Undergraduate Texts in Mathematics

Undergraduate Texts in Mathematics

continued after Index

Rudolf Lidl
Günter Pilz

Applied Abstract Algebra

With 175 Illustrations

Springer-Verlag
New York Berlin Heidelberg Tokyo

Rudolf Lidl
Department of Mathematics
University of Tasmania
Hobart, Tasmania, 7001
Australia

Günter Pilz
Institut für Mathematik
Universität Linz
A-4040 Linz
Austria

AMS Subject Classifications: 05-01, 06-01, 08-01, 13-01, 15-01, 94A24, 05BXX, 20B25, 68B10, 12C99, 94CXX, 68D30, 68F10, 20M35

Library of Congress Cataloging in Publication Data
Lidl, Rudolf.
    Applied abstract algebra.
    (Undergraduate texts in mathematics)
    Bibliography: p.
    Includes indexes.
    1. Algebra, Abstract.   I. Pilz, Günter.
II. Title.   III. Series.
QA162.L53   1984   512′.02   84-10576

Typeset by J. W. Arrowsmith Ltd., Bristol, England.

9 8 7 6 5 4 3 2 1

ISBN 978-0-387-96166-8   ISBN 978-1-4615-6465-2 (eBook)
DOI 10.1007/978-1-4615-6465-2

To Pamela and Gerti

Preface

There is at present a growing body of opinion that in the decades ahead discrete mathematics (that is, "noncontinuous mathematics"), and therefore parts of applicable modern algebra, will be of increasing importance. Certainly, one reason for this opinion is the rapid development of computer science, and the use of discrete mathematics as one of its major tools.

The purpose of this book is to convey to graduate students or to final-year undergraduate students the fact that the abstract algebra encountered previously in a first algebra course can be used in many areas of applied mathematics. It is often the case that students who have studied mathematics go into postgraduate work without any knowledge of the applicability of the structures they have studied in an algebra course.

In recent years there have emerged courses and texts on discrete mathematics and applied algebra. The present text is meant to add to what is available, by focusing on three subject areas. The contents of this book can be described as dealing with the following major themes:

Applications of Boolean algebras (Chapters 1 and 2).
Applications of finite fields (Chapters 3 to 5).
Applications of semigroups (Chapters 6 and 7).

Each of these three themes can be studied independently. We have not tried to write a comprehensive book on applied algebra, rather we have tried to highlight some algebraic structures which seem to have most useful applications. Each of these topics is relevant to and has strong connections with computer science.

We assume that the reader has the mathematical maturity of a beginning graduate student or of a last-year undergraduate student at a North American university or of a third or final-year Bachelor or Honors student in the United Kingdom or Australia. Thus the text is addressed mainly to a mature mathematics student and should also be useful to a computer scientist or a computer science student with a good background in algebra. Some students and lecturers might also be interested in seeing some not-so-well-known applications of selected algebraic structures. The reader is expected to be familiar with basic ideas about groups, rings, fields and linear algebra as the prerequisites for this book. All these requirements 'are met in a first course on linear algebra and an introductory course on abstract algebra.

The first topic, treated in Chapters 1 and 2, deals with properties and applications of Boolean algebras and their use in switching circuits and simplification methods. The next three chapters, which form the core of the text, comprise properties and applications of finite fields. Considerable emphasis is given to computational aspects. Chapter 3 contains the basic properties of finite fields and polynomials over finite fields; these will be used in the following two chapters. Chapter 4 contains topics from algebraic coding theory with a decoding procedure for BCH codes as its climax. Chapter 5 is devoted to other areas of applications of rings and finite fields, such as combinatorics, cryptography and linear recurring sequences. The third major topic, applications of semigroups to automata, formal languages, biology and sociology, is covered in Chapters 6 and 7.

Throughout the text, great emphasis is put on computational examples in the belief that most readers learn to do mathematics by solving numerical problems. A number of problems is given at the end of each section. Each paragraph ends with a number of exercises which are solved in Chapter 8 of this book. It is hoped that the reader will work through these problems and exercises and use the solutions in Chapter 8 only as a check of their understanding of the material.

The appendix consists of two parts. Part A contains fundamental definitions and properties of sets, logical symbols, relations, functions and algebraic operations. More on that can be found in almost every introductory text. Part B contains some computer programs to perform some of the algorithms presented in the text. The advent of microcomputers and the wide and rapidly increasing availability of desk-top computers prompted us to do so. In certain areas of applied mathematics, the computer is an indispensible tool.

The chapters are divided into sections; larger sections are subdivided into subsections A, B, etc. References in the text are organized such that 1.3.5 refers to item (Theorem, Definition, ...) number 5 in section 3 of Chapter 1. Within one chapter we use the abbreviation 3.5 to refer to item number 5 in section 3 of the present chapter. We refer to items in the Bibliography by writing the author's name in small capitals. The symbol □ denotes the end of a proof or an example. Some of the more difficult or not quite straightforward problems, exercises or whole sections are marked with an asterisk *. The notes at the end of each chapter provide some historical comments and references for further reading. Parts of the material of this book appeared (along with some other applications of algebra) in the authors' German Text *Angewandte Abstrakte Algebra*, Vols. I, II (Bibliographisches Institut, Mannheim, 1982).

It is with pleasure that we thank friends and colleagues for helpful suggestions after critically reviewing parts of the manuscript. We gratefully acknowledge contributions to the final draft by: Elizabeth J. Billington (Brisbane, Australia); Donald W. Blackett (Boston, Massachusetts); Henry E. Heatherly (Lafayette, Louisiana); Carlton J. Maxson (College Station, Texas); John D. P. Meldrum (Edinburgh, Scotland); Ken Miles (Melbourne, Australia); Alan Oswald (Teesside, England) and Peter G. Trotter (Hobart, Australia). Finally, we wish to thank the editorial and production staff of Springer-Verlag for their kind cooperation throughout the preparation of this book.

March 1984   R. L. and G. P.

Contents

List of Symbols

Symbols which are frequently used are listed in order of appearance in the text.

$\times$	cartesian product	2, 506
$:=$	colon indicating definition	2, 506
$\emptyset$	empty set	2
$\mathscr{P}$	the power set	2, 506
$\subseteq$	set-theoretic inclusion	2, 506
$\leq$	less than or equal	2
$<$	strictly less than	2
$\forall$	for all	4, 506
$\Leftrightarrow$	if and only if	4, 506
$\mathbb{R}$	the set of real numbers	4, 505
$\mathbb{N}$	the set of natural numbers	4, 505
sup	supremum	4
inf	infimum	4
$\sqcap$	meet	5
$\sqcup$	join	5
$|L|$	cardinality of L	6
$\Rightarrow$	implication	6, 506
$[x, y]$	interval	11
$\mathbb{C}$	the set of complex numbers	14, 505
$M \backslash A$	set-theoretic difference	19, 506
x'	complement	20
$\mathbb{B}$	the Boolean algebra of 0 and 1	23
$\cong_b$	Boolean isomorphic	27
$\hookrightarrow$	embedding monomorphism	30

$h \circ x^q$	substitution of polynomials	170
C_S	cyclotomic coset	177
$(\mathbf{A}, \mathbf{I})$	matrix consisting of a matrix $\mathbf{A}$ and the identity matrix $\mathbf{I}$	194
$d(\mathbf{x}, \mathbf{y})$	Hamming distance	197
$w(\mathbf{x})$	Hamming weight	197
$d_{\min}$	minimum distance	197
$S_r(\mathbf{x})$	the sphere of radius r about $\mathbf{x}$	198
$C^\perp$	dual code	199
$S(\mathbf{y})$	syndrome	201, 219
C_m	the Hamming code of length m	203
χ	character	207
V_n	the vector space of polynomials over $\mathbb{F}_q$ of degree $< n$	212
$Z(v)$	cyclic shift	212
lcm	least common multiple	217
DFT	discrete Fourier transform	222
BCH	Bose, Chaudhuri, Hocquenghem	229
RS	Reed, Solomon	231
QR	quadratic residue	238
$RM(m, d)$	Reed–Muller code	241
J_n	$n \times n$ matrix of ones	251
$S(t, k, v)$	Steiner system	260
$PG(2, q)$	the projective geometry	260
$J(a, n)$	the Jacobi symbol	296
$\lceil x \rceil$	smallest natural number $\geq x$	312
S_n	the symmetric group on $\{1, 2, \ldots, n\}$	314
$Z(S_n)$	the center of S_n	315
$\text{sign}(\pi)$	signature of π	315
A_n	alternating groups	316
$K(G)$	commutator group	316
$\text{Orb}(x)$	orbit of x	317
$\text{Stab}(x)$	stabilizer of x	317
$\text{Fix}(g)$		319
δ	next-state function	331
λ	output function	332
Δ	symmetric difference	338, 506
$\mathscr{R}(M)$	set of relations on M	339
G_s	group kernel	342
R^t	transitive hull	343
$\bar{S}$	least common multiple of semigroups	352
$\underline{S}$	greatest common divisor of semigroups	352
$d(S_1, S_2)$	distance between S_1 and S_2	353
$\mathscr{A}_1 \leq \mathscr{A}_2$	$\mathscr{A}_1$ is subautomaton of $\mathscr{A}_2$	361
$\mathscr{A}_1 \times \mathscr{A}_2$	direct product of automata	364
$\mathscr{A}_1 \| \mathscr{A}_2$	series composition of automata	365
$\mathscr{A}_1 \times_\varphi^\eta \mathscr{A}_2$	cascade of automata	366

CHAPTER 1

Lattices

In 1854, George Boole (1815–1864) introduced an important class of algebraic structures in connection with his research in mathematical logic. In his honor these structures have been called Boolean algebras. These are a special type of lattices. It was E. Schröder, who about 1890, considered the lattice concept in today's sense. At approximately the same time, R. Dedekind developed a similar concept in his work on groups and ideals. Dedekind defined in modern terminology modular and distributive lattices, which are types of lattices of importance in applications. The rapid development of lattice theory proper started around 1930 when G. Birkhoff made major contributions to the theory.

We could say that Boolean lattices or Boolean algebras are the simplest and at the same time the most important lattices for applications. Since they are defined as distributive and complemented lattices it is logical to consider some properties of distributive and complemented lattices first. Any distributive lattice is modular; therefore we introduce modular lattices before studying distributive ones.

§1. Properties of Lattices

A. Lattice Definitions

One of the important concepts in all of mathematics is that of a relation. Of particular interest are equivalence relations, functions and order relations. Here we concentrate on the latter concept and recall from an introductory mathematics course:

Let A and B be nonempty sets. A *relation* R from A to B is a subset of $A \times B$. Relations from A to A are called relations on A, for short. If $(a, b) \in R$ then we write $a\,R\,b$ and say that "a is in relation R to b". Also, if a is not in relation R to b, we write $a\,R\!\!\!/\,b$.

A relation R on a nonempty set A may have some of the following properties:

> R is *reflexive* if for all a in A we have $a\,R\,a$.
>
> R is *symmetric* if for all a and b in A: $a\,R\,b$
> implies $b\,R\,a$.
>
> R is *antisymmetric* if for all a and b in A: $a\,R\,b$ and $b\,R\,a$
> imply $a = b$.
>
> R is *transitive* if for all a, b, c in A: $a\,R\,b$
> and $b\,R\,c$ imply $a\,R\,c$.

A relation R on A is an *equivalence relation* if R is reflexive, symmetric and transitive. In this case $[a] := \{b \in A | a\,R\,b\}$ is called the *equivalence class of a*, for any $a \in A$.

1.1 Definition. A relation R on a set A is called a *partial order* (*relation*) if R is reflexive, antisymmetric and transitive.

In this case (A, R) is called a *partially ordered set* or *poset*.

Partial order relations are "hierarchical" relations, usually we write $\leq$ or $\subseteq$ instead of R. Partially ordered finite sets $(A, \leq)$ can be graphically represented by *Hasse diagrams*. Here the elements of A are represented as points in the plane and if $a \leq b$, $a \neq b$, we draw b higher up than a and connect a and b with a line segment. For example, the Hasse diagram of the poset $(\mathcal{P}(\{1, 2, 3\}), \subseteq)$ is

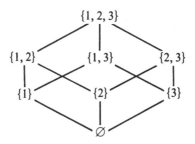

Figure 1.1

Here we do not draw a line from $\emptyset$ to $\{1, 2\}$, because this line already exists via $\{1\}$ or $\{2\}$, etc. The Hasse diagram of $(\{1, 2, 3, 4, 5\}, \leq)$, where $\leq$ means less than or equal to, is:

Figure 1.2

The difference between these two examples can be expressed by the follow-ing definition.

1.2 Definition. A partial order relation $\leq$ on A is called a *total order* (or *linear order*) if for each $a, b \in A$ either $a \leq b$ or $b \leq a$. $(A, \leq)$ is then called a *chain*, or *totally ordered set*.

For example, $(\{1, 2, 3, 4, 5\}, \leq)$ is a total order, $(\mathcal{P}(\{1, 2, 3\}), \subseteq)$ is not a total order.

If R is a relation from A to B then R^{-1}, defined by $(a, b) \in R^{-1} :\Leftrightarrow (b, a) \in R$ is a relation from B to A, called the *converse relation* of R.

If $(A, \leq)$ is a partially ordered set then $(A, \geq)$ is a partially ordered set and $\geq$ is the converse relation to $\leq$. In $(A, \leq)$ the following principle holds:

"Duality Principle". Every "statement" (formula, law, expression) on an ordered set $(A, \leq)$ remains correct, if everywhere in the statement the relation $\leq$ is replaced by its converse relation $\geq$. $(A, \geq)$ is called "dual" to $(A, \leq)$.

Let $(A, \leq)$ be a poset. We say, "a is a greatest element" if "all other elements are smaller". More precisely, $a \in A$ is called a *greatest element* of A if for all $x \in A$ we have $x \leq a$. The element b in A is called a *smallest element* of A if $b \leq x$ for all $x \in A$. The element $c \in A$ is called a *maximal element* of A if $c \leq x$ implies $c = x$ for all $x \in A$; similarly, $d \in A$ is called a *minimal element* of A if $x \leq d$ implies $x = d$ for all $x \in A$. It can be shown that $(A, \leq)$ has at most one greatest and one smallest element. However, there may be none, one, or several maximal or minimal elements. Every greatest element is maximal and every smallest element is minimal. For

instance, in the poset of Figure 1.3

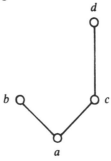

Figure 1.3

a is a minimal element and is a smallest element. b and d are maximal, but there is no greatest element.

1.3 Definition. Let $(A, \leq)$ be a poset and $B \subseteq A$.

(i) $a \in A$ is called an *upper bound* of $B :\Leftrightarrow \forall\, b \in B: b \leq a$.
(ii) $a \in A$ is called a *lower bound* of $B :\Leftrightarrow \forall\, b \in B: a \leq b$.
(iii) The greatest amongst the lower bounds, whenever it exists, is called the *infimum* of B, and is denoted by inf B.
(iv) The least upper bound of B, whenever it exists, is called the *supremum* of B, and is denoted by sup B.

For instance, let $(A, \leq) = (\mathbb{R}, \leq)$ and $B = [0, 3)$ then inf $B = 0$ and sup $B = 3$. Thus the infimum (supremum) of B may be an element of B, but does not have to be. If $B' = \mathbb{N}$ then in $(\mathbb{R}, \leq)$ we have inf $B' = 1$, but sup B' does not exist.

The following statement can neither be proved nor can it be refuted (it is undecidable). It is regarded as an additional axiom that may be used without comment in mathematical arguments.

1.4 Axiom ("Zorn's Lemma"). *If $(A, \leq)$ is a poset such that every chain of elements in A has an upper bound in A then A has at least one maximal element.*

The basic idea for a concept which is more general than that of a chain is as follows: If $x \leq z$ and $y \leq z$ hold then z is an upper bound for a and b. Whenever the least of all upper bounds of x and y exists it is the uniquely determined supremum of x and y, $\sup(x, y)$ for short. Hence $x \leq \sup(x, y)$ and $y \leq \sup(x, y)$. If z is any upper bound of x and y then $\sup(x, y) \leq z$. Similarly for the infimum of x and y, $\inf(x, y)$, we have $\inf(x, y) \leq x$ and $\inf(x, y) \leq y$. For any lower bound v of x and y we have $v \leq \inf(x, y)$.

In general, not every subset of a poset $(L, \leq)$ has a supremum or an infimum. We study more closely those posets which are axiomatically required to have a supremum and infimum for certain families of subsets.

1.5 Definition. A poset $(L, \leq)$ is called *lattice ordered* if for every pair x, y of elements of L the $\sup(x, y)$ and $\inf(x, y)$ exist.

1.6 Remark. (i) Every ordered set is lattice ordered.
(ii) In a lattice ordered set $(L, \leq)$ the following statements are equivalent for all x and y in L:
   (a) $x \leq y$;
   (b) $\sup(x, y) = y$;
   (c) $\inf(x, y) = x$.

There is another (yet equivalent) approach, that does not use order relations, but algebraic operations.

1.7 Definition. An (algebraic) *lattice* $(L, \sqcap, \sqcup)$ is a nonempty set L with two binary operations $\sqcap$ (*meet*) and $\sqcup$ (*join*) (also called intersection or product and union or sum, respectively), which satisfy the following conditions for all $x, y, z \in L$:

(L1)	$x \sqcap y = y \sqcap x,$	$x \sqcup y = y \sqcup x;$
(L2)	$x \sqcap (y \sqcap z) = (x \sqcap y) \sqcap z,$	$x \sqcup (y \sqcup z) = (x \sqcup y) \sqcup z;$
(L3)	$x \sqcap (x \sqcup y) = x,$	$x \sqcup (x \sqcap y) = x.$

Two applications of (L3), namely $x \sqcap x = x \sqcap (x \sqcup (x \sqcap x)) = x$, lead to the additional condition

(L4) $x \sqcap x = x, \qquad x \sqcup x = x.$

(L1) is the *commutative law*, (L2) is the *associative law*, (L3) is the *absorption law*, and (L4) is the *idempotent law*.

The connection between lattice ordered sets and algebraic lattices is as follows.

1.8 Theorem. (i) *Let* $(L, \leq)$ *be a lattice ordered set. If we define*

$$x \sqcap y := \inf(x, y), \qquad x \sqcup y := \sup(x, y),$$

then $(L, \sqcap, \sqcup)$ *is an algebraic lattice.*
(ii) *Let* $(L, \sqcap, \sqcup)$ *be an algebraic lattice. If we define*

$$x \leq y :\Leftrightarrow x \sqcap y = x \quad (or\ x \leq y :\Leftrightarrow x \sqcup y = y),$$

then $(L, \leq)$ *is a lattice ordered set.*

PROOF. Let $(L, \leq)$ be a lattice ordered set and define $x \sqcap y := \inf(x, y)$ and $x \sqcup y := \sup(x, y)$. Clearly L is a nonempty set with the above two binary operations.

(L1) $x \sqcap y = \inf(x, y) = \inf(y, x) = y \sqcap x \qquad \forall\, x, y \in L;$

$x \sqcup y = \sup(x, y) = \sup(y, x) = y \sqcup x \qquad \forall\, x, y \in L.$

(L2) $x \sqcap (y \sqcap z) = x \sqcap \inf(y, z) = \inf(x, \inf(y, z)) = \inf(x, y, z)$

$$= \inf(\inf(x, y), z) = \inf(x, y) \sqcap z = (x \sqcap y) \sqcap z,$$

$$\forall\, x, y, z \in L;$$

$$x \sqcup (y \sqcup z) = x \sqcup \sup(y, z) = \sup(x, \sup(y, z)) = \sup(x, y, z)$$

$$= \sup(\sup(x, y), z) = \sup(x, y) \sqcup z = (x \sqcup y) \sqcup z,$$

$$\forall\, x, y, z \in L.$$

(L3) $x \sqcap (x \sqcup y) = x \sqcap \sup(x, y) = \inf(x, \sup(x, y)) = x \quad \forall\, x, y \in L;$

$x \sqcup (x \sqcap y) = x \sqcup \inf(x, y) = \sup(x, \inf(x, y)) = x \quad \forall\, x, y \in L.$

(L4) $x \sqcap x = \inf(x, x) = x \quad \forall\, x \in L;$

$x \sqcup x = \sup(x, x) = x \quad \forall\, x \in L.$

Let $(L, \sqcap, \sqcup)$ be an algebraic lattice and define

$$x \le y :\Leftrightarrow x \sqcap y = x \quad (\text{or } x \le y :\Leftrightarrow x \sqcup y = y),$$

i.e. $x \le y :\Leftrightarrow \inf(x, y) = x$ (or $x \le y :\Leftrightarrow \sup(x, y) = y$). Clearly for all x, y, z in L:

   (i) $x \sqcap x = x$ and $x \sqcup x = x$ by (L4); so $x \le x$, i.e. $\le$ is reflexive.
  (ii) If $x \le y$ and $y \le x$, then $x \sqcap y = x$ and $y \sqcap x = y$ by (L1) $x \sqcap y = y \sqcap x$; so $x = y$, i.e. $\le$ is antisymmetric. Also $x \sqcap y = x$ and $x \sqcup y = y$ are equivalent, since, by (L3), e.g. $x \sqcap y = x$ implies

$$x \sqcup y = (x \sqcap y) \sqcup y = y.$$

 (iii) If $x \le y$ and $y \le z$ then $x \sqcap y = x$ and $y \sqcap z = y$. Therefore

$$x = x \sqcap y = x \sqcap (y \sqcap z) = (x \sqcap y) \sqcap z = x \sqcap z \text{ so } x \le z \quad \text{by (L2)},$$

i.e. $\le$ is transitive.

Let $x, y \in L$. Then $x \sqcap (x \sqcup y) = x$ implies $x \le x \sqcup y$ and similarly $y \le x \sqcup y$. If $z \in L$ with $x \le z$ and $y \le z$ then $(x \sqcup y) \sqcup z = x \sqcup (y \sqcup z) = x \sqcup z = z$ and so $x \sqcup y \le z$. Thus $\sup(x, y) = x \sqcup y$. Similarly $\inf(x, y) = x \sqcap y$. Hence $(L, \le)$ is a lattice ordered set. $\square$

1.9 Remark. It can be verified that Theorem 1.8 yields a one-to-one relationship between lattice ordered sets and algebraic lattices. Therefore we shall use the term lattice for both concepts. $|L|$ denotes the order (i.e. cardinality) of the lattice L.

   Following 1.8 we define two operations in posets: join (also called sum or union and meet (also called product or intersection). The supremum of two elements x and y is denoted by $x \sqcup y$ and is called the *join* of x and y. The infimum of x and y is denoted by $x \sqcap y$ and is called the *meet* of x

and y. We use the symbols $\sqcup$ and $\sqcap$ to distinguish the operations join and meet from the corresponding set-theoretic operations $\cup$ and $\cap$. More generally, if N is a subset of a poset then $\bigsqcup_{x \in N} x$ and $\bigsqcap_{x \in N} x$ denote the supremum and infimum of N, respectively, whenever they exist. We say that the supremum of N is the join of all elements of N and the infimum is the meet of all elements of N.

In Definition 1.7 we saw that for the conditions (L1) to (L4) two equations are given. The duality principle for posets is also valid for lattices.

1.10 "Duality Principle". *Any "formula" in a lattice* $(L, \sqcap, \sqcup)$ *involving the operations* $\sqcap$ *and* $\sqcup$ *remains valid if we replace* $\sqcap$ *by* $\sqcup$ *and* $\sqcup$ *by* $\sqcap$ *everywhere in the formula.* This process of replacing is called *dualizing*. $\square$

The validity of this assertion follows from the fact that in a lattice any formula, which can be derived by using (L1) to (L4), remains correct if we interchange $\sqcap$ with $\sqcup$ everywhere in the formula. So every dual of a condition in (L1)–(L4) holds too.

1.11 Definition. If a lattice L contains a smallest (greatest) element with respect to $\leq$ then this uniquely determined element is called the *zero element* (*one element*), denoted by 0 (by 1). 0 and 1 are called *universal bounds*.

Every finite lattice L has a 0 and a 1. If a lattice has a 0 and a 1 then every x in L satisfies $0 \leq x \leq 1$, $0 \sqcap x = 0$, $0 \sqcup x = x$, $1 \sqcap x = x$, $1 \sqcup x = 1$. We consider some examples of lattices.

1.12 Examples. Let M and M_i, $i \in I$, be linearly ordered sets with smallest element 0 and greatest element 1 and let G be a group with unit 1.

No	Set	$\leq$	$x \sqcap y$	$x \sqcup y$	0	1
1	M	linear order	$\min(x, y)$	$\max(x, y)$	smallest element	greatest element
2	$\underset{i \in I}{\times} M_i$	componentwise	componentwise	componentwise	$(\ldots, 0, \ldots)$	$(\ldots, 1, \ldots)$
3	$\mathcal{P}(M)$	$\subseteq$	$X \cap Y$	$X \cup Y$	$\varnothing$	M
4	$\mathbb{N}$	"divides"	$\gcd(x, y)$	$\text{lcm}(x, y)$	1	does not exist
5	$\{X \mid X \leq G\}$	$\subseteq$	$X \cap Y$	subgroup generated by $X \cup Y$	$\{1\}$	G

Theorem 1.8 and Remark 1.9 enable us to represent any lattice as a special poset or as an algebraic structure using operation tables. We present the Hasse diagrams of all lattices with at most six elements. V_i^n denotes the ith lattice with n elements.

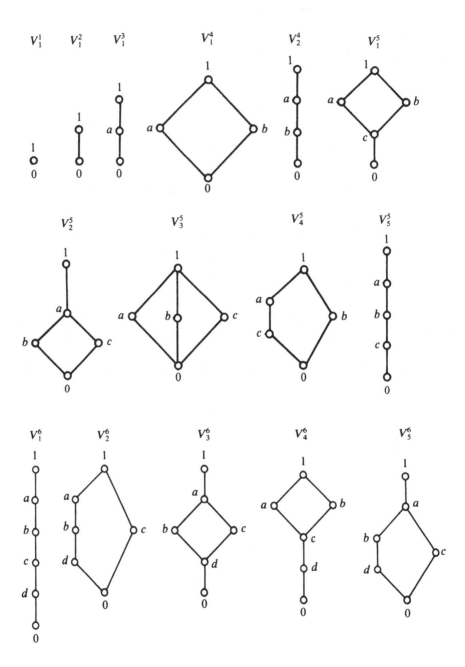

Figure 1.4. Hasse diagrams of all lattices with at most six elements.

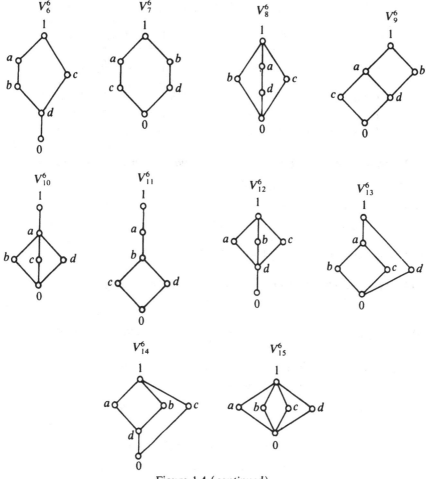

Figure 1.4 (*continued*)

The following diagram is an example of a poset which is not a lattice (since sup(*a*, *b*) does not exist).

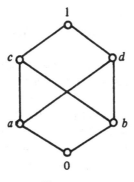

Figure 1.5

Next we give the operation tables for the lattice V_4^5.

$\sqcap$	0	a	b	c	1
0	0	0	0	0	0
a	0	a	0	c	a
b	0	0	b	0	b
c	0	c	0	c	c
1	0	a	b	c	1

$\sqcup$	0	a	b	c	1
0	0	a	b	c	1
a	a	a	1	a	1
b	b	1	b	1	1
c	c	a	1	c	1
1	1	1	1	1	1

$\square$

1.13 Lemma. *In every lattice L the operations $\sqcap$ and $\sqcup$ are isotone, i.e.*
$y \le z \Rightarrow x \sqcap y \le x \sqcap z$ and $x \sqcup y \le x \sqcup z$.

PROOF.

$$y \le z \Rightarrow x \sqcap y = (x \sqcap x) \sqcap (y \sqcap z) = (x \sqcap y) \sqcap (x \sqcap z) \Rightarrow x \sqcap y \le x \sqcap z.$$

The second formula is verified by duality. $\square$

1.14 Theorem. *The elements of an arbitrary lattice satisfy the following inequalities*:

(i) $\begin{cases} x \sqcap (y \sqcup z) \ge (x \sqcap y) \sqcup (x \sqcap z), \\ x \sqcup (y \sqcap z) \le (x \sqcup y) \sqcap (x \sqcup z), \end{cases}$ (*"Distributive inequalities"*).

(ii) $\begin{cases} x \ge z \Rightarrow x \sqcap (y \sqcup z) \ge (x \sqcap y) \sqcup z = (x \sqcap y) \sqcup (x \sqcap z), \\ x \le z \Rightarrow x \sqcup (y \sqcap z) \le (x \sqcup y) \sqcap z = (x \sqcup y) \sqcap (x \sqcup z), \end{cases}$

(*"Modular inequalities"*).

PROOF.

$$x \sqcap y \le x, x \sqcap y \le y \le y \sqcup z \Rightarrow x \sqcap y \le x \sqcap (y \sqcup z),$$

$$x \sqcap z \le x, x \sqcap z \le z \le y \sqcup z \Rightarrow x \sqcap z \le x \sqcap (y \sqcup z).$$

Thus $x \sqcap (y \sqcup z)$ is an upper bound for $x \sqcap y$ and $x \sqcap z$; therefore $x \sqcap (y \sqcup z) \ge (x \sqcap y) \sqcup (x \sqcap z)$. The second inequality in (i) follows from duality. (ii) is a special case of (i). $\square$

As usual, we can construct "new" lattices from given ones by forming substructures, homomorphic images and products.

1.15 Definition. A nonempty subset S of a lattice L is called *sublattice* of L if S is a lattice with respect to the restriction of $\sqcap$ and $\sqcup$ of L onto S.

It is obvious that $S \subseteq L$ is a sublattice of the lattice L if and only if S is "closed" with respect to $\sqcap$ and $\sqcup$ (i.e. $s_1, s_2 \in S \Rightarrow s_1 \sqcap s_2 \in S$ and $s_1 \sqcup s_2 \in S$). We note that a subset S of a lattice L can be a lattice with respect to the partial order of L without being a sublattice of L (see Example 1.16(iii) below).

1.16 Examples. (i) Every singleton of a lattice L is a sublattice of L.

(ii) For any two elements x, y in a lattice L the "interval" $[x, y] :=$ $\{a \in L | x \leq a \leq y\}$ is either $\varnothing$ or a sublattice of L.

(iii) Let L be the lattice of all subsets of a group G and let S be the set of all subgroups of G, then S is a lattice with respect to inclusion but not a sublattice of L.   □

1.17 Definition. Let L, M be lattices. A mapping $f: L \to M$ is called a

(i) *join-homomorphism*, if $x \sqcup y = z \Rightarrow f(x) \sqcup f(y) = f(z)$;

(ii) *meet-homomorphism*, if $x \sqcap y = z \Rightarrow f(x) \sqcap f(y) = f(z)$;

(iii) *order-homomorphism*, if $x \leq y \Rightarrow f(x) \leq f(y)$.

f is a *homomorphism* (or lattice homomorphism) if it is both a join- and a meet-homomorphism. Injective, surjective or bijective lattice homomorphisms are called lattice *monomorphisms, epimorphisms, isomorphisms*, respectively.

It can be shown that every join- (or meet-) homomorphism is an order-homomorphism. However, the converse is not true. The relationship between the different homomorphisms can be symbolized as follows:

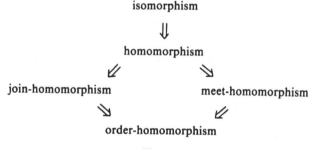

Figure 1.6

1.18 Example. Let L_1, L_2, L_3 be lattices with the Hasse diagrams

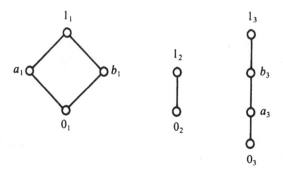

Figure 1.7

respectively. We define

$$f: L_1 \to L_2, \quad f(0_1) = f(a_1) = f(b_1) = 0_2, \quad f(1_1) = 1_2,$$
$$g: L_1 \to L_2, \quad g(1_1) = g(a_1) = g(b_1) = 1_2, \quad g(0_1) = 0_2,$$
$$h: L_1 \to L_3, \quad h(0_1) = 0_3, \quad h(a_1) = a_3, \quad h(b_1) = b_3, \quad h(1_1) = 1_3;$$

the three mappings are order-homomorphisms.

f is a meet-homomorphism, since

$$f(a_1 \sqcap b_1) = f(0_1) = 0_2 = f(a_1) \sqcap f(b_1) \text{ etc.}$$

However, f is not a homomorphism, since

$$f(a_1 \sqcup b_1) = f(1_1) = 1_2 \quad \text{and} \quad f(a_1) \sqcup f(b_1) = 0_2.$$

Dually, g is a join-homomorphism, but not a homomorphism.

h is neither a meet- nor a join-homomorphism, since

$$h(a_1 \sqcap b_1) = h(0_1) = 0_3 \quad \text{and} \quad h(a_1) \sqcap h(b_1) = a_3 \sqcap b_3 = a_3,$$

and

$$h(a_1 \sqcup b_1) = h(1_1) = 1_3 \quad \text{and} \quad h(a_1) \sqcup h(b_1) = a_3 \sqcup b_3 = b_3. \qquad \square$$

1.19 Definition. Let L and M be lattices. The set of ordered pairs

$$\{(x, y) | x \in L, y \in M\}$$

with operations $\sqcup$ and $\sqcap$ defined by:

$$(x_1, y_1) \sqcup (x_2, y_2) := (x_1 \sqcup x_2, y_1 \sqcup y_2),$$
$$(x_1, y_1) \sqcap (x_2, y_2) := (x_1 \sqcap x_2, y_1 \sqcap y_2),$$

is the *direct product* of L and M, in symbols $L \times M$, also called the *product lattice*.

It is easily verified that $L \times M$ is a lattice in the sense of Definition 1.7. The partial order of $L \times M$ which results from the definition in 1.8(ii) satisfies

$$(x_1, y_1) \le (x_2, y_2) \Leftrightarrow x_1 \le x_2 \quad \text{and} \quad y_1 \le y_2.$$

1.20 Example. The direct product of the lattices L and M can graphically be described in terms of Hasse diagrams:

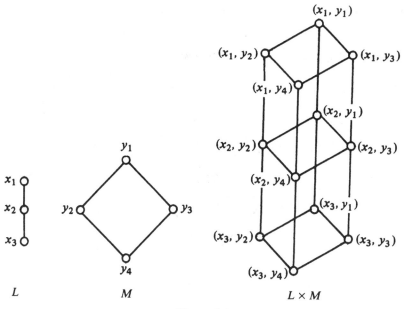

Figure 1.8

PROBLEMS

1. Determine all the partial orders and their Hasse diagrams on the set $L = \{a, b, c\}$. Which of them are chains?

2. Give an example of a poset which has exactly one maximal element but does not have a greatest element.

3. Let $(\mathbb{Q}, \leq)$ be the poset of the rational numbers and let $A = \{x \mid x \in \mathbb{Q}, x^3 < 3\}$. Is there an upper bound (or lower bound) or a supremum (or infimum) in A?

4. Let $(L, \leq)$ be a poset. Show that if $\inf(\inf(a, b), c)$ exists for $a, b, c \in L$ then also $\inf(a, b, c)$ exists and both are equal. Moreover, show that $\inf(a, b, c)$ may exist even when $\inf(\inf(a, b), c)$ or $\inf(a, \inf(b, c))$ do not exist. Finally, show that $\inf(\inf(a, b), c)$ may exist even when $\inf(a, \inf(b, c))$ does not exist.

5. Let $(L, \leq)$ be a poset with the following properties:
   (i) $A \subseteq L$ and $A \neq \varnothing$ implies that there exists an $a \in L$ such that $a = \inf A$.
   (ii) $B \subseteq L$ implies that there is a $b \in L$ such that $x \leq b$ for all $x \in B$.
   Prove that L is a lattice.

6. Prove that any finite lattice has a zero and a one.

7. Give an example of an infinite lattice without a zero and a one.

8. Let L be the set of complex numbers $z = x + iy$ where x and y are rationals. Define a partial order $\subseteq$ on L by: $x_1 + iy_1 \subseteq x_2 + iy_2$ if and only if $y_1 \leq y_2$. Is there a minimal or a maximal element in $(L, \subseteq)$? What additional condition is needed in order to make $(L, \subseteq)$ into a chain?

9. Prove that in any lattice L, for all x, y, z, $u \in L$:

$$(x \sqcap z) \sqcup (y \sqcap u) \leq (x \sqcup y) \sqcap (z \sqcup u).$$

10. The Hasse diagram of a lattice is as follows:

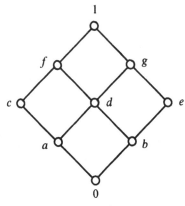

Figure 1.9

Find all sublattices of this lattice.

11. Prove: If f is an isomorphism of a poset L onto a poset M and if L is a lattice, then M is also a lattice and f is an isomorphism of the lattices.

12. Let $D(k)$ denote the lattice of all positive divisors of k. Construct the Hasse diagrams of the lattices $D(20)$ and $D(21)$ and show that $D(20) \times D(21)$ is isomorphic to $D(420)$.

13. Show that the direct product $D(120) \times D(432)$ is isomorphic to

$$D(2^3 \cdot 3 \cdot 5 \cdot 7^4 \cdot 11^3)$$

and also to the lattice

$$(\{(a_1, a_2, a_3, a_4, a_5) | 0 \leq x_1 \leq 3, 0 \leq x_2 \leq 1, 0 \leq x_3 \leq 1,$$

$$0 \leq x_4 \leq 4, 0 \leq x_5 \leq 3\}, \sqcup, \sqcap).$$

14. Let $(C([a, b]), \max, \min)$ be the lattice of continuous real-valued functions on a closed interval $[a, b]$, let $D((a, b))$ be the set of all differentiable functions on (a, b). Show by example that $D((a, b))$ is not a sublattice of $C([a, b])$.

B. Modular and Distributive Lattices

We now turn to special types of lattices with the aim of defining a very "rich" type of algebraic structure, a Boolean algebra.

1.21 Definition. A lattice L is called *modular* if $\forall$ x, y, $z \in L$

(M) $x \leq z \Rightarrow x \sqcup (y \sqcap z) = (x \sqcup y) \sqcap z$ ("*modular equation*").

1.22 Examples. An important example of a modular lattice is the lattice of all subspaces of a vector space.

Let A, B, C be subspaces of a vector space V and let $A \subseteq C$. Then $(A + B) \cap C \supseteq A + (B \cap C)$. Conversely, if $c \in (A + B) \cap C$, then there are $a \in A$ and $b \in B$ with $c = a + b$. Hence $b = c - a \in B \cap C$, i.e. $c \in A + (B \cap C)$. Therefore $(A + B) \cap C \subseteq A + (B \cap C)$.

The lattice of all subgroups of a group is in general not modular. The set of all normal subgroups of a group forms a modular lattice. Any chain and also the first eight lattices with Hasse diagrams given in Figure 1.4 are modular. V_4^5 is not modular. $\qquad\square$

We describe important characterizations of modular lattices.

1.23 Theorem. *A lattice L is modular if and only if $\forall\, x, y, z \in L$*

$$x \sqcup (y \sqcap (x \sqcup z)) = (x \sqcup y) \sqcap (x \sqcup z).$$

PROOF. If L is modular, then $x \le x \sqcup z$ yields the given equations. Conversely, the given equation implies the condition (M) of 1.21. $\qquad\square$

1.24 Theorem. *A lattice L is modular if and only if none of its sublattices is isomorphic to the "pentagon lattice" V_4^5, whose Hasse diagram is*

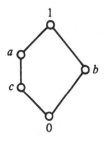

Figure 1.10

PROOF. The pentagon lattice is not modular since $c \le a$ but $c \sqcup (b \sqcap a) \ne (c \sqcup b) \sqcap a$, because of $c \sqcup (b \sqcap a) = c \sqcup 0 = c$ and $(c \sqcup b) \sqcap a = 1 \sqcap a = a$. Thus any lattice having a pentagon as a sublattice cannot be modular. To prove the converse we show that if a lattice is not modular, then it has a sublattice which is isomorphic to the pentagon lattice V_4^5. Let L be nonmodular. Then there are x, y, z in L such that

$$x \le z \quad \text{and} \quad x \sqcup (y \sqcap z) < (x \sqcup y) \sqcap z.$$

We shall show that the subset S of elements $u = y \sqcap z$, $a = (x \sqcup y) \sqcap z$, $b = y$, $c = x \sqcup (y \sqcap z)$, $v = x \sqcup y$ forms a sublattice of L which is isomorphic to the pentagon lattice.

We have

$$u \leq c < a \leq v \quad \text{and} \quad u \leq b \leq v. \tag{$*$}$$

Therefore

$$u \leq c \sqcap b \leq a \sqcap b = (x \sqcup y) \sqcap z \sqcap y = y \sqcap z = u,$$

$$v \geq a \sqcup b \geq c \sqcup b = x \sqcup (y \sqcap z) \sqcup y = x \sqcup y = v,$$

i.e.

$$c \sqcap b = a \sqcap b = u, \qquad c \sqcup b = a \sqcup b = v.$$

This shows that S is a sublattice of L. We verify that all elements of S are distinct. Suppose $u = b$, then

$$c \sqcap b = b \Rightarrow c \sqcup b = c \Rightarrow v = c,$$

which is a contradiction to $(*)$. So $u \neq b$. Suppose $v = a$, then

$$a \sqcup b = a \Rightarrow a \sqcap b = b \Rightarrow u = b$$

contradicting the foregoing. So $v \neq a$. Suppose $c = b$, then

$$u = b \sqcap b = b,$$

a contradiction. So $c \neq b$. The remaining cases, to show $v \neq b$, $u \neq c$ and $a \neq b$, are treated similarly. $\qquad\qquad\qquad\qquad\qquad\qquad\qquad\qquad\square$

1.25 Definition. A lattice L is called *distributive* if either of the following conditions hold for all x, y, z in L:

$$x \sqcup (y \sqcap z) = (x \sqcup y) \sqcap (x \sqcup z),$$
or $\qquad\qquad\qquad\qquad\qquad\qquad\qquad$ ("*distributive equations*")
$$x \sqcap (y \sqcup z) = (x \sqcap y) \sqcup (x \sqcap z).$$

1.26 Examples. (i) $(\mathcal{P}(M), \cap, \cup)$ is a distributive lattice.

(ii) Every chain is a distributive lattice.

(iii) If I and J are two ideals of a ring R, then we say I divides J, if $I \supseteq J$. Thus the $\gcd(I, J)$ is the ideal generated by the set $I \cup J$ in R, i.e. the set $\{a + b | a \in I, b \in J\}$. Also $\text{lcm}(I, J) = I \cap J$. The set of ideals of R is a lattice with respect to $\gcd(I, J), \text{lcm}(I, J)$. The product of the ideals I and J is the ideal generated by the elements ab, $a \in I$, $b \in J$, i.e. $IJ = \{\sum_{i=1}^{n} a_i b_i | a_i \in I, b_i \in J\}$. If every ideal of R can be uniquely expressed as a product of prime ideals, then the lattice of ideals is distributive. $\qquad\qquad\qquad\qquad\qquad\qquad\qquad\qquad\qquad\qquad\square$

1.27 Theorem. *A lattice L is distributive if and only if* $\forall\, x, y, z \in L$

$$(D) \qquad (x \sqcap y) \sqcup (y \sqcap z) \sqcup (z \sqcap x) = (x \sqcup y) \sqcap (y \sqcup z) \sqcap (z \sqcup x).$$

PROOF. Exercise.

1.28 Corollary. *Every distributive lattice is modular.*   □

Using mathematical induction it can be shown that the following formulas hold in a distributive lattice:

$$x \sqcup \left(\bigcap_{i=1}^{n} y_i \right) = \bigcap_{i=1}^{n} (x \sqcup y_i), \qquad x \sqcap \left(\bigsqcup_{i=1}^{n} y_i \right) = \bigsqcup_{i=1}^{n} (x \sqcap y_i).$$

1.29 Theorem. *A modular lattice is distributive if and only if none of its sublattices is isomorphic to the "diamond lattice"* V_3^5, *whose Hasse diagram is*

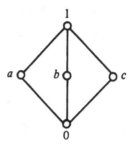

Figure 1.11

PROOF. The lattice V_3^5 is not distributive, since for instance

$$a \sqcup (b \sqcap c) \neq (a \sqcup b) \sqcap (a \sqcup c),$$

because

$$a \sqcup (b \sqcap c) = a \sqcup 0 = a \quad \text{and} \quad (a \sqcup b) \sqcap (a \sqcup c) = 1 \sqcap 1 = 1.$$

Thus any lattice having a diamond as a sublattice cannot be distributive.

Conversely, let L be a modular but not distributive lattice. By Theorem 1.27 there are elements x, y, z such that

$$(x \sqcap y) \sqcup (y \sqcap z) \sqcup (z \sqcap x) < (x \sqcup y) \sqcap (y \sqcup z) \sqcap (z \sqcup x). \qquad (*)$$

We shall show that the elements

$$u = (x \sqcap y) \sqcup (y \sqcap z) \sqcup (z \sqcap x),$$
$$v = (x \sqcup y) \sqcap (y \sqcup z) \sqcap (z \sqcup x),$$
$$a = u \sqcup (x \sqcap v),$$
$$b = u \sqcup (y \sqcap v),$$
$$c = u \sqcup (z \sqcap v),$$

form a sublattice of L which is isomorphic to the diamond lattice.

First, because of the modular law, we have

$$u \sqcup (x \sqcap v) = (u \sqcup x) \sqcap v \quad \text{for } u \leq v.$$

We have

$$a \sqcup b = u \sqcup (x \sqcap v) \sqcup (y \sqcap v)$$

$$= u \sqcup [(x \sqcap (y \sqcup z)) \sqcup (y \sqcap (z \sqcup x))], \quad \text{by the absorption law.}$$

Also $x \sqcap (y \sqcup z) \leq x \leq z \sqcup x$. Therefore

$$a \sqcup b = u \sqcup [((x \sqcap (y \sqcup z)) \sqcup y) \sqcap (z \sqcup x)] \quad \text{by modularity.}$$

Because $(x \sqcap (y \sqcup z)) \sqcup y = (x \sqcup y) \sqcap (y \sqcup z)$ by modularity, we have

$$a \sqcup b = u \sqcup v = v.$$

By duality $a \sqcap b = u$. In a similar way we can prove

$$b \sqcap c = c \sqcap a = u \quad \text{and} \quad b \sqcup c = c \sqcup a = v.$$

To show that the elements u, v, a, b, c are distinct, we proceed as follows. For instance, assume $u = a$. Then

$$a \sqcap b = a \Rightarrow a \sqcup b = b \Rightarrow v = b,$$

$$a \sqcap c = a \Rightarrow a \sqcup c = c \Rightarrow v = c,$$

and therefore

$$v = v \sqcap v = b \sqcap c = u,$$

a contradiction to $(*)$, which says $u < v$.

We leave the remainder of the proof as an exercise. $\square$

1.30 Corollary. *A lattice is distributive if and only if none of its sublattices is isomorphic to the pentagon lattice or the diamond lattice.* $\square$

Theorems 1.24 and 1.29 and Corollary 1.30 enable us to observe modularity or distributivity of the lattice from its Hasse diagram. Indeed, if we notice that somewhere in the diagram the pentagon lattice appears as a sublattice then we know that the lattice cannot be modular.

1.31 Example. The lattice with Hasse diagram

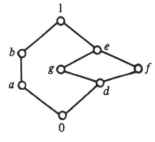

Figure 1.12

cannot be modular (and therefore not distributive) since it contains the
pentagon $\{0, a, f, b, 1\}$ as a sublattice.   □

1.32 Theorem. *A lattice L is distributive if and only if*

$$\forall\, x, y, z \in L: \quad (x \sqcap y = x \sqcap z, x \sqcup y = x \sqcup z) \Rightarrow y = z$$

                                                        (*"cancellation rule"*).

PROOF. Exercise.

1.33 Definition. A lattice L with 0 and 1 is called *complemented* if for each
$x \in L$ there is at least one element y such that $x \sqcap y = 0$, $x \sqcup y = 1$. y is
called a *complement* of x.

1.34 Examples. (i) Let $L = \mathcal{P}(M)$. Then $B = M \setminus A$ is a uniquely determined
  complement of A.
 (ii) In a bounded lattice 1 is a complement of 0 and 0 is a complement of 1.
(iii) Not every lattice with 0 and 1 is complemented. For instance, c in V_1^5
  does not have a complement:

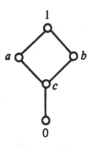

Figure 1.13

(iv) The complement does not have to be unique: e.g. in V_4^5, b has the two
  complements a and c.

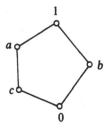

Figure 1.14

(v) $L = \{S \mid S$ is subspace of $\mathbb{R}^2\}$ is modular. If dim $S = 1$ then S has
  infinitely many complements, namely all subspaces T such that
  $S \oplus T = \mathbb{R}^2$. Therefore L cannot be distributive, as the following
  theorem shows.   □

1.35 Theorem and Definition. *If L is a distributive lattice then each $x \in L$ has at most one complement. We denote it by x'.*

PROOF. Suppose $x \in L$ has two complements y_1 and y_2. Then $x \sqcup y_1 = 1 = x \sqcup y_2$ and $x \sqcap y_1 = 0 = x \sqcap y_2$; thus $y_1 = y_2$ because of 1.32. $\square$

Complemented distributive lattices will be studied extensively in the following paragraphs.

PROBLEMS

1. Let A and B be two convex regions. Let $A \cdot B$ denote the largest convex region contained in A and B, $A + B$ denotes the smallest convex region that contains A and B. (Here $A + B$ is not the set-theoretic union.) Show that these operations on the set of all convex regions define a lattice which is not distributive.

2. Consider the subgroups of the group of the prime residue classes (mod m). Draw the Hasse diagram of the lattice of these subgroups. Is this lattice a distributive and complemented lattice?

3. Prove 1.28.

4. Prove the generalized distributive laws as stated after 1.28.

*5. Prove that a lattice is modular if and only if for all x, y, z:

$$(x \sqcup (y \sqcap z)) \sqcap (y \sqcap z) = (x \sqcap (y \sqcup z)) \sqcup (y \sqcap z).$$

*6. Prove: If for some elements x, y and z of a modular lattice

$$x \sqcap (y \sqcup z) = (x \sqcap y) \sqcup (x \sqcap z)$$

holds then

$$y \sqcap (x \sqcup z) = (y \sqcap x) \sqcup (x \sqcap z)$$

and

$$x \sqcup (y \sqcap z) = (x \sqcup y) \sqcap (x \sqcup z)$$

also hold.

7. Devise a formal algorithm for testing whether a given finite lattice is distributive.

*8. The elements $a_1, \ldots, a_n$ of a modular lattice with zero are called *independent*, if

$$(a_1 \sqcup \ldots \sqcup a_{i-1} \sqcup a_{i+1} \sqcup \ldots \sqcup a_n) \sqcap a_i = 0 \quad \text{for all } i = 1, \ldots, n.$$

Prove: If $a_1, \ldots, a_n$ are such that $(a_1 \sqcup \ldots \sqcup a_{i-1}) \sqcap a_i = 0$ for all $i = 1, \ldots, n$, then they are independent.

*9. In a modular lattice with zero, prove that the equality $(a_1 \sqcup \ldots \sqcup a_n) \sqcap b = 0$ implies

$$(a_1 \sqcup b) \sqcap \ldots \sqcap (a_n \sqcap b) = (a_1 \sqcap a_2 \sqcap \ldots \sqcap a_n) \sqcup b.$$

10. In a distributive lattice prove that $a \sqcap b \leq x \leq a \sqcup b$ and $x = (a \sqcap x) \sqcup (b \sqcap x) \sqcup (a \sqcap b)$ are equivalent.

EXERCISES (Solutions in Chapter 8, p. 409)

1. Let G be the group of quaternion units, $G = \{\pm 1, \pm i, \pm j, \pm k\}$ (see Chapter 3, §1). Draw the Hasse diagram for the lattice of all subgroups of G.

2. Prove the generalized distributive inequality for lattices:

$$y \sqcap \left(\bigsqcup_{i=1}^{n} x_i \right) \geq \bigsqcup_{i=1}^{n} (y \sqcap x_i).$$

3. Determine the operation tables for $\sqcap$ and $\sqcup$ for the lattice with Hasse diagram

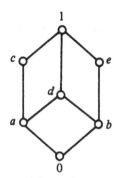

Figure 1.15

*4. A lattice L is called metric if there is a real-valued function $v : L \to \mathbb{R}$ such that $\forall\, x, y \in L$

$$v(x) + v(y) = v(x \sqcup y) + v(x \sqcap y),$$

$$y \leq x \Rightarrow v(y) \leq v(x).$$

(a) Prove that a metric lattice is modular.
(b) Define in a metric lattice L a "distance" d by $d : L^2 \to \mathbb{R}$, $(x, y) \mapsto v(x \sqcup y) - v(x \sqcap y)$. Prove that (L, d) is then a metric space.

*5. Prove: (i) The intervals $[x, x \sqcup y]$ and $[x \sqcap y, y]$ are isomorphic in a modular lattice.

(ii) If for all x, y in a lattice L the intervals $[x, x \sqcup y]$ and $[x \sqcap y, y]$ are isomorphic under $f : a \mapsto a \sqcap y$ then L is modular.

6. Prove: In any lattice L we have

$$[(x \sqcap y) \sqcup (x \sqcap z)] \sqcap [(x \sqcap y) \sqcup (y \sqcap z)] = x \sqcap y \quad \text{for all } x, y, z \in L.$$

7. Determine the lattice of all subgroups of the alternating group A_4 and show that this lattice is not modular.

8. Let C_1 and C_2 be the finite chains $\{0, 1, 2\}$ and $\{0, 1\}$, respectively. Draw the Hasse diagram of the product lattice $C_1 \times C_2 \times C_2$.

9. Show that the set of all normal subgroups of a group form a modular lattice.

10. Prove: If a, b, c are elements of a modular lattice with the property $(a \sqcup b) \sqcap c = 0$, then $a \sqcap (b \sqcup c) = a \sqcap b$.

*11. Prove: The following properties for a lattice L are equivalent:
   (i) L is modular.
   (ii) $\forall\ a, b, c \in L: a \sqcap ((a \sqcap b) \sqcup c) = (a \sqcap b) \sqcup (a \sqcap c)$.
   (iii) If $a \geq b$ and if $\exists\ c \in L$ with $a \sqcup c = b \sqcup c$ and $a \sqcap c = b \sqcap c$ then $a = b$.

*12. Prove: In a lattice $(L, \leq)$ every finite, nonempty subset S has a least upper bound.

13. Let S be an arbitrary set and D a distributive lattice. Show that the set of all functions from S to D is a distributive lattice, where $f \leq g$ means $f(x) \leq g(x)$ for all x.

14. Let L be a distributive lattice with 0 and 1. Prove: If a has a complement a', then

$$a \sqcup (a' \sqcap b) = a \sqcup b.$$

15. Is the lattice with Hasse diagram

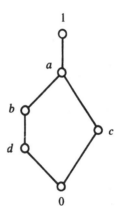

Figure 1.16

distributive? complemented? modular?

*16. Prove that a lattice is distributive if and only if

$$(x \sqcap y) \sqcup (x \sqcap z) \sqcup (y \sqcap z) = (x \sqcup y) \sqcap (x \sqcup z) \sqcap (y \sqcup z) \qquad \forall\ x, y, z \in L.$$

17. Show that the set of positive integers, $\mathbb{N}$, ordered by divisibility is a distributive lattice.

18. For a distributive lattice L show that $a \sqcup b = a \sqcup c$ and $a \sqcap b = a \sqcap c$ imply $b = c$.

19. Let L be a lattice. Show without the use of duality that the following conditions are equivalent:
   (i) $\forall\ a, b, c \in L: (a \sqcup b) \sqcap c = (a \sqcap c) \sqcup (b \sqcap c)$;
   (ii) $\forall\ a, b, c \in L: (a \sqcap b) \sqcup c = (a \sqcup c) \sqcap (b \sqcup c)$.

§2. Boolean Algebras

Boolean algebras are special lattices which are useful in the study of logic, both digital computer logic and that of human thinking, and of switching circuits. This latter application was initiated by C. E. Shannon, who showed that fundamental properties of electrical circuits of bistable elements can be represented by using Boolean algebras. We shall consider some such applications in Chapter 2.

A. Basic Properties

2.1 Definition. A complemented distributive lattice is called a *Boolean algebra* (or a *Boolean lattice*).

Distributivity in a Boolean algebra guarantees the uniqueness of complements (see 1.35), which of course exist because of the complementarity of each element (see Definition 1.33). Since every distributive lattice is modular (Corollary 1.28) all properties of distributive, modular and complemented lattices hold in Boolean algebras.

2.2 Notation. From now on, B denotes a set with the two binary operations $\sqcap$ and $\sqcup$, with a "zero" and a "one" element "0" and "1" and the unary operation complement "'", in short $B = (B, \sqcap, \sqcup, 0, 1, ')$ or $B = (B, \sqcap, \sqcup)$.

2.3 Examples. (a) $B = (\mathscr{P}(M), \cap, \cup, \varnothing, M, ')$ is the Boolean algebra of the power set of a set M. Here $\cap$ and $\cup$ are the set-theoretic operations intersection and union and the complement is the set-theoretic complement, namely $M \backslash A = A'$ is the complement of $A \in P(M)$; $\varnothing$ and M are the "universal" bounds. If M has n elements then B consists of 2^n elements.

(b) $\mathbb{B}^n := (\{0, 1\}^n, \sqcap, \sqcup, 0, 1, ')$ where the operations are defined as in 1.19. For $i_k, j_k \in \{0, 1\}$,

$$(i_1, \ldots, i_n) \sqcap (j_1, \ldots, j_n) := (\min(i_1, j_1), \ldots, \min(i_n, j_n)),$$

$$(i_1, \ldots, i_n) \sqcup (j_1, \ldots, j_n) := (\max(i_1, j_1), \ldots, \max(i_n, j_n)),$$

$$(i_1, \ldots, i_n)' := (i_1', \ldots, i_n') \quad \text{with } 0' := 1, 1' := 0,$$

and

$$0 := (0, \ldots, 0), 1 := (1, \ldots, 1). \qquad \square$$

2.4 Theorem ("De Morgan's Laws"). *In every Boolean algebra B we have* $\forall x, y \in B$:

$$(x \sqcap y)' = x' \sqcup y' \quad \text{and} \quad (x \sqcup y)' = x' \sqcap y'.$$

PROOF.

$$(x \sqcap y) \sqcup (x' \sqcup y') = (x \sqcup x' \sqcup y') \sqcap (y \sqcup x' \sqcup y')$$
$$= (1 \sqcup y') \sqcap (1 \sqcup x') = 1 \sqcap 1 = 1,$$
$$(x \sqcap y) \sqcap (x' \sqcup y') = (x \sqcap y \sqcap x') \sqcup (x \sqcap y \sqcap y')$$
$$= (0 \sqcap y) \sqcup (0 \sqcap x) = 0 \sqcup 0 = 0.$$

This implies that $x' \sqcup y'$ is the complement of $x \sqcap y$. The second formula follows dually. □

2.5 Corollary. *In a Boolean algebra B we have*

$$\forall\, x, y \in B \colon x \leq y \Leftrightarrow x' \geq y'.$$

PROOF.

$$x \leq y \Leftrightarrow x \sqcup y = y \Leftrightarrow x' \sqcap y' = (x \sqcup y)' = y'$$
$$\Leftrightarrow x' \geq y'. \qquad \qquad \square$$

2.6 Theorem. *In a Boolean algebra B we have*

$$\forall\, x, y \in B \colon x \leq y \Leftrightarrow x \sqcap y' = 0 \Leftrightarrow x' \sqcup y = 1 \Leftrightarrow x \sqcap y = x \Leftrightarrow x \sqcup y = y.$$

PROOF. Exercise.

2.7 Definition. A ring $R = (R, +, \cdot)$ with 1 is called *Boolean ring* if all elements $x \in R$ are idempotent, i.e. $x^2 = x$.

2.8 Theorem. *Every Boolean ring R is commutative and of characterististic 2.*

PROOF. Let $x, y \in R$. Then

$$x + y = (x + y)(x + y) = x^2 + xy + yx + y^2 = x + xy + yx + y.$$

Hence $xy + yx = 0$. For $x = y$ we have $2x = x + x = x^2 + x^2 = 0$. Also for any x, y we have $xy = xy + xy + yx = yx$. □

Every Boolean ring with identity can be given the structure of a Boolean algebra and conversely.

2.9 Example. The basic example for a Boolean ring is the set $\mathscr{P}(S)$ of all subsets of a set S, for which the operations $+$ and $\cdot$ are defined in terms of the set-theoretic union, $\cup$, intersection $\cap$ and complementation $\bar{}$ as follows:

$$\forall\, M, N \in \mathscr{P}(S) \colon \quad M + N := (M \cap \bar{N}) \cup (\bar{M} \cap N),$$
$$MN := M \cap N. \qquad \qquad \square$$

2.10 Theorem. *If we define the operations $+$ and $\cdot$ on a Boolean algebra $B = (B, \sqcap, \sqcup)$ by*

$$x + y := (x \sqcap y') \sqcup (x' \sqcap y), \qquad x \cdot y := x \sqcap y,$$

then we obtain a Boolean ring $R(B) = (B, +, \cdot)$ with identity.

Conversely, if we define the operation $\sqcup$ and $\sqcap$ on a Boolean ring $R = (R, +, \cdot)$ with 1 by

$$x \sqcup y := x + y + xy, \qquad x \sqcap y := x \cdot y$$

and the complement a' of a by $a' := a + 1$, then we obtain a Boolean algebra $B(R) = (R, \sqcap, \sqcup)$.

Furthermore, the Boolean algebra defined on the Boolean ring corresponding to the algebra is the algebra itself, i.e.

$$B(R(B)) = B, \qquad R(B(R)) = R.$$

PROOF. Given a Boolean algebra $(B, \sqcap, \sqcup, 0, 1, ')$, define

$$x + y := (x \sqcap y') \sqcup (x' \sqcap y),$$

and

$$xy := x \sqcap y.$$

Clearly $+$ is commutative.

Associativity:

$$(x + y) + z = ((x \sqcap y') \sqcup (x' \sqcap y)) + z$$
$$= [((x \sqcap y') \sqcup (x' \sqcap y)) \sqcap z'] \sqcup [((x \sqcap y') \sqcup (x' \sqcap y))' \sqcap z]$$
$$= [-] \sqcup [(x \sqcap y')' \sqcap (x' \sqcap y)' \sqcap z]$$
$$= [-] \sqcup [(x' \sqcup y) \sqcap (x \sqcup y') \sqcap z]$$
$$= [-] \sqcup [((x \sqcap y) \sqcup (x' \sqcap y')) \sqcap z]$$
$$= (x \sqcap y' \sqcap z') \sqcup (x' \sqcap y \sqcap z') \sqcup (x \sqcap y \sqcap z) \sqcup (x' \sqcap y' \sqcap z)$$
$$= x + (y + z),$$

by symmetry in x and z and using commutativity.

Also $x + 0 = (x \sqcap 0') \sqcup (x' \sqcap 0) = x$ and $x + x = (x \sqcap x') \sqcup (x' \sqcap x) = 0$. So $(B, +)$ is a commutative group. We know that $\cdot$ is associative and commutative and $x \cdot 1 = (x \sqcap 1) = x$.

Distributivity:

$$xy + xz = (x \sqcap y) + (x \sqcap z)$$
$$= [(x \sqcap y) \sqcap (x \sqcap z)'] \sqcup [(x \sqcap y)' \sqcap (x \sqcap z)]$$
$$= [x \sqcap y \sqcap (x' \sqcup z')] \sqcup [(x' \sqcup y') \sqcap x \sqcap z]$$
$$= (x \sqcap y \sqcap z') \sqcup (x \sqcap z \sqcap y')$$
$$= x \sqcap ((y \sqcap z') \sqcup (y' \sqcap z)$$
$$= x(y + z).$$

Finally, $x^2 = x \sqcap x = x$. Hence $(B, +, \cdot, 0, 1)$ is a Boolean ring.

Conversely, suppose $(B, +, \cdot, 0, 1)$ is a Boolean ring; define $x \sqcup y :=$
$x + y + xy$, $x \sqcap y := xy$, $x' := x + 1$. (L1) and (L4) follow by Theorem 2.8.

(L2)
$$x \sqcup (y \sqcup z) = x \sqcup (y + z + yz)$$
$$= x + y + z + yz + xy + xz + xyz$$
$$= (x + y + xy) \sqcup z$$
$$= (x \sqcup y) \sqcap z.$$

$x \sqcap (y \sqcap z) = (x \sqcap y) \sqcap z$ is immediate.

$x \sqcap (x \sqcup y) = x \sqcap (x + y + xy) = x^2 + xy + x^2y = x + xy + xy = x.$

Also
$$x \sqcup (x \sqcap y) = x \sqcup (xy) = x + xy + xxy = x.$$

So $(B, \sqcap, \sqcup)$ is a lattice.

Distributivity:
$$x \sqcap (y \sqcup z) = x(y + z + yz)$$
$$= xy + xz + xyz$$
$$= xy + xz + xyxz$$
$$= xy \sqcup xz$$
$$= (x \sqcap y) \sqcup (x \sqcap z).$$

Clearly 0 and 1 are lower and upper bounds.
$$x \sqcap x' = x(1 + x) = x + x = 0,$$
$$x \sqcup x' = x + x' + xx' = x + 1 + x + x + x^2 = 1.$$

So x' is the complement of x. So $(B, \sqcap, \sqcup, 0, 1, ')$ is a Boolean algebra. Finally, given a Boolean algebra B, we then define on $R(B)$ the operations $\wedge$ and $\vee$ on $R(B)$.

$$x \wedge y := xy$$

and

$$x \vee y := x + y + xy.$$

Then $x \wedge y = xy = x \sqcap y$ and
$$x \vee y = x + y + xy = 1 + (1 + x)(1 + y)$$
$$= 1 + x'y'$$
$$= (x' \sqcap y')'$$
$$= x \sqcup y.$$

Also $1 + x = (1 \sqcap x') \sqcup (1' \sqcap x) = x' \sqcup 0 = x'$. This proves $B(R(B)) = B$. Similarly, $R(B(R)) = R$. □

2.11 Definition. Let B_1 and B_2 be Boolean algebras. Then the mapping $f: B_1 \rightarrow B_2$ is called a (*Boolean*) *homomorphism* from B_1 into B_2 if f is a (lattice) homomorphism and for all $x \in B$ we have $f(x') = (f(x))'$.

Analogously we can define Boolean monomorphisms and isomorphisms by using 1.17. If there is a Boolean isomorphism between B_1 and B_2 we write $B_1 \cong_b B_2$. The simple proofs of the following properties are left as exercises.

2.12 Theorem. *Let $f: B_1 \rightarrow B_2$ be a Boolean homomorphism. Then*:

(i) $f(0) = 0, f(1) = 1$;
(ii) $\forall\, x, y \in B_1: x \le y \Rightarrow f(x) \le f(y)$;
(iii) $f(B_1)$ *is a Boolean algebra and a subalgebra of B_2.*

2.13 Examples. (i) If $M \subset N$ then the map $f: \mathscr{P}(M) \rightarrow \mathscr{P}(N)$, $A \mapsto A$ is a lattice monomorphism but not a Boolean homomorphism, since for $A \in \mathscr{P}(M)$ the complements in M and N are different. Also, $f(1) = f(M) = M \ne N =$ one in $\mathscr{P}(N)$.
(ii) If $M = \{1, \ldots, n\}$, then $\{0, 1\}^n$ and $\mathscr{P}(M)$ are Boolean algebras and the map $f: \{0, 1\}^n \rightarrow \mathscr{P}(M)$, $(i_1, \ldots, i_n) \mapsto \{k \,|\, i_k = 1\}$ is a Boolean isomorphism. □

The following terms are defined for any lattice V rather than just for Boolean algebras.

2.14 Definition. Let V be a lattice with zero. $a \in V$ is called an *atom* if for all $b \in V: 0 < b \le a \Rightarrow b = a$.

2.15 Definition. $a \in V$ is called *join-irreducible* if for all $b, c \in V$

$$a = b \sqcup c \Rightarrow a = b \quad \text{or} \quad a = c.$$

Otherwise a is called *join-reducible*.

2.16 Lemma. *Every atom of a lattice with zero is join-irreducible.*

PROOF. Let a be an atom and let $a = b \sqcup c$, $a \ne b$. Then $a = \sup(b, c)$; so $b \le a$. Therefore $b = 0$ and $a = c$. □

2.17 Lemma. *Let V be a distributive lattice with $p \in V$ join-irreducible and $p \le a \sqcup b$. Then $p \le a$ or $p \le b$.*

PROOF. $p \le a \sqcup b$ means $p = p \sqcap (a \sqcup b) = (p \sqcap a) \sqcup (p \sqcap b)$. Since p is join-irreducible; $p = p \sqcap a$ or $p = p \sqcap b$, i.e. $p \le a$ or $p \le b$. □

2.18 Definition. If $x \in [a, b] = \{v \in V \mid a \leq v \leq b\}$ and $y \in V$ with $x \sqcap y = a$ and $x \sqcup y = b$, then y is called a *relative complement* of x with respect to $[a, b]$. If all intervals $[a, b]$ in a lattice V are complemented, then V is called *relatively complemented*. If V has a zero element and all $[0, b]$ are complemented, then V is called *sectionally complemented*.

2.19 Theorem. *Let V be a lattice. Then the following implications hold:*

(1) (V *is a Boolean algebra*) $\Rightarrow$ (V *is relatively complemented*);
(2) (V *is relatively complemented*) $\Rightarrow$ (V *is sectionally complemented*);
(3) (V *is finite and sectionally complemented*) $\Rightarrow$ (*every* $0 \neq a \in V$ *is a join of finitely many atoms*).

PROOF. (1) First we show: If V is distributive and complemented then V is relatively complemented. Let $a \leq x \leq b$. Then there is a $c \in V$ such that $x \sqcap c = 0$, $x \sqcup c = 1$. Assume $y = b \sqcap (a \sqcup c)$. y is a complement of x in $[a, b]$, since

$$x \sqcap y = x \sqcap (b \sqcap (a \sqcup c)) = x \sqcap (a \sqcup c) = (x \sqcap a) \sqcup (x \sqcap c) = x \sqcap a = a$$

and

$$x \sqcup y = x \sqcup (b \sqcap (a \sqcup c)) = x \sqcup ((b \sqcap a) \sqcup (b \sqcap c)) = x \sqcup (b \sqcap c)$$

$$= (x \sqcup b) \sqcap (x \sqcup c) = x \sqcup b = b.$$

Thus V is relatively complemented.

(2) If V is relatively complemented then every $[a, b]$ is complemented; thus every interval $[0, b]$ is complemented, i.e. V is sectionally complemented.

(3) Let $\{p_1, \ldots, p_n\}$ be the set of atoms a and let $b = p_1 \sqcup \ldots \sqcup p_n$. Now $b \leq a$, and if we suppose that $b \neq a$ then b has a nonzero complement, say c, in $[0, a]$.

Let p be an atom less than c then $p \in \{p_1, \ldots, p_n\}$ and thus $p = p \sqcap b \leq c \sqcap b = 0$ which is a contradiction. Hence $a = b = p_1 \sqcup \ldots \sqcup p_n$. $\square$

The finite Boolean algebras can be characterized as follows.

2.20 Theorem (Representation Theorem). *Let B be a finite Boolean algebra, and let A denote the set of all atoms in B. Then B is isomorphic to $\mathscr{P}(A)$, i.e.*

$$(B, \sqcap, \sqcup) \cong_b (\mathscr{P}(A), \cap, \cup).$$

PROOF. Let $v \in B$ be an arbitrary element and let $A(v) := \{a \in B \mid a \text{ atom}, a \leq v\}$, and $A(v) = \varnothing$ if $v = 0$. Then $A(v) \subseteq A$. Define

$$h: B \to \mathscr{P}(A), \qquad v \mapsto A(v).$$

We show that h is a Boolean isomorphism. h is a Boolean homomorphism: let $x \in A(v \sqcap w)$ then x is an atom and $x \leq v \sqcap w$; also $x \leq v$ and $x \leq w$, so $x \in A(v)$ and $x \in A(w)$, hence $x \in A(v) \cap A(w)$. Thus $h(v \sqcap w) = h(v) \cap h(w)$. The proofs of $h(v \sqcup w) = h(v) \cup h(w)$ and $h(v') = A \setminus h(v)$ are similar. Since B is finite we are able to use Theorem 2.19 to verify that h is bijective. We know that every $v \in B$ can be expressed as a join of finitely many atoms: $v = a_1 \sqcup \ldots \sqcup a_n$, all atoms $a_i \le v$. Let $h(v) = h(w)$, i.e. $A(v) = A(w)$. Then $a_i \in A(v)$ and $a_i \in A(w)$. Therefore $a_i \le w$ and thus $v \le w$. Reversing the roles of v and w yields $v = w$ and this shows that h is injective.

To show that h is surjective we verify that for each $C \in \mathscr{P}(A)$ there is a $v \in B$ such that $h(v) = C$. Let $C \subseteq A$, $C = \{c_1, \ldots, c_n\}$. Let $v = c_1 \sqcup \ldots \sqcup c_n$. Then $A(v) \supseteq C$, hence $h(v) \supseteq C$.

Conversely, if $a \in h(v)$, then a is an atom with $a \le v = c_1 \sqcup \ldots \sqcup c_n$. Therefore $a \le c_i$ for some $1 \le i \le n$, by 2.16 and 2.17. So $a = c_i \in C$. Altogether this implies $h(v) = A(v) = C$. $\qquad\qquad\qquad\qquad\qquad\square$

2.21 Theorem. *The cardinality of a finite Boolean algebra is always of the form 2^n and any two Boolean algebras with the same cardinality are isomorphic.*

PROOF. We know that $|\mathscr{P}(A)| = 2^n$ if $|A| = n$. To show that two Boolean algebras B_1, B_2 with the same number of elements are isomorphic, we use Theorem 2.20 and let $B_i \cong_b \mathscr{P}(A_i)$, where $|B_i| = |\mathscr{P}(A_i)| = 2^{|A_i|}$, where A_i is the set of atoms of B_i, $i = 1, 2$. If $|A_1| = |A_2|$, then there exists a bijection $f: A_1 \to A_2$. Then we define a mapping

$$g: \mathscr{P}(A_1) \to \mathscr{P}(A_2), \qquad A_1 \mapsto f(A_1)$$

and leave it as an exercise to verify that g is a Boolean isomorphism. $\qquad\square$

A different proof of the representation theorem can be obtained from the following theorem. We note that in any Boolean algebra B, any interval $[a, b]$ is also a Boolean algebra but not a sub-Boolean algebra of B unless $a = 0$ and $b = 1$. We also note that direct products of Boolean algebras are again Boolean algebras.

2.22 Theorem. *For every finite Boolean algebra $B \ne \{0\}$ there is some $n \in \mathbb{N}$ with*

$$B \cong_b \{0, 1\}^n.$$

PROOF. We use induction on $|B|$. If $|B| = 2$, the result is obvious. So suppose that $|B| > 2$ and let $a \in B$, $a \ne 0$ and $a \ne 1$.

The mapping $f: B \to [0, a] \times [a, 1]$, via $b \mapsto (a \sqcap b, a \sqcup b)$, is injective. Because of $f(b \sqcap c) = (a \sqcap (b \sqcap c), a \sqcup (b \sqcap c)) = f(b) \sqcap f(c)$, $f(b \sqcup c) = (a \sqcap (b \sqcup c), a \sqcup (b \sqcup c)) = f(b) \sqcup f(c)$, and $f(b') = (a \sqcap b', a \sqcup b') = (a \sqcap b, a \sqcup b)' = f(b)'$ (in $[0, a] \times [a, 1]$) for all $b, c \in B$, we know that f is a Boolean homomorphism.

Finally, if $(x, y) \in [0, a] \times [a, 1]$ then $(x, y) = f(b)$ for $b := y \sqcap (a' \sqcup x)$. Hence f is a Boolean isomorphism and $B \cong_b [0, a] \times [a, 1]$. By induction, there are $n, m \in \mathbb{N}$ with $[0, a] \cong_b \{0, 1\}^n$ and $[a, 1] \cong_b \{0, 1\}^m$. Hence $B \cong_b \{0, 1\}^{n+m}$. $\qquad\square$

Then Theorem 2.20 follows from 2.13(ii).

2.23 Example. The lattice of the divisors of 30, i.e. the Boolean algebra $B = (\{1, 2, 3, 5, 6, 10, 15, 30\},$ gcd, lcm, 1, 30, complement with respect to 30), is isomorphic to the lattice of the power set $\mathscr{P}(\{a, b, c\})$. $\qquad\square$

We sketch the Hasse diagrams of some small Boolean algebras:

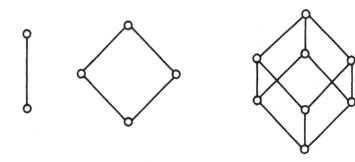

Figure 1.17

2.24 Remark. The identification of B with a power set as in 2.20 is not always possible. It can be shown that for every (not necessarily finite) Boolean algebra B there is a set M with $B \hookrightarrow \mathscr{P}(M)$. This is called Stone's Representation Theorem.

2.25 Definition and Theorem. *Let B be a Boolean algebra and B^n be its n-fold cartesian product. For mappings f and g from B^n into B we define*

$$f \sqcap g: B^n \to B, \qquad x \mapsto f(x) \sqcap g(x);$$
$$f \sqcup g: B^n \to B, \qquad x \mapsto f(x) \sqcup g(x);$$
$$f': B^n \to B, \qquad x \mapsto (f(x))';$$
$$f_0: B^n \to B, \qquad x \mapsto 0;$$
$$f_1: B^n \to B, \qquad x \mapsto 1;$$

for all $x \in B^n$. Then the set $F_n(B)$ of all mappings of B^n into B is a Boolean algebra. $\qquad\square$

PROBLEMS

1. Prove a generalization of Theorem 2.4 involving the meet of n elements of B instead of only two elements.

2. Give three examples of lattices which are not Boolean algebras.

3. How many Boolean algebras are there with four elements 0, 1, a and b?

*4. Consider the set $\mathcal{M}$ of $n \times n$ matrices $\mathbf{X} = (x_{ij})$ whose entries x_{ij} belong to a Boolean algebra $B = (B, \sqcap, \sqcup, 0, 1, ')$. Define two operations on $\mathcal{M}$:

$$\mathbf{X} \sqcup \mathbf{Y} = (x_{ij} \sqcup y_{ij}), \qquad \mathbf{X} \sqcap \mathbf{Y} = (x_{ij} \sqcap y_{ij}),$$

and two matrices $\mathbf{0}$ and $\mathbf{1}$, the all zeros and all ones matrix, respectively. Let $\mathbf{X}' = (x'_{ij})$. Show that $\mathcal{M}$ is a Boolean algebra. Furthermore, let

$$\mathbf{E} = \begin{pmatrix} 1 & 0 & \cdots & 0 \\ 0 & 1 & \cdots & 0 \\ \cdot & \cdot & \cdots & \cdot \\ 0 & 0 & \cdots & 1 \end{pmatrix}$$

and consider the subset $\mathcal{N}$ of $\mathcal{M}$ consisting of all $\mathbf{X} \in \mathcal{M}$ with the property $\mathbf{X} \geq \mathbf{E}$. (Here $\mathbf{X} \geq \mathbf{Y} \Leftrightarrow x_{ij} \geq y_{ij}$ for all i, j.) Show that $\mathcal{N}$ is a sublattice of $\mathcal{M}$ and that it is a filter of $\mathcal{M}$ (see Problem 12 in Subsection B). Verify also that $\mathcal{N}$ is the interval $[\mathbf{E}, \mathbf{I}]$ of $\mathcal{M}$.

*5. Let $\mathcal{N}$ be as in Problem 4, define matrix multiplication and show that $\mathcal{N}$ is closed with respect to multiplication, i.e. $\mathbf{X}, \mathbf{Y} \in \mathcal{N} \Leftrightarrow \mathbf{XY} \in \mathcal{N}$. Prove: If $\mathbf{X} \in \mathcal{N}$, then $\mathbf{X} \leq \mathbf{X}^2 \leq \ldots \leq \mathbf{X}^{n-1} = \mathbf{X}^n$.

*6. A *Newman algebra* N is a generalization of a Boolean algebra, obtained by dropping the commutative and associative axioms. Let N be a set closed under $+$ and $\cdot$ such that:

$$a(b + c) = ab + ac, \qquad (a + b)c = ac + bc \quad \text{for all } a, b, c \in N,$$

there exists a 1 such that $a1 = a$ for all $a \in N$, there exists a 0 such that $a + 0 = a = 0 + a$ for all $a \in N$. To each a there corresponds at least one "complement" a' such that

$$aa' = 0, \qquad a + a' = 1.$$

Prove: (i) $aa = a$; (ii) $(a')' = a$; (iii) $1a = a$; (iv) complements are unique; (v) addition is commutative; (vi) addition is associative.

7. (i) Show by example that relative complements are not always unique.
   (ii) Prove that a complemented modular lattice is relatively complemented.

B. Boolean Polynomials, Ideals

We introduce Boolean polynomials and polynomial functions in a form which is well suited for applications described in Chapter 2.

2.26 Definition. The notion of a Boolean polynomial is defined recursively. Let $X_n = \{x_1, \ldots, x_n\}$ be a set of n symbols (called indeterminates or variables), which does not contain the symbols 0 and 1. The *Boolean* *polynomials* over X_n are the objects which can be obtained by finitely many successive applications of:

(i) $x_1, x_2, \ldots, x_n$, and 0, 1 are Boolean polynomials;
(ii) if p and q are Boolean polynomials, then so are

$$(p) \sqcap (q), (p) \sqcup (q), (p)'.$$

We denote this set of all Boolean polynomials by P_n.

2.27 Remark. Two polynomials are equal if their sequences of symbols are identical. It is necessary to use brackets in 2.26(ii) in order to obtain the sequence of symbols $((x_1) \sqcap (x_2))'$. Instead of $(x_i)'$, $(0)'$ and $(1)'$ we shall write x_i', $0'$ and $1'$, for short.

2.28 Example. Some examples of Boolean polynomials over $\{x_1, x_2\}$ are $0, 1, x_1, x_2, x_1 \sqcap x_2, x_1 \sqcup x_2, x_1', x_1' \sqcap x_2, \ldots$. $\qquad\qquad\qquad\square$

Since every Boolean polynomial over $x_1, \ldots, x_n$ can be regarded as a Boolean polynomial over $x_1, \ldots, x_n, x_{n+1}$, we have

$$P_1 \subset P_2 \subset \ldots \subset P_n \subset P_{n+1} \subset \ldots.$$

Note that P_n is not a Boolean algebra. Of course, we want $x_1 \sqcap x_2$ and $x_2 \sqcap x_1$ to be related. Therefore we introduce the concept of polynomial functions as follows.

2.29 Definition. Let B be a Boolean algebra, B^n be the direct product of n copies of B, and p be a Boolean polynomial in P_n. Then

$$\bar{p}_B : B^n \to B, (a_1, \ldots, a_n) \mapsto \bar{p}_B(a_1, \ldots, a_n)$$

is called the *Boolean polynomial function* induced by p on B. Here $\bar{p}_B(a_1, \ldots, a_n)$ is the element in B which is obtained from p by replacing each x_i by $a_i \in B$, $1 \le i \le n$.

The following example shows that two different Boolean polynomials can have the same Boolean polynomial function. Again, $\mathbb{B}$ denotes the Boolean algebra $\{0, 1\}$ with the usual operations $\sqcup, \sqcap, '$.

2.30 Example. Let $n = 2$, $p = x_1 \sqcap x_2$, $q = x_2 \sqcap x_1$. Then

$$\bar{p}_B : \mathbb{B}^2 \to \mathbb{B}, (0,0) \mapsto 0, (0,1) \mapsto 0, (1,0) \mapsto 0, (1,1) \mapsto 1,$$

$$\bar{q}_B : \mathbb{B}^2 \to \mathbb{B}, (0,0) \mapsto 0, (0,1) \mapsto 0, (1,0) \mapsto 0, (1,1) \mapsto 1.$$

Therefore $\bar{p}_\mathbb{B} = \bar{q}_\mathbb{B}$. $\qquad\qquad\qquad\qquad\qquad\qquad\qquad\qquad\square$

Let B be a Boolean algebra. Using the notation introduced in 2.29, we define

2.31 Definition. $P_n(B) := \{\bar{p}_B | p \in P_n\}$.

2.32 Theorem. *Let B be a Boolean algebra; then the set $P_n(B)$ is a Boolean algebra and a subalgebra of the Boolean algebra $F_n(B)$ of all functions from B^n into B.*

PROOF. We have to verify that $P_n(B)$ is closed with respect to join, meet and complement of functions (as defined in 2.25) and that $P_n(B)$ contains f_0 and f_1. For $\sqcap$,

$$(\bar{p}_B \sqcap \bar{q}_B)(a_1, \ldots, a_n) = \bar{p}_B(a_1, \ldots, a_n) \sqcap \bar{q}_B(a_1, \ldots, a_n)$$

$$= \overline{p_B \sqcap q_B}(a_1, \ldots, a_n) \qquad \forall \, a_i \in B,$$

implies that $\forall \, \bar{p}_B, \bar{q}_B \in p_n(B)$, $\bar{p}_B \sqcap \bar{q}_B = \overline{p \sqcap q_B} \in p_n(B)$. For $\sqcup$ and $'$ we proceed similarly. Also $\bar{0} = f_0$, $\bar{1} = f_1$. $\qquad \square$

We partition the set of Boolean polynomials by using equality of Boolean polynomial functions.

2.33 Definition. Two Boolean polynomials $p, q \in P_n$ are called *equivalent* (in symbols $p \sim q$), if their Boolean polynomial functions on $\mathbb{B}$ are equal, i.e.

$$p \sim q :\Leftrightarrow \bar{p}_{\mathbb{B}} = \bar{q}_{\mathbb{B}}.$$

We shall show that $\sim$ is an equivalence relation on P_n which partitions Boolean polynomials into classes of equivalent polynomials with equal polynomial function. We formulate this precisely.

2.34 Theorem. (a) *The relation $\sim$ in 2.33 is an equivalence on P_n.*
(b) *$P_n/\sim$ is a Boolean algebra with respect to the usual operations on equivalence classes and*

$$P_n/\sim \, \cong_b P_n(\mathbb{B})$$

as Boolean algebras.

PROOF. (a) We have $p \sim p$ for all $p \in P_n$, since $\bar{p}_{\mathbb{B}} = \bar{p}_{\mathbb{B}}$. For all p, q, r in P_n we have

$$p \sim q \quad \text{and} \quad q \sim r \Rightarrow p \sim r,$$

since $p \sim q \Rightarrow \bar{p}_{\mathbb{B}} = \bar{q}_{\mathbb{B}}$ and $q \sim r \Rightarrow \bar{q}_{\mathbb{B}} = \bar{r}_{\mathbb{B}}$ imply $\bar{p}_{\mathbb{B}} = \bar{r}_{\mathbb{B}}$. Also $p \sim q \Rightarrow q \sim p$, for all $p, q \in P_n$, since $\bar{p}_{\mathbb{B}} = \bar{q}_{\mathbb{B}} \Rightarrow \bar{q}_{\mathbb{B}} = \bar{p}_{\mathbb{B}}$.
(b) We define the mapping $h: P_n(\mathbb{B}) \to P_n/\sim$, which maps $\bar{p}_{\mathbb{B}}$ to the equivalence class of p, denoted by C_p. h is well defined, since $\bar{p}_{\mathbb{B}} = \bar{q}_{\mathbb{B}} \Rightarrow p \sim q \Rightarrow C_p = C_q$. It is easily verified that h is a Boolean isomorphism. We leave the details as an exercise. $\qquad \square$

Moreover, it can be shown with the help of 2.24 that for equivalent polynomials the corresponding polynomial functions coincide on any Boolean algebra, not only on $\mathbb{B}$. That is:

2.35 Theorem. *Let $p, q \in P_n$; $p \sim q$ and let B be an arbitrary Boolean algebra. Then $\bar{p}_B = \bar{q}_B$.*

PROOF. From Remark 2.24 we assume $B \hookrightarrow \mathcal{P}(M)$; from 2.25 we know that the set of all functions from a set M into $\mathbb{B}$ is a Boolean algebra. We define

$$h: \mathcal{P}(M) \to \mathbb{B}^M, \qquad C \mapsto \chi_C$$

and claim that h is a Boolean isomorphism. Here $\mathbb{B}^M$ denotes the algebra of all mappings from M into $\mathbb{B}$, χ_C is the characteristic function. We leave it as an exercise to verify the properties of a Boolean isomorphism for h. Thus $B \hookrightarrow \mathbb{B}^M$ and it suffices to prove the theorem for $\mathbb{B}^M$. We know (from the definition) that

$$p \sim q \Leftrightarrow \bar{p}_\mathbb{B} = \bar{q}_\mathbb{B} \Leftrightarrow \bar{p}_\mathbb{B}(i_1, \ldots, i_n) = \bar{q}_\mathbb{B}(i_1, \ldots, i_n) \quad \text{for all } i_1, \ldots, i_n \in \mathbb{B}.$$

Let $f_1, \ldots, f_n \in \mathbb{B}^M$ and let $m \in M$. Then

$$(\bar{p}_{\mathbb{B}^M}(f_1, \ldots, f_n))(m) = \bar{p}_\mathbb{B}(f_1(m), \ldots, f_n(m))$$
$$= \bar{q}_\mathbb{B}(f_1(m), \ldots, f_n(m))$$
$$= (\bar{q}_{\mathbb{B}^M}(f_1, \ldots, f_n))(m).$$

Hence $\bar{p}_{\mathbb{B}^M} = \bar{q}_{\mathbb{B}^M}$. $\qquad\qquad\qquad\qquad\qquad\qquad\qquad\qquad\qquad\qquad$ $\square$

One frequently wants to replace a given polynomial p by an equivalent polynomial which is of simpler or more systematic form. This is achieved by considering so-called normal forms. The collection of normal forms provides a representative system for the equivalence classes of P_n.

2.36 Definition. $N \subseteq P_n$ is called a *system of normal forms* if

(i) Every $p \in P_n$ is equivalent to some $q \in N$;
(ii) $\forall\, q_1, q_2 \in N: q_1 \neq q_2 \Rightarrow q_1 \not\sim q_2$.

In the following theorem we describe two systems of normal forms. They represent Boolean polynomials by equivalent "join of meet" polynomials (or "meet of join" polynomials). We use the notation:

$$x_i^1 := x_i, \quad x_i^{-1} := x_i', \quad 0^1 := 0, \quad 0^{-1} := 1, \quad 1^1 := 1, \quad 1^{-1} := 0.$$

2.37 Theorem. *The following two sets are systems of normal forms in P_n:*

(i) $\quad N_d := \left\{ \bigsqcup_{(i_1, \ldots, i_n) \in \{1, -1\}^n} d_{i_1 \ldots i_n} \sqcap x_1^{i_1} \sqcap x_2^{i_2} \sqcap \ldots \sqcap x_n^{i_n} \text{ where } d_{i_1 \ldots i_n} \in \{0, 1\} \right\}.$

(ii) $N_c := \left\{ \displaystyle\bigsqcap_{(i_1,\dots,i_n)\in\{1,-1\}^n} c_{i_1\dots i_n} \sqcup x_1^{i_1} \sqcup x_2^{i_2} \sqcup \dots \sqcup x_n^{i_n} \text{ where } c_{i_1\dots i_n} \in \{0,1\} \right\}.$

PROOF. If

$$p = \bigsqcup_{(i_1,\dots,i_n)\in(1,-1)^n} d_{i_1\dots i_n} \sqcap x_1^{i_1} \sqcap \dots \sqcap x_n^{i_n} \in N_d,$$

then

$$\bar p_{\mathbb{B}}(1^{k_1},\dots,1^{k_n}) = \bigsqcup_{(i_1,\dots,i_n)\in\{1,-1\}^n} d_{i_1\dots i_n} \sqcap 1^{i_1 k_1} \sqcap \dots \sqcap 1^{i_n k_n}.$$

Now

$$d_{i_1\dots i_n} \sqcap 1^{i_1 k_1} \sqcap \dots \sqcap 1^{i_n k_n} = \begin{cases} 1 & \text{if } i_1 = k_1,\dots, i_n = k_n \text{ and } d_{i_1\dots i_n} = 1; \\ 0 & \text{otherwise.} \end{cases}$$

Therefore $\bar p_{\mathbb{B}}(1^{k_1},\dots,1^{k_n}) = d_{k_1\dots k_n}$.

Thus the function $\bar p_{\mathbb{B}}$ is completely determined by the values of the $d_{i_1\dots i_n}$'s in $p \in N_d$. We may regard the values of the $d_{i_1\dots i_n}$'s in p as a function from $\{-1,1\}^n$ into $\{0,1\}$. Thus there is a bijection between $\{\bar p_{\mathbb{B}} | p \in N_d\}$ and the set of functions from $\{-1,1\}^n$ into $\{0,1\}$. That is

$$|\{\bar p_{\mathbb{B}} | p \in N_d\}| = |\{0,1\}^{\{-1,1\}^n}| = 2^{2^n}.$$

$$2^{2^n} = |\{\bar p_{\mathbb{B}} | p \in N_d\}| \le |P_n/\!\sim| = |P_n(\mathbb{B})| \le |F_n(\mathbb{B})| = 2^{2^n}.$$

Hence there are precisely 2^{2^n} equivalence classes in P_n. This proves the assertion for (i), (ii) is shown similarly (or by "duality").

Alternatively, we give a more detailed proof that N_d is a system of normal forms. First we show that each equivalence class contains at most one element of N_d, i.e. for all $p, q \in N_d$ with $p \ne q \Rightarrow p \not\sim q$, i.e. $\bar p_{\mathbb{B}} \ne \bar q_{\mathbb{B}}$. Let p be as in 2.37(i), q be of the form

$$q = \bigsqcup_{(i_1,\dots,i_n)\in\{1,-1\}^n} e_{i_1,\dots,i_n} \sqcap x_1^{i_1} \dots \sqcap x_n^{i_n}$$

and let $d_{j_1,\dots,j_n} \ne e_{j_1,\dots,j_n}$ for suitable $j_1,\dots,j_n$. We define $(v_1,\dots,v_n) = (1^{j_1},\dots,1^{j_n}) \in \mathbb{B}^n$. Then

$$\bar p_{\mathbb{B}}(1^{j_1},\dots,1^{j_n}) = \overline{(\bigsqcup d_{i_1,\dots,i_n} \sqcap (x_1^{i_1} \sqcap \dots \sqcap x_n^{i_n}))}(1^{j_1},\dots,1^{j_n})$$

$$= \bigsqcup d_{i_1,\dots,i_n} \sqcap ((1^{j_1})^{i_1} \sqcap \dots \sqcap (1^{j_n})^{i_n}).$$

Denote this last expression by y. If $j_k \ne i_k$, $k = 1,\dots, n$, then $(1^{j_k})^{i_k} = 0$ and then the whole expression y will be 0. Therefore $j_k = i_k$ and then $y = d_{j_1,\dots,j_n} \sqcap 1 \ne e_{j_1,\dots,j_n} = \bar q_{\mathbb{B}}(1^{j_1},\dots,1^{j_n})$, with the same argument as for $\bar p_{\mathbb{B}}$. Therefore $\bar p_{\mathbb{B}} \ne \bar q_{\mathbb{B}}$.

Next we show that each equivalence class contains at least one element of N_d, i.e. for every Boolean polynomial p there is an $n \in N_d$ such that $p \sim n$. We show that $\bar N_d = \{\bar n_{\mathbb{B}} | n \in N_d\}$ equals $P_n(\mathbb{B}) = \{\bar p_{\mathbb{B}} | p \in P_n\}$. Here $P_n(\mathbb{B})$ is regarded as the set of all mappings f from $\mathbb{B}^n$ into $\mathbb{B}$, where f

consists of expressions obtained by applying $\sqcap$, $\sqcup$ and $'$ onto $\bar{0}$, $\bar{1}$ and $\bar{x}_1, \ldots, \bar{x}_n$ finitely many times. $\bar{x}_i : \mathbb{B}^n \to \mathbb{B}$ is the ith projection $(v_1, \ldots, v_n) \to v_i$. To show that there is a $\bar{n}_0 \in \bar{N}_d$ for $\bar{0}$ such that $\bar{0} = \bar{n}_0$, consider $0 \sim p$, where p is as in 2.37. Replace all coefficients $d_{i_1,\ldots,i_n}$ by 0, then $\bar{0} = \bar{n}_0$. To show that for $\bar{1}$ there is a $\bar{n}_1 \in \bar{N}_d$ such that $\bar{1} = \bar{n}_1$, put all coefficients $d_{i_1,\ldots,i_n}$ equal to 1. Then use mathematical induction to verify

$$ 1 \sim \bigsqcup_{(i_1,\ldots,i_n)\in\{1,-1\}^n} x_1^{i_1} \sqcap \ldots \sqcap x_n^{i_n}; $$

so $\bar{1} = \bar{n}_1$.

Next we have to show that for every $\bar{x}_k$ there is a $\bar{n}_k \in \bar{N}_d$ such that $\bar{x}_k = \bar{n}_k$. It is easily seen that

$$ \bigsqcup_{(i_1,\ldots,i_{k-1},i_{k+1},\ldots,i_n)\in\{1,-1\}^n} 1 \sqcap (x_1^{i_1} \sqcap \ldots \sqcap x_k' \sqcap \ldots \sqcap x_n^{i_n}) \sqcup $$

$$ 0 \sqcap (x_1^{i_1} \sqcap \ldots \sqcap x_k^{-1} \sqcap \ldots \sqcap x_n^{i_n}) $$

is equivalent to x_k.

Finally we have to verify that if for every function $\bar{p}$ and $\bar{q}$ there are functions $\bar{n}$ and $\bar{m}$ in $\bar{N}_d$, then for $\bar{p} \sqcap \bar{q}$, $\bar{p} \sqcup \bar{q}$ and $\bar{p}'$ there are equivalent functions in $\bar{N}_d$. We demonstrate this in the case $\bar{p} \sqcap \bar{q}$. Let p and q be as above. Then

$$ p \sqcap q \sim \left(\bigsqcup_{(i_1,\ldots,i_n)\in\{1,-1\}^n} d_{i_1,\ldots,i_n} \sqcap (x_1^{i_1} \sqcap \ldots \sqcap x_n^{i_n}) \right) $$

$$ \sqcap \left(\bigsqcup_{(j_1,\ldots,j_n)\in\{1,-1\}^n} e_{j_1,\ldots,j_n} \sqcap (x_1^{j_1} \sqcap \ldots \sqcap x_n^{j_n}) \right) $$

$$ \sim \bigsqcup_{(r_1,\ldots,r_n)\in\{1,-1\}^n} f_{r_1,\ldots,r_n} \sqcap (x_1^{r_1} \sqcap \ldots \sqcap x_n^{r_n}), $$

where $d_{i_1,\ldots,i_n} \sqcap e_{j_1,\ldots,j_n} =: f_{r_1,\ldots,r_n} \in \{0,1\}$.

The remaining cases are treated similarly. $\square$

2.38 Definition. Let $p \in P_n$.

(i) The uniquely determined polynomial $p_d \in N_d$ with $p \sim p_d$ is called the *disjunctive normal form* of p.

(ii) The uniquely determined polynomial $p_c \in N_c$ with $p \sim p_c$ is called the *conjunctive normal form* of p.

In the disjunctive normal form we only write down the terms $x_1^{i_1} \sqcap \ldots \sqcap x_n^{i_n}$ for which $d_{i_1\ldots i_n} = 1$ (and omit this coefficient). A similar convention applies to the conjunctive normal form, where we list only those terms $x_1^{i_1} \sqcup \ldots \sqcup x_n^{i_n}$ with coefficients $c_{i_1,\ldots,i_n} = 1$. Thus for $(1 \sqcap x_1 \sqcap x_2) \sqcup (0 \sqcap x_1 \sqcap x_2') \sqcup (1 \sqcap x_1' \sqcap x_2) \sqcup (0 \sqcap x_1' \sqcap x_2')$ we write $(x_1 \sqcap x_2) \sqcup (x_1' \sqcap x_2)$ and this is in disjunctive normal form, while $(x_1 \sqcap x_2) \sqcup x_1' \sqcup (x_2 \sqcap x_1' \sqcap x_2)'$ is not.

We note that in the systems N_d and N_c of normal forms defined above the order of the meets and joins, respectively, is not prescribed. Thus N_d and N_c are not uniquely determined as "normal forms" should be. This can be overcome by a formal device (see DORNHOFF and HOHN, p. 136). However, we omit this approach since it is not essential for the purpose of this chapter.

2.39 Example. We demonstrate the second part of the detailed proof of Theorem 2.37 for the special case:

$$n = 2, \quad \text{and} \quad p = (x_1 \sqcap x_2) \sqcup (x_1 \sqcap x_2')$$

$$q = (x_1 \sqcap x_2') \sqcup (x_1' \sqcap x_2'), \quad (p, q \in N_d).$$

(a) $\quad p \sqcap q = [(x_1 \sqcap x_2) \sqcup (x_1 \sqcap x_2')] \sqcap [(x_1 \sqcap x_2') \sqcup (x_1' \sqcap x_2')]$

$\qquad \sim \{(x_1 \sqcap x_2) \sqcap [(x_1 \sqcap x_2') \sqcup (x_1' \sqcap x_2')]\}$

$\qquad \quad \sqcup \{(x_1 \sqcap x_2') \sqcap [(x_1 \sqcap x_2') \sqcup (x_1' \sqcap x_2')]\}$

$\qquad \sim [(x_1 \sqcap x_2) \sqcap (x_1 \sqcap x_2')] \sqcup [(x_1 \sqcap x_2) \sqcap (x_1' \sqcap x_2')]$

$\qquad \quad \sqcup [(x_1 \sqcap x_2') \sqcap (x_1 \sqcap x_2')] \sqcup [(x_1 \sqcap x_2') \sqcap (x_1' \sqcap x_2')]$

$\qquad \sim [(x_1 \sqcap x_1) \sqcap (x_2 \sqcap x_2')] \sqcup [(x_1 \sqcap x_1') \sqcap (x_2 \sqcap x_2')]$

$\qquad \quad \sqcup [(x_1 \sqcap x_1) \sqcap (x_2' \sqcap x_2')] \sqcup [(x_1 \sqcap x_1') \sqcap (x_2' \sqcap x_2')]$

$\qquad \sim (x_1 \sqcap 0) \sqcup (0 \sqcap 0) \sqcup (x_1 \sqcap x_2') \sqcup (0 \sqcap x_2')$

$\qquad \sim 0 \sqcup 0 \sqcup (x_1 \sqcap x_2') \sqcup 0$

$\qquad \sim x_1 \sqcap x_2' \in N_d.$

(b) $\quad p \sqcup q = (x_1 \sqcap x_2) \sqcup (x_1 \sqcap x_2') \sqcup (x_1 \sqcap x_2') \sqcup (x_1' \sqcap x_2') \in N_d$

$\qquad \sim (x_1 \sqcap x_2) \sqcup (x_1 \sqcap x_2') \sqcup (x_1' \sqcap x_2') \in N_d.$

(c) $\quad p' = [(x_1 \sqcap x_2) \sqcup (x_1 \sqcap x_2')]' \sim (x_1 \sqcap x_2)' \sqcap (x_1 \sqcap x_2')'$

$\qquad \sim (x_1' \sqcup x_2') \sqcap (x_1' \sqcup x_2)$

$\qquad \sim [x_1' \sqcap (x_1' \sqcup x_2)] \sqcup [x_2' \sqcap (x_1' \sqcup x_2)]$

$\qquad \sim (x_1' \sqcap x_1') \sqcup (x_1' \sqcap x_2) \sqcup (x_2' \sqcap x_1') \sqcup (x_2' \sqcap x_2)$

$\qquad \sim x_1' \sqcup (x_1' \sqcap x_2) \sqcup (x_2' \sqcap x_1') \sqcup 0$

$\qquad \sim x_1' \sqcup (x_1' \sqcap x_2) \sqcup (x_2' \sqcap x_1')$

$\qquad \sim x_1' \sqcup (x_1' \sqcap x_2) \sqcup (x_1' \sqcap x_2')$

$\qquad \sim (x_1' \sqcap x_2) \sqcup (x_1' \sqcap x_2') \sqcup (x_1' \sqcap x_2) \sqcup (x_1' \sqcap x_2')$

$\qquad \sim (x_1' \sqcap x_2) \sqcup (x_1' \sqcap x_2') \in N_d.$ $\qquad\qquad\qquad \square$

2.40 Remark. The disjunctive normal form can be obtained by applying the following rules. Let p be a Boolean polynomial.

Step 1. Apply de Morgan's laws to bring complementation $'$ immediately to the x_i and the constants.

Step 2. Apply the distributive laws to express p as a join of meet expressions. Generally, the meet expressions will lack some x_i's; for each i we have in each meet either x_i or x_i' or none of both.

Step 3. If x_i is missing, insert $x_i \sqcup x_i'$.

Step 4. Now apply the distributive law until a join of meet expression is obtained. This will be the disjunctive normal form, after application of the idempotent law and suitable reordering.

2.41 Example. Let

$$p = [(a \sqcap x_1) \sqcup (b \sqcap x_2)']' \sqcup (x_1 \sqcup b)' \quad \text{with } a, b \in \mathbb{B}.$$

Step 1.

$$[(a \sqcap x_1) \sqcup (b \sqcap x_2)']' \sqcup (x_1 \sqcup b)' \sim [(a \sqcap x_1)' \sqcap ((b \sqcap x_2)')')] \sqcup (x_1' \sqcap b')$$

$$\sim [(a' \sqcup x_1') \sqcap (b \sqcap x_2)] \sqcup (x_1' \sqcap b').$$

Step 2.

$$[(a' \sqcup x_1') \sqcap (b \sqcap x_2)] \sqcup (x_1' \sqcap b')$$

$$\sim [(a' \sqcap b \sqcap x_2) \sqcup (x_1' \sqcap b \sqcap x_2)] \sqcup (x_1' \sqcap b')$$

$$\sim (a' \sqcap b \sqcap x_2) \sqcup (b \sqcap x_1' \sqcap x_2) \sqcup (x_1' \sqcap b').$$

Step 3.

$$\underbrace{(a' \sqcap b \sqcap x_2)}_{1} \sqcup \underbrace{(b \sqcap x_1' \sqcap x_2)}_{2} \sqcup \underbrace{(b' \sqcap x_1')}_{3} =: y$$

In Step 1: since x_1 is missing we insert $x_1 \sqcup x_1'$. In Step 2: x_1' and x_2 occur, so nothing is inserted. In Step 3: since x_2 is missing we insert $x_2 \sqcup x_2'$. Therefore:

$$y \sim [a' \sqcap b \sqcap (x_1 \sqcup x_1') \sqcap x_2] \sqcup (b \sqcap x_1' \sqcap x_2) \sqcup [(b' \sqcap x_1' \sqcap (x_2 \sqcup x_2'))].$$

Step 4.

$$[a' \sqcap b \sqcap (x_1 \sqcup x_1') \sqcap x_2] \sqcup (b \sqcap x_1' \sqcap x_2) \sqcup [b' \sqcap x_1' \sqcap (x_2 \sqcup x_2')]$$

$$\sim [a' \sqcap b \sqcap x_2 \sqcap (x_1 \sqcup x_1')] \sqcup (b \sqcap x_1' \sqcap x_2)$$

$$\sqcup [b' \sqcap x_1' \sqcap (x_2 \sqcup x_2')]$$

$$\sim [(a' \sqcap b \sqcap x_2 \sqcap x_1) \sqcup (a' \sqcap b \sqcap x_2 \sqcap x_1')] \sqcup (b \sqcap x_1' \sqcap x_2)$$

$$\sqcup [(b' \sqcap x_1' \sqcap x_2) \sqcup (b' \sqcap x_1' \sqcap x_2')]$$

$$\sim (a' \sqcap b \sqcap x_2 \sqcap x_1) \sqcup (a' \sqcap b \sqcap x_2 \sqcap x_1') \sqcup (b \sqcap x_1' \sqcap x_2)$$

$$\sqcup (b' \sqcap x_1' \sqcap x_2) \sqcup (b' \sqcap x_1' \sqcap x_2')$$

$$\sim (a' \sqcap b \sqcap x_1 \sqcap x_2) \sqcup (a' \sqcap b \sqcap x_1' \sqcap x_2) \sqcup (b \sqcap x_1' \sqcap x_2)$$

$$\sqcup (b' \sqcap x_1' \sqcap x_2) \sqcup (b' \sqcap x_1' \sqcap x_2'). \tag{$*$}$$

We put together the three expressions in the middle, where $x_1' \sqcap x_2$ occurs. Then:

$$(a' \sqcap b \sqcap x_1' \sqcap x_2) \sqcup (b \sqcap x_1' \sqcap x_2) \sqcup (b' \sqcap x_1' \sqcap x_2)$$

$$\sim [((a' \sqcap b) \sqcup b) \sqcap (x_1' \sqcap x_2)] \sqcup (b' \sqcap x_1' \sqcap x_2)$$

$$\sim [b \sqcap (x_1' \sqcap x_2)] \sqcup (b' \sqcap x_1' \sqcap x_2)$$

$$\sim (b \sqcup b') \sqcap (x_1' \sqcap x_2)$$

$$\sim 1 \sqcap (x_1' \sqcap x_2)$$

$$\sim 1 \sqcap x_1' \sqcap x_2.$$

Thus we have $(*)$ in the normal form

$$(a' \sqcap b \sqcap x_1 \sqcap x_2) \sqcup (1 \sqcap x_1' \sqcap x_2) \sqcup (b' \sqcap x_1' \sqcap x_2'). \qquad \square$$

Actually, the proof of 2.30 told us much more:

2.42 Corollaries. (i) $|P_n/\!\sim| = 2^{2^n}$ and $P_n(\mathbb{B}) = F_n(\mathbb{B})$.
(ii) If $|B| = m > 2$ then $|P_n(B)| = |P_n/\!\sim| = 2^{2^n} < m^{2^m} = |F_n(B)|$; so $P_n(B) \subset F_n(B)$.
(iii) If

$$p = \bigsqcup_{(i_1,\ldots,i_n) \in \{1,-1\}^n} d_{i_1 \ldots i_n} \sqcap x_1^{i_1} \sqcap \ldots \sqcap x_n^{i_n} \in N_d,$$

then $d_{i_1 \ldots i_n} = \bar{p}_{\mathbb{B}}(1^{i_1}, \ldots, 1^{i_n})$.
(iv) If

$$p = \bigsqcap_{(i_1,\ldots,i_n) \in \{1,-1\}^n} c_{i_1 \ldots i_n} \sqcup x_1^{i_1} \sqcup \ldots \sqcup x_n^{i_n} \in N_c,$$

then $c_{i_1 \ldots i_n} = \bar{p}_{\mathbb{B}}(0^{i_1}, \ldots, 0^{i_n})$. $\qquad \square$

The result (i) means that every function from $\mathbb{B}^n$ into $\mathbb{B}$ is a Boolean polynomial function. One therefore says that $\mathbb{B}$ is "*polynomially complete*". Corollary 2.42(ii) tells us that $\mathbb{B}$ is the only polynomially complete Boolean algebra; and (iii) and (iv) in 2.42 tell us how to find the disjunctive and conjunctive normal form of a given polynomial. Also, these results are very frequently used to find a polynomial (in normal form) which induces a given function from $\mathbb{B}^n$ into $\mathbb{B}$ (see 2.2.9–2.2.14).

2.43 Example. Let

$$p = ((x_1 \sqcup x_2) \sqcap x_1') \sqcup ((x_2')' \sqcap (x_1 \sqcup x_2')) \in P_2.$$

From $\bar{p}_{\mathbb{B}}(0,0) = 0$, $\bar{p}_{\mathbb{B}}(0,1) = 1$, $\bar{p}_{\mathbb{B}}(1,0) = 0$, $\bar{p}_{\mathbb{B}}(1,1) = 1$ we get $p \sim (x_1' \sqcap x_2) \sqcup (x_1 \sqcap x_2) = p_d$, the latter in disjunctive normal form. $\qquad\square$

Of course, p_d can sometimes be shortened by applying the rules of Boolean algebras. In the Example 2.43 above we can write

$$p_d \sim (x_1' \sqcup x_1) \sqcap x_2 \sim 1 \sqcap x_2 \sim x_2.$$

Reductions to "shortest forms" will be discussed in §3.

Finally, we note that the disjunctive (conjunctive) normal form of a Boolean polynomial p is "simpler" (i.e. shorter) than the conjunctive (disjunctive) normal form of p if there are more zeros (ones) in the function value table of $\bar{p}_{\mathbb{B}}$. This follows from 2.42(iii), (iv).

Often the values of polynomial functions are tabulated. We use this in an example.

2.44 Example. Given the Boolean polynomial function $\bar{p}_{\mathbb{B}}$ in terms of its values

v_1	v_2	v_3	$\bar{p}_{\mathbb{B}}(v_1, v_2, v_3)$
1	1	1	1
1	1	0	1
1	0	1	0
1	0	0	0
0	1	1	1
0	1	0	1
0	0	1	0
0	0	0	1

Then the disjunctive normal form of p is

$$(x_1 \sqcap x_2 \sqcap x_3) \sqcup (x_1 \sqcap x_2 \sqcap x_3') \sqcup (x_1' \sqcap x_2 \sqcap x_3)$$

$$\sqcup (x_1' \sqcap x_2 \sqcap x_3') \sqcup (x_1' \sqcap x_2' \sqcap x_3'),$$

since $\bar{p}_{\mathbb{B}}(1^1, 1^1, 1^1) = 1$, $\bar{p}_{\mathbb{B}}(1^1, 1^1, 1^{-1}) = 1$, $\bar{p}_{\mathbb{B}}(1^{-1}, 1^1, 1^1) = 1$, etc. The conjunctive normal form of p is

$$(x_1' \sqcup x_2 \sqcup x_3') \sqcap (x_1' \sqcup x_2 \sqcup x_3) \sqcap (x_1 \sqcup x_2 \sqcup x_3'). \qquad\square$$

For applications in Chapter 2, §3 we shall need a few further concepts in the theory of Boolean algebras. Some of the following terms may also be defined and studied in more general lattices. In the case of Boolean algebras, however, we obtain some simplifications; therefore we restrict our considerations to this case.

2.45 Definition. Let B be a Boolean algebra. $I \subseteq B$ is called an *ideal* in B, in symbols $I \trianglelefteq B$, if I is nonempty and if

$$\forall i, j \in I, \forall b \in B: (i \sqcap b \in I) \wedge (i \sqcup j \in I).$$

If we set $b = i'$ we see that 0 must be in I. Next we consider some useful characterizations of ideals. As one would expect the kernel of a Boolean homomorphism $h: B_1 \to B_2$ is defined as $\ker h := \{b \in B_1 \mid h(b) = 0\}$.

2.46 Theorem. *Let B be a Boolean algebra and let I be a nonempty subset of B. Then the following conditions are equivalent.*

(i) $I \trianglelefteq B$.
(ii) $I \trianglelefteq (B, +, \cdot)$. *This is the Boolean ring (see 2.10) corresponding to B.*
(iii) *I is the kernel of a Boolean homomorphism from B into another Boolean algebra.*
(iv) $\forall i, j \in I, \forall b \in B: (i \sqcup j \in I) \wedge (b \le i \Rightarrow b \in I)$.

PROOF. (i) $\Rightarrow$ (ii) is trivial.

(ii) $\Rightarrow$ (iii) follows from the homomorphism theorem of ring theory (see 3.1.11) and (2.10).

(iii) $\Rightarrow$ (iv). Let $h: B \to C$ be a Boolean homomorphism and $I = \ker h$. Moreover, let $i, j \in I$, $b \in B$, $b \le i$. Then $h(i \sqcup j) = h(i) \sqcup h(j) = 0 \sqcup 0 = 0$ and $h(b) \le h(i) = 0$, therefore $h(b) = 0$. Thus $i \sqcup j \in I$ and $b \in I$.

(iv) $\Rightarrow$ (i) is trivial. □

2.47 Examples. Let B be a Boolean algebra:

(i) $\{0\}$ and B are ideals of B, all other ideals are called *proper ideals* of B.
(ii) For $B = \mathcal{P}(M)$ the set $\{A \subseteq M \mid A \text{ finite}\}$ forms an ideal of B.
(iii) For any $A \in \mathcal{P}(M)$, $\{N \mid N \subseteq A\}$ is an ideal in $B = \mathcal{P}(M)$. □

The examples in 2.47(iii) are generated by a single element A. In general if b is an element in a Boolean algebra B, the set (b) of all "multiples" $x \sqcap b$ of b, for any $x \in B$, is called a *principal ideal*. An ideal M in B is *maximal* when the only ideals of B containing M are M and B itself. See also Chapter 3, §1.

If $I \trianglelefteq B = (B, \sqcap, \sqcup, 0, 1, ')$ then we have $I \trianglelefteq (B, +, \cdot)$, because of 2.46(ii). Therefore we can form the factor ring B/I. This is again a Boolean ring, which by 2.10 induces a Boolean algebra, denoted again by B/I.

2.48 Theorem. *Let B be a Boolean algebra and $b \in B$. Then the principal ideal generated by b is*

$$(b) = \{a \in B \mid a \le b\}.$$

The proof follows from 2.46(iv). The example given in 2.47(ii) is not a principal ideal, not even a "finitely generated" ideal, if M is an infinite set. The ideals in 2.47(iii) are principal ideals.

So far everything has been very similar to the concepts in ring theory. In the case of Boolean algebras, however, we have the possibility of dualizing. The dual of an ideal is a filter in the following sense.

2.49 Definition. Let B be a Boolean algebra and $F \subset B$. F is called a *filter* (or dual ideal) if

$$\forall f, g \in F, \forall b \in B: (f \sqcap g \in F) \wedge (f \sqcup b \in F).$$

In $\mathscr{P}(\mathbb{N})$, $F = \{A \subseteq \mathbb{N} \,|\, A' \text{ finite}\}$ is the filter of cofinite subsets of $\mathbb{N}$; this filter is widely used in convergence studies in analysis. The connection with 2.47(ii) is motivation for the following theorem, the proof of which is left to the reader.

2.50 Theorem. *Let B be a Boolean algebra and $I, F \subseteq B$.*

(i) *If $I \trianglelefteq B$ then $\{i' \,|\, i \in I\}$ is a filter in B.*
(ii) *If F is a filter in B, then $\{f' \,|\, f \in F\}$ is an ideal in B.*

This theorem enables us to dualize 2.46.

2.51 Theorem. *Let B be a Boolean algebra and $F \subseteq B$. Then the following conditions are equivalent:*

(i) *F is a filter in B.*
(ii) *There is a Boolean homomorphism h from B into another Boolean algebra such that $F = \{b \in B \,|\, h(b) = 1\}$.*
(iii) *$\forall f, g \in F, \forall b \in B: (f \sqcap g \in F) \wedge (b \geq f \Rightarrow b \in F)$.* $\square$

Maximal ideals (filters) can be characterized in a very simple way.

2.52 Theorem. *Let B be a Boolean algebra. An ideal (filter) M in B is maximal if and only if for any $b \in B$ either $b \in M$ or $b' \in M$, but not both, hold.*

PROOF. It is sufficient to prove the theorem for ideals. We use the usual notation I instead of M. Suppose for every $b \in B$ we have $b \in I$ or $b' \in I$ but not both. If J were an ideal in B which properly contained I, then $j \in J \setminus I \Rightarrow j, j' \in J$, which would imply $1 \in J$ and then $J = B$. Conversely, let I be a maximal ideal in B and $b_0 \in B$, such that neither b_0 nor b_0' are in I. Then $J := \{b \sqcup i \,|\, i \in I, b \in B, b \leq b_0\}$ is an ideal, generated by $I \cup \{b_0\}$, which contains I properly. Since I is maximal, we have $J = B$. Then there exists an $i \in I$, $b \leq b_0$ with $b \sqcup i = b_0'$. This implies $b_0' \sqcap (b \sqcup i) = b_0' \sqcap b_0'$, which means $b_0' \sqcap i = b_0'$; therefore $b_0' \leq i$ and by 2.46(iv) $b_0' \in I$, a contradiction. $\square$

We can show (using Zorn's Lemma) that any proper ideal (filter) is contained in a maximal ideal (filter). Maximal filters are called *ultrafilters*. Ultrafilters in $\mathscr{P}(M)$ include all filters of the form $F_m := \{A \subseteq M \mid m \in A\}$, for a fixed $m \in M$. Other ultrafilters must exist according to this remark. If M is finite all ultrafilters are of this form. If M is infinite, $F_c := \{A \subseteq M \mid M \setminus A \text{ is finite}\}$ is a proper filter of $\mathscr{P}(M)$. From the remark above there is an ultrafilter containing F_c and this ultrafilter is clearly not an F_m for any $m \in M$.

PROBLEMS

1. Find disjunctive and conjunctive normal forms of

$$f = (x_1' \sqcap x_2') \sqcup (x_3' \sqcap ((x_1 \sqcap x_2') \sqcup (x_1' \sqcap x_2))).$$

2. Simplify $f = (x_1 + x_2' + x_3')(x_1' + x_2)(x_2' + x_3)$.

3. Determine the disjunctive normal form of $p(x_1, x_2) = ((a \sqcup x_1) \sqcap (b \sqcup x_2)') \sqcup x_2$ with a and b in the Boolean algebra $\mathbb{B}$.

4. Find the disjunctive normal form of

$$f(x_1, x_2, x_3, x_4) = (x_1' \sqcap x_2) \sqcup (x_1 \sqcap x_2' \sqcap x_3) \sqcup (x_2' \sqcap x_3' \sqcap x_4') \sqcup (x_1' \sqcap x_3 \sqcap x_4).$$

5. Fill in the remaining details in the proof of Theorem 2.35.

*6. Prove part (ii) of Theorem 2.37 without dualizing part (i).

7. Demonstrate in detail how an interval $[a, b]$ in a Boolean algebra B can be made into a Boolean algebra.

8. Prove: A nonempty subset I of a Boolean algebra is an ideal if and only if

$$a \in I, b \in I \Leftrightarrow a \sqcup b \in I.$$

9. Ideals and principal ideals of lattices are defined just as those for Boolean algebras B by replacing B by a lattice L. If (a) and (b) are principal ideals in the lattice of ideals of a lattice L, prove that $(a) \sqcap (b) = (a \sqcap b), (a) \sqcup (b) = (a \sqcup b)$. (Here the operation $\sqcap$ on two ideals is just their common part and $\sqcup$ of two ideals is the set of elements c of L such that $c \leq x \sqcup y$, for $x \in (a)$, $y \in (b)$.)

10. Prove that the ideals of a lattice, ordered by set-theoretic inclusion, form a lattice.

*11. Prove that in a finite lattice every ideal is a principal ideal.

12. A nonempty subset J of a lattice L is called a filter of L if

  (i) $a \in J, b \in J \Rightarrow a \sqcap b \in J$;
  (ii) $a \in J, x \in L \Rightarrow a \sqcup x \in L$.

  Prove that in a filter J of L the following conditions hold: If $a \in J, a \leq j$ then $j \in J$. Also, if $a \sqcap b \in J$, then $a \in J$ and $b \in J$.

13. Find all ideals and filters of the lattice

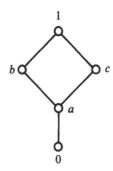

Figure 1.18

*14. Show that the correspondence between the ideals of a Boolean algebra and the ideals of the corresponding ring is one-to-one.

*15. Prove that any ideal of a Boolean algebra equals the intersection of all prime ideals containing it. ($I \vartriangleleft B$ is prime if $a, b \in B$, $a \sqcap b \in I \Rightarrow a \in I \vee b \in I$.)

*16. Prove: A distributive lattice is a Boolean algebra if and only if each of its prime ideals is maximal.

EXERCISES (Solutions in Chapter 8, p. 416)

1. Show that $(\{1, 2, 3, 6, 9, 18\}, \gcd, \operatorname{lcm})$ does not form a Boolean algebra for the set of positive divisors of 18.

2. Show that $(B, \gcd, \operatorname{lcm})$ is a Boolean algebra if B is the set of all positive divisors of 110.

*3. Prove that the lattice of all positive divisors of $n \in \mathbb{N}$, is a Boolean algebra with respect to lcm and gcd if and only if the prime factor decomposition of n does not contain any squares.

4. Prove Theorem 2.6.

5. Show that for any Boolean algebra B there are 2^{2^n} different Boolean functions for n variables.

6. Simplify the following terms in a Boolean algebra

   (a) $(x \sqcap y) \sqcup (x \sqcap y') \sqcup (x' \sqcup y)$;
   (b) $(x \sqcap y') \sqcup (x \sqcap (y \sqcap z)'] \sqcup z$.

7. Find the disjunctive normal form of

   (a) $(x_1 \sqcap (x_2 \sqcup x_3)') \sqcup ((x_1 \sqcap x_2) \sqcup x_3') \sqcap x_1)$;
   (b) $(x_2 \sqcup (x_1 \sqcap x_3)) \sqcap ((x_1 \sqcup x_3) \sqcap x_2)'$.

8. Find the conjunctive normal form of

$$(x_1 \sqcup x_2 \sqcup x_3) \sqcap ((x_1 \sqcap x_2) \sqcup (x_1' \sqcap x_3))'.$$

9. Let B be a Boolean algebra. Prove: F is a filter in B if and only if $F' := \{x' | x \in F\}$ is an ideal of B.

*10. Let S be a subset of a Boolean algebra B. Prove

    (i) The intersection J of all ideals I containing S is an ideal containing S.

    (ii) J in (i) consists of all elements of the form

$$(b_1 \sqcap s_1) \sqcup \ldots \sqcup (b_n \sqcap s_n), \quad n \geq 1,$$

    where $s_1, \ldots, s_n \in S$ and $b_1, \ldots, b_n$ are arbitrary elements of B.

    (iii) J in (i) consists of the set A of all a such that

$$a \leq s_1 \sqcup \ldots \sqcup s_n, \quad n \geq 1,$$

    where $s_1, \ldots, s_n$ are arbitrary elements of S.

    (iv) J in (i) is a proper ideal of B if and only if

$$s_1 \sqcup \ldots \sqcup s_n \neq 1$$

for all $s_1, \ldots, s_n$ in S.

*11. By the use of Zorn's Lemma prove that each proper filter in a Boolean algebra B is contained in an ultrafilter. (This is called the *Ultra Filter Theorem*.)

12. Let $h : B \to B'$ be a homomorphism of Boolean algebras. Show that $H = h^{-1}(1)$ is a filter. This is called the *hull* of h.

13. Show that for any distinct elements x and y in a Boolean algebra B an ultrafilter exists containing one but not the other element.

§3. Minimal Forms of Boolean Polynomials

We have seen in the previous section that it is possible to simplify a given Boolean polynomial by using the axioms of Boolean algebra. For this process of simplification it is often difficult to decide which axioms should be used and in which order they should be used. There are several systematic methods to simplify Boolean polynomials. Many of these methods have the disadvantage that the practical implementation is impossible when the number of indeterminates of the polynomial is too large. This problem area in the theory of Boolean algebras is called the *optimization* or *minimization problem* for Boolean polynomials; it is of importance in applications such as the simplification of switching circuits (see Chapter 2, §2).

Instead of $x \sqcap y$ and $x \sqcup y$ we shall write xy and $x + y$, respectively. We shall discuss the simplification of Boolean polynomials, especially the reduction of polynomials to a "minimal form" with respect to a suitably chosen minimal condition. Our considerations will be restricted to a special minimal condition for sum-of-product expressions or disjunctive normal forms. We define a *literal* to be any variable x_i either complemented or not complemented, and 0 and 1. Let d_f denote the total number of literals in a sum-of-product expression of f. Let e_f be the number of disjuncts (or product terms) in f. We say such an f is *simpler* than a sum-of-product expression g if $d_f \le d_g$, $e_f \le e_g$, and one of these inequalities is strict. f is called *minimal* if there is no simpler sum-of-product expression which is equivalent to f. In other words we are looking for the "shortest" expression with the smallest possible number of literals, which is equivalent to f. Such a minimal form does not always exist uniquely (for examples see p. 52). We shall describe one method amongst several methods for the simplification process. It is based on work by Quine and has been improved by McCluskey, so it is called the *Quine–McCluskey method*. A Boolean polynomial in this section is also called an "*expression*".

3.1 Definition. An expression p *implies* an expression q if for all

$$b_1, \ldots, b_n \in B, \qquad \bar{p}_B(b_1, \ldots, b_n) = 1 \quad \text{implies} \quad \bar{q}_B(b_1, \ldots, b_n) = 1;$$

p is called an *implicant* of q (cf. 2.2.9 and 2.2.10).

3.2 Definition. A *product expression* (briefly a *product*) α is an expression in which $+$ does not occur. A *prime implicant* for an expression p is a product expression α which implies p, but which does not imply p if one element in α is deleted. A product whose factors form a subset of the factors of another product is called *subproduct* of the latter, i.e. q is a subproduct of $p \in P_n$ if $\exists\, p_1, p_2 \in P_n \cup \{\Lambda\}$ such that $p = p_1 q p_2$. (Λ denotes the empty word.)

3.3 Example. $x_1 x_3$ is a subproduct of $x_1 x_2 x_3$ and also of $x_1 x_2' x_3$ and implies the expression

$$p = x_1 x_2 x_3 + x_1 x_2' x_3 + x_1' x_2' x_3'$$

because $(\overline{x_1 x_3})(1, i_2, 1) = 1$. But then $\bar{p}(1, i_2, 1) = 1$ as well. Neither x_1 nor x_3 imply p; therefore $x_1 x_3$ is a prime implicant. x_1 cannot be an implicant of p, since for all $(1, i_2, i_3) \in \{0, 1\}^3$ we have $\bar{x}_1(1, i_2, i_3) = 1$ but $\bar{p}(1, i_2, i_3) = 0$. $\qquad\square$

3.4 Theorem. *A polynomial $p \in P_n$ is equivalent to the sum of all prime implicants of p.*

PROOF. Let I_p be the set of all prime implicants of p and $q := \sum_{p_\alpha \in I_p} p_\alpha$. If $\bar{q}(b_1, \ldots, b_n) = 1$ for $(b_1, \ldots, b_n) \in \mathbb{B}^n$, then there is a $p_\alpha \in I_p$ with $\bar{p}_\alpha(b_1, \ldots, b_n) = 1$. Since p_α implies p, $\bar{p}(b_1, \ldots, b_n) = 1$ is a consequence.

Conversely, let $\bar{p}(b_1, \ldots, b_n) = 1$ and $s := x_1^{e_1} \ldots x_n^{e_n}$ with $x_i^{e_i} = x_i$ for $b_i = 1$ and $x_i^{e_i} = x_i'$ for $b_i = 0$. s is an implicant of p. In s we remove all those $x_i^{e_i}$ for which $\bar{p}(b_1, \ldots, b_{i-1}, b_i', b_{i+1}, \ldots, b_n) = 1$. The remaining product r still implies p, but does not imply p anymore if another factor is removed. Therefore r is a prime implicant for p with $\bar{r}(b_1, \ldots, b_n) = 1$. Hence $\bar{q}(b_1, \ldots, b_n) = 1$. $\qquad\square$

A sum of prime implicants of p is called *irredundant* if it has the smallest possible number of product expressions amongst all those sums of prime implicants, which are equivalent to p. A minimal sum-of-product expression must be irredundant. In order to determine a minimal expression we therefore determine the set of irredundant expressions and amongst them we look for the one with least number of literals. Here is a method due to Quine for determining the prime implicants.

Prime implicants are obtained by starting with the disjunctive normal form d for the Boolean polynomial p and applying the rule

$$yz + yz' \sim y,$$

wherever possible in d. More generally, we use

$$(\alpha\beta) + (\alpha\beta') \sim \alpha, \qquad (*)$$

where α and β are product expressions. The set of all expressions which cannot be simplified any more by this procedure is the set of prime implicants. The sum of these prime implicants gives the minimal form for p.

3.5 Example. Let p be the Boolean polynomial whose disjunctive normal form d is given by:

$$d = wxyz' + wxy'z' + wx'yz + wx'yz' + w'x'yz + w'x'yz' + w'x'y'z.$$

We use the idempotent law and $(*)$ for the $\binom{7}{2} = 21$ pairs of products in d (as far as this is possible) and by doing this we "shorten" these products. For instance, the first and second product expression of d yields wxz' by using $(*)$. If a product expression is used once or more for the simplification, it is ticked. Since $+$ stands for $\sqcup$, an expression can be used any number of times and one tick suffices. In this way all the expressions are ticked which contain other product expressions and therefore cannot be prime implicants. Altogether this process leads us

from	$wxyz'$	and	$wxy'z'$	to	wxz',
from	$wx'yz$	and	$wx'yz'$	to	$wx'y$,
from	$wxyz'$	and	$wx'yz'$	to	wyz',
from	$w'x'yz$	and	$w'x'yz'$	to	$w'x'y$,
from	$w'x'yz$	and	$w'x'y'z$	to	$w'x'z$,
from	$wx'yz$	and	$w'x'yz$	to	$x'yz$,
from	$wx'yz'$	and	$w'x'yz'$	to	$x'yz'$.

In here, all seven summands are used and therefore ticked.

In general, this procedure is repeated over and over again using only the ticked expressions (which become shorter and shorter). The other ones are prime implicants and remain unchanged.

In our example, the second round of simplifications yields:

$$\text{from} \quad wx'y \quad \text{and} \quad w'x'y \quad \text{to} \quad x'y,$$

$$\text{from} \quad x'yz \quad \text{and} \quad x'yz' \quad \text{to} \quad x'y.$$

These four expressions $wx'y$, $w'x'y$, $x'yz$ and $x'yz'$ are ticked. The remaining ones, namely wxz', wyz' and $w'x'z$ cannot be simplified. Hence p can be written as a sum of prime implicants:

$$p \sim wxz' + wyz' + w'x'z + x'y. \qquad \square$$

McCluskey improved this method, which leads us to the general *Quine–McCluskey algorithm*. We use the polynomial of Example 3.5 to describe the procedure.

Step 1. Represent all product expressions in terms of 0–1 sequences, such that x_i' and x_i are denoted by 0 and 1, respectively. Missing variables are indicated by a dash, e.g. $w'x'y'z$ is 0001, $w'x'z$ is 00–1.

Step 2. The product expressions, regarded as binary n-tuples, are partitioned into equivalence classes according to their number of ones. We order the classes according to increasing numbers of ones. In our example

$w'x'y'z$	0	0	0	1
$w'x'yz'$	0	0	1	0
$w'x'yz$	0	0	1	1
$wx'yz'$	1	0	1	0
$wxy'z'$	1	1	0	0
$wx'yz$	1	0	1	1
$wxyz'$	1	1	1	0.

Step 3. Each expression with r ones is added to each expression containing $r + 1$ ones. If we use (*), we can simplify expressions in neighboring equivalence classes. We have to compare expressions in neighboring classes with dashes in the same position. If two of these expressions differ in exactly one position, then they are of the form $p = i_1 i_2 \ldots i_r \ldots i_n$ and $q = i_1 i_2 \ldots i_r' \ldots i_n$, where all i_k are in $\{0, 1, -\}$, and i_r is in $\{0, 1\}$.

Then (*) reduces p, q to $i_1 i_2 \ldots i_{r-1} - i_{r+1} \ldots i_n$, and p and q are ticked. This yields in our example

0	0	–	1	
0	0	1	–	$\sqrt{}$
–	0	1	0	$\sqrt{}$
–	0	1	1	$\sqrt{}$
1	0	1	–	$\sqrt{}$
1	–	1	0	
1	1	–	0.	

The expressions with ticks are not prime implicants and will be subject to further reduction. They yield the single expression

$$- \quad 0 \quad 1 \quad -$$

Thus we found all prime implicants, namely

$$
\begin{array}{cccc@{\qquad}l}
0 & 0 & - & 1 & w'x'z \\
1 & - & 1 & 0 & wyz' \\
1 & 1 & - & 0 & wxz' \\
- & 0 & 1 & - & x'y.
\end{array}
$$

Since the sum of all prime implicants is not necessarily in minimal form we perform the last step in the procedure.

Step 4. Since the sum of all the prime implicants of p is equivalent to p by 3.4, for each product expression in the disjunctive normal form d of p there must be a prime implicant which is a subproduct of this product expression. This is determined by establishing a table of prime implicants. The heading elements for the columns are the product expressions in d at the beginning of the rows we have the prime implicants calculated in step 3. A cross $\times$ is marked off at the intersection of the ith row and jth column if the prime implicant in the ith row is a subproduct of the product expression in the jth column. A product expression is said to *cover* another product expression if it is subproduct of the latter one. In order to find the sum of prime implicants, which then is equivalent to d, we choose a subset of the set of prime implicants in such a way that each product expression in d is covered by at least one prime implicant of the subset. Then a minimal form is a sum of prime implicants with the least number of terms and the least number of letters. A prime implicant is called a *main term* if it covers a product expression which is not covered by any other prime implicant; the sum of the main terms is called the *core*. First we find the core, then we denote by $q_1, \ldots, q_k$ those product expressions which are not covered by prime implicants in the core; the prime implicants not in the core are denoted by $p_1, \ldots, p_m$. We form a second table with index elements q_j for the columns and index elements p_i for the rows. An $\times$ is placed in the entry (i, j) indicating that p_i covers q_j.

Next we form a product-of-sums. Each factor corresponds to one of the q_j and consists of a sum of those p_i's which cover that q_j. Using the Boolean algebra laws we convert this to the simplest possible sum of products. Each of these products represents a subset of the p_i's which covers all the q_j's. We now concentrate on those products with the least number of factors. From these shortest products we select those with the least total number of literals in their constituent prime implicants. Each of these last when written as a sum of its prime implicant factors and added to the core will give a minimal sum of products representation of p.

3.6 Example. Determine the minimal form of p, which is given in disjunctive normal form

$$d = v'w'x'y'z' + v'w'x'yz' + v'w'xy'z' + v'w'xyz' + v'wx'y'z + v'wx'yz'$$
$$+ v'wxy'z' + v'wxyz' + v'wxyz + vw'x'y'z' + vw'x'y'z + vw'xy'z$$
$$+ vwx'yz' + vwxy'z' + vwxyz' + vwxyz.$$

Steps 1 *and* 2.

							row numbers
0 ones	0	0	0	0	0	$\checkmark$	(1)
1 one	0	0	0	1	0	$\checkmark$	(2)
	0	0	1	0	0	$\checkmark$	(3)
	1	0	0	0	0	$\checkmark$	(4)
2 ones	0	0	1	1	0	$\checkmark$	(5)
	0	1	0	0	1	$\checkmark$	(6)
	0	1	0	1	0	$\checkmark$	(7)
	1	0	0	0	1	$\checkmark$	(8)
3 ones	0	1	1	0	1	$\checkmark$	(9)
	0	1	1	1	0	$\checkmark$	(10)
	1	0	1	0	1	$\checkmark$	(11)
	1	1	0	1	0	$\checkmark$	(12)
	1	1	1	0	0	$\checkmark$	(13)
4 ones	0	1	1	1	1	$\checkmark$	(14)
	1	1	1	1	0	$\checkmark$	(15)
5 ones	1	1	1	1	1	$\checkmark$	(16)

Step 3. Combination of rows (i) and (j) yields the following simplifications:

(1) (2)	0	0	0	–	0	$\checkmark$
(1) (3)	0	0	–	0	0	$\checkmark$
(1) (4)	–	0	0	0	0	J
(2) (5)	0	0	–	1	0	$\checkmark$
(2) (7)	0	–	0	1	0	$\checkmark$
(3) (5)	0	0	1	–	0	$\checkmark$
(4) (8)	1	0	0	0	–	I

(5) (10)	0	–	1	1	0	√
(6) (9)	0	1	–	0	1	H
(7) (10)	0	1	–	1	0	√
(7) (12)	–	1	0	1	0	√
(8) (11)	1	0	–	0	1	G

(9) (14)	0	1	1	–	1	F
(10) (14)	0	1	1	1	–	√
(10) (15)	–	1	1	1	0	√
(12) (15)	1	1	–	1	0	√
(13) (15)	1	1	1	–	0	E

(14) (16)	–	1	1	1	1	√
(15) (16)	1	1	1	1	–	√

Repeating this step by combining the rows as indicated gives

(1) (2), (3) (5)	0	0	–	–	0	D
(2) (5), (7) (10)	0	–	–	1	0	C
(7) (10), (12) (15)	–	1	–	1	0	B
(10) (15), (14) (16)	–	1	1	1	–	A

The marking of the expressions by √ or letters $A, B, \ldots$ is, of course, done after the simplification processes. Having found the prime implicants we denote them by $A, B, \ldots, J$.

Step 4. We give the table of prime implicants, where the first "row" represents the product expressions of d as binary 5-tuples in column form.

		(1)	(2)	(3)	(4)	(5)	(6)	(7)	(8)	(9)	(10)	(11)	(12)	(13)	(14)	(15)	(16)
		0	0	0	1	0	0	0	1	0	0	0	1	1	1	0	1	1
		0	0	0	0	0	1	1	0	1	1	0	1	1	1	1	1
		0	0	1	0	1	0	0	0	1	1	1	0	1	1	1	1
		0	1	0	0	1	0	1	0	0	1	0	1	0	1	1	1
		0	0	0	0	0	1	0	1	1	0	1	0	0	1	0	1
– 1 1 1 –	A										×				×	×	×
– 1 – 1 0	B							×			×		×		×		
0 – – 1 0	C		×			×		×			×						
0 0 – – 0	D	×	×	×		×											
1 1 1 – 0	E													×		×	
0 1 1 – 1	F									×					×		
1 0 – 0 1	G								×			×					
0 1 – 0 1	H						×			×							
1 0 0 0 –	I				×				×								
– 0 0 0 0	J	×			×												

The core, i.e. the sum of the main terms, is $D + H + G + B + E + A$ (in our short notation). (4) is the only product expression which is not covered by the core; it is denoted by q_i. C, F, I, J are the prime implicants p_i which are not in the core. The new table is of the form

$$
\begin{array}{cccccc}
 & & & & & (4) \\
 & & & & & 1 \\
 & & & & & 0 \\
 & & & & & 0 \\
 & & & & & 0 \\
 & & & & & 0 \\
\hline
\end{array}
$$

0	–	–	1	0	C	
0	1	1	–	1	F	
1	0	0	0	–	I	×
–	0	0	0	0	J	×

This means that the minimal form is

(i) $D + H + G + B + E + A + I$

if we use I; it is

(ii) $D + H + G + B + E + A + J$

if we choose J. In our usual notation the minimal form (i) of p is

$$v'w'z' + v'wy'z + vw'y'z + wyz' + vwxz' + wxy + vw'x'y'. \qquad \square$$

We refer to HOHN for a detailed description of this method, proofs included.

PROBLEMS

1. Determine all prime implicants of $f(x, y, z, u) = xyz + xyz' + x'y'u + yzu$.

2. Find the minimal form for $x_3(x_2 + x_4) + x_2x_4' + x_2'x_3'x_4$ using the Quine–McCluskey procedure.

3. Repeat Problem 2 for $f = x_1'x_2' + x_1x_3x_4 + x_1x_2x_4' + x_2'x_3$.

4. Simplify:

$$f = xyzuv + xyz'uv + xy'zu'v + x'yz'uv + xy'zu'v + x'yz'u'v' + x'yzuv.$$

5. Simplify the following disjunctive normal form by using the Quine–McCluskey procedure:

$$f = xy'zu + xy'zu' + xy'z'u' + x'yzu + x'yzu$$

$$+ x'yz'u + x'yz'u' + x'y'zu + x'y'x'u'.$$

6. Five different locks L_i, $i = 1, 2, 3, 4, 5$, secure a door. The electrical locking mechanism opens the door only if at least three of the locks are operated, amongst them either L_1 or L_2 and L_3. Find the disjunctive normal form of the Boolean polynomial that represents unlocking the door and simplify it.

EXERCISES (Solutions in Chapter 8, p. 420)

1. Find all prime implicants of $xy'z + x'yz' + xyz' + xyz$ and form the corresponding prime implicant table.

2. Find three prime implicants of $xy + xy'z + x'y'z$.

3. Use Quine–McCluskey's method to find the minimal form of

$wx'y'z + w'xy'z' + wx'y'z' + w'xyz + w'x'y'z' + wxyz + wx'yz + w'xyz' + w'x'yz'$.

4. Determine the prime implicants of

$$f = w'x'y'z' + w'x'yz' + w'xy'z + w'xyz' + w'xyz + wx'y'z' + wx'yz$$

$$+ wxy'z + wxyz + wxyz'$$

by using Quine's procedure. Complete the minimizing process of f by using the Quine–McCluskey method.

5. Find the disjunctive normal form of f and simplify it:

$$f = x'y + x'y'z + xy'z' + xy'z.$$

NOTES

Standard reference books on lattice theory are BIRKHOFF, RUTHERFORD, SZÁSZ, GRÄTZER$_1$, HALMOS, GRÄTZER$_2$. The latter book is a recent one and also a more advanced book on the subject. Several of the books on applied algebra also contain sections on lattices; most have a chapter on Boolean algebras, we mention a few of them: BIRKHOFF and BARTREE, DORNHOFF and HOHN, GILBERT, GILL, FISHER, STREET and WALLIS, and PRATHER. We shall refer again in the notes to Chapter 2 to some of these books which include lattices, Boolean algebras and their applications.

The history of the lattice concept and the development of lattice theory from the early beginnings in the nineteenth century up to the concept of universal algebra is beautifully traced in MEHRTENS. Here we indicate some of the highlights in the history of lattices.

In 1847, G. Boole wrote an epoch making little book *The Mathematical Analysis of Logic* in which logic is treated as a purely formal system and the interpretation in ordinary language comes afterwards. Boole wrote that mathematics is characterized by its form, not its contents. His next book *Investigation of the Law of Thought* (1854) contains the concept of Boolean algebra.

George Boole's calculus of logic centred on the formal treatment of logic by means of mathematical (especially algebraic) methods and on the description of logical equations. Following Boole, a school of English mathematicians, Schröder, and also Whitehead developed the axiomatization of operations (conjunction, disjunction, negation); on the other hand Peirce and Schröder created the axiomatics of order, with inclusion as the fundamental term. In 1904 E. V. Huntington studied the two systems of axioms and thus started the treatment of Boolean algebras as mathematical structures apart from logic.

Another approach to lattices was taken by R. Dedekind, who transferred the divisibility relation on $\mathbb{N}$ to ideals, modules and even fields, and reformulated gcd and lcm as set-theoretic operations. Thus the lattice structure appeared in several concrete applications. In 1897, Dedekind arrived at the abstract concept of a lattice, which he called "Dualgruppe"; but this abstract axiomatic foundation of lattice theory remained unnoticed.

Some 30 years later, around 1930, several mathematicians formulated the lattice concept anew. The axiomatic method was accepted by then and the time was ripe for lattices. The most important root for lattices was algebra starting with group theory. Garrett Birkhoff published his very important paper on lattice theory in 1933, in which he introduced "lattices". In following papers he deepened some aspects and widened the area of applications. Other authors or contributors to lattice theory at that time were Karl Menger, Fritz Klein, Oystein Ore. Schröder's lattice concepts stemmed from his work on the algebra of logic and abstract arithmetic, Dedekind's lattice concepts originated in the structure of his algebraic number theory.

Boole uses distributivity of meet with respect to join, which had been noted by J. Lambert before him. He worked with sets and denoted the meet of x and y by xy, the join by $x + y$, if x and y are disjoint. Similar to Leibniz he interpreted the inclusion relation as $xy = x$, which easily gave him the classical rules of syllogism. Jevons then extended the operation join to arbitrary x and y; de Morgan and later Peirce proved the duality relations called De Morgan's laws.

Most of the nineteenth century logicians did not show much interest in applying their findings to mathematics. One reason for this was the lack of the use of variables and quantifiers, which were introduced by Frege and C. S. Peirce. Peano, among others, introduced the symbols $\cup$, $\cap$, $-$ for join, meet and difference of sets. After Van der Waerden's book on modern algebra, the concept of universal algebra was not far away. Birkhoff developed the concepts of an "algebra" from the approach of Van der Waerden and took the name "universal algebra" from Whitehead's book. In 1934, MacLane also stated some ideas on universal algebra influenced by his stay in Göttingen, but did not publish them. Ore published a paper in 1935, which was one of the fundamental papers on lattice theory. The following years saw many contributions to the subject and work on varied applications of lattices, e.g. in group theory, projective geometry, quantum mechanics, functional analysis, measure and integration theory (see MEHRTENS, pp. 202–203).

In the years 1933–1937, M. H. Stone developed important results on Boolean algebras, which he interpreted as special rings, namely Boolean rings, made amenable to ideal theory. Other fundamental questions tackled by Stone were the representation of Boolean algebras and applications of Boolean algebras in topology. From then on lattice theory expanded steadily into a healthy and vigorous discipline of its own. It is, however, not completely accepted that lattices are part of algebra. For instance, several of the most influential or most popular algebra texts do not include lattices at all, e.g. VAN DER WAERDEN, FRALEIGH, HERSTEIN, LANG, RÉDEI; on the other hand some important texts do include lattices, e.g. BIRKHOFF and MACLANE, MACLANE and BIRKHOFF, JACOBSON, KUROSH.

There are several other algebraic structures which can be mentioned in connection with lattices. For instance, a *semilattice* is an ordered set in which any two elements have an infimum. An example of a semilattice is the set of a man and all his male descendants, where the order relation is defined as "is ancestor of". Semilattices are used in the developmental psychology of Piaget. A semilattice could also be defined as an idempotent and commutative semigroup.

A *quasilattice* is a set with a reflexive and transitive relation, where to any two elements in the set there exist infimum and supremum. It can be shown that quasilattices can also be described as a set with two binary operations ⊔ and ⊓ which satisfy the associative and absorption laws, but not the commutative law. JORDAN, MATSUSHITA and others studied these structures in the context of quantum physics (and called them skew lattices, Schrägverbände).

There are other methods available for finding minimal forms. DORNHOFF and HOHN describe a method for finding all prime implicants and for finding minimal forms which is based on work by REUSCH. REUSCH and DETERING's paper is a tutorial on this topic.

CHAPTER 2

Applications of Lattices

One of the most important practical applications and also one of the oldest applications of modern algebra, especially lattice theory, is the use of Boolean algebras in modeling and simplifying switching or relay circuits. This application will be described in §1. It should be noted that the algebra of switching circuits is not described because of its primary importance today but rather for historical reasons since it represented one of the first applications in this field and also because of its elegant mathematical formulation. The same theory will also describe other systems, e.g. plumbing systems, road systems with blocks, etc. The second section considers propositional logic and the third section indicates applications in probability.

§1. Switching Circuits

A. Basic Definitions

The main aspect of the algebra of switching circuits is to describe electrical or electronic switching circuits in a mathematical way or to design a diagram of a circuit with given properties. Here we combine electrical or electronic switches into series or parallel circuits. Such switches or contacts are switching elements with two states (open/closed), e.g. mechanical contacts, relays, semiconductors, photocells, or transistors. The type of the two states depends on these switching elements; we can consider conductor–nonconductor elements, charged–uncharged, positively magnetized–negatively magnetized, etc. We shall use the notation introduced in Chapter 1. Again we replace $x_1 \sqcap x_2$ by $x_1 x_2$ and $x_1 \sqcup x_2$ by $x_1 + x_2$ and call them product and sum. Again, P_n denotes the set of Boolean polynomials; $\mathbb{B}$ is the Boolean algebra of two elements, $\{0, 1\}$; $\bar{p}_\mathbb{B} =: \bar{p}$ is the Boolean polynomial function associated with $p \in P_n$.

Electrical switches or contacts can be symbolized in a *switching* or *circuit diagram* or *contact sketch*:

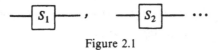

Figure 2.1

Such a switch can be bi-stable, either "open" or "closed". Sometimes open and closed switches are symbolized as

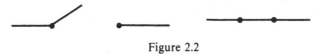

Figure 2.2

The basic assumption is that for current to flow through a switch it is necessary that the switch be closed.

The symbol $-\boxed{S_1'}-$ (complementation) indicates a switch, which is

open, if $-\boxed{S_1}-$ (a switch appearing elsewhere in the circuit) is closed

and is closed if $-\boxed{S_1}-$ is open. In other words, S_1 and S_1' constitute

two switches which are linked, in the sense that their states are related in this way. Similarly, if S_1 appears in two separate places in a circuit it means that there are two separate switches linked so as to ensure that they are always either both open or both closed. In the diagram

Figure 2.3

we have "current" if and only if S_1 and S_2 are both closed. In

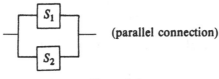

Figure 2.4

we have "current", if and only if either or both of S_1 and S_2 are closed. These properties of electrical switches motivate the following definitions, which give a connection between electrical switches and the elements of a Boolean algebra. As usual, $X_n := \{x_1, \ldots x_n\}$.

1.1 Definition. (a) The elements $x_1, \ldots, x_n \in X_n$ are called *switches*.

(b) Every $p \in P_n$ is called *switching circuit*.

(c) x_i' is called the *complementation switch* of x_i.

(d) $x_i x_j$ is called the *series connection* of x_i and x_j.

(e) $x_i + x_j$ is called the *parallel connection* of x_i and x_j.

(f) For $p \in P_n$ the corresponding $\bar{p} \in P_n(\mathbb{B})$ is called the *switching function* of p.

(g) $\bar{p}(a_1, \ldots, a_n)$ is called the *value* of the switching circuit p at $a_1, \ldots, a_n \in \mathbb{B}$. The a_i are called *input variables*.

Switches and switching circuits in the sense of 1.1, the mathematical models of circuits, can be graphically represented by using contact diagrams. Instead of S_i we use x_i according to 1.1.

The polynomial (i.e. the circuit) $x_1 x_2 + x_1 x_3$ can be represented as:

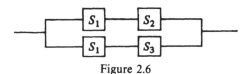

Figure 2.5

The electrical realization would be:

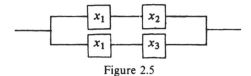

Figure 2.6

Another method of representation is as a switching or circuit diagram. These show the circuit in terms of "boxes", which convert input variables into values:

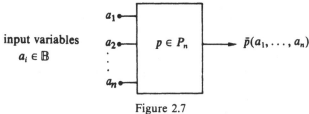

Figure 2.7

For the example given above we have the diagram

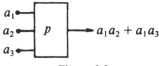

Figure 2.8

$\bar{p}(a_1, \ldots, a_n) = 1$ (or $= 0$) means that in the circuit p the electrical circuit has current (or does not have current).

Thus it is possible to model electrical circuits by using Boolean polynomials. Here different electrical circuits operate "identically" if their values are the same for all possible combinations of the input variables. This means for the corresponding polynomials p, q that $\bar{p}_B = \bar{q}_B$; that is, $p \sim q$.

In order to find a possible simplification of an electrical circuit retaining its original switching properties we can look for a "simple" Boolean polynomial which is equivalent to the original polynomial. This can be done by transposing the given polynomial into disjunctive normal form and then applying the Quine–McCluskey algorithm (see Chapter 1, §3).

At this point we mention that in this way we can construct and simplify a wide variety of flow diagrams with "barriers", not only electrical circuits, e.g. pipe systems (for water, oil or gas pipes) and traffic diagrams (with streets as circuits and traffic lights as switches). We only describe some aspects of the electrical interpretation of the situation as given above.

1.2 Examples. (i) We draw the diagram for the switching circuit

$$p = x_1(x_2(x_3 + x_4) + x_3(x_5 + x_6)).$$

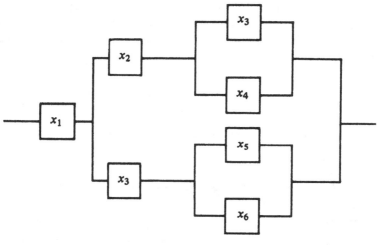

Figure 2.9

(ii) Next, determine the switching circuit p given by the diagram overleaf (Figure 2.10):

$$p = x_1(x_2'(x_6 + x_3(x_4 + x_5')) + x_7(x_3 + x_6)x_8'). \qquad \square$$

Nowadays, electrical switches are of less importance than semiconductor elements. These elements are types of electronic blocks which are predominant in the logical design of digital building components of electronic computers. In this context the switches are represented as so-called *gates*, or combinations of gates. We call this the "*symbolic representation*". Thus a gate (or combination of gates) is a polynomial p which has as value in

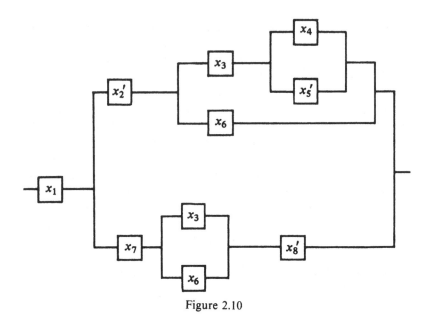

Figure 2.10

$\mathbb{B}$ the element obtained by replacing x_i by a_i in $\mathbb{B}$ for each i. We also say that the gate is a realization of a switching function. $\bar{p}(a_1, \ldots, a_n) = 1$ (or 0) means that we have current (or no current) in the switching circuit p. We define some special gates.

1.3 Definition.

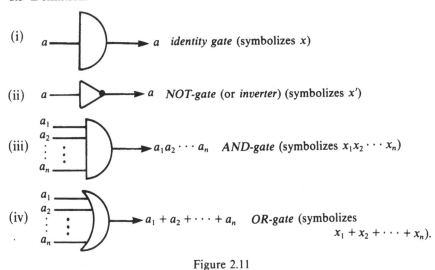

Figure 2.11

In the propositional logic (see §2) the three polynomials $x_1' + x_2$ (*subjunction*), $(x_1 + x_2)'$ (*Pierce-operation*) and $(x_1 x_2)'$ (*Sheffer-operation*) are of importance.

A briefer notation for the NOT-gate is to draw a black disc immediately before or after one of the other gates to indicate an inverter, e.g.

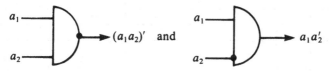

Figure 2.12

1.4 Definition.

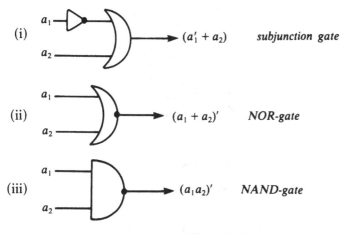

(i) $(a_1' + a_2)$ *subjunction gate*

(ii) $(a_1 + a_2)'$ *NOR-gate*

(iii) $(a_1 a_2)'$ *NAND-gate*

Figure 2.13

1.5 Examples. (i) The symbolic representation of $p = (x_1' x_2)' + x_3$ is

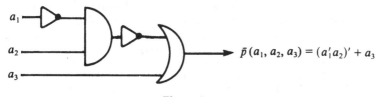

$$\bar{p}(a_1, a_2, a_3) = (a_1' a_2)' + a_3$$

Figure 2.14

Here we used the gates of 1.3. The representation of p by using 1.4 as well is

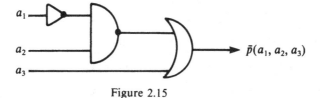

$$\bar{p}(a_1, a_2, a_3)$$

Figure 2.15

(ii) The polynomial p which corresponds to the diagram

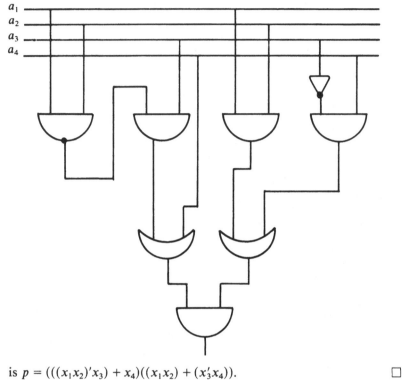

is $p = (((x_1 x_2)' x_3) + x_4)((x_1 x_2) + (x_3' x_4))$.   □

Figure 2.16

1.6 Example. In 1.2.42(i) we noted that there are $2^{(2^2)} = 16$ Boolean polynomial functions on $\mathbb{B}$ in the case $n = 2$. The function value table for all these polynomial functions, written as columns, is as follows:

a_1	a_2	$\bar{p}_1$	$\bar{p}_2$	$\bar{p}_3$	$\bar{p}_4$	$\bar{p}_5$	$\bar{p}_6$	$\bar{p}_7$	$\bar{p}_8$	$\bar{p}_9$	$\bar{p}_{10}$	$\bar{p}_{11}$	$\bar{p}_{12}$	$\bar{p}_{13}$	$\bar{p}_{14}$	$\bar{p}_{15}$	$\bar{p}_{16}$
1	1	1	1	1	1	1	1	1	1	0	0	0	0	0	0	0	0
1	0	1	1	1	1	0	0	0	0	1	1	1	1	0	0	0	0
0	1	1	1	0	0	1	1	0	0	1	1	0	0	1	1	0	0
0	0	1	0	1	0	1	0	1	0	1	0	1	0	1	0	1	0

Minimal forms of the polynomials $p_1, \ldots, p_{16}$ inducing $\bar{p}_1, \ldots, \bar{p}_{16}$ are as follows:

$p_1 = 1$, $p_5 = x_1' + x_2$, $p_9 = x_1' + x_2'$, $p_{13} = x_1'$,

$p_2 = x_1 + x_2$, $p_6 = x_2$, $p_{10} = x_1 x_2' + x_1' x_2$, $p_{14} = x_1' x_2$,

$p_3 = x_1 + x_2'$, $p_7 = x_1 x_2 + x_1' x_2'$, $p_{11} = x_2'$, $p_{15} = x_1' x_2'$,

$p_4 = x_1$, $p_8 = x_1 x_2$, $p_{12} = x_1 x_2'$, $p_{16} = 0$.   □

From these sixteen polynomial functions eight are very important in the algebra of switching circuits and get special names.

1.7 Definition.

$\bar{p}_8 \ldots$ *AND-function*, $\bar{p}_5 \ldots$ *implication function*,

$\bar{p}_{12} \ldots$ *inhibit function*, $\bar{p}_7 \ldots$ *equivalence function*,

$\bar{p}_{10} \ldots$ *antivalence function*, $\bar{p}_9 \ldots$ *NAND-function*,

$\bar{p}_2 \ldots$ *OR-function*, $\bar{p}_{15} \ldots$ *NOR-function*.

1.8. Remark. Especially in the algebra of switching circuits it is usual to call the products (or sums) in the disjunctive (or conjunctive) normal form of a Boolean polynomial the *minterms* (or *maxterms*). Each minterm (maxterm) has the value 1 (or 0) for exactly one assignment $x_i = a_i$ and has value 0 (or 1) otherwise.

Before giving applications to switching circuits we briefly describe a different way of representing switching functions, namely *Karnaugh diagrams* (also called Veitch diagrams). We explain them by using the AND-function.

row	a_1	a_2	minterm	$\bar{p}(a_1, a_2) = a_1 a_2$
(1)	1	1	$x_1 x_2$	1
(2)	1	0	$x_1 x_2'$	0
(3)	0	1	$x_1' x_2$	0
(4)	0	0	$x_1' x_2'$	0

The fourth column is the unique minterm which has the value 1 for the given assignment of input variables. The Karnaugh diagram consists of an a_1 and a_1' column and an a_2 and a_2' row for two input variables a_1, a_2.

	a_1	a_1'
a_2	(1)	(3)
a_2'	(2)	(4)

Figure 2.17

Each section in the intersection of a row and a column corresponds to a minterm. In the case of the AND-function the shaded section has value 1; the others have values 0.

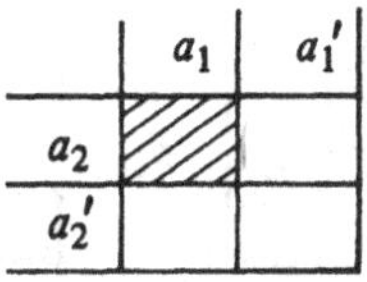

	a_1	a_1'
a_2	▨	
a_2'		

Figure 2.18

Karnaugh diagrams with three input variables a_1, a_2, a_3 can be presented as follows:

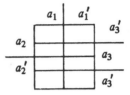

Figure 2.19

Karnaugh diagrams for four input variables are of the following form (called the *standard square* SQ):

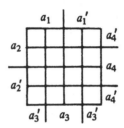

Figure 2.20

The standard square enables us to construct Karnaugh diagrams with more than four input variables.

5 Variables:

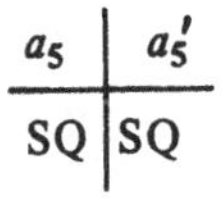

	a_5	a_5'
	SQ	SQ

6 Variables:

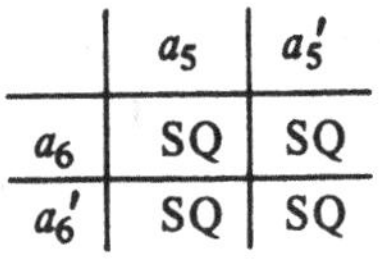

	a_5	a_5'
a_6	SQ	SQ
a_6'	SQ	SQ

Figure 2.21

We give examples of the Karnaugh diagrams of some of the functions introduced in 1.7.

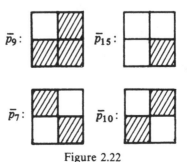

Figure 2.22

Karnaugh diagrams can be used to simplify Boolean polynomials. The following idea is fundamental; we try to collect as many portions of the diagram as possible to form a block; these represent simple polynomials or polynomial functions. Here we may use part of the diagram more than once, since the polynomials corresponding to blocks are connected by +.

As an example we consider simplifying the circuit

$$p = (x_1 + x_2)(x_1 + x_3) + x_1 x_2 x_3.$$

Its Karnaugh diagram is:

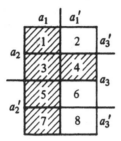

Figure 2.23

The diagram consists of the block formed by squares (1), (3), (5), (7) and the block formed by (3), (4). The first block represents x_1, the second $x_2 x_3$. Thus $p \sim x_1 + x_2 x_3$.

PROBLEMS

1. Simplify and represent in terms of gates:

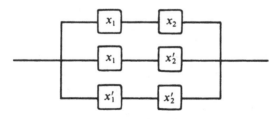

Figure 2.24

2. As in Problem 1

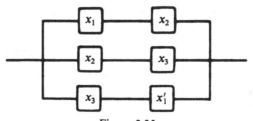

Figure 2.25

3. Simplify

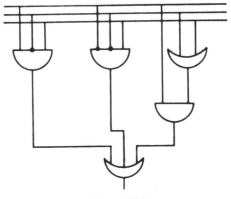

Figure 2.26

4. Find the gate representation corresponding to the function

$$f(x, y, z, u) = xyzu + x'y'zu + xy'z'u + xyz'u'$$
$$+ x'yzu' + x'yz'u + xy'zu' + x'y'z'u'.$$

5. Find the Boolean polynomial for

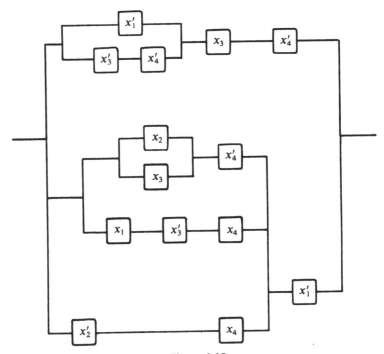

Figure 2.27

6. Use NAND-gates repeatedly to represent AND, OR and NOT gates.

7. Use NOR-gates repeatedly to represent AND, OR and NOT gates.

8. Use Karnaugh diagrams to simplify all polynomials of Example 1.2.

9. Give the Karnaugh diagrams of all polynomial functions $\bar{p}_i$ of Example 1.6.

B. Applications of Switching Circuits

We describe some examples of applications.

1.9 Example. In a large room there are electrical switches next to the three doors to operate the central lighting. The three switches operate alternatively, i.e. each switch can switch on or switch off the lights. We wish to determine the switching circuit p, its symbolic representation and contact diagram. Each switch has two positions: either on or off. We denote the switches by x_1, x_2, x_3 and the two possible states of the switches x_i by $a_i \in \{0, 1\}$.

The light situation in the room is given by the value $\bar{p}(a_1, a_2, a_3) = 0$ ($=1$) if the lights are off (are on, respectively). We arbitrarily choose $\bar{p}(1, 1, 1) = 1$.

(a) If we operate one or all three switches then the lights go off, i.e. we have $\bar{p}(a_1, a_2, a_3) = 0$ for all (a_1, a_2, a_3) which differ in one or in three places from $(1, 1, 1)$.

(b) If we operate two switches, the lights stay on, i.e. we have $\bar{p}(a_1, a_2, a_3) = 1$ for all those (a_1, a_2, a_3) which differ in two places from $(1, 1, 1)$.

This yields the following table of function values:

a_1	a_2	a_3	minterms	$\bar{p}(a_1, a_2, a_3)$
1	1	1	$x_1 x_2 x_3$	1
1	1	0	$x_1 x_2 x_3'$	0
1	0	1	$x_1 x_2' x_3$	0
1	0	0	$x_1 x_2' x_3'$	1
0	1	1	$x_1' x_2 x_3$	0
0	1	0	$x_1' x_2 x_3'$	1
0	0	1	$x_1' x_2' x_3$	1
0	0	0	$x_1' x_2' x_3'$	0

From this table we can derive the disjunctive normal form for the switching circuit p as in 1.2.42(iii):

$$p = x_1 x_2 x_3 + x_1 x_2' x_3' + x_1' x_2 x_3' + x_1' x_2' x_3.$$

Thus we obtain the symbolic representations:

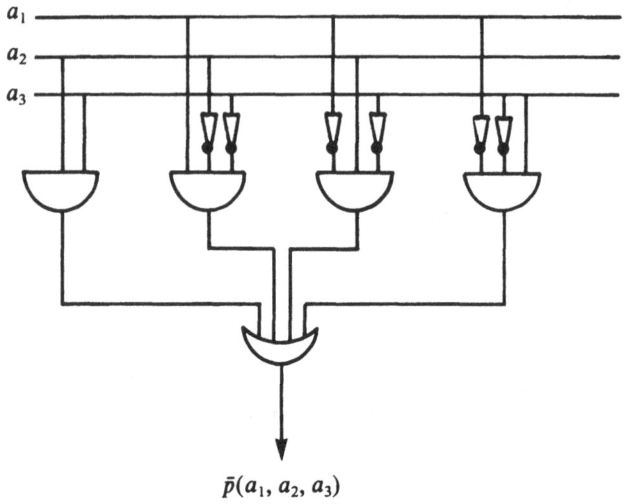

$$\bar{p}(a_1, a_2, a_3)$$

Figure 2.28

This switching circuit can also be represented in terms of antivalence and equivalence switches (see 1.6 and 1.7):

$$p = x_1 x_2 x_3 + x_1 x_2' x_3' + x_1' x_2 x_3' + + x_1' x_2' x_3$$

$$\sim [x_1 \underbrace{[x_2 x_3 + x_2' x_3']}] + [x_1' \underbrace{[x_2 x_3' + x_2' x_3]}].$$

$$\text{equivalence of } x_2, x_3 \qquad \text{antivalence of } x_2, x_3$$

This solution is symbolically represented as in Figure 2.29. A circuit diagram is as in Figure 2.30.   □

1.10 Example. In fast printers for computers, and in machines for paper production or in machines with high speed of paper transport careful control of the paper movements is essential. We draw a schematic model of the method of paper transportation and the control mechanism (see Figure 2.31). The motor operates the pair of cylinders (1), which transports the paper strip (2). This paper strip forms a light barrier for lamp (3). If the paper strip breaks, the photo cell (4) receives light and passes on an impulse which switches off the motor. The light in lamp (3) can vary in its brightness or it can fail, therefore a second photo cell (5) supervises the brightness of lamp (3). The lamp works satisfactorily as long as its brightness is above a given value a. If the brightness falls below a, but remains above a minimum value b, then the diminished brightness is indicated by a warning lamp (6). In this case the transportation mechanism still operates. If the brightness

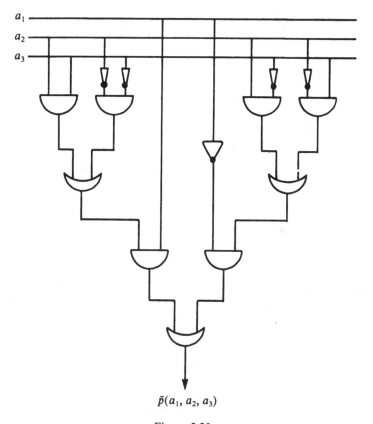

$$\bar{p}(a_1, a_2, a_3)$$

Figure 2.29

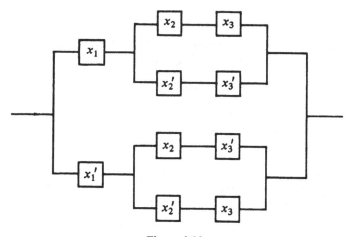

Figure 2.30

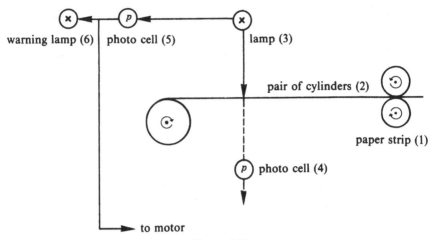

Figure 2.31

of the lamp goes below b then the photo cell (4) cannot work satisfactorily and the motor is switched off. We obtain the switching circuit and its symbolic representation, using the following notation:

$$a_1 = 1 \text{ if brightness of (3)} > a;$$
$$a_1 = 0 \text{ if brightness of (3)} < a;$$
$$a_2 = 1 \text{ if brightness of (3)} > b;$$
$$a_2 = 0 \text{ if brightness of (3)} < b;$$
$$a_3 = 1 \text{ if paper strip is broken;}$$
$$a_3 = 0 \text{ if paper strip is unbroken. Note that } b < a.$$

Thus we need a function value table for the state of the motor (say $\bar{p}_1(a_1, a_2, a_3)$) and one for the warning lamp (say $\bar{p}_2(a_1, a_2, a_3)$). We define

$$\bar{p}_1(a_1, a_2, a_3) = 1; \quad \text{motor operates;}$$
$$\bar{p}_1(a_1, a_2, a_3) = 0: \quad \text{motor is switched off;}$$
$$\bar{p}_2(a_1, a_2, a_3) = 1: \quad \text{warning lamp (6) operates;}$$
$$\bar{p}_2(a_1, a_2, a_3) = 0: \quad \text{warning lamp (6) does not operate.}$$

Therefore the values of the functions can be summarized

a_1	a_2	a_3	$\bar{p}_1(a_1, a_2, a_3)$	$\bar{p}_2(a_1, a_2, a_3)$
1	1	1	0	0
1	1	0	1	0
1	0	1		
1	0	0		
0	1	1	0	1
0	1	0	1	1
0	0	1	0	0
0	0	0	0	0

According to our definitions the case $a_1 = 1$, $a_2 = 0$ cannot occur. This yields as the disjunctive normal form of the polynomial expressing the switching circuit $p_1 = x_1 x_2 x_3' + x_1' x_2 x_3' \sim x_2 x_3'$. For p_2 we obtain the disjunctive normal form $p_2 = x_1' x_2 x_3 + x_1' x_2 x_3' \sim x_1' x_2$. We see that the state of the motor is independent of a_1 and the state of the warning lamp is independent of a_3. The symbolic representation is

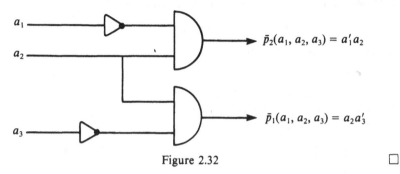

Figure 2.32

1.11 Example. A motor is supplied by three generators. The operation of each generator is monitored by a corresponding switching element which closes a circuit as soon as a generator fails. We demand the following conditions from the electrical monitoring system:

(i) A warning lamp lights up if one or two generators fail.
(ii) An acoustic alarm is initiated if two or all three generators fail.

We determine a symbolic representation as a mathematical model of this problem. Let $a_i = 0$ denote that generator i is operating, $i \in \{1, 2, 3\}$; $a_i = 1$ denotes that generator i does not operate. The table of function values has two parts $\bar{p}_1(a_1, a_2, a_3)$, $\bar{p}_2(a_1, a_2, a_3)$, defined by:

$$\bar{p}_1(a_1, a_2, a_3) = 1; \quad \text{acoustic alarm sounds};$$
$$\bar{p}_1(a_1, a_2, a_3) = 0: \quad \text{acoustic alarm does not sound};$$
$$\bar{p}_2(a_1, a_2, a_3) = 1: \quad \text{warning lamp lights up};$$
$$\bar{p}_2(a_1, a_2, a_3) = 0: \quad \text{warning lamp is not lit up}.$$

Then we obtain the following table for the function values:

a_1	a_2	a_3	$\bar{p}_1(a_1, a_2, a_3)$	$\bar{p}_2(a_1, a_2, a_3)$
1	1	1	1	0
1	1	0	1	1
1	0	1	1	1
1	0	0	0	1
0	1	1	1	1
0	1	0	0	1
0	0	1	0	1
0	0	0	0	0

For p_1 we choose the disjunctive normal form, namely

$$p_1 = x_1 x_2 x_3 + x_1 x_2 x_3' + x_1 x_2' x_3 + x_1' x_2 x_3.$$

This can be simplified by using rules of a Boolean algebra:

$$p_1 \sim x_1 x_2 + x_2 x_3 + x_1 x_3.$$

For p_2 we choose the conjunctive normal form:

$$p_2 = (x_1 + x_2 + x_3)(x_1' + x_2' + x_3').$$

The symbolic representation is

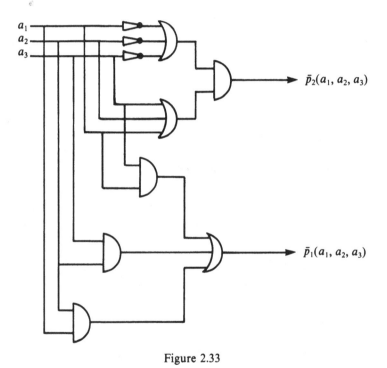

Figure 2.33   □

One of the applications of Boolean algebras is the simplification of electromechanical or electronic switching circuits. In order to economize it is often useful to construct switching circuits in such a way that the costs for their technical realization are as small as possible, e.g. that a minimal number of gates is used. Unfortunately, it is often difficult to decide from the diagram of a switching circuit whether its technical implementation is simple. Also, the simplest and most economical switching circuit may not necessarily be a series-parallel connection, in which case switching algebra is not of much help. Some methods of simplification are discussed in DORNHOFF and HOHN and also in HOHN.

1.12 Remark. A switching circuit p can be simplified as follows:

1. It can be simplified according to the laws of a Boolean algebra (e.g., by applying the distributive, idempotent, absorption and de Morgan laws).
2. Sometimes calculating the dual $d(p)$ of p and simplifying the dual yields a simple expression.
3. We can also determine the minimal form of p (e.g. by using the method of Quine and McCluskey, see Chapter 1, §3).

1.13 Example. We give an example for each of the methods mentioned in 1.12.

1.
$$p = (x_1' + x_2 + x_3 + x_4)(x_1' + x_2 + x_3 + x_4')(x_1' + x_2' + x_3 + x_4')$$
$$\sim (x_1' + x_2 + x_3 + x_4 x_4')(x_1' + x_3 + x_4' + x_2 x_2')$$
$$\sim (x_1' + x_3 + x_2)(x_1' + x_3 + x_4')$$
$$\sim x_1' + x_3 + x_2 x_4'.$$

2.
$$p = ((x_1 + x_2)(x_1 + x_3)) + (x_1 x_2 x_3)$$
$$\sim \underbrace{((x_1 + x_2) + (x_1 x_2 x_3))}_{:= p_1} \underbrace{((x_1 + x_3) + (x_1 x_2 x_3))}_{:= p_2}.$$

Let d denote "dual of". We have $d(p_1) = (x_1 x_2)(x_1 + x_2 + x_3) \sim x_1 x_2$. Therefore $d(d(p_1)) \sim x_1 + x_2$, $d(p_2) = (x_1 x_3)(x_1 + x_2 + x_3) \sim x_1 x_3$. Thus $d(d(p_2)) \sim x_1 + x_3$. Altogether we have $p \sim p_1 p_2 \sim (x_1 + x_2)(x_1 + x_3) \sim x_1 + (x_2 x_3)$.

3. We apply the Quine–McCluskey algorithm to
$$p = x_1' x_2' x_3' x_4' + x_1' x_2' x_3' x_4 + x_1' x_2' x_3 x_4' + x_1' x_2' x_3 x_4$$
$$+ x_1' x_2 x_3 x_4 + x_1 x_2' x_3' x_4' + x_1 x_2 x_3' x_4'.$$

This yields the minimal form
$$x_1 x_3' x_4' + x_1' x_3 x_4 + x_1' x_2'.$$
□

We consider two more examples of applications (due to DOKTER and STEINHAUER).

1.14 Example. An elevator services three floors. On each floor there is a call-button C to call the elevator. It is assumed that at the moment of call the cabin is stationary at one of the three floors. Using these six input variables we want to determine a control which moves the motor M in the right direction for the current situation. One, two, or three call-buttons may be pressed simultaneously; so there are eight possible combinations of calls, the cabin being at one of the three floors. Thus we have to consider $8 \cdot 3 = 24$ combinations of the total of $2^6 = 64$ input variables. We use the following notation: $a_i := c_i$ (for $i = 1, 2, 3$) for the call-signals. $c_i = 0$ (or 1) indicates that no call (or a call) comes from floor i. $a_4 := f_1$, $a_5 := f_2$, $a_6 := f_3$ are position signals; $f_i = 1$ means the elevator cabin is on floor i. $\bar{p}_1(a_1, \ldots, a_6) := M\uparrow$, $\bar{p}_2(a_1, \ldots, a_6) := M\downarrow$ indicate the direction of movement to be given to the motor; then the signal $\bar{M}\uparrow = 1$ means movement of the motor upwards. The output signals (function values) of the motor are determined as follows. If there is no call for the cabin the motor does not operate. If a call comes from the floor where the cabin is at present, again the motor does not operate. Otherwise the motor follows the direction of the call. The only exception is the case when the cabin is at the second floor and there are two simultaneous calls from the third and first floor. We agree that the cabin goes down first. Here is the table of function values:

Call			Floor			Direction of motor	
c_1	c_2	c_3	f_1	f_2	f_3	$\bar{M}\uparrow$	$\bar{M}\downarrow$
1	1	1	1	0	0	0	0
1	1	0	1	0	0	0	0
1	0	1	1	0	0	0	0
1	0	0	1	0	0	0	0
0	1	1	1	0	0	1	0
0	1	0	1	0	0	1	0
0	0	1	1	0	0	1	0
0	0	0	1	0	0	0	0
1	1	1	0	1	0	0	0
1	1	0	0	1	0	0	0
1	0	1	0	1	0	0	1
1	0	0	0	1	0	0	1
0	1	1	0	1	0	0	0
0	1	0	0	1	0	0	0
0	0	1	0	1	0	1	0
0	0	0	0	1	0	0	0
1	1	1	0	0	1	0	0
1	1	0	0	0	1	0	1
1	0	1	0	0	1	0	0
1	0	0	0	0	1	0	1
0	1	1	0	0	1	0	0
0	1	0	0	0	1	0	1
0	0	1	0	0	1	0	0
0	0	0	0	0	1	0	0

From this table we derive the switching circuits p_1 for $M\uparrow$ and p_2 for $M\downarrow$ in disjunctive normal form. Here x_i are replaced by C_i, for $i = 1, 2, 3$ and by F_{i-3} for $i = 4, 5, 6$.

$$p_1 = C_1'C_2C_3F_1F_2'F_3' + C_1'C_2C_3'F_1F_2'F_3'$$
$$+ C_1'C_2'C_3F_1F_2'F_3' + C_1'C_2'C_3F_1'F_2F_3'.$$

The first and third minterms are complementary with respect to C_2 and can be combined. This gives:

$$p_1 = C_1'C_2C_3'F_1F_2'F_3' + C_1'C_3F_1F_2'F_3' + C_1'C_2'C_3F_1'F_2F_3'.$$

For $M\downarrow$ we obtain

$$p_2 = C_1C_2'C_3F_1'F_2F_3' + C_1C_2'C_3'F_1'F_2F_3' + C_1C_2C_3'F_1'F_2'F_3$$
$$+ C_1C_2'C_3'F_1'F_2'F_3 + C_1'C_2C_3'F_1'F_2'F_3.$$

The first two minterms are complementary with respect to C_3, the third and fourth minterm are complementary with respect to C_2. Simplification gives

$$p_2 = C_1C_2'F_1'F_2F_3' + C_1C_3'F_1'F_2'F_3 + C_1'C_2C_3'F_1'F_2'F_3.$$

The two switching circuits enable us to design the symbolic representation of Figure 2.34 (we have six NOT-gates, six AND-gates and two OR-gates).

□

1.15 Example. We consider a simplified model of a container for chemical reactions and design a circuit which involves four function values depending on two input variables (temperature and pressure). In a container, in which chemical reactions can take place, we have a thermometer T and a manometer (pressure-gauge) P for monitoring purposes. Both instruments have upper and lower contacts: $a_1 := t_l$, $a_2 := t_u$ (lower and upper temperature contacts) and $a_3 := p_l$, $a_4 := p_u$ (lower and upper pressure contacts). We want to control the reaction in a certain way, involving a mixing motor $\bar{p}_1(a_1, a_2, a_3, a_4) = m$, a cooling-water valve $\bar{p}_2(a_1, a_2, a_3, a_4) = c$, a heating device $\bar{p}_3(a_1, a_2, a_3, a_4) = h$, and a safety valve $\bar{p}_4(a_1, a_2, a_3, a_4) = s$ (Figure 2.35).

We use the following interpretation:

$$t_l = 0 \text{ and } t_u = 0, \quad \text{temperature is too low,}$$

(e.g., $t_l = 0$ means that the temperature is less than the lower temperature contact).

$t_l = 1$ and $t_u = 0$,	temperature is correct;
$t_l = 1$ and $t_u = 1$,	temperature is too high;
$p_l = 0$ and $p_u = 0$,	pressure is too low;
$p_l = 1$ and $p_u = 0$,	pressure is correct;
$p_l = 1$ and $p_u = 1$,	pressure is too high;
$\bar{m} = 0/\bar{m} = 1$,	mixing motor off/on;
$\bar{c} = 0/\bar{c} = 1$.	cooling-water valve off/on;
$\bar{h} = 0/\bar{h} = 1$,	heating off/on;
$\bar{s} = 0/\bar{s} = 1$,	safety valve closed/open.

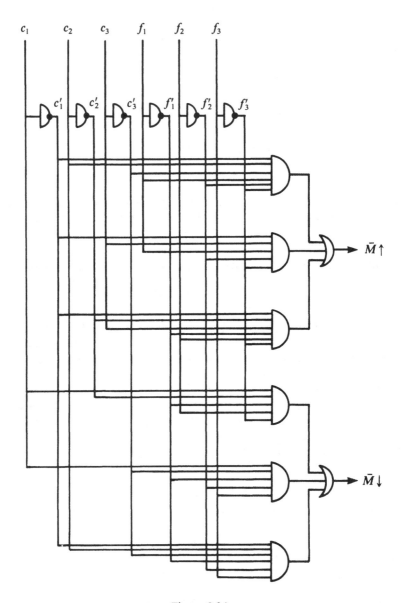

Figure 2.34

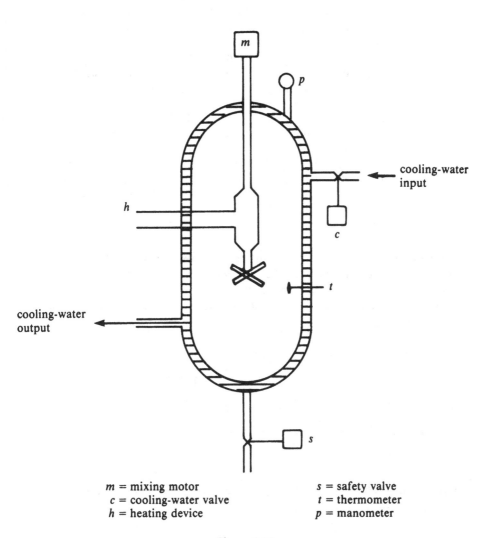

m = mixing motor   s = safety valve
c = cooling-water valve t = thermometer
h = heating device   p = manometer

Figure 2.35

The following function table shows the desired operation of the input and output variables:

t_l	t_u	p_l	p_u	$\bar{h}$	$\bar{c}$	$\bar{m}$	$\bar{s}$	
0	0	0	0	1	0	1	0	
1	0	0	0	1	0	0	0	initial state
1	1	0	0	0	0	1	0	
0	0	1	0	1	0	0	0	
1	0	1	0	0	0	0	0	normal state
1	1	1	0	0	1	1	0	
0	0	1	1	0	0	1	0	
1	0	1	1	0	1	1	0	danger state
1	1	1	1	0	1	1	1	

We denote the corresponding switches and switching circuits by using corresponding capital letters. The table enables us to represent H, C, M and S in disjunctive normal forms, which can be simplified by using the laws of Boolean algebra (some of the steps of simplification have not been written down explicitly):

$$H = T_l'T_u'P_l'P_u' + T_lT_u'P_l'P_u' + T_l'T_u'P_lP_u' \sim \ldots \sim T_u'P_u'(P_l' + T_l'),$$

$$C = T_lT_uP_lP_u' + T_lT_u'P_lP_u + T_lT_uP_lP_u \sim \ldots \sim T_lP_l(T_u + P_u),$$

$$M = T_l'T_u'P_l'P_u' + T_lT_uP_l'P_u' + T_lT_uP_lP_u' + T_l'T_u'P_lP_u + T_lT_u'P_lP_u + T_lT_uP_lP_u,$$

$$\sim T_l'T_u'P_l'P_u' + T_lT_uP_lP_u + T_lT_uP_u' + T_u'P_lP_u,$$

$$S = T_lT_uP_lP_u.$$

Using four NOT-gates, six AND-gates and three OR-gates we can represent the switching circuit in Figure 2.36. □

   As a final example of applications of this type we consider the addition of binary numbers with halfadders and adders. Decimals can be represented in terms of quadruples of binary numbers; such a quadruple is called a *tetrad*. Each digit of a decimal gets assigned a tetrad; thus we use ten different tetrads corresponding to $0, 1, 2, \ldots, 9$. Using four binary positions we can form $2^4 = 16$ tetrads. Since we need only ten tetrads to represent $0, 1, \ldots, 9$, there are six superfluous tetrads, which are called *pseudotetrads*. A binary coded decimal then uses the following association between

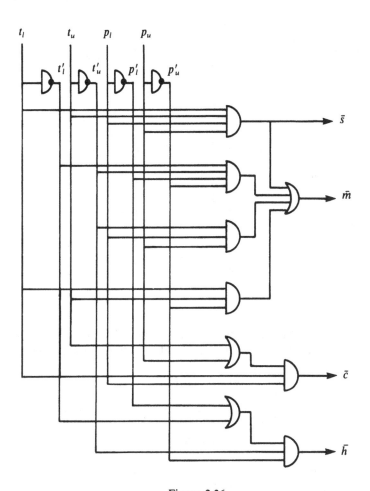

Figure 2.36

$0, 1, \ldots, 9$ and tetrads:

		a_3	a_2	a_1	a_0	$\bar{p}(a_0, a_1, a_2, a_3)$
pseudotetrads		1	1	1	1	1
		1	1	1	0	1
		1	1	0	1	1
		1	1	0	0	1
		1	0	1	1	1
		1	0	1	0	1
decimals	9	1	0	0	1	0
	8	1	0	0	0	0
	7	0	1	1	1	0
	6	0	1	1	0	0
	5	0	1	0	1	0
	4	0	1	0	0	0
	3	0	0	1	1	0
	2	0	0	1	0	0
	1	0	0	0	1	0
	0	0	0	0	0	0

$\bar{p}(a_0, a_1, a_2, a_3) = 1$ denotes the pseudotetrads. We have to evaluate $\bar{p}(a_0, a_1, a_2, a_3)$ to find out if the result of a computing operation is a pseudotetrad.

We represent p in disjunctive normal form:

$$p = x_3x_2x_1x_0 + x_3x_2x_1x_0' + x_3x_2x_1'x_0 + x_3x_2x_1'x_0'$$
$$+ x_3x_2'x_1x_0 + x_3x_2'x_1x_0'.$$

The pairs of minterms 1 and 2, 3 and 4, 5 and 6 are complementary with respect to x_0 and can be simplified:

$$p \sim x_3x_2x_1 + x_3x_2x_1' + x_3x_2'x_1$$
$$\sim x_3x_2x_1 + x_3x_2x_1 + x_3x_2x_1' + x_3x_2'x_1$$
$$\sim (x_3x_2x_1 + x_3x_2x_1') + (x_3x_2x_1 + x_3x_2'x_1)$$
$$\sim x_3x_2 + x_3x_1 \sim x_3(x_2 + x_1).$$

This result indicates that determining if a tetrad with the four positions a_0, a_1, a_2, a_3 is a pseudotetrad is independent of a_0. If we use the a_i as inputs, then Figure 2.37 indicates the occurrence of a pseudotetrad.

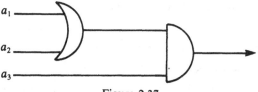

Figure 2.37

1.16 Example ("*Half-adders*"). We describe the addition of two binary numbers. In order to add two single digit binary numbers a_1 and a_2 we have to consider a carry $\bar{c}(a_1, a_2)$, in case $a_1 = a_2 = 1$. The table of the function values for the sum $\bar{s}(a_1, a_2)$ is as follows:

a_1	a_2	$\bar{s}(a_1, a_2)$	$\bar{c}(a_1, a_2)$
1	1	0	1
1	0	1	0
0	1	1	0
0	0	0	0

s and c have the disjunctive normal forms $x_1 x_2' + x_1' x_2$ and $x_1 x_2$, respectively. Thus the corresponding circuit is

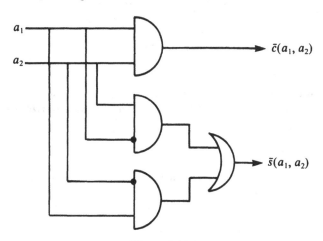

Figure 2.38

To obtain a simpler circuit we modify s according to the axioms of a Boolean algebra.

$$s = x_1 x_2' + (x_1' x_2) \sim ((x_1' + x_2)' + (x_1 + x_2')')$$

$$\sim ((x_1' + x_2)(x_1 + x_2'))' \sim (x_1' x_1 + x_2 x_1 + x_1' x_2' + x_2 x_2')'$$

$$\sim (x_1 x_2 + x_1' x_2')' \sim (x_1 x_2)'(x_1' x_2')' \sim c'(x_1 + x_2).$$

This leads to a circuit called the *half-adder*.

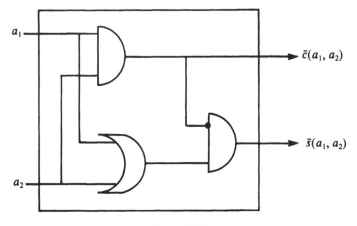

Figure 2.39

Symbolically we write

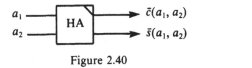

Figure 2.40   □

1.17 Example. *Full-adders* can add three one-digit binary numbers. Let a_1, a_2, a_3 denote the three numbers. Then we can summarize all possible cases in the following table:

a_1	a_2	a_3	$\bar{s}(a_1, a_2, a_3)$	$\bar{c}(a_1, a_2, a_3)$
1	1	1	1	1
1	1	0	0	1
1	0	1	0	1
1	0	0	1	0
0	1	1	0	1
0	1	0	1	0
0	0	1	1	0
0	0	0	0	0

Next we consider partial sums of two summands.

a_2	a_3	$\bar{s}_1(a_2, a_3)$	$\bar{c}_1(a_2, a_3)$
1	1	0	1
1	0	1	0
0	1	1	0
0	0	0	0
1	1	0	1
1	0	1	0
0	1	1	0
0	0	0	0

a_1	$\bar{s}_1(a_2, a_3)$	$\bar{s}(a_1, a_2, a_3)$	$\bar{c}_2(a_1, \bar{s}_1(a_2, a_3))$
1	0	1	0
1	1	0	1
1	1	0	1
1	0	1	0
0	0	0	0
0	1	1	0
0	1	1	0
0	0	0	0

$\bar{c}_1(a_2, a_3)$	$\bar{c}_2(a_1, \bar{s}_1(a_2, a_3))$	$\bar{c}(a_1, a_2, a_3)$
1	0	1
0	1	1
0	1	1
0	0	0
1	0	1
0	0	0
0	0	0
0	0	0

From these tables we derive: a_2 and a_3 are inputs of a half-adder with outputs $\bar{s}_1(a_2, a_3)$ and $\bar{c}_1(a_2, a_3)$. The output $\bar{s}_1(a_2, a_3)$ together with a_1 forms inputs of a second half-adder, whose outputs are $\bar{s}(a_1, a_2, a_3)$ and $\bar{c}_2(a_1, \bar{s}_1(a_2, a_3))$. Here $\bar{s}(a_1, a_2, a_3)$ is the final sum. Finally, disjoining

$\bar{c}_1(a_2, a_3)$ and $\bar{c}_2(a_1, \bar{s}_1(a_2, a_3))$ yields $\bar{c}(a_1, a_2, a_3)$. Hence a full-adder is composed of half-adders in the form:

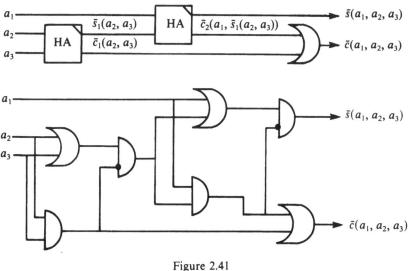

Figure 2.41

A symbol for the full-adder is:

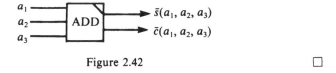

Figure 2.42   □

1.18 Example. Addition of two binary numbers $a_2a_1a_0$ and $b_2b_1b_0$ is performed according to three steps:

(i)   A half-adder calculates the provisional sums s_i and carry c_i', $i = 0, 1, 2$.
(ii) A second half-adder combines s_i with c_{i-1} $(i = 1, 2)$ to the final sum x_i in position i. There may be a carry c_i''.
(iii) Either c_i' or c_i'' is 1. An OR-gate generates the carry c_i for the following position $i + 1$.

The symbols are shown in Figure 2.43.   □

We have seen that an arbitrary electrical circuit can be obtained from suitable compositions of series and parallel circuits. Any function from $\{0, 1\}^n$ into $\{0, 1\}$, thus any switching function of an electrical circuit, is a Boolean polynomial; therefore any electrical circuit can be described by a Boolean polynomial in its mathematical model.

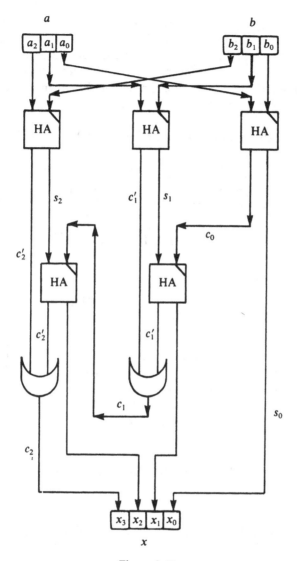

Figure 2.43

Different circuits (namely equivalent ones) can describe different electrical circuits with the same behavior. On the other hand there are electrical circuits which are not composed of series and parallel circuits. These are called *bridge circuits*. According to these remarks there is a corresponding equivalence class of Boolean polynomials, the polynomial functions which reflect the electrical behavior of the bridge circuit.

If we replace single switches in a series-parallel circuit by another series-parallel circuit we again obtain a series-parallel. We can use this to recognize bridge circuits as follows: In a given electrical circuit we look for

a series-parallel circuit which contains at least two switches and connects
any two endpoints of the given circuit. These are replaced by a single switch.
If we continue this process of simplification and obtain a single switch after
possibly several steps, then the given circuit must be a series-parallel circuit.
If we reach a point where we cannot further simplify, then the given circuit
is a bridge circuit.

1.19 Example. Is the following circuit a series-parallel circuit?

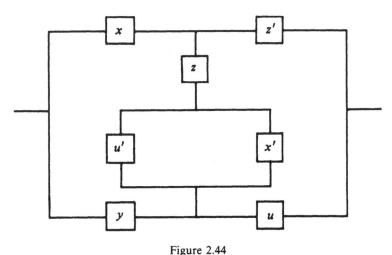

Figure 2.44

If we replace the "inner" series-parallel circuit by a single switch a, we obtain

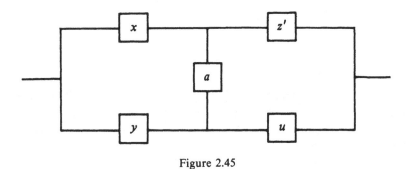

Figure 2.45

Since a further simplification is not possible, we recognize this circuit as a
bridge circuit.   □

For bridge circuits we can find a corresponding polynomial by determining all possible "paths" through the circuit. Thus we determine the switching function. Here is an example for illustration.

1.20 Example. Given

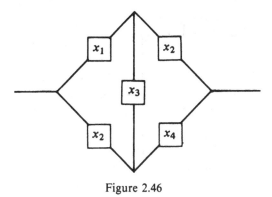

Figure 2.46

We have current in the circuit if and only if either

x_1 and x_2 are closed,   or
x_1, x_3 and x_4 are closed,   or
x_2 and x_4 are closed,   or
x_2 and x_3 (and x_2) are closed.

A switching circuit with this switching function is given by the polynomial

$$p = x_1 x_2 + x_1 x_3 x_4 + x_2 x_4 + x_2 x_3.$$

A series-parallel circuit with the same switching function is given by:

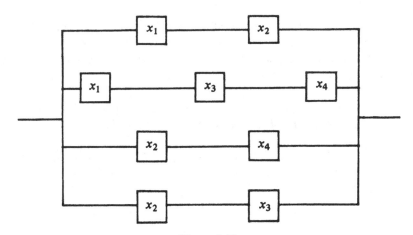

Figure 2.47

A simpler series-parallel circuit with the same "behavior" is:

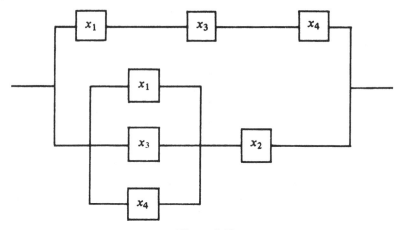

Figure 2.48

The corresponding polynomial $x_1x_3x_4 + (x_1 + x_3 + x_4)x_2$ is equivalent to p. We note that the bridge circuit has fewer contacts than the series-parallel circuit and could be preferred because of financial considerations.   □

At present there is no general method known to find as simple as possible a realization of a given Boolean polynomial as a bridge circuit.

So far we considered so-called two pole circuits or networks. Now we extend this concept.

1.21 Definition. An *n-pole circuit* (or *network*) is a configuration of electrical circuits which are connected by wires, where n points are denoted as *poles* P_i. The set $\{P_i, P_j\}$ of two poles is denoted by p_{ij}.

There are $\frac{1}{2}n(n-1)$ such pairs p_{ij}. If we study all connecting paths between two different poles P_i, P_j in the network, then we obtain a 2-pole network in each case, i.e. an electrical circuit in the sense above. The corresponding Boolean polynomials are also denoted by p_{ij}.

1.22 Definition. Two *n*-pole circuits are called *equivalent*, if corresponding Boolean polynomials p_{ij} of the two circuits are equivalent.

1.23 Example. The 3-pole network (see Figure 2.49) is a combination of $\frac{1}{2}3(3-1)=$ three 2-pole circuits p_{12}, p_{13}, p_{23} which have the corresponding polynomials

$$p_{12} = x_1 + x_2x_3, \qquad p_{13} = (x_2 + x_1x_3)(x_4 + x_2'x_4'),$$
$$p_{23} = (x_1x_2 + x_3)(x_4 + x_2'x_4').$$ □

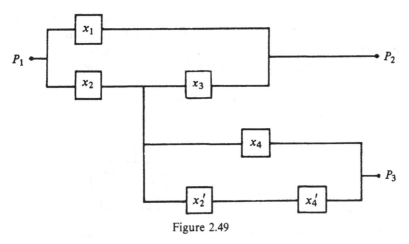

Figure 2.49

Now we show how to reduce arbitrary 3-pole networks to series-parallel circuits.

1.24 Definition. A *star-circuit* is a 3-pole circuit in which the three 2-pole circuits have one point in common which is not a pole. A *triangle circuit* is a 3-pole circuit in which the only connecting points of any two 2-pole circuits are the poles themselves.

The simple cases are:

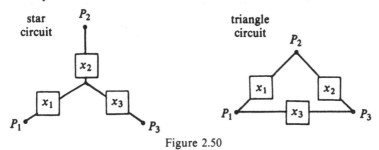

Figure 2.50

The following "transformation" is called a *star-triangle transformation*, because it transforms a given star circuit into an equivalent triangle circuit; all occurring 2-pole circuits p_{ij} are equivalent.

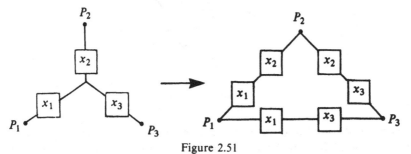

Figure 2.51

This transformation transforms a given circuit into an equivalent one. Let p_{ij} and q_{ij} denote the 2-pole circuits of the star and triangle circuit, respectively. Then $p_{12} = x_1 x_2 = q_{12}$, $p_{13} = x_1 x_3 = q_{13}$, $p_{23} = x_2 x_3 = q_{23}$.

The following transformation is called a *triangle-star transformation* and it transforms a given triangle circuit into an equivalent star circuit, as we readily see by applying the distributive law.

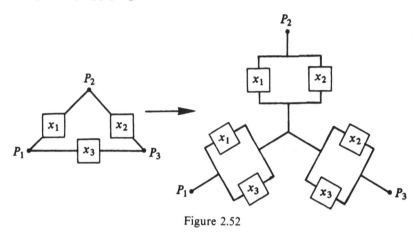

Figure 2.52

To show the equivalence of the two circuits let p_{ij} and q_{ij} be as before. Then

$$q_{12} = x_1 + x_3 x_2 \sim (x_1 + x_3)(x_1 + x_2) = p_{12},$$

$$q_{13} = x_3 + x_1 x_2 \sim (x_3 + x_1)(x_3 + x_2) = p_{13},$$

$$q_{23} = x_2 + x_1 x_3 \sim (x_2 + x_1)(x_2 + x_3) = p_{23}.$$

1.25 Example. The bridge circuit

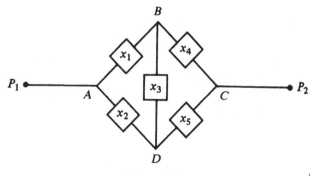

Figure 2.53

is to be transformed into an equivalent series-parallel circuit. Ignoring x_2 and x_5 the point B in the bridge circuit is the "center" of a star with points

A, D and *C.* Using the star-triangle transformation we obtain the following equivalent circuit.

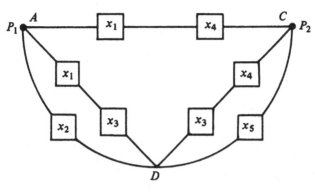

Figure 2.54

This circuit has the contact diagram

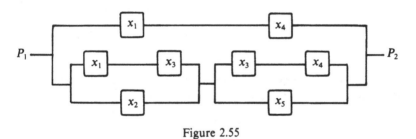

Figure 2.55

The contact diagram yields:

$$p_{12} = x_1 x_4 + (x_1 x_3 + x_2)(x_3 x_4 + x_5).$$ □

The star-to-triangle transformation may be generalized to a *knot-to-mesh transformation* where the knot represents an *n*-pole circuit with a common central point which then is eliminated to give the equivalent mesh circuit (see Figure 2.56). The *knot* of the *n*-pole circuit is the common central point which is to be eliminated. The arrow indicates that further poles with increasing index up to P_n may be added. The paths from P_n to P_2 in both circuits are equivalent. For let p_{ij} be a 2-pole circuit in the knot circuit, and let q_{ij} be a 2-pole circuit in the mesh circuit. Then

$$q_{n2} = x_n x_1 x_1 x_2 + x_n x_2 + x_n x_3 x_3 x_2 \sim x_n x_2 = p_{n2}.$$

We proceed similarly for the other paths.

In Example 1.20 we described one method of finding polynomials for given circuits. The following method is another trial-and-error method for

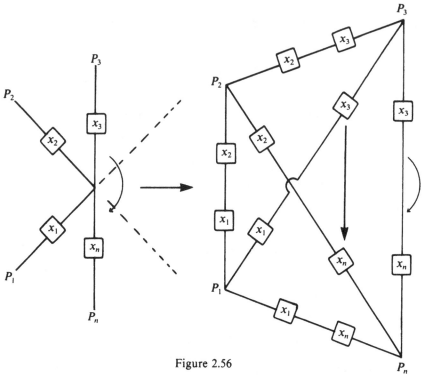

Figure 2.56

the same purpose. It requires one to draw lines through the circuit in all possible ways such that the circuit is broken. For example, consider

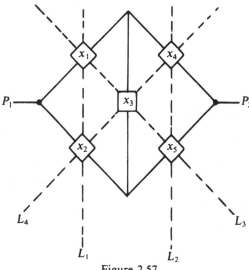

Figure 2.57

From the diagram we see that L_1 cuts x_1, x_2, L_2 cuts x_4, x_5, L_3 cuts x_1, x_3, x_5, L_4 cuts x_2, x_3, x_4. Thus $p_{12} = (x_1 + x_2)(x_4 + x_5)(x_1 + x_3 + x_5)(x_2 + x_3 + x_4)$.

1.26 Example. We wish to simplify the circuit

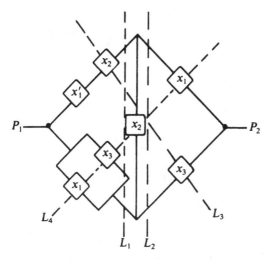

Figure 2.58

First we determine the corresponding Boolean polynomial. The lines are drawn in such a way that the following switches are involved:

$$L_1: x_1', x_2, x_1, x_3; \qquad L_2: x_1, x_3; \qquad L_3: x_1', x_2, x_2, x_3; \qquad L_4: x_1, x_3, x_2, x_1.$$

Therefore

$$p_{12} = (x_1'x_2 + x_1 + x_3)(x_1 + x_3)(x_1'x_2 + x_2 + x_3)(x_1 + x_3 + x_2 + x_1).$$

We simplify this according to the rules of a Boolean algebra.

$$p_{12} = (x_1'x_2 + x_1 + x_3)(x_1 + x_3)(x_1'x_2 + x_2 + x_3)(x_1 + x_3 + x_2 + x_1)$$

$$\sim (x_1x_1'x_2 + x_1x_1 + x_1x_3 + x_1'x_2x_3 + x_1x_3 + x_3x_3)$$

$$\times (x_1'x_2 + x_2 + x_3)(x_1 + x_2 + x_3 + x_1)$$

$$\sim (0 + x_1 + x_1x_3 + x_1'x_2x_3 + x_3)(x_1'x_2 + x_2 + x_3)(x_1 + x_2 + x_3)$$

$$\sim (x_1 + x_3)(x_1'x_2 + x_2 + x_3)(x_1 + x_2 + x_3)$$

$$\sim (x_1x_1'x_2 + x_1x_2 + x_1x_3 + x_1'x_2x_3 + x_2x_3 + x_3x_3)(x_1 + x_2 + x_3)$$

$$\sim (0 + x_1x_2 + x_1x_3 + x_1'x_2x_3 + x_2x_3 + x_3)(x_1 + x_2 + x_3)$$

$$\sim (x_1x_2 + x_3)(x_1 + x_2 + x_3)$$

$$\sim x_1x_1x_2 + x_1x_3 + x_1x_2x_2 + x_2x_3 + x_1x_2x_3 + x_3x_3$$

$$\sim x_1x_2 + x_1x_3 + x_2x_3 + x_1x_2x_3 + x_3$$

$$\sim x_1x_2 + x_3.$$

The corresponding diagrams are of the form:

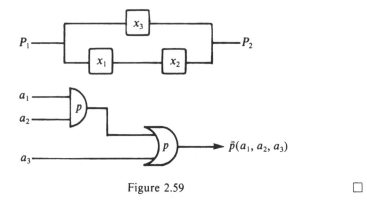

Figure 2.59   □

PROBLEMS

1. In a production process there are three motors operating, but only two are allowed to operate at the same time. Design a switching circuit which prevents that more than two motors can be switched on simultaneously.

2. Design a switching circuit that enables you to operate one lamp in a room from four different switches in that room.

*3. Design a circuit for the subtraction of two three-digit integers. (Use binary representation for the integers.)

*4. Design a circuit for the addition of three three-digit binary numbers.

5. Construct a circuit which helps the water level moving between the points A and B in the diagram below. Thus the motor M of a pump P should always be switched on, if the water level goes below B. The motor should be switched off whenever the water level goes above A.

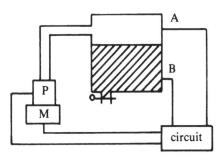

Figure 2.60

6. Let T_1 and T_2 be two telephones which are arranged in such a way that T_2 cannot be used unless T_1 is engaged but T_2 is not cut off when T_1 is not engaged.

A light L is switched on whenever both T_1 and T_2 are engaged but does not switch off until both telephones are disengaged. Construct suitable switching circuits and obtain Boolean polynomials for the circuits for T_2 and L.

*7. Find a 2-pole series-parallel switching circuit which is equivalent to the bridge circuit.

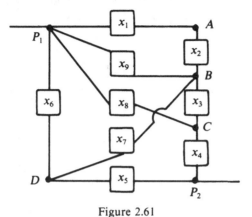

Figure 2.61

8. Find a series-parallel circuit which is equivalent to the following circuit:

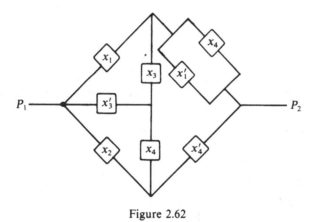

Figure 2.62

*9. Design a circuit for the addition of three two-digit binary numbers.

EXERCISES (Solutions in Chapter 8, p. 425)

1. Determine the symbolic representation of the circuit given by

$$p = (x_1 + x_2 + x_3)(x_1' + x_2)(x_1x_3 + x_1'x_2)(x_2' + x_3).$$

2. Determine the Boolean polynomial p of the circuit

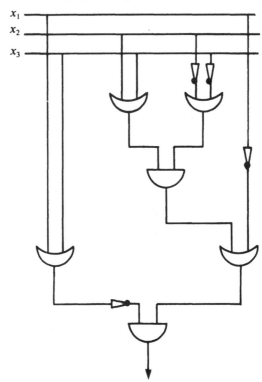

Figure 2.63

3. Find the symbolic representation of a simple circuit for which the binary polynomial function f in four variables is defined as follows: f is 0 at $(0, 0, 1, 0)$, $(0, 0, 1, 1)$, $(0, 1, 1, 0)$, $(0, 1, 1, 1)$, $(1, 0, 0, 0)$, $(1, 0, 0, 1)$, $(1, 1, 0, 0)$, $(1, 1, 0, 1)$, $(1, 1, 1, 1)$ and has value 1 otherwise.

4. Find the symbolic gate representation of the contact diagram (see Figure 2.64).

5. Simplify $p = ((x_1 + x_2)(x_1 + x_3)) + (x_1 x_2 x_3)$.

6. Determine which of the contact diagrams in Figure 2.65 give equivalent circuits.

7. A voting-machine for three voters has three YES–NO switches. Current is in the circuit precisely when YES has a majority. Draw the contact diagram and the symbolic representation by gates and simplify it.

8. An oil pipeline has three pipelines b_1, b_2, b_3 which feed it. Design a plan for switching off the pipeline at three points S_1, S_2, S_3 such that oil runs in the following two situations: S_1 and S_3 are both open or both closed but S_2 is open; S_1 is open and S_2, S_3 are closed.

9. A hall light is controlled by two switches, one upstairs and one downstairs. Design a circuit so that the light can be switched on or off from the upstairs or the downstairs.

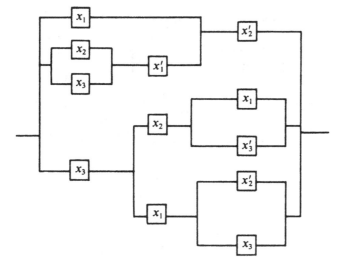

Figure 2.64

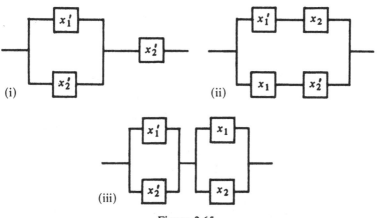

Figure 2.65

10. Determine a series-parallel circuit which is equivalent to the following bridge circuit:

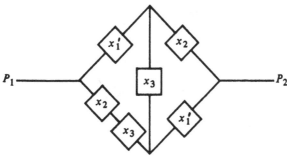

Figure 2.66

11. Find the Boolean polynomial for the bridge circuit:

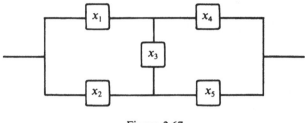

Figure 2.67

and draw an equivalent series-parallel circuit.

*12. Find a series-parallel circuit that is equivalent to the bridge circuit in the figure and simplify your circuit.

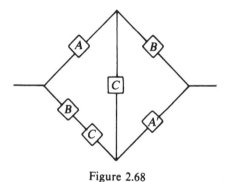

Figure 2.68

13. Find a series-parallel circuit equivalent to the following 4-pole network.

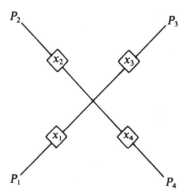

Figure 2.69

14. Simplify the following circuit.

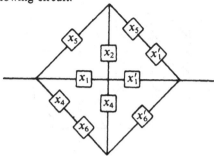

Figure 2.70

15. Determine the Karnaugh diagrams of $\bar{p}_{12}$, $\bar{p}_{14}$, $\bar{p}_5$ and $\bar{p}_3$.
16. Use a Karnaugh diagram to simplify
    (i) $a_1a_2a_3' + a_1'a_2a_3' + (a_1 + a_2' + a_3')' + (a_1 + a_2 + a_3)' + a_3(a_1' + a_2)$.
    (ii) $a_1a_2a_3 + a_2a_3a_4 + a_1'a_2a_4' + a_1'a_2a_3a_4' + a_1'a_2'a_4'$.

17. Simplify the following using Karnaugh diagrams.
    (i) $x_1x_3 + x_1'x_3x_4 + x_2x_3'x_4 + x_2'x_3x_4$.
    (ii) $(x_1' + x_2)(x_3 + x_4')(x_2 + x_4'x_1)$.

18. Find Karnaugh diagrams for the following polynomials.
    (i) $x_1x_2 + x_1'x_2x_3 + x_1'x_2'x_3'$.
    (ii) $x_1x_2x_4 + (x_1' + x_2)x_3$.
    (iii) $x_3x_4 + x_1x_2x_5 + x_1'x_4'x_5'$.
    (iv) $(x_1 + x_2)(x_2 + x_3)(x_3 + x_4)$.

*19. Find the minimal forms for $x_3(x_2 + x_4) + x_2x_4' + x_2'x_3'x_4$ using Karnaugh diagrams.

*20. Simplify

$$f = x_1'x_2' + x_1x_3x_4 + x_1x_2x_4' + x_2'x_3.$$

21. Find simple functions for the following Karnaugh diagrams
    (i)

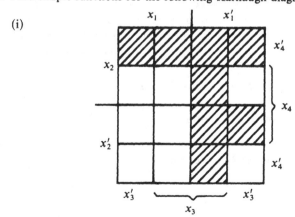

Figure 2.71

(ii)

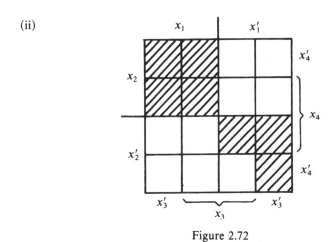

Figure 2.72

§2. Propositional Logic

The algebra of symbolic logic represents one of the early applications of Boolean algebras. Symbolic logic is concerned with studying and analyzing modes of thoughts, arguments and conclusions by making extensive use of symbols. This was the historical origin and the initial purpose for the foundation of the algebra named after G. Boole. The term "proposition" is central in this section and the algebra of truth values will serve as a tool for studying them.

2.1 Informal Definition. A *proposition* is a meaningful sentence (in the natural language) that can properly be assigned the notion true or false.

The careful reader will have noticed that 2.1 is not really a definition but rather a description of the term proposition. At the basis of the concept of propositions we have the two-value-principle (also called "*principle of the excluded middle*" or "*tertium non datur*"), a principle which goes back to the classical propositional logic of Aristotle. It means that each proposition must be either true or false, there is no other possibility.

2.2. Examples. The following are propositions:

(a) 7 is a prime number.
(b) Addition of 2 and 3 gives 4 as the sum.

The following are not propositions:

(c) Be quiet!
(d) Two dogs on the way to yesterday.   □

Propositions will be denoted by capital letters $A, B, C, \ldots$. They can be compounded in several ways, e.g. by "and", "or", "if... then...". We can also obtain a new proposition from a given one by negating it. The truth value "true" of a proposition will be denoted by 1 and the truth value "false" will be 0. Truth tables will be used to define the compound proposition by describing its truth value according to the truth values of the propositions involved in the combination.

2.3. Definition. Let A and B be propositions. The *negation* (or *complementation*) of A (in symbols: $\neg A$) is the proposition "not A", i.e. $\neg A$ is true if and only if A is false. The *conjunction* of A and B (in symbols: $A \wedge B$) is the proposition "A and B", i.e. $A \wedge B$ is true if and only if A as well as B are true. The *disjunction* of A and B (in symbols: $A \vee B$) is the proposition "A or B", i.e. $A \vee B$ is true if and only if either A or B or both are true.

We shall use the following notation in the context of propositional algebra, also used in Chapter 1, §2. As before, $\mathbb{B}$ is the Boolean algebra $\{0, 1\}$.

2.4 Definition. (i) $x_1, \ldots, x_n$ are called *propositional variables*.
(ii) Each $p \in p_n$ is called a *propositonal form*.
(iii) Each $\bar{p}_\mathbb{B} \in P_n(\mathbb{B})$ is called the *truth function* of p.

2.5 Definition. Let $A_1, \ldots, A_n$ be propositions and let A be composed of $A_1, \ldots, A_n$ by means of the operations $\wedge, \vee$ and $\neg$ with suitable use of brackets. If we replace $A_1, \ldots, A_n, \wedge, \vee, \neg$ by $x_1, \ldots, x_n, \sqcap, \sqcup, '$, respectively, then we obtain a polynomial in P_n, which is called the *polynomial corresponding to A with respect to $A_1, \ldots, A_n$*. If B is another proposition formed from $A_1, \ldots, A_n$, then A and B are called *equivalent* (or *logically equivalent*) if the polynomials corresponding to A and B are equivalent (in the sense of 1.2.33), in symbols $A \sim B$.

2.6 Theorem and Definition. Let M be the set of all propositions which can be formed according to 2.5 using propositions $A_1, \ldots, A_n$. Then $M/\sim$ together with the induced operations is a Boolean algebra called the *propositional algebra* over $A_1, \ldots, A_n$. $A_1, \ldots, A_n$ are the *atomic propositions* of M and all other propositions of M are called *compound*.

2.7 Example. Let A_1 be the proposition "It rains" and A_2 be "The sun is shining". M as defined in 2.6 can have at most as many elements as there are nonequivalent polynomials over $\{x_1, x_2\}$, namely $|P_2/\sim| = |P_2(\mathbb{B})| = |\mathbb{B}^{\mathbb{B}^2}| = 2^4 = 16$, by 1.2.34 and 1.2.42. In fact, M has 16 elements (i.e. equivalence classes of propositions) which we can give by the representatives "It rains and it does not rain", "It rains", "The sun is shining", "The sun is shining or the sun is not shining", etc. Therefore, in short,

$$M = \{[A_1 \wedge \neg A_1], [A_1], [A_2], [A_1 \wedge A_2], [A_1 \vee A_2], \ldots, [A_2 \vee \neg A_2]\}.$$

Here $[A_1 \wedge \neg A_1]$ is the 0 and $[A_2 \vee A_2]$ is the 1 of M.   □

We noted above that the compound propositions can be characterized by truth tables.

2.8 Examples. We give some examples of truth tables

A	$\neg A$		A	B	$A \wedge B$	$A \vee B$	$\neg A \vee B$	$(\neg A \vee B) \wedge (\neg B \vee A)$
1	0		1	1	1	1	1	1
0	1		1	0	0	1	0	0
			0	1	0	1	1	0
			0	0	0	0	1	1

Here $A \wedge B$, $A \vee B$, $\neg A \vee B$ and $(\neg A \vee B) \wedge (\neg B \vee A)$ are the polynomial functions induced by $x_1 \wedge x_2$, $x_1 \vee x_2$, $x_1' \wedge x_2$, $(x_1' \vee x_2) \wedge (x_2' \vee x_1)$, respectively. Therefore we use polynomials which have the values of the corresponding polynomial functions. We use the expression "truth table" which gives the truth values of the polynomial function. Two compound propositions are equivalent according to Definition 2.5 if their columns in the truth tables are equal.   □

2.9 Definition. $p \in P_n$ is called a *tautology* if the Boolean polynomial function $\bar{p}$ is always 1, i.e. $\bar{p} = f_1$. A propositional form p is called *contradictory* if the Boolean polynomial function $\bar{p}$ is always 0, i.e. $\bar{p} = f_0$.

The normal forms of Boolean polynomials enable us to decide whether a given proposition is a tautology. Corollary 1.2.42 shows:

2.10 Theorem. (i) *A propositional form is a tautology if and only if its disjunctive normal form has all coefficients equal to 1.*

(ii) *A proposition is contradictory if and only if its conjunctive normal form has all coefficients equal to 0.*   □

The two last propositions in 2.8 are of special importance in symbolic logic and get special names.

2.11 Definition. $\neg A \vee B =: A \to B$ is called the *subjunction* of A and B (or "if-then" operation).

$(\neg A \vee B) \wedge (\neg B \vee A) =: A \leftrightarrow B$ is called the *bijunction* of A and B (or "if and only if" operation).

2.12 Theorem. *Two propositional forms p and q are equivalent if and only if the bijunction p ↔ q is a tautology.* □

The proof is obvious.

One of the problems of propositional logic is to study and simplify propositions and to determine their truth values. In practical applications it is often useful to describe the logical framework in complicated treaties, rules, or laws and to determine their truth values by applying methods of propositional algebra. A simple example will suffice to demonstrate a possible use.

2.13 Example. We determine if the following argument is correct by representing the sentences as propositions and by checking whether the conjunction of the assumptions implies the conclusion. This could be done by writing down truth tables, but we demonstrate a quicker procedure: "If the workers in a company do not go on strike, then a necessary and sufficient condition for salary increases is that the hours of work increase. In case of a salary increase there will be no strikes. If working hours increase there will be no salary increase. Therefore salaries will not be increased." The assumptions can be formulated as follows:

$$\neg S \rightarrow (I \leftrightarrow W), \qquad I \rightarrow \neg S, \qquad W \rightarrow \neg I,$$

where S denotes "strike", I "salary increase" and W "increase of working hours". Then the conclusion is denoted as $\neg I$. We want to determine if the assumptions imply the conclusion. We assume there is a truth assignment making the conjunction of the assumptions true and the conclusion false. For such an assignment $\neg S \rightarrow (I \leftrightarrow W)$, $I \rightarrow \neg S$, $W \rightarrow \neg I$ are true and $\neg I$ is false; therefore I is true. $I \rightarrow \neg S$ is true implies that $\neg S$ is true. Hence by the truth of the first assumption above we know that $I \leftrightarrow W$ is true. I is true; therefore W must be true. But since $W \rightarrow \neg I$ is true, $\neg I$ is true, which is a contradiction to the assumption. Therefore the original argument is correct. □

Propositions such as "x is a natural number", where we can substitute "things" for x which make this proposition true or false (e.g. $x = 3$ or $x = \sqrt{3}$, respectively), are called *predicates*. These predicates can be introduced formally. We need the following ingredients. Let A be a set, called the alphabet, let X be a set disjoint with A, and let V be an arbitrary set and B a propositional algebra whose propositions are elements of the free semigroup (formal language) over A. Those elements of the "free semigroup" on $A \cup X \cup \{0, 1, '\}$ which are propositions in B if the elements in X are replaced by those from V are called predicates over A, B, X and V. We are not considering this formal approach but refer to the literature.

Next we show how equations in Boolean algebras can be used in logic. In order to do this we have to clarify the term "equation". We wish to know under which conditions two Boolean polynomials p, q have the same value. We cannot, in general, speak of the equation $p = q$, since $p = q$ holds by definition only if p and q are identical. Therefore we have to be a little more careful (extensive treatment of this topic can be found in LAUSCH and NÖBAUER).

2.14 Definition. Let p and q be Boolean polynomials in P_n. Then the pair (p, q) is called an *equation*; $(b_1, \ldots, b_n) \in B^n$ is called a *solution* of this equation in a Boolean algebra B if $\bar{p}_B(b_1, \ldots, b_n) = \bar{q}_B(b_1, \ldots, b_n)$. A *system of equations* is a set of equations $\{(p_i, q_i) | i \in I\}$; a *solution of a system* is a common solution of all equations (p_i, q_i).

Often we shall write $p = q$ instead of (p, q) in case no confusion can arise. For instance, $x_1' x_2 + x_3 = x_1 (x_2 + x_3)$ is an equation and $(1, 0, 1)$ is a solution of the equation since the polynomial functions corresponding to the polynomials have value 1 at $(1, 0, 1)$. However, $x_1 + x_1' = 0$ is an equation, which does not have a solution. For the following it is convenient to transform an equation $p = q$ into an equation of the form $r = 0$.

2.15 Theorem. *The equations $p = q$ and $pq' + p'q = 0$ have the same solutions.*

PROOF. Let B be a Boolean algebra and let $(b_1, \ldots, b_n) \in B^n$. Then for $a := \bar{p}_B(b_1, \ldots, b_n)$ and $b := \bar{q}_B(b_1, \ldots, b_n)$ we have

$$a = b \Leftrightarrow (a + b)(a' + b') = aa' + ab' + a'b' + bb' = ab' + a'b$$

which proves the theorem. □

Using this theorem we are able to transform the system of equations $\{(p_i, q_i) | \leq i \leq n\}$ into a single equation

$$p_1 q_1' + p_1' q_1 + p_2 q_2' + p_2' q_2 + \ldots + p_n q_n' + p_n' q_n = 0. \tag{$*$}$$

If we express the left-hand side in conjunctive normal form we see that $(*)$ has a solution if at least one factor has value 0, since for this n-tuple the whole expression on the left of $(*)$ has value 0. In this way we obtain all solutions of the original system of equations.

2.16 Example. We wish to solve the system $\{(x_1 x_2, x_1 x_3 + x_2), (x_1 + x_2', x_3)\}$. Or, written more loosely, the system

$$x_1 x_2 = x_1 x_3 + x_2,$$

$$x_1 + x_2' = x_3.$$

Using 2.15 we can transform this system into a single equation of the form

$$(x_1 x_2)(x_1 x_3 + x_2)' + (x_1 x_2)'(x_1 x_3 + x_2) + (x_1 + x_2')x_3' + (x_1 + x_2')'x_3 = 0.$$

If we express the left-hand side in conjunctive normal form we obtain the equation

$$(x_1 + x_2 + x_3')(x_1' + x_2' + x_3') = 0,$$

which has the same set of solutions as the original system. The solutions over $B = \mathbb{B}$ are the zeros of the first or second factor, namely all $(a_1, a_2, a_3) \in \mathbb{B}^3$, such that

$$a_1 + a_2 + a_3' = 0, \quad \text{i.e.} \quad a_1 = a_2 = 0, \quad a_3 = 1,$$

or with

$$a_1' + a_2' + a_3' = 0, \quad \text{i.e.} \quad a_1 = a_2 = a_3 = 0.$$

Our system of equations therefore has exactly two solutions in $\mathbb{B}$, namely $(0, 0, 1)$ and $(0, 0, 0)$. $\qquad\square$

Knowing the behavior of solutions of Boolean equations we can find applications in logic (see Exercises).

PROBLEMS

1. Represent the propositional form

$$(A \wedge B \wedge \neg C) \vee (\neg A \wedge (B \vee \neg C)) \qquad (*)$$

   as an electrical circuit. Then construct a circuit which has current if and only if the circuit corresponding to $(*)$ does not have current. Give a table which indicates the behavior of the switches in $(*)$ such that current can flow through the circuit.

2. If A wants to see a particular movie, so will B. C and D do not want to see the movie at the same time. B and C either want to see the movie together or neither of them sees it. If A does not see the movie then B and D want to see it. Who is watching the movie?

3. Three people A, B and C are eligible to serve as members of a committee. It is desirable that as many of the three as possible should serve, but the following restrictions hold: A and B should not serve on the committee together, A should serve if and only if C serves; B serves on the committee only if C serves. Who will be members of the committee?

4. Of three women A, B, C, one is young, one is middle-aged and one is elderly. Of the three following statements one is true and two are false: "A is young", "B is not young", "C is not middle aged". Determine the age of each woman.

EXERCISES (Solutions in Chapter 8, p. 439)

1. Determine the truth function of $((x_1 \vee x_2) \vee x_3) \vee (x_1 \wedge x_3)$.

2. Which of the following are tautologies:
   (a) $x_1 \to (x_2 \wedge x_1 \to x_2)$;
   (b) $x_1 \to (x_1 \vee x_2)$;
   (c) $(x_1 \to x_2) \vee (x_2 \to x_3)$.

3. Determine whether the following statements are consistent by representing the sentences as compound propositions: "An economist says that either devaluation will occur or if exports increase, then price controls will have to be imposed. If devaluation does not occur, then exports will not increase. If price controls are imposed then exports will increase".

4. A politician says in four of his election speeches:
   "Either full employment will be maintained or taxes must not be increased";
   "Since politicians have to worry about the people, taxes have to be increased";
   "Either politicians worry about people or full employment cannot be maintained";
   "It is not true that full employment has increased taxes as a consequence".
   (i) Are these four statements consistent?
   (ii) Are the first three statements consistent or are these three statements put together nonsensical?

*5. Each of the objects A, B, C is either green or red or white. Of the following statements one is true and four are false.
   (1) B is not green and C is not white.
   (2) C is red and (4) is true.
   (3) Either A is green or B is red.
   (4) Either A is red or (1) is false.
   (5) Either A is white or B is green.
   Determine the color of each object.

*6. Solve the system of equations
$$x_1 x_3 = x_2 + x_1 x_2 x_3',$$
$$x_1' + x_3 = x_1 x_2.$$

7. Show that the equation $a \sqcup x = 1$ in a Boolean algebra B has the general solution $x = a' \sqcup u$, where u is an arbitrary element in B.

8. Prove: The general solution of $x \sqcup y = c$ is of the form
$$x = c \sqcap (u \sqcup v'), \qquad y = c \sqcap (u' \sqcup v).$$
[Hint: show that the given forms of x, y satisfy the equation and, conversely, if x and y are solutions, then they must be of the given form.]

9. Show that the equation $a \sqcup x = b$ has a solution iff $a \le b$, and if this condition holds show that the general solution is of the form
$$x = (u \sqcup a') \sqcap b.$$

*10. Prove: The equation $x \sqcup (a \sqcap y) = b$ has the general solution
$$x = (u \sqcup (a' \sqcup v')) \sqcap b,$$
$$y = (a' \sqcup b) \sqcap v,$$
for arbitrary u and v in B.

*§3. Further Applications

This final section will show how lattices also occur in other parts of mathematics and outside mathematics in a natural way. We begin with the connection between lattices and topology. If the reader wants to omit the applications of lattice theory in topology he/she should start reading after 3.13. If the reader is more interested in the applications of lattices in probability theory, we recommend the reading of BIRKHOFF (Ch. 9) and HALMOS. Firstly, we need some algebraic preparations.

3.1 Definition. A lattice L is called *complete* if arbitrary (not only finite) subsets of L have a supremum and an infimum in L.

If L has a zero element, then $a \in L$ is called an *atom* if $a \neq 0$, but $b \in L$, $b < a \Rightarrow b = 0$. Thus atoms are the minimal elements in $L\backslash\{0\}$ (see 1.2.14. Equivalently, an element which covers 0 is an atom.) L is called *atomic* if for any $v \in L$ there is an atom $a \in L$ such that $a \leq v$. L is called *atom free* if L does not have any atoms.

A lattice L is called *Brouwer lattice* if, for any two elements $a, b \in L$, the set $\{v \in L | av \leq b\}$ always has a greatest element.

3.2 Examples. $\mathscr{P}(M)$ is always complete and $\{A \subset \mathbb{N} | A \text{ finite}\}$ is not complete. Complete lattices are bounded by $\inf L$ and $\sup L$. In $\mathscr{P}(M)$ all one-element subsets (called singletons) are atoms. The one-dimensional subspaces are the atoms in the lattice of all subspaces of a vector space. In $(\mathbb{N}, \gcd, \text{lcm})$ the atoms are exactly the primes. All lattices mentioned so far are atomic. However, $\{A \subseteq \mathbb{N} | A' \text{ finite}\}$ does not contain any atoms; thus it is atom free. Complete atomic Boolean algebras are precisely those which are Boolean isomorphic to $\mathscr{P}(M)$.

Brouwer lattices are distributive (see BIRKHOFF, p. 45). Any Boolean algebra is a Brouwer lattice, since $a' + b$ is the greatest element in $\{v \in L | av \leq b\}$. Every finite distributive lattice and every chain are Brouwer lattices. $\qquad\square$

Next we need a few topological terms. A *topological space* is a pair $(X, \mathscr{A})$, which consists of a nonempty set X and a subset $\mathscr{A}$ of $\mathscr{P}(X)$ such that $\varnothing$ and X belong to $\mathscr{A}$ and $\mathscr{A}$ is closed with respect to finite unions and arbitrary intersections. The elements of $\mathscr{A}$ are called *closed sets*. $M \subseteq X$ is *open* if the complement M' is closed. The sets which are both open and closed are called *clopen sets*, e.g. $\varnothing$ and X are clopen sets. A topological space $(X, \mathscr{A})$ is called *discrete* if $\mathscr{A} = \mathscr{P}(X)$. In this case all subsets of X are clopen. These definitions and 3.2 imply

3.3 Theorem. *Let $(X, \mathscr{A})$ be a topological space. Then the set $\mathscr{A}$ of all closed sets forms a complete distributive lattice and the set $\mathscr{A}^*$ of all open sets is a Brouwer lattice.* $\qquad\square$

However, in general, $\mathscr{A}$ is not a Brouwer lattice, since, e.g. in $]0, 1[$ with the usual topology, $\{V \in \mathscr{A} | \{1/2\} \cap V = \varnothing\}$ does not have a greatest element. A topological space $(X, \mathscr{A})$ is traditionally called a T_1-*space* if all singletons are in $\mathscr{A}$, i.e. all singletons are closed. T_1-spaces are obviously atomic. The following theorem says that a T_1-space can be characterized by the lattice of its closed subsets.

We say that a mapping f from a topological space $(X_1, \mathscr{A}_1)$ into a topological space $(X_2, \mathscr{A}_2)$ is *continuous* if the pre-images of all sets in $\mathscr{A}_2$ are in $\mathscr{A}_1$. f is called a *homeomorphism* if f is bijective and f and f^{-1} are continuous. In that case $(X_1, \mathscr{A}_1)$ and $(X_2, \mathscr{A}_2)$ are homeomorphic or topologically isomorphic. Thus homeomorphisms are those bijective mappings from X_1 to X_2 which induce a bijective mapping from $\mathscr{A}_1$ to $\mathscr{A}_2$.

3.4 Theorem. *Let* $(X_i, \mathscr{A}_i)$, $i = 1, 2$, *be two* T_1-*spaces.* $(X_1, \mathscr{A}_1)$ *and* $(X_2, \mathscr{A}_2)$ *are homeomorphic if and only if* $\mathscr{A}_1$ *and* $\mathscr{A}_2$ *are isomorphic lattices.*

PROOF. If $(X_1, \mathscr{A}_1)$ and $(X_2, \mathscr{A}_2)$ are homeomorphic, then $\mathscr{A}_1$ and $\mathscr{A}_2$ are isomorphic.

Conversely, any lattice isomorphism from $\mathscr{A}_1$ onto $\mathscr{A}_2$ induces a bijective mapping between the singletons of $\mathscr{A}_1$ and $\mathscr{A}_2$ and hence a bijective mapping between X_1 and X_2. Obviously this must be a homeomorphism.   □

This also tells us which lattices occur as lattices of closed sets of a topological space.

3.5 Theorem. *A lattice* $L = (L, \sqcap, \sqcup)$ *is isomorphic to the lattice of all closed subsets of a* T_1-*space* $(X, \mathscr{A})$ *if and only if* L *is a complete atomic lattice whose dual lattice* $(L, \sqcup, \sqcap)$ *is a Brouwer lattice.*   □

Here we may assume that X is the set of the atoms of L. The last two theorems imply that one may identify parts of topology and lattice theory. This gives rise to hope that one may transfer theorems and results from one theory into the other theory. We shall see that this is indeed possible. A topological space is *compact* if any family $(A_i)_{i \in I}$ of closed sets with the property $\bigcap_{i \in I} A_i = \varnothing$ contains a finite subfamily $(A_k)_{k \in K}$ with $\bigcap_{k \in K} A_k = \varnothing$ (*finite intersection property*). For instance, in $\mathbb{R}^n$ with the usual topology the closed bounded subset are compact spaces (theorem of Borel–Lebesgue).

3.6 Theorem. *A* T_1-*space* $(X, \mathscr{A})$ *is compact if and only if in the lattice* $\mathscr{A}$ *each ultrafilter is generated by one element, i.e. it is of the form* $\{A \in \mathscr{A} | x \in A\}$ *for some* $x \in X$.

PROOF. Let $(X, \mathscr{A})$ be compact. If F is an ultrafilter in $\mathscr{A}$, then the intersection of all sets in F cannot be the empty set, since in that case there is a finite subfamily of sets in F with intersection $\varnothing$ so that $\varnothing \in F$ and thus

$F = \mathscr{A}$, a contradiction. Let x be in this intersection. Then $F \subseteq \{A \in \mathscr{A} | x \in A\} = \{A \in \mathscr{A} | \{x\} \subseteq A\}$. The latter is a proper filter in $\mathscr{A}$ and so the maximality of F implies equality. F is also generated by $\{x\}$.

Conversely, suppose every ultrafilter in $\mathscr{A}$ is of the form as given in the theorem. Let A_i, $i \in I$, be closed sets whose intersection is $\varnothing$. If the filter generated by all A_i were not equal to $\mathscr{A}$ then it would be contained in an ultrafilter F which is generated by $\{x\}$, say. x is an element of all A_i, which contradicts $\bigcap_{i \in I} A_i = \varnothing$. Therefore the A_i generate a filter equal to $\mathscr{A}$, and we can obtain $\varnothing$ in finitely many steps as an intersection of A_i's. $\qquad\square$

Further connections between topology and lattice theory are obtained as follows. A toplogical space $(X, \mathscr{A})$ is called a *Hausdorff space* if any two different points $x_1, x_2 \in X$ are contained respectively in open sets $0_1, 0_2$ with $0_1 \cap 0_2 = \varnothing$. Any Hausdorff space is a T_1-space. If $(X_i, \mathscr{A}_i)$ are topological spaces, then we can build up a system $\mathscr{A}$ of closed sets on $X := \bigtimes_{i \in I} X_i$ from the sets $\mathscr{A}_i$. Here $\mathscr{A}$ is the smallest possible system of closed sets on X such that all projections $X \to X_i$ are continuous. $(X, \mathscr{A})$ is the product of the spaces $(X_i, \mathscr{A}_i)$. A theorem of Tychonoff states that $(X, \mathscr{A})$ is compact if all $(X_i, \mathscr{A}_i)$ are compact. If the spaces $(X_i, \mathscr{A}_i)$ are Hausdorff spaces then $(X, \mathscr{A})$ is a Hausdorff space. If all the factors $(X_i, \mathscr{A}_i)$ are equal to $(X, \mathscr{A})$, say, we write the product as a power $(X^I, \mathscr{A}^I)$.

3.7 Definition. Let B be a Boolean algebra. Then $B^* := \{h | h$ is a Boolean homomorphism from B to $\mathbb{B}\}$ with the topology it inherits as a subspace of the product $\mathbb{B}^B$. Here $\mathbb{B} = (\mathbb{B}, \mathscr{P}(\mathbb{B}))$ has the discrete topology.

Every finite topological space and in particular $(\mathbb{B}, \mathscr{P}(\mathbb{B}))$ are compact; $(\mathbb{B}, \mathscr{P}(\mathbb{B}))$ is also a Hausdorff space. Therefore B^* is a compact Hausdorff space. B^* is completely disconnected, which means that any open set is the union of clopen sets. B^* is nonempty, since for any $b \in B$ there is a maximal ideal I which does not contain b. So B/I is simple (see Exercise 10 in Chapter 3, §2), therefore it is isomorphic to $\mathbb{B}$. Hence we have a Boolean epimorphism $B \to \mathbb{B}$.

3.8 Definition. A compact Hausdorff space is called a *Boolean space* if any open set is the union of clopen sets.

Obviously, finite discrete topological spaces are Boolean spaces. From the above we obtain:

3.9 Theorem. *If B is a Boolean algebra, then B^* is a Boolean space.* $\qquad\square$

The converse is true, as can be verified by elementary calculations.

3.10 Theorem. *If $(X, \mathscr{A})$ is a Boolean space, then the set X^* of all clopen sets is a Boolean algebra.* $\qquad\square$

HALMOS gives proofs of the following two theorems which show a connection between Boolean algebras and Boolean spaces.

3.11 Theorem. (i) *Let B be a Boolean algebra. Then* $(B^*)^* =: B^{**}$ *is Boolean isomorphic to B.*
(ii) *Let* $(X, \mathscr{A})$ *be a Boolean space. Then* $(X^*)^* =: X^{**}$ *is homeomorphic to X.* $\qquad\qquad\square$

3.12 Definition. If $(X, \mathscr{A}) = B^*$ then (B, X) is called a *dual pair.*

Results on Boolean pairs reflect the connections between topology and lattice theory. We mention a few of these results. Here we call $x \in X$ an *isolated point* if $\{x\}$ is open ($\{x\}$ is also closed since X is a T_1 space).

3.13 Theorem. *Let* (B, X) *be a dual pair. Then:*

(i) *B is finite* $\Leftrightarrow$ *X is discrete.*
(ii) *B is countable* $\Leftrightarrow$ *X is metrizable (i.e. the topology can be generated by a metric).*
(iii) *B is atomic* $\Leftrightarrow$ *the isolated points are dense in X (i.e. X is the smallest closed set which contains all isolated points).*
(iv) *B is atom free* $\Leftrightarrow$ *X is perfect (i.e. does not contain isolated points).*
(v) *The ideals (filters) in B correspond to the open (closed) sets in X; here the clopen sets in C correspond to the principal ideals, the minimal closed subsets correspond to the ultrafilters, and the maximal open sets to the maximal ideals.* $\qquad\qquad\square$

Moreover, if (X_i, B_i), $i = 1, 2$ are two dual pairs, then to a Boolean homomorphism $h: B_1 \to B_2$ there corresponds a continuous mapping $f: X_1 \to X_2$, the kernel of h corresponds to the complement of the image of f, etc.

We leave this topic and consider next the connections between lattices and probability theory. At the foundations of probability theory and statistics we have the analysis of "experiments with random outcome". Here we mean "experiments" or observations of experiments whose outcome is not predetermined. This may be the case because of lack of knowledge of the circumstances of the experiment, because of too complicated a situation (see 3.14(i)), or because of real chance (see 3.14(ii)); we could also have a situation with seemingly chance outcome (see 3.14(iii)).

3.14 Example. (i) Let an experiment consist of casting a die once. If the die is completely homogeneous and symmetrical, then we may assume that any of the numbers $1, 2, \ldots, 6$ have the same chance of occurring. If the die is not "ideal", then one (or more) numbers will occur more often and would be less random.

(ii) The experiment consists of counting the number of α-particles which are emitted by a radioactive substance during one second. The outcome of this experiment is truly random.

(iii) Consider the values of the first, second, third ... decimal place of a real number, given in decimal notation. If x is rational, then x has a periodic decimal expansion and in the sequence of the numbers in the various decimal places there is nothing random. However, if x is irrational then the numbers in the decimal places may be randomly distributed.   □

These simple examples show that the term randomness is not an easy one. The question if and how much the outcome of an experiment is random can be answered by statistical methods in such a way that we try to confirm or contradict the assumption of randomness by using a series of tests, e.g. homogeneity of a die tested by a series of casts. In general an experiment has many possible outcomes not all of which are of interest in a special situation. Moreover, some or all combinations of outcomes may be of interest, e.g. "the die shows more than four points", etc.

We shall construct a mathematical model to study random experiments similar to the model constructed for switching circuits. This mathematical model will depend on the main aspects of the situation we are interested in.

3.15 Model for Random Experiments I. In a random experiment let Ω be the set of all (interesting) outcomes of the experiment. The elements of Ω are called *samples* and Ω is called the *sample space* of the experiment. Combinations of outcomes of an experiment can be modeled as subsets of the sample space. For example, if $\Omega = \{1, 2, \ldots, 6\}$ is the sample space for tossing a die once, then the combination "the die shows more than four points" of possible outcomes can be described by $\{5, 6\}$. If K_1, K_2 denote two combinations of outcomes (described by the subsets A_1, A_2 of Ω), then "K_1 and K_2" (described by $A_1 \cap A_2$), "K_1 or K_2" (described by $A_1 \cup A_2$) and "not K_1" (described by A_1') are also combinations of outcomes. These three "operations" on outcome combinations (which interest us) should again be interesting outcome combinations. This leads to the concept of Boolean algebras and to the second stage of our model building process.

3.16 Model for Random Experiments II. A *random experiment* is modeled by the pair $(\Omega, \mathscr{A})$. Here Ω is the sample space and $\mathscr{A}$ is a Boolean subalgebra of $\mathscr{P}(\Omega)$. The elements of $\mathscr{A}$ (corresponding to the interesting combinations of outcomes) are called *events*. $\mathscr{A}$ is called the *algebra of events* over the sample space Ω.

3.17 Examples. (i) In tossing a die let all outcomes and all combinations be of interest. Then a mathematical model is given as $(\Omega, \mathscr{P}(\Omega))$, where $\Omega = \{1, 2, \ldots, 6\}$.

(ii) In tossing a die assume we are only interested whether the points are less than 3 or greater than or equal to 3. In this case a model is given by $(\Omega, \mathcal{P}(\Omega))$, where $\Omega = \{a, b\}$, a means "points value <3", b means "points value ≥ 3".

(iii) In tossing two dice a suitable mathematical model could be $(\Omega, \mathcal{P}(\Omega))$, where $\Omega = \{1, 2, \ldots, 6\} \times \{1, 2, \ldots, 6\}$. The event $\{(4, 6), (5, 5), (6, 4)\}$ can be interpreted as the outcome combination "the sum of points is 10".

(iv) If we consider the experiment of counting α-particle emissions during one second (see 3.14(ii)) then $(\mathbb{N}_0, \mathcal{A})$ would be a suitable model where $\mathcal{A} = \{A \subseteq \mathbb{N}_0 | A$ is finite or $\mathbb{N}_0 \backslash A$ is finite$\}$.   □

3.18 Definition. Let $\mathcal{A}$ be the algebra of events over the sample space Ω. If $\mathcal{A}$ (as a Boolean algebra) is generated by a subset $\mathcal{E}$, then the elements of $\mathcal{E}$ are called *elementary events.*

In the examples 3.17 we can take the one-element subsets as elementary events. If we chose $(\mathbb{N}_0, \mathcal{P}(\mathbb{N}_0))$ as a model in 3.17(iv), we would have to choose a much more complicated system of elementary events.

So far we have not mentioned "probability". It is useful to restrict the definition of an algebra of events. There are good reasons why we would like to have unions and intersections in $\mathcal{A}$ for countably many sets $A_1, A_2, \ldots$. Thus we have to compromise between arbitrary Boolean algebras (in which any two-element set, therefore also any finite set, has a supremum and an infimum) and those Boolean algebras, which are complete lattices. Boolean algebras of this type are called σ-algebras.

3.19 Definition. A Boolean algebra B is called a σ-*algebra* if every countable subset of B has a supremum and an infimum in B.

For example, $\mathcal{P}(M)$ is a σ-algebra for any set M. $\mathcal{A}$ in 3.17(iv) is not a σ-algebra since the family $\{\{0\}, \{1\}, \{2\}, \ldots\}$ has no supremum in $\mathcal{A}$. We refer to HALMOS (pp. 97–103) for connections with σ-spaces (see 3.9–3.12), and for the representability of any σ-algebra as a σ-algebra in a suitable $\mathcal{P}(M)$, factorized by a suitable σ-ideal (theorem of Loomis); the theory of free σ-algebras over a set is also described there (these are given by the σ-algebra of Baire sets generated by the clopen sets in the Boolean σ-space B^M).

3.20 Definition. Let $\mathcal{A}$ be an algebra of events on the sample space Ω. If $\mathcal{A}$ is a σ-algebra then $(\Omega, \mathcal{A})$ is called a *measurable space.*
Now we are able to "measure".

3.21 Definition. Let B be a σ-algebra. A *measure* on B is a mapping μ from B into $\{x \in \mathbb{R} | x \geq 0\} \cup \{\infty\} =: [0, \infty]$ with the following properties:

(i) $\mu(b) < \infty$ for at least one $b \in B$.

(ii) If $b_1, b_2, \ldots$ are countably many elements in B with $b_i b_j = 0$ for $i \neq j$ and if b is their supremum, then

$$\mu(b) = \sum_{i=1}^{\infty} \mu(b_i) \quad (\sigma\text{-additivity}).$$

Moreover

(iii) if $\mu(1) = 1$, then μ is called a *probability measure* and $\mu(b)$ is the *probability* of $b \in B$.

The pair (B, μ) is called a *measure space*. If μ is a probability measure then (B, μ) is also called a *probability measure space*.

3.22 Examples. Let $M \neq \emptyset$ be a finite set. Then μ, defined by $\mu(A) := |A|$, is a measure on $\mathcal{P}(M)$. If $|M| \geq 2$ then μ is not a probability measure. $P(A) := |A|/|M|$ defines a probability measure P on $\mathcal{P}(M)$. If $B = \mathcal{P}(\{1, 2, \ldots, 6\})$, see 3.17(i), then we have, e.g.

$$P(\{5, 6\}) = \frac{|\{5, 6\}|}{|\{1, 2, \ldots, 6\}|} = \frac{2}{6} = \frac{1}{3}. \qquad \square$$

Now we are able to conclude our model.

3.23 Model for Random Experiments III. The triplet $(\Omega, \mathcal{A}, P)$ is called a *probabilistic model of a random experiment* if Ω is the set of all outcomes (which interest us), $\mathcal{A}$ is a suitable σ-algebra in $\mathcal{P}(\Omega)$ and P is a probability measure on $\mathcal{A}$.

In this case $(\Omega, \mathcal{A})$ is a measurable space; $(\mathcal{A}, P)$ is a probability space. The question of whether a given probability measure P is the "correct" measure to model a random experiment is one of the central questions of mathematical statistics. We cannot go into these problems here and refer the reader to the literature.

In case of a finite sample space Ω one is usually best served by the σ-algebra $\mathcal{A} = \mathcal{P}(\Omega)$. However, for infinite sample spaces, there are some problems: if we choose $\mathcal{A}$ "to small" (e.g. as in 3.17(iv)) then we do not obtain a σ-algebra; if we choose $\mathcal{A}$ "too large" as, for instance, in $\mathcal{A} = \mathcal{P}(\Omega)$, it often happens that we cannot have a probability measure on $\mathcal{A}$, as happens for $\Omega = \mathbb{R}$. Therefore we have to compromise. For $\Omega = \mathbb{R}$ and, more generally, for $\Omega = \mathbb{R}^n$ and its subsets, we obtain a solution to this problem as follows.

3.24 Definition. Let $\Omega \subseteq \mathbb{R}^n$. We consider the σ-algebra $\mathcal{B}$ in $\mathcal{P}(\mathbb{R}^n)$ which is generated by the set of all products of open intervals. $\mathcal{B}$ is called the *σ-algebra of Borel sets in $\mathbb{R}^n$*. The σ-algebra of reduced Borel sets on Ω is defined as $\mathcal{B}_\Omega := \{B \cap \Omega \mid B \in \mathcal{B}\}$.

We obtain a measure space $(\mathcal{B}, \mu)$ if we define μ on the product of open sets as follows:

$$\mu((a_1, b_1) \times (a_2, b_2) \times \ldots \times (a_n, b_n)) := (b_1 - a_1)(b_2 - a_2) \ldots (b_n - a_n)$$

and extend μ to all of $\mathcal{B}$ in an obvious way. If we define

$$\mu((a_1, b_1) \times (a_2, b_2) \times \ldots \times (a_n, b_n))$$
$$:= g_1(b_1 - a_1)g_2(b_2 - a_2) \ldots g_n(b_n - a_n)$$

with suitable weights g_i, then we can obtain a probability space. For more information we have to refer to the literature on probability theory.

We summarize some of the basic properties of probability measures.

3.25 Theorem. *Let* (B, P) *be a probability space. Then*:

 (i) $b_1, b_2 \in B$, $b_1 b_2 = 0$ *imply* $P(b_1 + b_2) = P(b_1) + P(b_2)$;
 (ii) $\forall\, b_1, b_2 \in B$: $P(b_1 + b_2) = P(b_1) + P(b_2) - P(b_1 b_2)$;
 (iii) $\forall\, b_1, b_2 \in B$: $b_1 \leq b_2 \Rightarrow P(b_1) \leq P(b_2)$;
 (iv) $\forall\, b \in B$: $P(b) \in [0, 1]$;
 (v) $\forall\, b \in B$: $P(b') = 1 - P(b)$.

PROOF. (i) Is a special case of σ-additivity,
    (ii) We have $b_1 + b_2 = b_1 b_2' + b_1 b_2 + b_1' b_2$ and the terms on the right-hand side have intersection (product) 0. Therefore

$$P(b_1 + b_2) = P(b_1 b_2' + b_1 b_2 + b_1' b_2) = P(b_1 b_2') + P(b_1 b_2) + P(b_1' b_2)$$
$$= P(b_1 b_2') + P(b_1 b_2) + P(b_1' b_2) + P(b_1 b_2) - P(b_1 b_2)$$
$$= P(b_1 b_2' + b_1 b_2) + P(b_1' b_2 + b_1 b_2) - P(b_1 b_2)$$
$$= P(b_1) + P(b_2) - P(b_1 b_2).$$

 (iii) $b_1 \leq b_2 \Rightarrow b_2 = b_1 + b_1' b_2$ (with $b_1(b_1' b_2) = 0$). Therefore $P(b_2) = P(b_1) + P(b_1' b_2) \geq P(b_1)$.
    (iv) Follows from (iii), since $P(b) \leq P(1) = 1$ for all $b \in B$.
    (v) $1 = b + b'$ with $bb' = 0$. Therefore $1 = P(1) = P(b) + P(b')$.   □

Next we introduce and study the term "conditional probability".

3.26 Definition. Let (B, P) be a probability space and $b_1, b_2 \in B$ where $P(b_2) > 0$. Then $P(b_1|b_2) := P(b_1 b_2)/P(b_2)$ is called *conditional probability* of b_1 under the condition b_2. If $P(b_1|b_2) = P(b_1)$, then b_1 and b_2 are said to be *independent*.

Simple calculations verify the following theorem.

3.27 Theorem. *Let* (B, P) *and* b_1, b_2 *be as in 3.26. Then*

(i) $P(b_1 b_2) = P(b_1|b_2)P(b_2)$;
(ii) $P(b_1 b_2) = P(b_1)P(b_2)$, *if* b_1, b_2 *are independent.*   □

Finally we mention a completely different situation where lattices appear in "nature". We consider a "classical mechanical system", such as our system of nine planets. Each planet can be described by its position (three local coordinates). Thus the system can be described as a "point" in $\mathbb{R}^{27}$. $\mathbb{R}^{27}$ is called the *phase space* of the system. Any property of the system (e.g. "the distance between Jupiter and Saturn is less than k kilometres") determines a subset of $\mathbb{R}^{27}$ (in the example this is $\{(x_1, \ldots, x_{27})|(x_{13} - x_{16})^2 + (x_{14} - x_{17})^2 + (x_{15} - x_{18})^2 \leq k^2\}$ if (x_{13}, x_{14}, x_{15}) gives the position of Jupiter and (x_{16}, x_{17}, x_{18}) the position of Saturn). Conversely, there is the question of whether one can assign a relevant physical property to any of the subsets of $\mathbb{R}^{27}$, or more general phase spaces. It seems to make sense to assume that we can assign sensible physical properties to the Borel sets in $\mathbb{R}^{27}$. Thus the physical system of our planets can be studied by means of the σ-algebra of the Borel sets on $\mathbb{R}^{27}$.

In microcosms we have some problems, since not all observables like place, impulse, spin, energy, etc. of a quantum-theoretical system can be precisely measured at the same time. We know this from Heisenberg's uncertainty principle. In this case it is advisable to choose an infinite-dimensional separable Hilbert space as our phase space. This is a vector space H with an inner product $\langle \, , \, \rangle$ such that any Cauchy sequence (h_n), which is characterized by $\lim_{n,m\to\infty}\langle h_n - h_m, h_n - h_m \rangle = 0$, converges to an $h \in H$. Moreover H has a countable subset B such that any $h \in H$ is limit of a sequence of elements in B.) Then the observable properties of a quantum system correspond to the closed subspaces of H. Here lattices of the following type arise:

3.28 Definition. Let L be a lattice with zero element 0. L is called an *orthocomplemented lattice* if for any $v \in L$ there is a $v^\perp \in L$ (called the *orthocomplement* of v) such that:

(i) $(v^\perp)^\perp = v$;
(ii) $v \leq w \Rightarrow v^\perp \geq w^\perp$;
(iii) $vv^\perp = 0$.

An orthocomplemented lattice L with 0, 1 is called *orthomodular* if the orthomodular identity

$$v \leq w \Rightarrow w = v + wv^\perp$$

is satisfied.

3.29 Example. Let L be the lattice of all subspaces of an inner product space. L is orthocomplemented and the orthocomplement of a subspace U

of L is the orthocomplemented space $U^{\perp} := \{x | \langle x, u \rangle = 0 \text{ for all } u \in U\}$. That is, the set of all vectors x orthogonal to U.   ☐

We verify immediately:

3.30 Theorem. *Let H be a separable Hilbert space. Then the closed subspaces of H form a complete, atomic and orthocomplemented lattice.*   ☐

The fact that some of the observables of a quantum-theoretic system can be measured simultaneously and others not, however, can be expressed in the following definition.

3.31 Definition. Let L be an orthomodular lattice and $v, w \in L$. v and w are called *orthogonal*, if $v \le w^{\perp}$. v and w are called *simultaneously verifiable* if there are pairwise orthogonal elements $a, b, c \in L$ such that $v = ac$, $w = bc$. We write $v \leftrightarrow w$.

The relation $\leftrightarrow$ is reflexive and symmetric, but in general is not transitive.

3.32 Definition. Let L be an orthomodular lattice. $z \in L$ is called *central* if $z \leftrightarrow v$ for all $v \in L$.

In all classical mechanical systems all observables are simultaneously measurable and therefore we obtain Boolean algebras. The next and last theorem shows that quantum mechanics in this sense can also be regarded as an extension of classical mechanics and quantum logic as an extension of classical logic.

3.33 Theorem. *Let L be an orthomodular lattice. Then the central elements in L form a Boolean algebra.*   ☐

This theory goes deep into lattice theory and quantum mechanics.

PROBLEMS

*1. Prove: a lattice is complete unless it has a subset which forms an infinite chain.

2. Show that the existence of a 0 and a 1 is a necessary but not a sufficient condition for a lattice to be complete.

3. Prove that for any homomorphism f of a complete lattice L into itself there is at least one element $a \in L$ such that $a = f(a)$.

*4. Determine if the Cartesian product of a family of Boolean spaces is a Boolean space with respect to the product topology.

*5. Prove: Every closed subset Y of a Boolean space $(X, \mathscr{A})$ is a Boolean space with respect to the topology on X. Every clopen set in Y is the intersection of Y with some clopen subset of X.

6. Two gamblers A and B have $\$a$ and $\$b$, respectively, at the start of a game. The game is repeated until one of them is "ruined". Let p and q, $p + q = 1$, be the probability for A and B, respectively, to win in each game. In each game the win of a player (and the loss of the other player) is $\$1$. Find the probability for the loss of each player.

7. Let $(\Omega, \mathscr{A})$ be an experiment and $\varnothing \neq B \in \mathscr{A}$ be an event. The experiment $(\Omega, \mathscr{A})$ conditioned by the event B is $(B, B\mathscr{A})$, where $B\mathscr{A}$ denotes the family of those $A \in \mathscr{A}$ for which $A \subseteq B$. When is $(B, B\mathscr{A})$ the trivial experiment? Let $\mathscr{E}_i = (\Omega_i, \mathscr{A}_i)$, $i = 1, 2, \ldots$, be a finite or denumerably infinite sequence of experiments such that Ω_i and Ω_j, $i \neq j$, are disjoint. Define a sum and product of the experiments $\mathscr{E}_i$ and give an interpretation to these new experiments.

*8. Let Ω be a nondenumerable set and let $\mathscr{A}$ be a family of subsets A of Ω such that either A or $\bar{A}$ is denumerable. Show that:
   (i) $\mathscr{A}$ is a σ-algebra;
   (ii) $\mathscr{A}$ contains nondenumerably many atoms;
   (iii) not every union of atoms belongs to $\mathscr{A}$.

9. A positive integer I is selected with probability $P(I = n) = (\frac{1}{2})^n$, $n = 1, 2, \ldots$. If I takes the value n, a coin with probability e^{-n} of heads is tossed once. Find the probability that the resulting toss is a head.

10. Prove finite additivity for the measure μ, in Definition 3.21(ii), i.e. $\mu(b) = \sum_1^k \mu(b_i)$, where the b_i are disjoint and b is their supremum.

*11. Verify that the lattice of all subspaces of three-dimensional real Euclidean space is an orthomodular lattice.

12. Construct an example of an orthocomplemented lattice with six elements that is not orthomodular.

13. If $a < b$ in an orthomodular lattice L, prove that the sublattice $[a, b]$ is also orthomodular.

14. Let L be an orthocomplemented lattice. Verify in L that $(a + b)^\perp = a^\perp b^\perp$. Also show that L is orthomodular if and only if in case $a \leq b$ there exists $c \in L$ such that a and c are orthogonal and $a + c = b$.

EXERCISES (Solutions in Chapter 8, p. 443)

1. Justify the propositions in 3.2.

2. Which of the lattice in and following 1.1.12 are Boolean lattices? Which are atomic, or complete or Brouwer lattices?

3. Which compact Hausdorff spaces are Boolean spaces?

4. A random experiment consists of going to the doctor's surgery to find out how long one has to wait to get attended to. Give a model for this experiment.

5. An experiment consists of tossing a die on a table and measuring the distance of the die from the edge of the table. Give a model.

6. In the models of Exercises 4 and 5 state a system of elementary events for each experiment.

*7. Justify the statements in 3.22.

8. Let $\mathscr{B}$ be the σ-algebra of Borel sets on $\mathbb{R}$. Is $\mathscr{B}$ atomic? Are all finite subsets elements of $\mathscr{B}$? Is the set of positive real numbers in $\mathscr{B}$? Is $\mathbb{Q} \in \mathscr{B}$?

9. Give a probability measure P for 3.17(iii). Interpret $A = \{1\} \times \{1, 2, \ldots, 6\}$ and $B = \{(1, 1), \ldots, (6, 6)\}$. Calculate $P(A|B)$ and $P(B|A)$. Are A and B independent?

10. A deck of 52 cards is shuffled and two cards are dealt in succession face up. Describe the sample space for (y_1, y_2), where y_i is the designation on the ith card dealt, e.g., AS, ace of spades, etc. How many points are in the sample space? Describe the subset A for the event "both are spades". How many points are in A?

Let A_j be the occurrence of an ace on the jth draw and let B_j be the occurrence of exactly j aces. Express B_j in terms of the A_j and calculate $p(B_0), p(B_1), p(B_2)$.

NOTES

Some standard introductory textbooks on Boolean algebras and applications are HOHN, MENDELSON, WHITESITT.

The collection of survey articles by ABBOTT contains applications of lattices to various areas of mathematics. It includes a paper on orthomodular lattices, one on geometric lattices, a general survey on "what lattices can do for you" within mathematics, and a paper on universal algebra. Most of the books on applied algebra consider Boolean algebras in an introductory way and have applications to switching circuits, simplification methods, logic: BIRKHOFF and BARTEE, DORNHOFF and HOHN, FISHER, GILBERT, PRATHER and PREPARATA.

Further texts describing the applications given in this chapter and several additional examples are DOKTER and STEINHAUER, DWORATSCHEK, HARRISON, PERRIN, DENQUETTE and DALCIN, PESCHEL. A comprehensive book* by DAVIO, DESCHAMPS and THAYSE on discrete and switching functions (algebraic theory and applications) has been published recently. RUDEANU deals with polynomial equations over Boolean algebras.

As to the development of logic, the great merit of the fundamental work of Aristotle is that Aristotle succeeded in describing and systematizing in a set of rules the process of reaching logical conclusions from assumptions. Aristotle focused his attention mainly on a certain type of logical relations and chains, syllogisms. Leibniz tried to give Aristotelian logic an algebraic form. He used the notation AB for the conjunction of two terms,

---

* DAVIO, M., DESCHAMPS, J. P. and THAYSE, A. *Discrete and Switching Functions.* McGraw-Hill, New York, 1978.

noted the idempotent law, and knew that "every A is a B" can be written as $A = AB$. He noted that the calculus of logic applied to statements or expressions as well. However, he did not go far enough and only Boolean algebra was an adequate tool. Leibniz's interest in this context was mainly in transcribing the rules of syllogism into his notation. Boolean algebras reduced an important class of logical questions to a simple algebraic form and gave an algorithm of solving them mechanically.

Under the three Boolean operations the binary relations on a set X form a Boolean algebra isomorphic to that of the power set of all subsets of X^2. This was initially considered independently by C. S. Peirce and Schröder. The success of Boolean algebra and relation algebra in simplifying and clarifying many logical questions encouraged mathematical logicians to try to formalize all mathematical reasoning. Regarding texts on mathematical logic we mention HILBERT and ACKERMAN, a classic, and also BARNES and MACK, RENNIE and GIRLE.

In 1900 at the International Congress of Mathematicians in Paris, D. Hilbert gave an address entitled "Mathematical Problems" in which he listed 23 problems or problem areas in mathematics. Problem 6 refers to the question of the axiomatization of physics and probability theory, which prompted S. N. Bernstein (in 1917) and R. V. Mises (in 1919) to use algebra in the foundations of probability theory. Kolmogorov (in 1933) based probability theory on the concept of measure theory and σ-algebras. An introductory text on probability theory based on Boolean σ-algebras is FRASER. See also LOÈVE, GNEDENKO, CHUNG.

A survey article by Holland in ABBOTT's collection gives an excellent description of the history and development of orthomodular lattices and also includes a bibliography on the subject. VARADARAJAN is a modern introduction to quantum theory, VON NEUMANN is a classic on quantum mechanics.

Further applications of lattices and Boolean algebras are due to ZELMER and STANCU, who try to describe biosystems (e.g. organisms and their environment) axiomatically and interpret them in terms of lattice theory. FRIEDELL gives lattice theoretical interpretations to social hierarchies interpreted as partial orders.

CHAPTER 3

Finite Fields and Polynomials

Finite fields give rise to particularly useful and, in our view, beautiful examples of the applicability of rings and fields. Such applications arise both within mathematics and in other areas; for example, in communication theory, in computing and in statistics. In this chapter we present the basic properties of finite fields, with special emphasis on polynomials over these fields. The simplest finite field is the field $\mathbb{F}_2$ consisting of 0 and 1, with binary addition and multiplication as operations. Many of the results for $\mathbb{F}_2$ can be extended to more general finite fields.

Section 1 contains a summary of the basic properties of rings and fields. The core of the present chapter consists of §2 on finite fields and §3 on irreducible polynomials. We apply some of the results herein to the problem of factorization of polynomials over finite fields. Section 5 is an appendix, giving algorithms for finding null spaces.

§1. Rings and Fields

We assume that the reader has a basic knowledge of the ring concept and therefore we give only a brief summary of fundamental results on rings and fields. Most proofs can be found in standard introductory texts on abstract algebra, such as FRALEIGH or HERSTEIN.

A. Rings, Ideals, Homomorphisms

1.1 Definition. A *ring* is a set R together with two binary operations, $+$ and $\cdot$, called *addition* and *multiplication*, such that

(i) $(R, +)$ is an abelian group;

(ii) the product $r \cdot s$ of any two elements $r, s \in R$ is in R and multiplication is associative;

(iii) for all $r, s, t \in R$: $r \cdot (s + t) = r \cdot s + r \cdot t$ and $(r + s) \cdot t = r \cdot t + s \cdot t$ ("distributive laws").

We will then write $(R, +, \cdot)$ or simply R. In general, the neutral element in $(R, +)$ will be denoted by 0, the additive inverse of $r \in R$ by $-r$. Instead of $r \cdot s$ we shall write rs. Again, let $R^* := R \setminus \{0\}$. Rings according to 1.1 are also called *"associative rings"* in contrast to *"nonassociative rings"* (where associativity is not assumed). The "prototype" of a ring is $(\mathbb{Z}, +, \cdot)$.

1.2 Definition. Let R be a ring. R is said to be *commutative* if this applies to $\cdot$. If there is an element 1 in R such that $r \cdot 1 = 1 \cdot r = r$ for any $r \in R$ then 1 is called an *identity* (or unit) element. If $rs = 0$ then r is called a *left divisor* and s a *right divisor* of zero. If $rs = 0$ implies $r = 0$ or $s = 0$ for all $r, s \in R$, R is called *integral*. A commutative integral ring with identity is called an *integral domain*. If $(R^*, \cdot)$ is a group then R is called a *skew field* or a *division ring*. If, moreover, R is commutative, we speak of a *field*. The *characteristic* of R is the smallest natural number k with $kr := r + \ldots + r$ (k-times) equal to 0 for all $r \in R$. We then write $k = \operatorname{char} R$. If no such k exists, we put $\operatorname{char} R = 0$.

Now we list a series of examples of rings. The assertions contained in this list are partly obvious, some will be discussed in the sequel; the rest is left to the reader.

1.3 Examples. See table on page 122.

Remarks to this list: In $\mathbb{R}^R$, $+$ and $\cdot$ are defined pointwise; i.e.

$$(f + g)(x) := f(x) + g(x) \quad \text{and} \quad (f \cdot g)(x) := f(x) \cdot g(x) \quad \text{for } f, g \in \mathbb{R}^R.$$

$0'$ and $1'$ are the functions which have constant values 0 and 1, respectively. The operation $*$ in G is defined by $g * h := 0$ for all $g, h \in G$; hence every abelian group can be made into a (commutative) ring. □

We state how some of the concepts introduced in 1.2 are interrelated.

1.4 Theorem. (i) *Every field is an integral domain.*

(ii) *Every finite integral domain with more than one element is a field.*

(iii) *Every finite skew-field is a field (Wedderburn's theorem).*

(iv) *If R is an integral domain then* $\operatorname{char} R$ *is 0 or a prime.* □

The finite fields (by 1.4(ii) and (iii) hence also the finite integral domains and the finite skew-fields) will be explicitly described in §2. A first example of a proper skew-field is given by

Underlying set	Operation first	Operation second	Zero element	Identity	Commutative?	Integral?	Integral domain	Field?	char R
Z	$+$	$\cdot$	0	1	yes	yes	yes	no	0
Q, R or C	$+$	$\cdot$	0	1	yes	yes	yes	yes	0
$nZ = \{0, \pm n, \pm 2n, \ldots\}$ $(n > 1)$	$+$	$\cdot$	0	$(-)$	yes	yes	no	no	0
Z_n $(n \notin P)$	$+$	$\cdot$	$[0]$	$[1]$	yes	no	no	no	n
Z_p $(p \in P)$	$+$	$\cdot$	$[0]$	$[1]$	yes	yes	yes	yes	p
R_n^n $(n > 1)$	$+$	$\cdot$	0	I	no	no	no	no	0
R^R	$+$	$\cdot$	0	$1'$	yes	no	no	no	0
$R[x]$	$+$	$\cdot$	0	$1'$	yes	yes	yes	no	0
$\mathscr{P}(M)$ $(M \neq \varnothing)$	$\triangle$	$\cap$	$\varnothing$	M	yes	no	no	no	2
$(G, +)$ (Abelian group)	$+$	$*$	0	$(-)$	yes	no	no	no	max ord(g) or 0
$\{0\}$	$+$	$\cdot$	0	0	yes	yes	yes	no	1

1.5 Example. Let $Q := (\mathbb{R}^4, +, \cdot)$ with component-wise addition and multi-plication defined by

$$(a, b, c, d) \cdot (a', b', c', d')$$
$$:= (aa' - bb' - cc' - dd', ab' + ba' + cd' - dc', ac' - bd' + ca'$$
$$+ db', ad' + bc' - cb' + da').$$

With $1 := (1, 0, 0, 0)$, $i := (0, 1, 0, 0)$, $j := (0, 0, 1, 0)$, $k := (0, 0, 0, 1)$ we get $i^2 = j^2 = k^2 = -1$, $ij = k$, $jk = i$, $ki = j$, $ji = -k$, $kj = -i$, $ik = -j$ and

$$Q = \{a1 + bi + cj + dk \,|\, a, b, c, d \in \mathbb{R}\}.$$

Hence Q can be considered as a four-dimensional vector space over $\mathbb{R}$. $(Q, +, \cdot)$ is a skew-field; it is not a field and can be viewed as an "extension" of $\mathbb{C}$ to a system with three "imaginary units" i, j, k. $(\{\pm 1, \pm i, \pm j, \pm k\}, \cdot)$ is a group of order 8, isomorphic to the quaternion group (4.1.4(iii)). Q is called the *skew-field of quaternions.* □

For our purposes the following example will be of great interest.

1.6 Example. Let n be a positive integer and let $\equiv_n$ denote the following relation on $\mathbb{Z}$: $a \equiv_n b :\Leftrightarrow n$ divides $a - b$ (denoted by $n \,|\, a - b$). Then $\equiv_n$ is an equivalence relation on $\mathbb{Z}$, called the *congruence modulo n*. The equivalence classes are denoted by $[a]$, or by $[a]_n$. We also write $a \equiv b \pmod{n}$. Thus $[a]_n$ is the set of all integers z which on division by n give the same remainder as a gives. Therefore $[a]$ is also called the *residue class of a modulo n*. We have

$$[0] = \{0, n, -n, 2n, -2n, 3n, -3n, \ldots\},$$
$$[1] = \{1, n + 1, -n + 1, 2n + 1, -2n + 1, -2n + 1, \ldots\},$$
$$[2] = \{2, n + 2, -n + 2, 2n + 2, -2n + 2, \ldots\},$$
$$\vdots$$
$$[n - 1] = \{n - 1, n + n - 1, -n + n - 1, 2n + n - 1, -3n + n - 1, \ldots\}.$$

There are no more: $[n] = [0]$, $[n + 1] = [1], \ldots$. In general, $[k] = [a]$ for $a \in \{0, 1, \ldots, n - 1\}$, if k divided by n gives remainder a. The equivalence relation $\equiv_n$ on $\mathbb{Z}$ satisfies

$$a \equiv_n b, c \equiv_n d \Rightarrow a + c \equiv_n b + d, ac \equiv_n bd.$$

Thus $\equiv_n$ is compatible with the operations $+$ and $\cdot$ on $\mathbb{Z}$ and $\equiv_n$ is a congruence relation on $\mathbb{Z}$. We can define two operations addition and multiplication on $\mathbb{Z}_n = \{[0], \ldots, [n - 1]\}$ by

$$[a] + [b] = [a + b],$$
$$[a][b] = [ab].$$

It can be verified that $(\mathbb{Z}_n, +, \cdot)$ is a commutative ring with identity and is called the *residue class ring* mod n. $\square$

The following theorem contains important information, particularly that $\mathbb{Z}_n$ is a field if n is prime. This is our first example of a field with finitely many elements.

1.7 Theorem. *For every $n > 1$, the following conditions are equivalent.*

   (i) $\mathbb{Z}_n$ *is an integral domain*;
  (ii) $\mathbb{Z}_n$ *is a field*;
 (iii) $n \in \mathbb{P}$. $\square$

Let R be a ring and let $S \subseteq R$. S is called a *subring* of R (denoted by $S \leq R$) if S is a ring with respect to the operations of R operating on the subset S of R.

A subgroup I of $(R, +)$ is called *ideal* of R and is denoted by $I \trianglelefteq R$ if

$$\forall\, i \in I, \forall\, r \in R: ir \in I \text{ and } ri \in I. \tag{$*$}$$

If only the first part $ir \in I$ (or the second part $ri \in I$) of $(*)$ holds then we speak of *right* (*left*) *ideals* of R. A left (right) ideal is thus a special subring. The condition $(*)$ is often written as "$IR \subseteq I \wedge RI \subseteq I$". We put $I \triangleleft R$ if $I \trianglelefteq R$, but $I \neq R$.

For every subring S of R we get an equivalence relation $\sim_S$ by $r_1 \sim_S r_2 :\Leftrightarrow r_1 - r_2 \in S$. Since $(R, +)$ is abelian, $\sim_S$ is automatically compatible with $+$ that is $r_1 \sim_S r_2, r_1' \sim_S r_2' \Rightarrow r_1 + r_1' \sim_S r_2 + r_2'$. However, $\sim_S$ is not necessarily compatible with $\cdot$. For instance, if $R = \mathbb{Q}$ and $S = \mathbb{Z}$ then $S \trianglelefteq R$ and $1 \sim_{\mathbb{Z}} 0$ as well as $\frac{3}{2} \sim_{\mathbb{Z}} \frac{1}{2}$ (since $1 - 0 \in \mathbb{Z}$ and $\frac{3}{2} - \frac{1}{2} \in \mathbb{Z}$), but $1 \cdot \frac{3}{2} - 0 \cdot \frac{1}{2} \notin \mathbb{Z}$ whence $1 \cdot \frac{3}{2} \not\sim_{\mathbb{Z}} 0 \cdot \frac{1}{2}$.

When is $\sim_S$ compatible with $\cdot$, i.e. when is $\sim_S$ a congruence relation in R?

1.8 Theorem. *Let $I \leq R$. Then $\sim_I$ is a congruence relation in $(R, +, \cdot)$ if and only if $I \trianglelefteq R$.* $\square$

One readily sees that $R/\sim_I$ then turns out to be a ring with respect to $(r + 1) + (s + I) = (r + s) + I$ and $(r + I)(s + I) = rs + I$. $R/\sim_I =: R/I$ is called the *factor ring* of R with respect to the ideal I.

In the case of groups, normal subgroups are intimately connected with homomorphisms. The same applies to rings where ideals are related to homomorphisms.

Let R, S be rings and let $h: R \to S$ be a function. h is called a (*ring-*) *homomorphism* provided that $h(r_1 + r_2) = h(r_1) + h(r_2)$ and $h(r_1 r_2) = h(r_1)h(r_2)$ hold for all r_1, r_2 in R. Ker $h := \{r \in R \,|\, h(r) = 0\}$ is then called its *kernel* and Im $h = h(R)$ its *image*. It is easy to see that Im $h \leq S$ (in the notation above). The kernel is more than just a subring:

1.9 Theorem. *Every kernel of a homomorphism leaving R is an ideal of R, and all ideals can be obtained in this way.* □

Injective homomorphisms are called *monomorphisms* (or *embeddings*), surjective homomorphisms are called *epimorphisms*, bijective homomorphisms are *isomorphisms*. We use the symbol ≅ to mean "is isomorphic to". If there exists a monomorphism from a ring R into a ring R' then R is embeddable in R', in symbols $R \hookrightarrow R'$. Let

$$\text{Hom}(R, R') := \{f \mid f: R \to R' \text{ homomorphism}\}.$$

1.10 Theorem. *Every ring R can be embedded in a ring $R^{(1)}$ with identity.* □

Such a ring $R^{(1)}$ may be of the form $(R \times \mathbb{Z}, +, \cdot)$, where addition of pairs is defined component-wise and multiplication is defined by $(r, z) \cdot (r', z') = (rr' + zr' + z'r, zz')$ for all $r, r' \in R$ and $z, z' \in \mathbb{Z}$.

It should be remarked that if R already has an identity 1 then 1 loses its role in the process of embedding R into the ring $R^{(1)}$ with identity.

1.11 Theorem. *Let R, S be rings and $h: R \to S$ a homomorphism.*

(i) $\{0\} \trianglelefteq R$ *and* $R \trianglelefteq R$.

(ii) *h is an epimorphism iff* Im $h = S$ *and a monomorphism iff* Ker $h = \{0\}$.

(iii) *("Homomorphism Theorem for Rings".)* $R/\text{Ker } h \cong \text{Im } h$.

(iv) *There is a 1–1 correspondence between the subrings (one- and two-sided ideals) of R containing* Ker h *and the ones of* Im h.

(v) *("First Isomorphism Theorem".) If $S \leq R$ and $I \trianglelefteq R$ then $(I + S)/I \cong S/(I \cap S)$.*

(vi) *("Second Isomorphism Theorem".) If $I \trianglelefteq R$ with* Ker $h \subseteq I$ *then $R/I \cong h(R)/h(I)$. With $J := \ker h$ this reads as $(R/J)/(I/J) \cong R/I$.* □

Intersections of subrings (ideals) of R are again subrings (ideals). We can also speak of the concept of a *generated subring* (ideal, respectively). Ideals which are generated by a single element deserve special interest. Let R be a ring and $a \in R$. The ideal generated by a will be denoted by (a) and is called the *principal ideal* generated by a. If R is a commutative ring with identity, then for all $a \in R$ we get $(a) = aR = \{ar \mid r \in R\}$. In a ring with identity, $\{0\} = (0)$ and $R = (1)$ are principal ideals. In $\mathbb{Z}$, $n\mathbb{Z} = (n)$ is a principal ideal for every $n \in \mathbb{N}_0$. An integral domain in which every ideal is principal is called a *principal ideal domain* (PID). For example, $\mathbb{Z}$ is a PID.

An ideal I in R is called a *maximal ideal* if $I \neq R$ and there is no ideal strictly between I and R. It can be shown that the ideal (n) is a maximal ideal in $\mathbb{Z}$ iff n is a prime. R is *simple* if $\{0\}$ and R are its only ideals.

An ideal P in a commutative ring R is called a *prime ideal* of R if $P \neq R$ and for all $r, s \in R$: $rs \in P$ implies $r \in P$ or $s \in P$.

1.12 Theorem. *Let $I \unlhd R$. R a commutative ring with identity. Then*:

(i) *I is maximal $\Leftrightarrow R/I$ is a field.*

(ii) *I is prime $\Leftrightarrow R/I$ is an integral domain.*

(iii) *I is maximal $\Rightarrow I$ is prime.*   □

PROBLEMS

1. Prove parts (i), (ii) and (iv) of Theorem 1.4 (for (iii) see, e.g. HERSTEIN).

2. In Example 1.3 is there a ring in which every element is either invertible or a divisor of zero?

3. Verify in detail that $\mathbb{Q}$ of Example 1.5 really is a skew-field.

4. Is the following true: "$n \le m \Leftrightarrow \mathbb{Z}_n$ is a subring of $\mathbb{Z}_m$"?

5. Determine all rings with two and three elements.

*6. Prove Theorem 1.7.

7. Find all rings with $(\mathbb{Z}_4, +)$ as their additive group.

8. Are there integral rings of characteristic 6?

9. Let R be a commutative ring of characteristic p, a prime. Show that

$$(x + y)^p = x^p + y^p \quad \text{and} \quad (xy)^p = x^p y^p$$

hold for all $x, y \in R$.

10. Show that char R divides n, if R has $n \in \mathbb{N}$ elements.

*11. Prove Theorems 1.8 and 1.9.

12. Prove Theorem 1.10.

13. Show that $n\mathbb{Z}$ is an ideal in $\mathbb{Z}$. What does one obtain for $n = 0$ and for $n = 1$?

14. Find all ideals in $\mathbb{Z}_3$, $\mathbb{Z}_4$ and $\mathbb{Z}_6$ and determine homomorphisms with these ideals as kernels.

15. Are the following rings PID: $\mathbb{Z}_n$; $n\mathbb{Z}$; every skew-field; every field; every simple ring?

16. Find all maximal ideals in the following rings: $\mathbb{R}$; $\mathbb{Q}$; $\mathbb{Z}_2$; $\mathbb{Z}_3$; $\mathbb{Z}_4$; $\mathbb{Z}_6$; $\mathbb{Z}_{24}$; $\mathbb{Z}_{p^e}$ (p a prime); $\mathbb{Z}_n$.

17. Is the intersection of prime ideals a prime ideal?

18. Find a prime ideal in a given ring which is not maximal.

*19. Show that if $h: R \to S$ is a homomorphism and if R is commutative then $h(I + J) = h(I) + h(J)$ and $h(I \cap J) = h(I) \cap h(J)$ hold for all ideals I, J in R.

20. Find, in $\mathbb{Z} \times \mathbb{Z}$, the subring and the ideal generated by $(2, 2)$. Is $\mathbb{Z} \times \mathbb{Z}$ a PID?

21. Find the ideal generated by $\{12, 14\}$ in $\mathbb{Z}$ and in $2\mathbb{Z} = \{0, \pm 2, \pm 4, \pm 6, \dots\}$.

22. Find all endomorphisms of $(\mathbb{Q}, +, \cdot)$.

23. Find a commutative ring R and some $r \in R$ such that $(a) \neq aR$.

24. Is $2\mathbb{Z}_n$ a maximal ideal in $\mathbb{Z}_n$?

*25. Prove Theorem 1.12.

B. Polynomials

Usually one thinks of polynomials as "formal expressions" $a_0 + a_1 x + \ldots + a_n x^n$ in the "indeterminate" x. But, if x is "not determined" or "unknown", one might raise the question, how this "undetermined" x can be squared, added to some a_0, and so on. We will overcome this difficulty by introducing polynomials in a way which at first glance might look much too formal. But exactly this "formalism" will prove very useful in applications. The whole matter starts by observing that a "polynomial $a_0 + a_1 x + \ldots + a_n x^n$" is already determined by the sequence $(a_0, a_1, \ldots, a_n)$ of its "coefficients".

1.13 Definition. Let R be a commutative ring with identity. All sequences of elements of R which have only finitely many nonzero elements are called *polynomials* over R, all sequences of elements of R are called *formal power series* over R. The set of polynomials over R is denoted by $R[x]$, the set of power series over R is $R[[x]]$. If $p = (a_0, a_1, \ldots, a_n, 0, 0, \ldots) \in R[x]$ we will also write $p = (a_0, \ldots, a_n)$. If $a_n \neq 0$ then we call n the *degree of p* $(n = \deg p)$; if $a_n = 1$ we call p *monic*. We put $\deg(0, 0, 0, \ldots) := -\infty$. Polynomials of degree ≤ 0 are called *constant*.

In $R[x]$ and $R[[x]]$ we define multiplication $(a_0, a_1, \ldots) \cdot (b_0, b_1, \ldots) = (c_0, c_1, \ldots)$ with $c_k := \sum_{i+j=k} a_i b_j = \sum_{i=0}^{k} a_i b_{k-i}$, and addition $(a_0, a_1, \ldots) + (b_0, b_1, \ldots) = (a_0 + b_0, a_1 + b_1, \ldots)$. Note that $\deg pq = \deg p + \deg q$, for $p, q \in R[x]$.

With respect to the operations of multiplication and addition the sets $R[x]$ and $R[[x]]$ are commutative rings with identity $(1, 0, 0, \ldots)$. If R is integral, the same applies to $R[x]$ and $R[[x]]$. $(R[x], +, \cdot)$ and $(R[[x]], +, \cdot)$ are called the *ring of polynomials* and the *ring of formal power series*, respectively. In $R[x]$ and $R[[x]]$ we define $x := (0, 1, 0, 0, \ldots) = (0, 1)$. We then get $x \cdot x = x^2 = (0, 0, 1)$, $x^3 = (0, 0, 0, 1)$, and so on. With $x^0 := (1, 0, 0, \ldots)$ and $a_i = (a_i, 0, 0, \ldots)$ we see that in $R[x]$ and $R[[x]]$ we can write

$$p = (a_0, a_1, a_2, \ldots) = a_0 + a_1 x + a_2 x^2 + \cdots =: \sum_{i \geq 0} a_i x^i.$$

This gives the familiar form of polynomials as $\sum_{i=0}^{n} a_i x^i$ and formal power series as $\sum_{i=0}^{\infty} a_i x^i$ (they are called "formal" since one is not concerned with questions of convergence). We see: x is *not* an "indeterminate", it is just a special polynomial.

Let $p, q \in R[x]$. p *divides* q (denoted by $p \mid q$) if $p = q \cdot r$ for some $r \in R[x]$. If $\deg q > \deg(p) > 0$ then p is called a *proper divisor* of q. A polynomial q with $\deg q \geq 1$ which has no proper divisors is called *irreducible*.

Every polynomial of degree 1 is, of course, irreducible. If $R = \mathbb{C}$ there are no more irreducible ones; if $R = \mathbb{R}$ then "half" of the quadratic polynomials are irreducible as well. This is the contents of the so-called "*Fundamental Theorem of Algebra*":

1.14 Theorem. (i) *The irreducible polynomials in $\mathbb{C}[x]$ are precisely the ones of degree* 1.

(ii) *The irreducible polynomials in $\mathbb{R}[x]$ are the ones of degree ≥ 1 and the polynomials $a_0 + a_1 x + a_2 x^2$ with $a_1^2 - 4a_0 a_2 < 0$.* $\qquad\square$

The best way to check if $g \mid f$ is to "divide" f by g and to see if there is some nonvanishing remainder. This is possible by the following theorem.

1.15 Theorem (The Division Algorithm). *Let R be a field and $f, g \in R[x]$. Then there exist $q, r \in R[x]$ with*

$$f = g \cdot q + r \quad \text{and} \quad \deg r < \deg g. \qquad\square$$

There are some properties of some polynomial rings which will prove useful.

1.16 Theorem. *Let R be a field and $p, f, g \in R[x]$. Then:*

(i) *$R[x]$ is a PID.*

(ii) *(p) is a maximal ideal $\Leftrightarrow (p)$ is a prime ideal $\neq R \Leftrightarrow p$ is irreducible and $p \neq 1$.*

(iii) *p irreducible $\wedge p \mid f \cdot g \Rightarrow p \mid f \vee p \mid g$.* $\qquad\square$

Theorem 1.16(i) has important consequences: we can define concepts like "greatest common divisors":

1.17 Theorem. *Let R be a field and $f, g \in R[x]$. Then there exists exactly one $d \in R[x]$ which enjoys the following properties:*

(i) *$d \mid f$ and $d \mid g$.*

(ii) *d is monic.*

(iii) *If $d' \mid f$ and $d' \mid g$ then $d' \mid d$.*

For this d there exist $p, q \in R[x]$ with $d = p \cdot f + q \cdot g$. $\qquad\square$

The polynomial d in 1.17 is called the *greatest common divisor* of f and g, denoted by $\gcd(f, g)$. f and g are called *relatively prime* (or *coprime*) if $\gcd(f, g) = 1$.

1.18 Theorem (The Unique Factorization Theorem). *Let R be a field. Then every $f \in R[x]$ has a representation (which is unique up to order) of the form $f = r \cdot p_1 \cdot p_2 \cdot \ldots \cdot p_k$ with $r \in R$, $p_1, \ldots, p_k$ monic and irreducible polynomials over R.*   □

We mention explicitly that the proof of this theorem does not indicate how to construct this "prime factor decomposition". In general, there is no efficient constructive way to do this. However, we will describe a (constructive) algorithm in §4, if R is a finite field.

In general, rings with a property analogous to 1.18 are called *unique factorization domains* (UFD). It can be shown that every PID is a UFD.

The penetrating similarity between $\mathbb{Z}$ and $R[x]$ (R a field) is enunciated by the following table. Let f, g be in $\mathbb{Z}$ or in $R[x]$, according to the left-hand side or right-hand side of the table.

$\mathbb{Z}$	$R[x]$
"Norm": absolute value of f.	"Norm": degree of f.
Invertible: numbers with value 1.	Invertible: polynomials of degree 0.
Every integer can be represented in the form $a_0 + a_1 \cdot 10 + a_2 \cdot 10^2 + \ldots + a_n \cdot 10^n$.	Every polynomial can be represented in the form $a_0 + a_1 x + a_2 x^2 + \ldots + a_n x^n$.
$f \mid g :\Leftrightarrow \exists\, q \in \mathbb{Z}: f = g \cdot q$.	$f \mid g :\Leftrightarrow \exists\, q \in R[x]: f = g \cdot q$.
$\exists\, q, r \in \mathbb{Z}: f = g \cdot q + r \wedge 0 \leq r < |g|$.	$\exists\, q, r \in R[x]: f = g \cdot q + r \wedge \deg r < \deg g$.
$\mathbb{Z}$ is an integral domain and a PID.	$R[x]$ is an integral domain and a PID.
f and g have a uniquely determined greatest common divisor d which can be written as $d = f \cdot a + g \cdot b$ with $a, b \in \mathbb{Z}$.	f and g have a uniquely determined greatest common divisor d which can be written as $d = f \cdot a + g \cdot b$ with $a, b \in R[x]$.
f is a prime $:\Leftrightarrow f$ has no proper divisors.	f is irreducible $\Leftrightarrow f$ has no proper divisors.
Every integer is a "unique" product of primes.	Every polynomial is a "unique" product of irreducible polynomials.

Many people think of "functions" when polynomials are discussed. In fact, every polynomial induces a "polynomial function", but not necessarily conversely.

If $p = (a_0, \ldots, a_n) = a_0 + a_1 x + \ldots + a_n x^n \in R[x]$ then $\bar{p}: R \to R$, $r \mapsto a_0 + a_1 r + a_2 r^2 + \ldots + a_n r^n$ is called the *polynomial function* induced by p. Here $a_0 + \ldots + a_n r^n = p(r)$. Let $P(R) := \{\bar{p} \mid p \in R[x]\}$ be the *set of all polynomial functions* induced by polynomials over R. If no confusion is to be expected, we will simply write p instead of $\bar{p}$.

1.19 Theorem. (i) $\forall\, p, q \in R[x]\colon \overline{p+q} = \bar{p} + \bar{q} \wedge \overline{p \cdot q} = \bar{p} \cdot \bar{q}$.

(ii) $P(R)$ is a subring of R^R. $\square$

1.20 Theorem. (i) The map $h\colon R[x] \to P(R),\, p \mapsto \bar{p}$, is a (ring-) epimorphism.

(ii) If all nonzero elements of $(R, +)$ are of infinite order then h is an isomorphism. $\square$

This result allows us to "identify" $R[x]$ and $P(R)$ if $R = \mathbb{R}$, for instance. This is, however, not allowed if R is a finite ring. In $\mathbb{Z}_n[x]$, the polynomial $p := x(x+1)\ldots(x+(n-1))$ has degree n, but $\bar{p}$ is the zero function. Hence p is in ker h of 1.20. If R is finite, the same applies to $P(R) \leq R^R$. On the other hand, $R[x]$ is infinite as long as $|R| > 1$. Hence the map h of 1.20 cannot be an isomorphism in this case.

An element $r \in R$ is called a *root* (or a *zero*) of the polynomial $p \in R[x]$ if $p(r) = 0$. There is an important link between roots and divisibility of polynomials.

1.21 Theorem. An element r of a field R is a root of the polynomial $p \in R[x]$ if and only if $x - r$ divides p. $\square$

Let $r \in R$ be a root of $p \in R[x]$. If k is a positive integer such that p is divisible by $(x - r)^k$, but not by $(x - r)^{k+1}$, then k is called the *multiplicity* of r. If $k = 1$, then r is called a *simple root* of p.

We also remark that by $(R[x])[y] =: R[x, y]$, etc., one gets polynomial rings in more "variables" (they are not variables at all). If one were very precise, one would have to write $(R[x])[x]$.

PROBLEMS

1. Let $f = x^6 + 3x^5 + 4x^2 - 3x + 2$, $g = x^2 + 2x - 3 \in \mathbb{Z}_7[x]$. Determine $g, r \in \mathbb{Z}_7[x]$ such that $f = gq + r$.

2. Let $f(x) = 2 + 2x + 2x^2 + 2x^3 + x^4 + x^7 + x^8$, $g(x) = 2 + x + x^4$ be two polynomials in $\mathbb{Z}_3[x]$. Is f divisible by g?

3. If $f = x + 2x^3 + 3x^4 + 3x^5 + 2x^6 + 3x^7 + x^8$ and $g = x + 3x^2 + x^3$ are polynomials over R, compute polynomials q and r (with deg $r < 3$) such that $f = g \cdot q + r$. Does $g \mid f$ hold? Answer these questions for $R = \mathbb{R}$, $R = \mathbb{Q}$ and finally $R = \mathbb{Z}_5$. Do you get the same answers?

4. With f, g of Problem 3, compute gcd(f, g) again for $R = \mathbb{R}$, $\mathbb{Q}$ and $\mathbb{Z}_5$.

*5. Show that $R[x]$ and $R[[x]]$ are integral domains if R is integral.

6. Show that $p \in R[x]$ (R a field and deg $p \leq 3$) is irreducible iff $\bar{p}$ has no zero (i.e. no $r \in R$ with $\bar{p}(r) = 0$).

7. Decompose $x^5 + x^4 + 3x^3 + 3x^2 + x + 1 \in \mathbb{Z}_5[x]$ into irreducible factors over $\mathbb{Z}_5$.

*8. If R is a field, show that $a_0 + a_1x + \ldots \in R[[x]]$ is invertible iff $a_0 \neq 0$.

9. Find the inverse of $1 - x$ in $\mathbb{R}[[x]]$.

10. List the elements of $P(\mathbb{Z}_2)$ and $P(\mathbb{Z}_3)$ explicitly.

11. Show that $P(R) \leq \{f \in R^R \mid r - s \in I \Rightarrow f(r) - f(s) \in I$ for each $I \trianglelefteq R\}$.

12. Find at least five distinct polynomials in $\mathbb{Z}_3[x]$ which induce the zero function.

13. Determine all homomorphic images of $F[x]$, F a field.

*14. Let F be a field, $a, b \in F$, $a \neq 0$. Show that there is exactly one automorphism ϕ of $F[x]$ with $\phi(x) = ax + b$ and ϕ maps each element of F into itself. Show that all automorphisms of $F[x]$ are of this form.

*15. Prove that for any ideal I in a ring R we have $(R/I)[x]$ is isomorphic to $R[x]/I[x]$.

16. Prove: Let p be a prime and let $a_0, \ldots, a_n$ be finitely many integers, $p \nmid a_n$. Then the equation $a_0 + a_1x + \ldots + a_nx^n = 0$ over $\mathbb{Z}_p$ has at most n distinct solutions x in $\mathbb{Z}_p$. (Theorem of Lagrange.)

17. Which of the following ideals in $\mathbb{Z}[x]$ are prime ideals, which are maximal: $(x + 1)$; $(2, x)$; $(3, x)$; $(6, x)$; (x^2); $(x^3 - 2x^2 - 2x - 3)$? (Here $(2, x)$ denotes the ideal generated by 2 and x.)

18. Show that $(x^2 + 1)$ is not a maximal ideal of $\mathbb{R}[x]$ and $(2, x^2)$ is not a principal ideal of $\mathbb{Z}[x]$.

19. Find three maximal ideals in $\mathbb{R}[x]$. Also for $\mathbb{Z}_2[x]$.

*20. Show that $\mathbb{R}[x, y]$ is a UFD but not a PID.

21. x and y are relatively prime in $\mathbb{R}[x, y]$. Show that there are no $a, b \in \mathbb{R}[x, y]$ such that $1 = ax + by$.

22. Show that the ideals (x), (x, y), $(2, x, y)$ are prime ideals in $\mathbb{Z}[x, y]$, but only (x, y) is maximal.

23. Determine all zeros in $\mathbb{Z}_5$ of $2x^{219} + 3x^{74} + 2x^{57} + 3x^{44} \in \mathbb{Z}_5[x]$.

24. Show that $f, g \in F[x]$, F a field, have a common zero b, if b is zero of $\gcd(f, g)$. Find the common zeros of $f = 2x^3 + x^2 - 4x + 1$ and $g = x^2 - 1$ in $\mathbb{R}$.

25. Let $f(x) = 1 + x + 4x^2 + 2x^3 + 4x^4 + 3x^6 \in \mathbb{Z}_5[x]$. Factor f into a product of irreducible polynomials over $\mathbb{Z}_5$.

26. (i) Let $f(x) = \sum_{i=0}^{n} a_ix^i$. Explain the following table due to W. G. Horner (in 1819) for the calculation of $f(x) = (x - c)(b_{n-1}x^{n-1} + \ldots + b_0) + f(c)$ and also for determining the value of f at c.

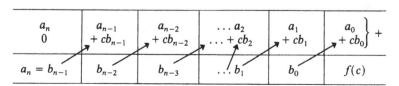

(ii) Generalize Horner's scheme to obtain the coefficients $b_{n-2}, b_{n-1}, \ldots, b_0$; r_1, r_0 of $b(x) = \Sigma\, b_i x^i$ and $r(x) = r_0 + r_1 x$ in the equation $f(x) = (x^2 - ax - c)b(x) + r(x)$.

(iii) Use (i) to determine $f(4)$ for $f(x) = 2x^6 - x^5 - 9x^4 + 10x^3 - 11x + 9$ over $\mathbb{R}$.

(iv) Determine the multiplicity of the zero 1 of $f(x) = x^5 - 2x^4 + x^3 + x^2 - 2x + 1$ over $\mathbb{R}$.

(v) Determine $f(-3), f'(-3), f''(-3)$ for $f(x) = x^4 + 3x^2 + 2x + 1$ over $\mathbb{R}$ by continued use of (i).

(vi) Use (i) to find $f(3)$ over $\mathbb{Z}_5$ for $f(x) = x^4 - x^3 - x + 1$.

C. Fields

Rings R which can be embedded in a field F (in symbols $R \leftarrow F$) obviously have to be commutative and integral, since F has these properties and every subring inherits them. We may assume that R is an integral domain.

1.22 Theorem. *For every integral domain $\neq \{0\}$ there is a field F with the following properties*:

(i) $R \leftarrow F$.
(ii) *If $R \leftarrow F'$, F' a field, then $F \leftarrow F'$.*

SKETCH OF PROOF (the details are left as an exercise). Let $S := R \times R^*$. We define on S $(a, b) + (c, d) := (ad + bc, bd)$ and $(a, b) \cdot (c, d) := (ac, bd)$, as well as $(a, b) \sim (c, d) :\Leftrightarrow ad = bc$. One has to check that $(S, +, \cdot)$ is a ring, $\sim$ is a congruence relation in S and $F := S/\sim$ is a field. The map h, sending r into the equivalence class of $(r, 1)$, is an embedding. If F' is another field with an embedding $h' : R \to F'$ then the map $g : F \to F'$, sending the equivalence class of (a, b) into $h'(a)h'(b)^{-1}$, is well defined and is an embedding as well. The equivalence class of (a, b), by the way, is usually denoted by a/b. $\square$

Thus every integral domain can be embedded in a "minimal" field.

The field of 1.22 is called the *quotient field* of R (or *field of quotients* of R).

1.23 Theorem. *Let $h : R_1 \to R_2$ be an isomorphism between the integral domains R_1 and R_2. Then h can be uniquely extended to an isomorphism $\bar{h}$ between the quotient fields F_1 and F_2 of R_1 and R_2.*

PROOF. $\bar{h} : F_1 \to F_2$, $[(u, b)] \to h(u)h(b)^{-1}$ does the job. $\square$

1.24 Corollary. *Any two quotient fields of an integral domain are isomorphic.*
$\square$

Thus we can speak of *the* quotient field of an integral domain. Applying this construction to our three standard examples of integral domains yields the following fields:

1.25 Examples and Definitions. (i) $\mathbb{Q}$ is the quotient field of $\mathbb{Z}$.

(ii) Let R be a field. The field of quotients of $R[x]$ is denoted by $R(x)$ and is called the *field of rational functions* over R. But the name is quite misleading: the elements in $R(x)$ are not functions at all. They consist of fractions p/q with $p, q \in R[x]$, $q \neq 0$ (in the sense of 1.22).

(iii) If R is a field then the quotient field of $R[[x]]$ is denoted by $R\langle x \rangle$ and is called the *field of formal Laurent series* over R. In an isomorphic copy, $R\langle x \rangle$ consists of sequences $(a_{-n}, \ldots, a_0, a_1, \ldots)$ of elements of R. $\qquad\Box$

1.26 Definition. A subset U of a field F is called a *subfield* of F, in symbols $U \leq F$, if U is a subring of F and U is a field with respect to the operations in F. If $U \neq F$ then $(U, +, \cdot)$ is called a *proper subfield* of $(F, +, \cdot)$, in symbols $U < F$. $(F, +, \cdot)$ is called an *extension field* (or *extension*) of the field $(U, +, \cdot)$ if $(U, +, \cdot)$ is a subfield of $(F, +, \cdot)$. A field P is called a *prime field* if it has no proper subfield.

The following theorem characterizes all prime fields.

1.27 Theorem. *Up to isomorphism, all distinct prime fields are given by $\mathbb{Q}$ and $\mathbb{Z}_p$, p prime.*

PROOF. Let P be a prime field and let 1 be its identity. It can be verified immediately that $C = \{n1 \mid n \in \mathbb{Z}\}$ is an integral domain in P. The mapping $\psi: \mathbb{Z} \to C, z \mapsto z1$, is an epimorphism of $\mathbb{Z}$ onto C. We distinguish between two cases:

(i) If $\ker \psi = \{0\}$, then ψ is an isomorphism. The quotient field of C is the smallest field containing C and is isomorphic to the quotient field of $\mathbb{Z}$, which is $\mathbb{Q}$. Therefore $P \cong \mathbb{Q}$.

(ii) If $\ker \psi \neq \{0\}$ then there is a $k \in \mathbb{N} \setminus \{1\}$ with $\ker \psi = (k)$. The homomorphism theorem 1.11(iii) implies $\mathbb{Z}_k = \mathbb{Z}/(k) \cong C$. C and $\mathbb{Z}_k$ are finite integral domains with more than one element, so they are fields and k must be a prime. In this case $P = C \cong \mathbb{Z}_k$. $\qquad\Box$

Let F be an arbitrary field. It is easily verified that the intersection of all subfields of F is a subfield of F as well as a prime field. Therefore the intersection P of all subfields of F is called the prime field of F.

1.28 Theorem. *Let P be the prime field of the field F. Then*

(i) *If* char $F = 0$ *then* $P \cong \mathbb{Q}$.

(ii) *If* char $F = p \in \mathbb{P}$ *then* $P \cong \mathbb{Z}_p$. $\qquad\Box$

A field with prime characteristic does not necessarily have to be finite, as the following example shows:

1.29 Example. The field $\mathbb{Z}_2(x)$ of all rational functions f/g, $f, g \in \mathbb{Z}_2[x]$, $g \neq 0$, has prime field $\mathbb{Z}_2$, i.e. its characteristic is 2, but it has infinitely many elements.   □

We now note that an element r lying in some extension field of a field F is called a *root* (or a *zero*) of $p \in F[x]$ if $p(r) = 0$.

We proceed to prove one of the basic theorems of modern algebra, Theorem 1.32, which is a result due to L. Kronecker (1821–1891). This theorem guarantees the existence of an extension containing a zero of an arbitrary polynomial over a field. The proof of the theorem also suggests a method of the construction of such an extension. Let F be a field. We already know that the ideals in $F[x]$ are principal ideals (see 1.16). Let $f \in F[x]$ be a polynomial of positive degree and let (f) denote the principal ideal generated by f. An important result for the factor ring $F[x]/(f)$ is that $F[x]/(f)$ is a field if and only if f is irreducible over F (see 1.12 and 1.16).

Suppose f is a monic polynomial of degree k over F. Let $g + (f)$ be an arbitrary element in $F[x]/(f)$. From the division algorithm (1.15) it follows that $g = hf + r$, where $\deg r < k$. Since $hf \in (f)$, it follows that $g + (f) = r + (f)$. Hence each element of $F[x]/(f)$ can be uniquely expressed in the form

$$a_0 + a_1 x + \ldots + a_{k-1} x^{k-1} + (f), \qquad a_i \in F. \tag{1}$$

If we identify F with the subfield $\{a + (f) \mid a \in F\}$ of $F[x]/(f)$, then $F[x]/(f)$ can be regarded as a vector space over F.

1.30 Remark. If f is an irreducible polynomial of degree $n > 0$ over F then $[1], [x], [x^2], \ldots, [x^{n-1}]$ is a basis for $F[x]/(f)$ over F.

Each element of $F[x]/(f)$ can be uniquely represented in the form

$$a_0 + a_1 \alpha + \ldots + a_{k-1} \alpha^{k-1}, \qquad a_i \in F, \quad \text{where } \alpha := x + (f). \tag{2}$$

Since $f + (f)$ is the zero element of $F[x]/(f)$, we have $\bar{f}(\alpha) = f + (f) = 0 + (f)$, i.e. α is a root of f. Clearly, α is an element in $F[x]/(f)$ but not in F. Thus the elements in $F[x]/(f)$ of the form (2) can be regarded so that α is a symbol with the property that $\bar{f}(\alpha) = 0$.

1.31 Example. Let F be the field $\mathbb{Z}_2 = \{0, 1\}$; then $f = x^2 + x + 1$ is an irreducible polynomial of degree 2 over $\mathbb{Z}_2$. $\mathbb{Z}_2[x]/(x^2 + x + 1)$ is a field whose elements can be represented in the form $a + b\alpha$, $a, b \in \mathbb{Z}_2$, where α satisfies $\bar{f}(\alpha) = 0$, i.e. $\alpha^2 + \alpha + 1 = 0$.

The product $(a + b\alpha)(c + d\alpha)$ of two elements in $\mathbb{Z}_2[x]/(x^2 + x + 1)$ can be evaluated. $ac + (ad + bc)\alpha + bd\alpha^2 = ac + (ad + bc)\alpha + bd(\alpha + 1) = (ac + bd) + (ad + bc + bd)\alpha.$ ☐

This example indicates that calculations in $F[x]/(f)$ and $\mathbb{Z}_n$ can be performed in a "similar" way.

1.32 Theorem (Kronecker). *Let F be a field and let g be an arbitrary polynomial of positive degree in $F[x]$. Then there is an extension field K of F such that g has a zero in K.*

PROOF. If g has a zero in F then the theorem is trivial. If this is not the case then there is a divisor f of g of degree at least 2, which is irreducible over F. Let $K := F[x]/(f)$ and consider g as a polynomial over K. Denoting the element $x + (f)$ of K by α we have $f(\alpha) = 0$, i.e. α is a zero of f and therefore also a zero of g. ☐

We now consider a field which is large enough to contain all zeros of a given polynomial.

1.33 Definition. A polynomial $f \in F[x]$ is said to *split* in an extension K of F if f can be expressed as a product of linear factors in $K[x]$. K is called a *splitting field* of f over F if f splits in K, but does not split in any proper subfield of K containing F.

1.34 Corollary. *Let F be a field and let $g \in F[x]$ be of positive degree. Then there is an extension K of F such that g splits into linear factors over K.*

PROOF. The polynomial g has $x - \alpha$ as a divisor in $K_1[x]$, where $F \le K_1$. If g does not split into linear factors over K_1, then we repeat the construction of 1.32 and construct extensions $K_2, K_3, \ldots$ until g splits completely into linear factors over K. ☐

The following notation will prove useful. Let F be a subfield of a field M and let A be an arbitrary set of elements in M. Then $F(A)$ denotes the intersection of all subfields of M, which contain both F and A. $F(A)$ is called the extension of F which is obtained by adjunction of the elements of A. If $A = \{a\}$, $a \notin F$, then $F(\{a\})$ is called a *simple extension* of F. We also write $F(a)$ in this case. We have $F(A) = \langle F \cup A \rangle$. For $F(\{a_1, \ldots, a_n\})$ we shall write $F(a_1, \ldots, a_n)$.

1.35 Theorem. *Let $f \in F[x]$ be of degree n and let K be an extension of F. If $f = c(x - a_1) \ldots (x - a_n)$ in $K[x]$ then $F(a_1, \ldots, a_n)$ is a splitting field of f over F.* ☐

Theorem 1.20 and Corollary 1.33 secure the existence of splitting fields. The proof of the uniqueness of the splitting field of a polynomial f over a field F is slightly more complicated and we omit it (see, e.g. FRALEIGH).

1.36 Theorem. *For any field F and polynomial $f \in F[x]$ of degree ≥ 1 all splitting fields of f over F are isomorphic.* $\qquad\qquad\square$

PROBLEMS

1. Let $R = \{a + b\sqrt{2} \mid a, b \in \mathbb{Z}\}$. Define operations $+$ and $\cdot$ and show that $(R, +, \cdot)$ is a commutative ring with identity. Determine the quotient field of R in the field $\mathbb{R}$.

2. Let G denote the set of complex numbers $\{a + bi \mid a, b \in \mathbb{Z}\}$. With the usual addition and multiplication of complex numbers G forms the domain of *Gaussian integers*. Show that its quotient field is isomorphic to the subring of $\mathbb{C}$ consisting of $\{p + qi \mid p, q \in \mathbb{Q}\}$.

3. Show that an automorphism of a field maps every element of its prime field into itself.

*4. Let Q be the quotient field of an integral domain R. Let I be an ideal of R. Prove or disprove that $\{ab^{-1} \mid a \in I, b \in R \setminus \{0\}\}$ is an ideal of Q and every ideal of Q can be obtained in this way.

*5. Let Q be the quotient field of an integral domain R. Show that any monomorphism of R into a field F has a unique extension to a monomorphism of Q into F.

6. According to 1.31 construct a field of nine elements, given $\mathbb{Z}_3$, and $f = x^2 + x + 2 \in \mathbb{Z}_3[x]$. Construct the operation tables for this field.

7. Show that $\mathbb{Q}((-1 + \sqrt{3}i)/2)$ is the splitting field of $x^4 + x^2 + 1$ over $\mathbb{Q}$.

8. Find the splitting field of $x^p - 1 \in \mathbb{Q}[x]$, p a prime.

9. Discuss all possible splitting fields of $x^3 + ax^2 + bx + c$ over a field F depending on the polynomial being reducible or irreducible over F.

10. If $(i_1, \ldots, i_n)$ is any permutation of $(1, \ldots, n)$ and $F(a_1, \ldots, a_n)$ is as in 1.35, prove that $F(a_1, \ldots, a_n) = F(a_{i_1}, \ldots, a_{i_n})$.

*D. Algebraic Extensions

We now introduce special types of extension fields of a given field F.

1.37 Definition. Let K be an extension of a field F. An element α of K is called *algebraic* over F, if there is a nonzero polynomial g with coefficients in F such that $\bar{g}(\alpha) = 0$. If α is not algebraic over F then α is said to be *transcendental* over F.

Using this definition of algebraic elements we can divide all extension fields into two classes:

1.38 Definition. An extension K of a field F is called *algebraic* if each element of K is algebraic over F. If K contains at least one element which is transcendental over F then K is a *transcendental extension*. The *degree* of K over F, in symbols $[K:F]$, is the dimension of K as a vector space over F.

1.39 Examples. If $\beta \in K$ is also an element of F then β is a zero of $x - \beta \in F[x]$. Thus any element in F is algebraic over F. The real number $\sqrt{2}$ is algebraic over $\mathbb{Q}$, since $\sqrt{2}$ is a zero of $x^2 - 2$. It can be shown, though with considerable difficulty, that π and e are transcendental over $\mathbb{Q}$. The result implies that $\mathbb{R}$ is for example, a transcendental extension of $\mathbb{Q}$. $\square$

The following two theorems determine all extensions up to isomorphisms. We state the first result without proof.

1.40 Theorem. *Let K be an extension of F, and let $\alpha \in K$ be transcendental over F. Then the extension $F(\alpha)$ is isomorphic to the field $F(x)$ of rational functions in x.* $\square$

For simplicity, we shall use $F[\alpha] := \{a_0 + a_1\alpha + \ldots + a_n\alpha^n \mid n \in \mathbb{N}_0, a_i \in F\}$.

1.41 Theorem. *Let K be an extension of F, and let $\alpha \in K$ be algebraic over F. Then:*

(i) $F(\alpha) = F[\alpha] \cong F[x]/(f)$, *where f is a uniquely determined, monic, irreducible polynomial in $F[x]$ with zero α in K.*

(ii) α *is a zero of a polynomial $g \in F[x]$ if and only if g is divisible by f.*

(iii) *If f in (i) is of degree n then $1, \alpha, \ldots, \alpha^{n-1}$ is a basis of $F(\alpha)$ over F. We have $[F(\alpha):F] = n$ and each element of $F(\alpha)$ can be uniquely expressed as $a_0 + a_1\alpha + \ldots + a_{n-1}\alpha^{n-1}$, $a_i \in F$.*

PROOF. (i) We consider $\psi: F[x] \to F[\alpha]$ defined by $g \mapsto \bar{g}(\alpha)$. Then $F[x]/\ker \psi \cong F[\alpha]$. Since α is algebraic over F, the kernel of ψ is not zero and not $F[x]$, i.e. it is a proper ideal. Ker ψ is a principal ideal, say Ker $\psi = (f)$, where f is irreducible. We may assume f is monic, since F is a field. The uniqueness of f is clear. By the irreducibility of f, (f) is maximal and $F[x]/(f)$ is a field. Consequently $F[\alpha]$ is a field and we have $F[\alpha] = F(\alpha)$, since $F(\alpha)$ is the smallest field which contains $F[\alpha]$.

(ii) This follows from Ker $\psi = (f)$.

(iii) This is a consequence of $[1], [x], \ldots, [x^{n-1}]$ being a basis of $F[x]/(f)$ over F. $\square$

The polynomial f in Theorem 1.41(i) plays an important role in field extensions.

1.42 Definition. Let $\alpha \in L$ be algebraic over a field F. The unique, monic, irreducible polynomial $f \in F[x]$ with α as a zero is called the *minimal polynomial* of α over F. The *degree* of α over F is defined as the degree of f.

1.43 Example. The minimal polynomial of $\sqrt[3]{2} \in \mathbb{Q}(\sqrt[3]{2})$ over $\mathbb{Q}$ is $x^3 - 2$. We have $\mathbb{Q}(\sqrt[3]{2}) = \mathbb{Q}[\sqrt[3]{2}]$ and $[\mathbb{Q}(\sqrt[3]{2}):\mathbb{Q}] = 3$. The elements $1, \sqrt[3]{2}, \sqrt[3]{4}$ form a basis of $\mathbb{Q}(\sqrt[3]{2})$ over $\mathbb{Q}$ such that any element of $\mathbb{Q}(\sqrt[3]{2})$ can be uniquely expressed in the form $a_0 + a_1\sqrt[3]{2} + a_2\sqrt[3]{4}$, $a_i \in \mathbb{Q}$.   □

1.44 Theorem. *An element α in an extension K of F is algebraic over F if and only if it is a zero of an irreducible polynomial $f \in F[x]$ of degree ≥ 1.*

PROOF. This follows from the fact that α is a zero of f if and only if it is a zero of an irreducible factor of f.   □

Next we describe a relationship between extensions K of a field F and vector spaces over F. Let K be an extension of F. Then K can be regarded as a vector space over F by considering the additive group $(F, +)$ together with scalar multiplication by elements in F.

1.45 Definition. Let $F \leq K$. K is called a *finite extension* of F if dim $K =:$ $[K:F]$ is finite. Otherwise K is called an *infinite extension* of F. $[K:F]$ is the *degree* of K over F. The degree of an element $\alpha \in K$ over F is $[F(\alpha):F]$.

If $\alpha_1, \ldots, \alpha_n$ is a basis of K over F then $F(\alpha_1, \ldots, \alpha_n) = \{c_1\alpha_1 + \ldots + c_n\alpha_n \,|\, c_i \in F\}$. If K is a finite extension of F of degree n then there is a subset $\{\alpha_1, \ldots, \alpha_n\}$ of K such that $K = F(\alpha_1, \ldots, \alpha_n)$.

1.46 Theorem. *Any finite extension K of F is an algebraic extension.*

PROOF. If $n = [K:F]$, then any set of $n + 1$ elements in K is linearly dependent. Let $\alpha \in K$. Then $1, \alpha, \alpha^2, \ldots, \alpha^n$ in K are linearly dependent over F, i.e. there are $c_i \in F$ not all zero, such that $c_0 + c_1\alpha + \ldots + c_n\alpha^n = 0$. Thus α is a zero of the polynomial $g = c_0 + \ldots + c_nx^n \in F[x]$ and therefore it is algebraic.   □

We mention briefly that there do exist algebraic extensions of a field which are not finite, although we restrict ourselves to finite extensions in Theorem 1.46. An important example of an infinite algebraic extension is the field of all algebraic numbers, which consists of all algebraic elements over $\mathbb{Q}$. For extensions it can be shown that if a field L is algebraic over K and K is algebraic over F then L is algebraic over F.

In a certain sense, the following theorem represents a generalization of the Theorem of Lagrange to the case of finite extensions.

1.47 Theorem. *Let L be a finite extension of K and K be a finite extension of F. Then $[L:K][K:F] = [L:F]$.*

PROOF. Let $\{\alpha_i \mid i \in I\}$ be a basis of L over K and $\{\beta_j \mid j \in J\}$ be a basis of K over F. One may verify that the $|I| \cdot |J|$ elements $\{\alpha_i\beta_j \mid i \in I$ and $j \in J\}$ form a basis of L over F. □

1.48 Corollary. *Let K be a finite extension of F.*

(i) *The degree of an element of K over F divides $[K:F]$.*
(ii) *An element in K generates the vector space K over F if and only if its degree over F is $[K:F]$.*
(iii) *If $[K:F] = 2^m$ and f is an irreducible polynomial over F of degree 3 then f is irreducible over K.* □

Part (iii) of this corollary enables us to give proofs of the impossibility of certain classical Greek construction problems, i.e. constructions with the use of ruler and compass only. We mention the problem of doubling the cube. Given a cube of volume 1 then the construction of a cube of twice this volume makes it necessary to solve $x^3 - 2 = 0$. This polynomial is irreducible over $\mathbb{Q}$. In general, equations of circles are of degree 2 and equations of lines are of degree 1, so that their intersection leads to equations of degree 2^m. This implies that Greek construction methods lead to fields of degree 2^m over $\mathbb{Q}$. The irreducibility of $x^3 - 2$ over $\mathbb{Q}$ implies that it is impossible to construct a side of a cube with twice the volume of a given cube by using ruler and compass alone.

The problem of trisecting an angle is similar. It is equivalent to determining the cosine of one third of a given angle. By analytic geometry this cosine c must satisfy the equation $4x^3 - 3x - c = 0$. In general, this is an irreducible polynomial over $\mathbb{Q}(c)$, so this implies the impossibility of trisecting an angle using only ruler and compass constructions.

1.49 Example. $\mathbb{Q}(\sqrt{2}, \sqrt{3})$ is a field of degree 4 over $\mathbb{Q}$. Since $(\sqrt{2} + \sqrt{3})^3 - 9(\sqrt{2} + \sqrt{3}) = 2\sqrt{2}$, we see that $\sqrt{2}$ is an element of $\mathbb{Q}(\sqrt{2} + \sqrt{3})$. Since $\sqrt{3} = (\sqrt{2} + \sqrt{3}) - \sqrt{2}$, we have $\sqrt{3} \in \mathbb{Q}(\sqrt{2} + \sqrt{3})$, and so $\mathbb{Q}(\sqrt{2}, \sqrt{3}) = \mathbb{Q}(\sqrt{2} + \sqrt{3})$. □

This example shows that at least in special cases it is possible to regard an algebraic extension $F(\alpha_1, \ldots, \alpha_n)$ of F as a simple algebraic extension $F(\alpha)$ for suitable α.

An irreducible polynomial f in $F[x]$ is called *separable* if there are no multiple zeros of f in its splitting field. An arbitrary polynomial in $F[x]$ is called *separable* if each of its irreducible factors is separable. An algebraic element α over F is called *separable* if its minimal polynomial is separable over F.

1.50 Theorem. *Let F be a field and let $\alpha_1, \ldots, \alpha_n$ be algebraic and separable over F. Then there is an element α in $F(\alpha_1, \ldots, \alpha_n)$ such that $F(\alpha) = F(\alpha_1, \ldots, \alpha_n)$.* $\quad\square$

This theorem is useful in the study of field extensions since simple extensions are more easily handled than multiple extensions. It is therefore important to be able to determine the separability or nonseparability of a polynomial. Here we need the concept of the *formal derivative*. Let $D: F[x] \to F[x]$ be defined by

$$D: f = a_0 + a_1 x + \ldots + a_n x^n \to f' := a_1 + \ldots + n a_n x^{n-1}.$$

We may verify immediately that

$$(af + bg)' = af' + bg',$$

i.e. D is an F-endomorphism of the vector space $F[x]$ with kernel F. D is called the *differential operator* and f' is called the *derivative* of f. We now state, without proof, important criteria for separability of an irreducible polynomial over F.

1.51 Theorem. *A polynomial f over F is separable if and only if $\gcd(f, f') = 1$.* $\quad\square$

1.52 Theorem. *An irreducible polynomial f over F is separable if and only if its derivative is nonzero.* $\quad\square$

A field F is called *algebraically closed* if any nonconstant polynomial in $F[x]$ splits into linear factors in $F[x]$. A field $\bar{F}$ is called an *algebraic closure* of a field F, if $\bar{F}$ is algebraically closed and is an algebraic extension of F. We note that this is equivalent to saying that F does not have any algebraic extension which properly contains F. It is easy to see that a polynomial of degree $n > 0$ in $F[x]$, F algebraically closed, can be expressed as product of n linear monic polynomials. In this context we repeat an important theorem (cf. 1.14) for which there are more than one hundred proofs, the first of which was given by C. F. Gauss in 1799.

1.53 Theorem (Fundamental Theorem of Algebra). *The field of complex numbers is algebraically closed.* $\quad\square$

PROBLEMS

1. Describe the smallest subfield of the real numbers containing $\sqrt{2}$ and $\sqrt{3}$. Find three proper subfields of this field.

2. Show that the given number $\alpha \in \mathbb{C}$ is algebraic over $\mathbb{Q}$ by finding $f(x) \in \mathbb{Q}[x]$

such that $f(\alpha) = 0$, for

(i) $\alpha = \sqrt{2} + \sqrt{3}$;

(ii) $\alpha = \sqrt{1 + \sqrt[3]{2}}$.

3. Does there exist a polynomial with rational coefficients of degree less than 4 such that $\sqrt{2} + \sqrt{3}$ is a root?

*4. Let L be a simple algebraic extension of a field F. Prove that H is also a simple algebraic extension of F if $L \supset H \supset F$.

*5. Let L be a finite extension of a field F. Prove that L is a simple extension if and only if there are only finitely many fields H such that $L \supset H \supset F$.

6. Let p_i, $i = 1, 2, 3$, be distinct primes. Determine the degree of $Q(\sqrt{p_1}, \sqrt{p_2}, \sqrt{p_3})$ over Q.

7. Let F be a field and $b \in F$, $b \neq 0$, be a zero of $f(x) = \sum_{i=0}^{n} a_i x^i \in F[x]$. Show that $1/b$ is a zero of $g(x) = \sum_{i=0}^{n} a_{n-i} x^i$.

8. Let $f(x) = \sum_{i=0}^{n} a_i x^i \in Z[x]$, and let $f(0)$ and $f(1)$ be odd. Show that f does not have integer zeros.

9. Show that $f(x) = x^3 + x + 1 \in Q[x]$ is irreducible over Q and determine the multiplicative inverse of $x^2 + x + 1 + (f(x))$ in the field $Q[x]/(f(x))$.

10. Find necessary and sufficient conditions on $a, b \in Q$ so that the splitting field of $x^3 + ax + b$ has degree exactly 3 over Q.

*11. Let L be an extension of F, let $f \in F[x]$ and let ϕ be an automorphism of L that maps every element of F into itself. Prove that ϕ must map a root of f in L into a root of f in L.

12. Determine a primitive element α of the splitting field of $x^3 - 7 \in Q[x]$ over Q and determine its minimal polynomial.

13. Let $K = F(x)$ and $L = F(x^3(x + 1)^{-1})$. Show that K is a simple algebraic extension of L and determine $[K : L]$.

14. Let G be the Gaussian integral domain $\{a + bi \mid a, b \in Z\}$. Show that $G/(7)$ is a finite field, determine its prime field P and an element $t \in G$ such that $G = P(t)$. Also determine the minimal polynomial of t over P.

*15. Show that the splitting field of $x^n - a \in Q[x]$ can be obtained by adjoining a primitive nth root of unity and a root of $x^n - a$.

16. Is it possible to divide the angle $\pi/3$ into five equal parts using ruler and compass constructions?

17. Show that the regular 9-gon cannot be constructed by ruler and compass.

*18. Prove: An algebraic element a over a field F of prime characteristic p is separable if and only if $F(a^p) = F(a)$.

*19. If F is a field of prime characteristic p, prove that $f \in F[x]$ has a multiple root only if it is of the form $f(x) = g(x^p)$ for a suitable polynomial g over F.

20. Prove that $x^{p^n} - x$ over a field of prime characteristic has no multiple roots.

EXERCISES (Solutions in Chapter 8, p. 446)

1. How many possibilities are there to define the multiplication operations on $R = \{0, 1, 2, 3\}$ to make R into a ring, if addition is defined as mod 4 addition?

*2. Prove that a ring $R \neq \{0\}$ is a skew-field if and only if for every nonzero $a \in R$ there is an $x \in R$ such that $axa = a$.

*3. Show that if a finite ring R has a nonzero element which is not a zero divisor then R has an identity.

4. Give an example of a ring R with identity 1 and a subring R' of R with identity $1'$ such that $1 \neq 1'$.

5. Let R be a finite ring of characteristic p, a prime. Show that the order of R is a power of p.

6. Show that $\mathbb{Z}$ is a PID.

*7. A ring R with more than one element is called *simple* if R has no ideals except $\{0\}$ and R itself. Prove that R is a field if and only if R is a simple commutative ring with identity.

8. Prove that an ideal $I \neq R$ of a ring R is maximal if and only if R/I is simple.

*9. Let R be a ring with identity and let $I \trianglelefteq R$. Prove that I is contained in a maximal ideal.

*10. Show that the ideals of $R[[x]]$, R a field, are precisely the members of the chain

$$R[[x]] = (1) \supset (x) \supset (x^2) \supset \ldots \supset (0).$$

Hence $R[[x]]$ is a PID with exactly one maximal ideal.

*11. Prove Theorem 1.16.

*12. Prove Theorem 1.17.

*13. Prove the Unique Factorization Theorem 1.18.

14. Prove Theorem 1.19.

*15. Prove Theorem 1.20.

16. Show that if I is an ideal of a ring R then $I[x]$ is an ideal of $R[x]$.

17. Determine all roots and their multiplicity of the polynomial $x^6 + 3x^5 + x^4 + x^3 + 4x^2 + 3x + 2$ over $\mathbb{Z}_5$.

18. Let $f(x) = 3 + 4x + 5x^2 + 6x^3 + x^6$ and $g(x) = 1 + x + x^2$ be two polynomials over $\mathbb{Z}_{11}$. Is f divisible by g? Determine $\bar{f}(3) + \bar{g}(3)$ and $\overline{f + g}(3)$.

19. Let $f(x) + 1 + x + x^3 + x^6$ and $g(x) = 1 + x + x^2 + x^4$ be polynomials over $\mathbb{R}$. If $R = \mathbb{R}$, is f divisible by g? If $R = \mathbb{Z}_2$, is f divisible by g?

*20. Show that an element α in an extension K of a field F is transcendental (algebraic) over F, if the map $\psi : F[x] \to F(\alpha)$, $f \mapsto \bar{f}(\alpha)$ is an isomorphism (not an isomorphism).

21. Show that $f = x^3 + x + 1$ is irreducible over $\mathbb{Z}_2$. Determine the elements of the field $\mathbb{Z}_2[x]/(f)$ and show that this field is the splitting field of f over $\mathbb{Z}_2$.

22. Show that $\alpha = \sqrt{2} + i$ is of degree 4 over $\mathbb{Q}$ and of degree 2 over $\mathbb{R}$. Determine the minimal polynomial of α in both cases.

23. Determine the multiplicative inverse of $1 + \sqrt[3]{2} + \sqrt[3]{4}$ in $\mathbb{Q}(\sqrt[3]{2})$.

24. Describe the elements of $\mathbb{Z}_2[x]/(x)$.

25. (i) Show that $x^2 + 1$ is irreducible in $\mathbb{Z}_3[x]$.
    (ii) Let α be a zero of $x^2 + 1$ in an extension of $\mathbb{Z}_3$. Give the addition and multiplication tables for the nine elements of $\mathbb{Z}_3(\alpha)$.

26. Find the degree and a basis for each of the given field extensions.

    (i) $\mathbb{Q}(\sqrt{2}, \sqrt{3}, \sqrt{5})$ over $\mathbb{Q}$.
    (ii) $\mathbb{Q}(\sqrt{2}, \sqrt{6})$ over $\mathbb{Q}(\sqrt{3})$.
    (iii) $\mathbb{Q}(\sqrt{2}, \sqrt[3]{2})$ over $\mathbb{Q}$.
    (iv) $\mathbb{Q}(\sqrt{2} + \sqrt{3})$ over $\mathbb{Q}(\sqrt{3})$.

27. (i) Find the splitting field of $(x^2 - 3)(x^3 + 1)$ over $\mathbb{Q}$.
    (ii) Find the splitting field of $(x^2 - 2x - 2)(x^2 + 1)$ over $\mathbb{Q}$.
    (iii) Find the splitting fields of $x^2 - 3$ and $x^2 - 2x - 2$ over $\mathbb{Q}$.
    (iv) Find the splitting field of $x^2 + x + 1$ over $\mathbb{Z}_2$.
    (v) Determine the splitting field of $x^3 + x + 2$ over $\mathbb{Q}$.

28. Show that $f = x^2 + x + 1$ is irreducible over $\mathbb{Z}_5$. Let $\alpha = x + (f)$ be in $\mathbb{Z}_5[x]/(f)$ and let β be another zero of f. Determine an isomorphism from $\mathbb{Z}_5(\alpha)$ onto $\mathbb{Z}_5(\beta)$.

29. Let α be a zero of $x^2 + x + 1$ and let β be a zero of $x^2 + 4$. Determine an element γ such that $\mathbb{Z}_5(\gamma) = \mathbb{Z}_5(\alpha, \beta)$.

30. Show that $(x^n - 1)/(x - 1)$ is a polynomial over $\mathbb{Q}$ for all positive integers n. Also determine precisely the set of values of n for which it is irreducible.

31. Deduce algebraically that by repeated bisection it is possible to divide an arbitrary angle into four equal parts. (Use a relationship between $\cos 4\theta$ and $\cos \theta$.)

32. (a) Can the cube be "trebled"?
    (b) Can the cube be "quadrupled"?

*33. A regular n-gon is constructible for $n \geq 3$ if and only if the angle $2\pi/n$ is constructible. $2\pi/n$ is constructible if and only if a line segment of length $\cos(2\pi/n)$ is constructible. Prove: If the regular n-gon is constructible and if the odd prime p divides n, then p is of the form $2^{2^k} + 1$.

34. Given a segment s, show that it is impossible to construct segments m and n such that $s : m = m : n = n : 2s$.

35. Determine whether the following polynomial has multiple roots:

$$x^4 - 5x^3 + 6x^2 + 4x - 8 \in \mathbb{Q}[x].$$

§2. Finite Fields

A field with m elements ($m \in \mathbb{N}$) is called a finite field of order m. One of the main aims of this section is to show that for any prime number p and positive integer n there is (up to isomorphism) exactly one finite field of order p^n. This field is the splitting field of $x^{p^n} - x$ over $\mathbb{Z}_p$. We know from §1 that a finite field F must be of prime characteristic p and that the prime field of F is isomorphic to $\mathbb{Z}_p$. We shall identify the prime field of a finite field with $\mathbb{Z}_p$, i.e. we shall regard any field of prime characteristic p as extension field of the field $\mathbb{Z}_p$ of the integers mod p.

2.1 Theorem. *Let F be a finite field of characteristic p. Then F contains p^n elements, where $n = [F : \mathbb{Z}_p]$.*

PROOF. F, considered as a vector space over its prime field $\mathbb{Z}_p$, contains a finite basis of n elements. Each element of F can be expressed as a unique linear combination of the n basis elements with coefficients in $\mathbb{Z}_p$. Therefore there are p^n elements in F.   □

For the proof of the following theorem we recall two results from an introductory course on group theory. Let G be a finite group of order $|G|$, and let g be an element of G. Then $g^{|G|} = 1$. We also require the "*Fundamental Theorem on Finite Abelian Groups*", which states that every finite abelian group is the direct product of cyclic groups. For proofs see e.g. HERSTEIN or FRALEIGH.

2.2. Theorem. *Let F be a finite field with p^n elements.*

(i) *The multiplicative group of the nonzero elements of F is cyclic and of order $p^n - 1$.*
(ii) *All elements a of F satisfy $a^{p^n} - a = 0$.*

PROOF. We first prove the theorem by using the fundamental theorem for finite abelian groups. The multiplicative group G of nonzero elements of F is a group of order $p^n - 1$. G is a direct product of the cyclic subgroups $U_1, \ldots, U_m$, where $|U_i|$ divides $|U_{i+1}|$. This implies that the order of each element in G divides the order r of U_m. For any element a in G we therefore have $a^r - 1 = 0$. The polynomial $x^r - 1$ over F can have at most r zeros in F, hence $|G| = p^n - 1 \le r$. Since $|U_m|$ divides $|G|$, we have $r \le p^n - 1$, which proves (i). Since U_m is of order $p^n - 1$, we have $G = U_m$. Part (ii) follows from the fact that for any nonzero element a we have $a^{p^{n-1}} - 1 = 0$ as mentioned above.   □

An elementary proof of Theorem 2.2(i), which does not rely on the fundamental theorem for finite abelian groups, goes as follows. Let $p^n = q$. We may assume $q \ge 3$. Let $h = p_1^{r_1} p_2^{r_2} \ldots p_m^{r_m}$ be the prime factor decomposition of $h = q - 1$, the order of the group $F\backslash\{0\}$. For every i, $1 \le i \le m$, the polynomial $x^{h/p_i} - 1$ has at most h/p_i roots in F. Since $h/p_i < h$, it follows that there are nonzero elements in F which are not roots of this polynomial. Let a_i be such an element and set $b_i = a_i^{h/p_i^{r_i}}$. We have $b_i^{p_i^{r_i}} = 1$, hence the order b_i is a divisor of $p_i^{r_i}$ and is therefore of the form $p_i^{s_i}$ with $0 \le s_i \le r_i$. On the other hand,

$$b_i^{p_i^{r_i-1}} = a_i^{h/p_i} \ne 1,$$

and so the order of b_i is $p_i^{r_i}$. We claim that the element $b = b_1 b_2 \dots b_m$ has order h. Suppose, on the contrary, that the order of b is a proper divisor of h and is therefore a divisor of at least one of the m integers h/p_i, $1 \le i \le m$, say of h/p_1. Then we have

$$1 = b^{h/p_1} = b_1^{h/p_1} b_2^{h/p_1} \dots b_m^{h/p_1}.$$

Now if $2 \le i \le m$, then $p_i^{r_i}$ divides h/p_1, and hence $b_i^{h/p_1} = 1$. Therefore $b_1^{h/p_1} = 1$. This implies that the order of b_1 must divide h/p_1, which is impossible since the order of b_1 is $p_1^{r_1}$. Thus $F\backslash\{0\}$ is a cyclic group with generator b. $\square$

In the special case $F = \mathbb{Z}_p$ we have elements $a \in \mathbb{Z}_p$ such that the powers $a, a^2, \dots, a^{p-1}$ represent all nonzero elements of $\mathbb{Z}_p$. Such an element a is called a *primitive root modulo p*. A generator of the cyclic group of a finite field F is called a *primitive element*.

2.3 Theorem. *Let F be a finite field and let $\alpha_1, \dots, \alpha_k$ be algebraic over F. Then $F(\alpha_1, \dots, \alpha_k) = F(\alpha)$ for some α in $F(\alpha_1, \dots, \alpha_k)$.* $\square$

It can be shown that the extension field $F(\alpha_1, \dots, \alpha_k)$ is finite over F, and that it is an algebraic extension of F. Therefore it is a finite field with cyclic multiplicative group. If α is a generating element of this group then the theorem follows.

2.4 Corollary. *Let F be a finite field of characteristic p and let $[F:\mathbb{Z}_p] = n$. Then there is an element α in F such that α is algebraic of degree n over $\mathbb{Z}_p$ and $F = \mathbb{Z}_p(\alpha)$.* $\square$

Theorem 2.2(ii) ensures that any finite field F consists of the roots of the polynomial $x^{p^n} - x$ for some n where $p = $ char F. The following theorem describes all finite fields and shows that there is a finite field for any prime power p^n.

2.5 Theorem. (i) *Any finite field is of order p^n where p is a prime and n is a positive integer.*
(ii) *For any prime p and any $n \in \mathbb{N}$ there is a field of order p^n.*

(iii) *Any field of order p^n is (up to isomorphism) the splitting field of $x^{p^n} - x \in$*
      $\mathbb{Z}_p[x]$.
(iv) *Any two fields of order p^n are isomorphic.*

PROOF. (i) follows from Theorem 2.1. Let K be the splitting field of
$x^{p^n} - x =: f$ over $\mathbb{Z}_p$ and let $\alpha \in K$ be a root of f so that $f = (x - \alpha)^k g$ in
$K[x]$, where α is not a root of g. Then $f' = -1$ is divisible by $(x - \alpha)^{k-1}$
and hence $k - 1 = 0$. This means that α is simple. All roots of f in K are
distinct and f has p^n roots in K. It is easily verified that sums, products
and inverses of roots of f in K are also roots of f. Thus the roots of f form
a field with p^n elements, which must be the splitting field of K of f over
$\mathbb{Z}_p$. This implies $[K:\mathbb{Z}_p] = n$, which proves (ii) and (iii). The uniqueness
(iv) follows from the uniqueness of the splitting field (see 1.36).   □

2.6 Corollary. *For any positive integer n there is an irreducible polynomial of
degree n in $\mathbb{Z}_p[x]$. The finite field of p^n elements is the splitting field of an
irreducible polynomial in $\mathbb{Z}_p[x]$ of degree n.*   □

Theorem 2.5 enables us to speak of *the* finite field with p^n elements. This
field is also called the *Galois field*, in honour of Evariste Galois (1811–1832)
and is denoted by GF(p^n) or $\mathbb{F}_{p^n}$. The multiplicative group of $\mathbb{F}_{p^n}$ is denoted
by $\mathbb{F}_{p^n}^*$. The prime field $\mathbb{F}_p$ is isomorphic to $\mathbb{Z}_p$.
   The results obtained so far make it possible to determine the elements
of a finite field. We know that $\mathbb{F}_{p^n}$ is a vector space of dimension n over $\mathbb{F}_p$.
Moreover, it is a simple extension of the prime field $\mathbb{F}_p$, say $\mathbb{F}_{p^n} = \mathbb{F}_p(\alpha)$,
and any $n + 1$ elements of $\mathbb{F}_{p^n}$ are linearly dependent, so that $a_0 + a_1\alpha +$
$\ldots + a_n\alpha^n = 0$. This means that α is a root of a polynomial in $\mathbb{F}_p(\alpha)$. Let
f be the minimal polynomial of α, then $\mathbb{F}_{p^n} = \mathbb{F}_p(\alpha) \cong \mathbb{F}_p[x]/(f)$. In order
to obtain the elements of $\mathbb{F}_{p^n}$ explicitly, we determine an irreducible monic
polynomial of degree n over $\mathbb{F}_p$ and form $\mathbb{F}_p[x]/(f)$. More generally, to
obtain $\mathbb{F}_{q^m}$, $q = p^n$, we find an irreducible, monic polynomial g of degree
m over $\mathbb{F}_q$ and form $\mathbb{F}_q[x]/(g)$, which is then isomorphic to $\mathbb{F}_{q^m}$.

2.7 Example. We determine the elements of $\mathbb{F}_{2^3}$. If we regard $\mathbb{F}_{2^3}$ as a simple
extension of degree 3 of the prime field $\mathbb{F}_2$ then this extension is obtained
by adjoining to $\mathbb{F}_2$ a root of an irreducible cubic polynomial over $\mathbb{F}_2$. It is
easily verified that $x^3 + x + 1$ and $x^3 + x^2 + 1$ are irreducible over $\mathbb{F}_2$. There-
fore $\mathbb{F}_{2^3} \cong \mathbb{F}_2[x]/(x^3 + x + 1)$ and also $\mathbb{F}_{2^3} \cong \mathbb{F}_2[x]/(x^3 + x^2 + 1)$. Let α be a
root of $f = x^3 + x + 1$, then $1, \alpha, \alpha^2$ form a basis of $\mathbb{F}_{2^3}$ over $\mathbb{F}_2$. The elements
of $\mathbb{F}_{2^3}$ are of the form

$$a + b\alpha + c\alpha^2 \quad \text{for all } a, b, c \in \mathbb{F}_2.$$

We can also use $g = x^3 + x^2 + 1$ to determine the elements of $\mathbb{F}_{2^3}$. Let β be a root of g, so $\beta^3 + \beta^2 + 1 = 0$. It can be easily verified that $\beta + 1$ is a root of f in $\mathbb{F}_2[x]/(g)$. The two fields $\mathbb{F}_2[x]/(f)$ and $\mathbb{F}_2[x]/(g)$ are splitting fields of $x^8 - x$ and are thus isomorphic. Therefore there is an isomorphism ψ such that $\psi(\alpha) = \beta + 1$ and ψ restricted to $\mathbb{F}_2$ is the identity mapping. The elements $1, \beta + 1, (\beta + 1)^2$ form a basis of $\mathbb{F}_2[x]/(g)$ over $\mathbb{F}_2$. Thus the isomorphism ψ is given by

$$\psi(a + b\alpha + c\alpha^2) = a + b(\beta + 1) + c(\beta + 1)^2 \quad \text{with } a, b, c \in \mathbb{F}_2.$$

The multiplication table of the multiplicative group $\mathbb{F}_2[x]/(x^3 + x^2 + 1)\backslash\{0\}$ is as follows (β is as above)

$\cdot$	1	β	$\beta+1$	β^2	$\beta^2+\beta$	β^2+1	$\beta^2+\beta+1$
1	1	β	$\beta+1$	β^2	$\beta^2+\beta$	β^2+1	$\beta^2+\beta+1$
β	β	β^2	$\beta^2+\beta$	β^2+1	1	$\beta^2+\beta+1$	$\beta+1$
$\beta+1$	$\beta+1$	$\beta^2+\beta$	β^2+1	1	$\beta^2+\beta+1$	β	β^2
β^2	β^2	β^2+1	1	$\beta^2+\beta+1$	β	$\beta+1$	$\beta^2+\beta$
$\beta^2+\beta$	$\beta^2+\beta$	1	$\beta^2+\beta+1$	β	$\beta+1$	β^2	β^2+1
β^2+1	β^2+1	$\beta^2+\beta+1$	β	$\beta+1$	β^2	$\beta^2+\beta$	1
$\beta^2+\beta+1$	$\beta^2+\beta+1$	$\beta+1$	β^2	$\beta^2+\beta$	β^2+1	1	β

If F is a subfield of order p^m in $\mathbb{F}_{p^n}$ then F is the splitting field of $x^{p^m} - x$ in $\mathbb{F}_{p^n}$ over $\mathbb{F}_p$. We shall describe all subfields of a finite field. The following lemma can be proved as an exercise.

2.8 Lemma. $x^m - 1$ *divides* $x^n - 1$ *over a field F if and only if m divides n.*

2.9 Theorem. *Let p be a prime and let m, n be natural numbers.*

(i) *If $\mathbb{F}_{p^m}$ is a subfield of $\mathbb{F}_{p^n}$ then $m | n$.*
(ii) *If $m | n$ then $\mathbb{F}_{p^m} \subsetneq \mathbb{F}_{p^n}$. There is exactly one subfield of $\mathbb{F}_{p^n}$ with p^m elements.*

PROOF. (i) Theorem 1.47 implies

$$[\mathbb{F}_{p^n} : \mathbb{F}_p] = [\mathbb{F}_{p^n} : \mathbb{F}_{p^m}][\mathbb{F}_{p^m} : \mathbb{F}_p].$$

Since the term on the left-hand side is n and the second factor on the right-hand side is m, we have $m | n$.

(ii) Now $m | n$ implies $p^m - 1 | p^n - 1$, thus (by 2.8) $x^{p^m-1} - 1 | x^{p^n-1} - 1$ and $x^{p^m} - x | x^{p^n} - x$. The roots of $x^{p^m} - x$ form a subfield of $\mathbb{F}_{p^n}$ of order p^m, which is isomorphic to $\mathbb{F}_{p^m}$. There can not be another subfield with p^m elements, because otherwise there would be more than p^m roots of $x^{p^m} - x$ in $\mathbb{F}_{p^n}$. $\square$

2.10 Example. We draw a diagram of all subfields of $\mathbb{F}_{2^{12}}$.

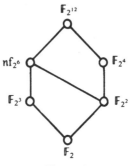

Figure 3.1

Because of Theorem 2.9 and the property $m!\mid n!$ for positive integers $m < n$, we have an ascending chain of fields

$$\mathbb{F}_p \subset \mathbb{F}_{p^{2!}} \subset \mathbb{F}_{p^{3!}} \subset \ldots .$$

We define $\mathbb{F}_{p^\infty}$ as $\bigcup_n \mathbb{F}_{p^{n!}}$ and note that $\mathbb{F}_{p^\infty}$ is a field, which contains $\mathbb{F}_{p^n}$ as a subfield for any positive integer n. Each element in $\mathbb{F}_{p^\infty}$ is of finite multiplicative order, but $\mathbb{F}_{p^\infty}$ itself is infinite. The field $\mathbb{F}_{p^\infty}$ is the algebraic closure of $\mathbb{F}_p$.

Of importance in field theory is the set of automorphisms of an extension K of F which fix the elements of F. It can be shown that this set G forms a group under composition of mappings. G is called the *Galois group* over F. In the case of automomorphisms of finite fields $\mathbb{F}_{p^n}$, all elements of $\mathbb{F}_p$ remain fixed. Thus G consists of all automorphisms of $\mathbb{F}_{p^n}$. Let $q = p^n$.

2.11 Definition. The mapping $\theta : \mathbb{F}_q \to \mathbb{F}_q$, $a \mapsto a^p$, is called the *Frobenius automorphism* of $\mathbb{F}_q$.

It can be verified that θ is an automorphism. If a is a generating element of $\mathbb{F}_q^*$ of order $q - 1$ then $\theta^n(a) = a^{p^n} = a$. For $i = 2, \ldots, n - 1$ we have $\theta^i(a) = a^{p^i} \neq a$; therefore θ is an automorphism of order n. We state without proof:

2.12 Theorem. *The group G of automorphisms of $\mathbb{F}_{p^n}$ is cyclic of order n. G consists of the elements $\theta, \theta^2, \ldots, \theta^{n-1}$ and $\theta^n = \iota$, where ι is the identity mapping.*

Finally, we consider a generalization of the well-known concept of complex roots of unity. In $\mathbb{C}$ the nth roots of unity are $z_k = e^{2\pi i k/n}$, $k = 0, 1, \ldots, n - 1$. Geometrically they can be represented by the n vertices of a regular polygon in the unit circle in the complex plane. All z_k with $(k, n) = 1$ are generators. They are again called *primitive nth roots of unity.*

These complex numbers z_k form a multiplicative group of order n, which is cyclic with generator $z_1 = e^{2\pi i/n}$. We define for an arbitrary field F:

2.13 Definition. Let F be a field. A root of $x^n - 1$ in $F[x]$ is called an nth *root of unity*. The *order* of an nth root α of unity is the least positive integer such that $\alpha^n = 1$. An nth root of unity of order n is called *primitive*. The splitting field S_n of $x^n - 1 \in F[x]$ is called the *associated cyclotomic field.*

2.14 Theorem. *Let n be a positive integer and let F be a field whose characteristic does not divide n.*

(i) *There is a finite extension K of F which contains a primitive nth root of unity.*

(ii) *If α is a primitive nth root of unity then $F(\alpha)$ is the splitting field of $f = x^n - 1$ over F.*

(iii) *$x^n - 1$ has exactly n distinct roots in $F(\alpha)$. These roots form a cyclic group. The order of an nth root α of unity is just the order of α in this group. The primitive nth roots of unity in $F(\alpha)$ are precisely the generators of this group. There are $\phi(n)$ primitive nth roots of unity, which can be obtained from one root by raising it to the powers $k < n$, $\gcd(k, n) = 1$.*

PROOF. The proofs of (i) and (ii) are similar to that of 2.5 and are omitted. We show that the set of nth roots of unity is a cyclic group and leave the rest of (iii) as an exercise. Let char $F = p$. Then the extension K of F contains the splitting field S_n of $x^n - 1$ over $\mathbb{F}_p$. S_n is finite and has a cyclic multiplicative group. The roots $\alpha_1, \ldots, \alpha_n$ of $x^n - 1$ in S_n form a subgroup S of G, which is cyclic. α is a generator of S if and only if it is of order n, so that indeed α is a primitive nth root of unity. The case where char $F = 0$ is treated separately. $\square$

In factoring $x^n - 1$ into irreducible factors, the so-called cyclotomic polynomials are useful.

2.15 Definition. The polynomial $Q_n := (x - \alpha_1) \ldots (x - \alpha_{\phi(n)})$ is called the nth *cyclotomic polynomial* over a field F, if $\alpha_1, \ldots, \alpha_{\phi(n)}$ are the $\phi(n)$ primitive nth roots of unity.

Let α be a primitive nth root of unity. Then it follows from 2.14 that $Q_n = \prod (x - \alpha^i)$, where the product is formed over all i with $\gcd(i, n) = 1$. The polynomial Q_n is of degree $\phi(n)$. Let $n = kd$ so that α^k is of order d and is a primitive dth root of unity. The dth cyclotomic polynomial is of the form $Q_d = \prod_{\gcd(i,d)=1} (x - \alpha^{ik})$.

Any nth root of unity is a primitive dth root of unity for exactly one d. Therefore we can group the nth roots of unity together and obtain

2.16 Theorem. $x^n - 1 = \prod_{d \mid n} Q_d.$

2.17 Theorem. *Let p be a prime number and m a positive integer. Then*

$$Q_{p^m} = 1 + x^{p^{m-1}} + \ldots + x^{(p-1)p^{m-1}}.$$

PROOF. Theorem 2.16 shows

$$Q_{p^m} = \frac{x^{p^m} - 1}{Q_1 Q_p \ldots Q_{p^{m-1}}} = \frac{x^{p^m} - 1}{x^{p^{m-1}} - 1}$$

which yields the result.   □

The decomposition of $x^n - 1$ in 2.16 does not necessarily give irreducible factors. It is called the *cyclotomic decomposition* of $x^n - 1$. Using the so-called Möbius inversion formula we can derive a formula for cyclotomic polynomials.

2.18 Definition. The mapping $\mu : \mathbb{N} \to \{0, 1, -1\}$, defined by

$$\mu(1) := 1,$$
$$\mu(p_1 \ldots p_t) := (-1)^t \quad \text{if } p_i \text{ are distinct primes},$$
$$\mu(n) := 0 \qquad \text{if } p^2 | n \text{ for some prime } p,$$

is called the *Möbius function* or μ-function.

There is a very simple and useful property of the μ-function, namely

$$\sum_{d|n} \mu(d) = \begin{cases} 1 & \text{if } n = 1, \\ 0 & \text{if } n > 1. \end{cases}$$

To verify this, for $n > 1$, we have to take into account only those positive divisors d of n for which $\mu(d) \neq 0$, i.e. for which $d = 1$ or d is a product of distinct primes. Thus, if $p_1, p_2, \ldots, p_k$ are the distinct prime divisors of n, we get

$$\sum_{d|n} \mu(d) = \mu(1) + \sum_{i=1}^{k} u(p_i) + \sum_{1 \le i_1 < i_2 \le k} \mu(p_{i_1} p_{i_2}) + \ldots + \mu(p_1 p_2 \ldots p_k)$$

$$= 1 + \binom{k}{1}(-1) + \binom{k}{2}(-1)^2 + \ldots + \binom{k}{k}(-1)^k = (1 + (-1))^k = 0.$$

The case $n = 1$ is trivial.

2.19 Theorem (Möbius Inversion Formula). (i) (Additive Form.) *Let $f : \mathbb{N} \to (A, +)$ and $g : \mathbb{N} \to (A, +)$ be mappings from $\mathbb{N}$ into an additive abelian group A; then*

$$g(n) = \sum_{d|n} f(d) \Leftrightarrow f(n) = \sum_{d|n} \mu\left(\frac{n}{d}\right) g(d).$$

(ii) (Multiplicative Form.) *Let $f : \mathbb{N} \to (A, \cdot)$ and $g : \mathbb{N} \to (A, \cdot)$ be mappings from $\mathbb{N}$ into a multiplicative abelian group A; then*

$$g(n) = \prod_{d|n} f(d) \Leftrightarrow f(n) = \prod_{d|n} g(d)^{\mu(n/d)}.$$

PROOF. We show the additive form of the inversion formula. Assuming $g(n) = \sum_{d|n} f(d)$ we get

$$\sum_{d|n} \mu\left(\frac{n}{d}\right) g(d) = \sum_{d|n} \mu(d) g\left(\frac{n}{d}\right) = \sum_{d|n} \mu(d) \sum_{c|n/d} f(c) = \sum_{c|n} \sum_{d|n/c} \mu(d) f(c)$$

$$= \sum_{c|n} f(c) \sum_{d|n/c} \mu(d) = f(n).$$

The converse is derived in a similar calculation. The multiplicative form follows from the additive form by replacing sums with products and products with powers. □

2.20 Corollary.

$$Q_n = \prod_{d|n} (x^d - 1)^{\mu(n/d)} = \prod_{d|n} (x^{n/d} - 1)^{\mu(d)}.$$ □

It can be verified that all cyclotomic polynomials over $\mathbb{Q}$ have integer coefficients, while in the case of polynomials over $\mathbb{Z}_p$, the coefficients are to be taken mod p. Cyclotomic polynomials are irreducible over $\mathbb{Q}$, but not necessarily over $\mathbb{Z}_p$. A curious property of Q_n is that the first 104 cyclotomic polynomials have coefficients in $\{0, 1, -1\}$ only. In Q_{105} we have 2 as one of the coefficients.

2.21 Examples.

n	Q_n
1	$x - 1$
2	$x + 1$
3	$x^2 + x + 1$
4	$x^2 + 1$
5	$x^4 + x^3 + x^2 + x + 1$
6	$x^2 - x + 1$
7	$x^6 + x^5 + x^4 + x^3 + x^2 + x + 1$
8	$x^4 + 1$
9	$x^6 + x^3 + 1$
10	$x^4 - x^3 + x^2 - x + 1$
11	$x^{10} + x^9 + x^8 + x^7 + x^6 + x^5 + x^4 + x^3 + x^2 + x + 1$
12	$x^4 - x^2 + 1$
13	$x^{12} + x^{11} + x^{10} + x^9 + x^8 + x^7 + x^6 + x^5 + x^4 + x^3 + x^2 + x + 1$
14	$x^6 - x^5 + x^4 - x^3 + x^2 - x + 1$
15	$x^8 - x^7 + x^5 - x^4 + x^3 - x + 1$

□

It follows from 2.3 and 2.14 that the cyclotomic field S_n can be constructed as a simple extension of $\mathbb{Z}_p$ by using a polynomial which divides Q_n. The finite field F_{p^n} is the cyclotomic field of the $(p^n - 1)$th roots of unity. In F_{p^n} we have

$$x^{p^n-1} - 1 = \prod_{d|p^n-1} Q_d.$$

In general the polynomial Q_d is not irreducible, but has as roots all elements of order d in F_{p^n}. An element of order d in an extension of F_p has a minimal polynomial of degree k over F_p, where k is the smallest integer such that $d|p^k - 1$. since there are $\phi(d)$ elements of order d, we have $\phi(d)/k$ irreducible polynomials of degree k over F_p with this property. The product of these polynomials is equal to Q_d.

2.22 Example. We want to factorize $x^{15} - 1$ over F_2. First we consider $x^{15} - 1$ as a product of cylotomic polynomials, namely

$$x^{15} - 1 = Q_{15}Q_5Q_3Q_1,$$

where

$$Q_1 = x + 1,$$
$$Q_3 = x^2 + x + 1,$$
$$Q_5 = x^4 + x^3 + x^2 + x + 1,$$
$$Q_{15} = x^8 - x^7 + x^5 - x^4 + x^3 - x + 1.$$

Q_1, Q_3 and Q_5 are irreducible over F_2. Since 15 divides $2^4 - 1$ and $\phi(15) = 8$, we conclude that Q_{15} is a product of irreducible polynomials of degree 4 over F_2.

$$Q_{15} = (x^4 + x + 1)(x^4 + x^3 + 1). \qquad \square$$

The procedure indicated in this example is useful for determining all elements of finite fields.

2.23 Example. We want to describe the elements of F_{3^2}. This is the eighth cyclotomic field. We determine Q_8 and factorize it over F_3

$$Q_8 = x^4 + 1 = (x^2 + x - 1)(x^2 - x - 1).$$

A root of $x^2 + x - 1$ is a primitive eighth root of unity over F_3. Let ζ be such a root, so $\zeta^2 + \zeta - 1 = 0$. Now all nonzero elements of F_{3^2} can be represented in the form ζ^i, $1 \le i \le 8$, so $F_{3^2} = \{0, \zeta, \zeta^2, \zeta^3, \zeta^4, \zeta^5, \zeta^6, \zeta^7, \zeta^8\}$. We may represent the elements of F_{3^2} also by using the approach of Example 2.7. First we choose any irreducible polynomial of degree 2 over F_3, say $x^2 + 1 = 0$. Let α be a root of this polynomial, so $\alpha^2 + 1 = 0$ in F_{3^2}. Then the nine elements of F_{3^2} can be represented in the form $F_{3^2} = \{a + b\alpha \,|\, a, b \in F_3\}$. In order to establish a connection between this representation and the one above we note that $\zeta = 1 + \alpha$ is a root of $x^2 + x - 1$. The nonzero elements of $\mathbb{F}_{3^2}$ can be represented in a table, called *index table* or *table of discrete logarithm*:

i	ζ^i
1	$1 + \alpha$
2	2α
3	$1 + 2\alpha$
4	2
5	$2 + 2\alpha$
6	α
7	$2 + \alpha$
8	1

More generally if $\mathbb{F}_q$ can be represented as $\{\zeta^i | 0 \le i < q - 1\} \cup \{0\}$ then the unique i is called the *index* or *discrete logarithm* of a if any $a \in \mathbb{F}_q$ can be written as ζ^i for a primitive root ζ in $\mathbb{F}_q$. The discrete logarithm of a is denoted as $i = \mathrm{ind}_\zeta(a)$ if $a = \zeta^i$. It satisfies the following basic rules:

$$\mathrm{ind}_\zeta(ab) \equiv \mathrm{ind}_\zeta(a) + \mathrm{ind}_\zeta(b) \pmod{q - 1},$$

$$\mathrm{ind}_\zeta(ab^{-1}) \equiv \mathrm{ind}_\zeta(a) - \mathrm{ind}_\zeta(b) \pmod{q - 1}.$$

The inverse function of the discrete logarithm is the antilogarithm that maps i onto ζ^i.

PROBLEMS

1. Define the operations $+$ and $\cdot$ on $\mathbb{Z}_p^2$ as follows

$$(a_1, b_1) + (a_2, b_2) = (a_1 + a_2, b_1 + b_2),$$

$$(a_1, b_1) \cdot (a_2, b_2) = (a_1 a_2 - b_1 b_2, a_1 b_2 + a_2 b_1).$$

Prove that $(\mathbb{Z}_p^2, +, \cdot)$ is a finite field $\mathbb{F}_{p^2}$ if and only if $p \equiv 3 \pmod 4$.

2. Let α_i, $i = 0, 1, \ldots, 7$, be the eight elements of $\mathbb{F}_{2^3}$, defined by the irreducible polynomial $x^3 + x + 1$ over $\mathbb{F}_2$. Find the operation tables for addition and multiplication of elements of $\mathbb{F}_{2^3}$.

3. Show each element of a finite field is the sum of two squares.

*4. Prove: If $x^3 + ax + b$ is irreducible over a finite field F then $-4a^3 - 27b^2$ is a square in F.

*5. The map $x \to x^3$ is never an automorphism of the additive group of $\mathbb{F}_{2^n}$, $n > 1$. For which p is the map $x \to x^p$ an automorphism of the multiplicative group of $\mathbb{F}_{2^n}$?

6. Let a, b be elements of $\mathbb{F}_{2^n}$, n odd. Prove that $a^2 + ab + b^2 = 0$ implies $a = b = 0$.

*7. Let a, b be elements of $\mathbb{F}_{2^m}$. Prove or disprove if $\mathbb{F}_2(a) \cap \mathbb{F}_2(b) = \mathbb{F}_2$ then $\mathbb{F}_2(a, b) = \mathbb{F}_2(a + b)$.

8. Prove that every element of $\mathbb{F}_q$ is the kth power of some element of $\mathbb{F}_q$ if and only if $\gcd(k, q - 1) = 1$.

9. Let α be a primitive element of $\mathbb{F}_q$. The Zech's logarithm Z is a function which is defined on the integers, for $0 \le n \le q - 1$, in such a way that $\alpha^{Z(n)} = \alpha^n + 1$. This can be used to add elements α^i and α^j in $\mathbb{F}_q$ by using the equation $\alpha^i + \alpha^j = \alpha^{j+Z(i-j)}$. Determine the Zech logarithm in $\mathbb{F}_{2^4}$ and evaluate $\alpha^3 + \alpha^5$ and also $\alpha^4 + \alpha^{13}$, where $\alpha^4 + \alpha + 1 = 0$.

*10. Let $\mathbb{F}_q$ be of characteristic p. Prove that there exists exactly one pth root for each element of $\mathbb{F}_q$.

11. For the cyclotomic polynomial $Q_n(x)$ prove:
    (i) $Q_n(0) = 1$ if $n \ge 2$;
    (ii) $Q_n(1) = \begin{cases} 0 & \text{if } n = 1, \\ p & \text{if } n \text{ is a power of the prime } p, \\ 1 & \text{if } n \text{ has at least two distinct prime factors.} \end{cases}$

12. Find the index table (or discrete logarithms) for a finite field with 27 elements.

13. Represent all elements of $\mathbb{F}_{25}$ as linear combinations of basis elements over $\mathbb{F}_5$. Then find a primitive element β of $\mathbb{F}_{25}$ and determine for each element α of $\mathbb{F}_{25}^*$ the least nonnegative integer n such that $\alpha = \beta^n$.

EXERCISES (Solutions in Chapter 8, p. 451)

1. Determine all elements of $\mathbb{F}_{16}$.

2. The polynomials $x^3 - x + 1$ and $x^3 - x - 1$ are irreducible over $\mathbb{F}_3$. Determine the isomorphism between their respective splitting fields over $\mathbb{F}_3$.

3. Prove that every mapping $f: \mathbb{F}_q \to \mathbb{F}_q$ can be expressed uniquely as a polynomial function $\bar{p}$ of degree $q - 1$ of the form

$$\bar{p}(\beta) = \sum_{\alpha \in \mathbb{F}_q} f(\alpha)(1 - (\beta - \alpha)^{q-1}), \qquad \beta \in \mathbb{F}_q.$$

This is called the *Lagrange Interpolation Formula* for finite fields.

Conversely, if R is a finite commutative ring such that any function from R into itself can be represented as a polynomial function, R is a finite field.

*4. Prove:

$$\sum_{a \in \mathbb{F}_q} a^m = \begin{cases} -1 & \text{if } (q - 1) | m \\ 0 & \text{otherwise} \end{cases} \quad \text{for any positive integer } m.$$

In particular $\sum_{a \in \mathbb{Z}_p} a = 0$ for $p \in \mathbb{P}$, $p \ne 2$.

5. For $\alpha \in \mathbb{F}_{q^m} = F$ and $K = \mathbb{F}_q$, the trace $\mathrm{Tr}_{F/K}(\alpha)$ of α over K is defined by

$$\mathrm{Tr}_{F/K}(\alpha) = \alpha + \alpha^q + \ldots + \alpha^{q^{m-1}}.$$

Prove:
  (i) $\mathrm{Tr}_{F/K}(\alpha + \beta) = \mathrm{Tr}_{F/K}(\alpha) + \mathrm{Tr}_{F/K}(\beta)$ for all $\alpha, \beta \in F$;
  (ii) $\mathrm{Tr}_{F/K}(c\alpha) = c\,\mathrm{Tr}_{F/K}(\alpha)$ for all $c \in K, \alpha \in F$;
  *(iii) $\mathrm{Tr}_{F/K}$ is a linear transformation from F onto K, where both F and K are viewed as vector spaces over K;
  (iv) $\mathrm{Tr}_{F/K}(a) = ma$ for all $a \in K$;
  (v) $\mathrm{Tr}_{F/K}(\alpha^q) = \mathrm{Tr}_{F/K}(\alpha)$ for all $\alpha \in F$.

*6. Let $a_0, \ldots, a_{q-1}$ be elements of $\mathbf{F}_q$. Prove

$$\sum_{i=0}^{q} a_i^t = \begin{cases} 0 & \text{for} \quad 1 \le t \le q - 2, \\ -1 & \text{for} \quad t = q - 1, \end{cases}$$

if all elements $a_0, \ldots, a_{q-1}$ are distinct.

7. Show as consequences of properties of finite fields that for $a \in \mathbb{Z}$ and $p \in \mathbb{P}$
  (i) $a^p \equiv a \bmod p$ ("*Little Fermat's Theorem*")
  (ii) $(p - 1)! \equiv -1 \bmod p$ ("*Wilson's theorem*").

8. Find all subfields of $\mathbf{F}_{15625}$ and $\mathbf{F}_{2^{30}}$.

9. Prove Lemma 2.8.

10. Determine all primitive elements of $\mathbf{F}_7$.

11. Prove $f(x)^q = f(x^q)$ for $f \in \mathbf{F}_q[x]$.

12. Let ζ be an nth root of unity over a field K. Prove that $1 + \zeta + \zeta^2 + \ldots + \zeta^{n-1} = 0$ or n according as $\zeta \ne 1$ or $\zeta = 1$.

13. Prove the following properties of cyclotomic polynomials:
  (i) If p is prime and $p \nmid m$ then

$$Q_{mp^k}(x) = Q_{pm}(x^{p^{k-1}}).$$

  (ii) If p is prime and $p \nmid m$ then

$$Q_{pm}(x) = \frac{Q_m(x^p)}{Q_m(x)}.$$

  (iii) If $n \ge 2$, $Q_n(x) = \prod_{d|n} (1 - x^{n/d})^{\mu(d)}$.
  (iv) If $n \ge 3$, n odd, then $Q_{2n}(x) = Q_n(-x)$.
  (v) If $n \ge 2$, $Q_n(x^{-1})x^{\phi(n)} = Q_n(x)$.

14. Find the cyclotomic polynomials Q_{36} and Q_{105}.

15. The companion matrix of a monic polynomial

$$f = a_0 + a_1 x + \ldots + a_{n-1}x^{n-1} + x^n$$

of degree $n \ge 1$ over a field is defined to be the $n \times n$ matrix

$$\mathbf{A} = \begin{pmatrix} 0 & 0 & \ldots & 0 & -a_0 \\ 1 & 0 & \ldots & 0 & -a_1 \\ \multicolumn{5}{c}{\cdots\cdots\cdots\cdots\cdots} \\ 0 & 0 & \ldots & 1 & -a_{n-1} \end{pmatrix}$$

$\mathbf{A}$ satisfies $f(\mathbf{A}) = a_0\mathbf{I} + a_1\mathbf{A} + \ldots + a_{n-1}\mathbf{A}^{n-1} + \mathbf{A}^n = 0$. If f is irreducible over $\mathbf{F}_p$ then $\mathbf{A}$ can play the role of a root of f and the polynomials in $\mathbf{A}$ over $\mathbf{F}_p$ of degree less than n yield a representation of the elements of $\mathbf{F}_q$, where $q = p^n$.

(i) Let $f = x^2 + 1 \in \mathbb{F}_3[x]$. Find the companion matrix $\mathbf{A}$ of f and a representation of $\mathbb{F}_9$ using $\mathbf{A}$. Establish the multiplication table for the elements of $\mathbb{F}_9$ given in terms of $\mathbf{A}$.

(ii) Let $f = x^2 + x + 2 \in \mathbb{F}_3[x]$ be an irreducible factor of the cyclotomic polynomial $Q_8 \in \mathbb{F}_3[x]$. Find the companion matrix $\mathbf{A}$ of f and a representation of the elements of $\mathbb{F}_9$ in terms of $\mathbf{A}$.

16. Let $\mathbb{F}_q$ be a finite field and $\mathbb{F}_r$ a finite extension. Show that $\mathbb{F}_r$ is a simple algebraic extension of $\mathbb{F}_q$ and that every primitive element of $\mathbb{F}_r$ can be adjoined to $\mathbb{F}_q$ to give $\mathbb{F}_r$.

17. Show that a finite field $\mathbb{F}_q$ is the $(q - 1)$th cyclotomic field over any one of its subfields.

18. Let $\mathbb{F}_{q^m}$ be an extension of $\mathbb{F}_q$ and let $\alpha \in \mathbb{F}_{q^m}$. The elements $\alpha, \alpha^q, \ldots, \alpha^{q^{m-1}}$ are called the *conjugates* of α with respect to $\mathbb{F}_q$. Find an element of $\mathbb{F}_{16}$ and its conjugates with respect to $\mathbb{F}_2$ and with respect to $\mathbb{F}_4$.

19. A basis of $\mathbb{F}_{q^m}$ over $\mathbb{F}_q$, which consists of a suitable element $\alpha \in \mathbb{F}_{q^m}$ and its conjugates with respect to $\mathbb{F}_q$, is called a *normal basis* of $\mathbb{F}_{q^m}$ over $\mathbb{F}_q$. Find a normal basis of $\mathbb{F}_8$ over $\mathbb{F}_2$.

*20. Let α and β be nonzero elements of $\mathbb{F}_q$. Show that there exist elements $a, b \in \mathbb{F}_q$ such that $1 + \alpha a^2 + \beta b^2 = 0$.

§3. Irreducible Polynomials over Finite Fields

We have seen in §1 and §2 that irreducible polynomials over a field are of fundamental importance in the theory of field extensions. This is true, in particular, in the case of extensions of finite fields. In this section we consider polynomials over $\mathbb{F}_q$, these have many applications in combinatorics, number theory and algebraic coding theory.

We recall that the splitting field of an irreducible polynomial of degree k over $\mathbb{F}_q$ is $\mathbb{F}_{q^k}$.

3.1 Theorem. *Let f be an irreducible polynomial over $\mathbb{F}_q$ of degree k. f divides $x^{q^n} - x$ if and only if k divides n.*

PROOF. Suppose $f | x^{q^n} - x$. Then f has its roots in $\mathbb{F}_{q^n}$ and its splitting field $\mathbb{F}_{q^k}$ must be contained in $\mathbb{F}_{q^n}$. Theorem 2.9 implies $k|n$. Conversely, let $k|n$, so that $\mathbb{F}_{q^k}$ is a subfield of $\mathbb{F}_{q^n}$. Since f and $x^{q^n} - x$ split into linear factors in $\mathbb{F}_{q^n}$, $f | x^{q^n} - x$ holds over $\mathbb{F}_{q^n}$ by 2.2(ii). $\qquad\square$

By 2.2(ii), $x^{q^n} - x \in \mathbb{F}_q[x]$ has only simple roots. Theorem 3.1 implies that this is so for all irreducible polynomials in $\mathbb{F}_q[x]$ as well. If α is any root of an irreducible polynomial, all other roots are given by $\alpha^q, \alpha^{q^2}, \ldots, \alpha^{q^{k-1}}$, called the *conjugates* of α.

3.2 Theorem. $x^{q^n} - x = \prod_i f_i$, where the product is extended over all distinct, monic, irreducible polynomials over $\mathbb{F}_q$, with degree a divisor of n.

PROOF. It is easily verified that if f_i and f_j are two distinct, monic, irreducible polynomials over $\mathbb{F}_q$ whose degrees divide n, then f_i and f_j are relatively prime and hence $f_i f_j | x^{q^n} - x$. The theorem follows from Theorem 3.1 and the fact that $x^{q^n} - x$ has only simple roots in its splitting field over $\mathbb{F}_q$. $\square$

Theorem 3.1 asserts that any element $\alpha \in \mathbb{F}_{q^n}$ is a root of an irreducible polynomial of degree $\leq n$ over $\mathbb{F}_q$. Let M be the minimal polynomial of α over $\mathbb{F}_q$, and let the degree of M be k where $k|n$. Then $\alpha, \alpha^q, \ldots, \alpha^{q^{k-1}}$ are the roots of M and M is also the minimal polynomial for each of these roots. The factorization of Theorem 3.2 can also be regarded as the product of all distinct minimal polynomials of elements of $\mathbb{F}_{q^n}$ over $\mathbb{F}_q$. Properties of minimal polynomials are summarized as follows; see also 4.8 and 4.11 in this chapter.

3.3. Theorem. Let $\alpha \in \mathbb{F}_{q^n}$. Suppose the degree of α over $\mathbb{F}_q$ is d and let M be the minimal polynomial of α over $\mathbb{F}_q$.

(i) M is irreducible over $\mathbb{F}_q$ and $\deg M = d$ divides n.
(ii) $f \in \mathbb{F}_q[x]$ satisfies $\bar{f}(\alpha) = 0$ if and only if $M|f$.
(iii) If α is primitive then $\deg M = n$.
(iv) $\alpha, \alpha^q, \ldots, \alpha^{q^{n-1}}$ all have M as minimal polynomial.
(v) If f is a monic irreducible polynomial of $\mathbb{F}_q[x]$ with $\bar{f}(\alpha) = 0$, then $f = M$.
(vi) M divides $x^{q^d} - x$ and $x^{q^n} - x$.
(vii) The roots of M are $\alpha, \alpha^q, \ldots, \alpha^{q^{d-1}}$, and M is the minimal polynomial over $\mathbb{F}_q$ of all these elements. $\square$

As we saw before, it is often important to find the minimal polynomial of an element in a finite field. A straightforward method of determining minimal polynomials is the following one. Let ζ be a defining element of $\mathbb{F}_{q^n}$ over $\mathbb{F}_q$, so that $\{1, \zeta, \ldots, \zeta^{n-1}\}$ is a basis of $\mathbb{F}_{q^n}$ over $\mathbb{F}_q$. If we wish to find the minimal polynomial g of $\beta \in \mathbb{F}_{q^n}^*$ over $\mathbb{F}_q$, we represent $\beta^0, \beta^1, \ldots, \beta^n$ in terms of the basis elements. Let

$$\beta^{i-1} = \sum_{j=1}^n d_{ij} \zeta^{j-1} \quad \text{for } 1 \leq i \leq n + 1.$$

Let g be of the form $g(x) = c_n x^n + \ldots + c_1 x + c_0$. In order that g be the monic polynomial of least positive degree with $g(\beta) = 0$ we proceed as follows. The condition $g(\beta) = c_n \beta^n + \ldots + c_1 \beta + c_0 = 0$ leads to the homogeneous system of linear equations.

$$\sum_{i=1}^{n+1} c_{i-1} d_{ij} = 0 \quad \text{for } 1 \leq j \leq n, \tag{*}$$

with unknowns $c_0, c_1, \ldots, c_n$. Let $\mathbf{D}$ be the matrix of coefficients of the system, i.e., $\mathbf{D}$ is the $(n+1) \times n$ matrix whose (i, j) entry is d_{ij}, and let r be the rank of $\mathbf{D}$. Then the dimension of the space of solutions of the system is $s = n + 1 - r$, and since $1 \le r \le n$, we have $1 \le s \le n$. Therefore we let s of the unknowns $c_0, c_1, \ldots, c_n$ take prescribed values and then the remaining ones are uniquely determined. If $s = 1$, we set $c_n = 1$, and if $s > 1$, we set $c_n = c_{n-1} = \ldots = c_{n-s+2} = 0$ and $c_{n-s+1} = 1$.

3.4 Example. Let $\zeta \in \mathbb{F}_{64}$ be a root of the irreducible polynomial $x^6 + x + 1$ in $\mathbb{F}_2[x]$. For $\beta = \zeta^3 + \zeta^4$ we have

$$\beta^0 = 1,$$
$$\beta^1 = \zeta^3 + \zeta^4,$$
$$\beta^2 = 1 + \zeta + \zeta^2 + \zeta^3,$$
$$\beta^3 = \zeta + \zeta^2 + \zeta^3,$$
$$\beta^4 = \zeta + \zeta^2 + \zeta^4,$$
$$\beta^5 = 1 + \zeta^3 + \zeta^4,$$
$$\beta^6 = 1 + \zeta + \zeta^2 + \zeta^4.$$

Therefore the matrix $\mathbf{D}$ is of the form

$$\mathbf{D} = \begin{pmatrix} 1 & 0 & 0 & 0 & 0 & 0 \\ 0 & 0 & 0 & 1 & 1 & 0 \\ 1 & 1 & 1 & 1 & 0 & 0 \\ 0 & 1 & 1 & 1 & 0 & 0 \\ 0 & 1 & 1 & 0 & 1 & 0 \\ 1 & 0 & 0 & 1 & 1 & 0 \\ 1 & 1 & 1 & 0 & 1 & 0 \end{pmatrix}$$

and its rank is $r = 3$. Hence $s = n + 1 - r = 4$, and we set $c_6 = c_5 = c_4 = 0$, $c_3 = 1$. The remaining coefficients are determined from (*), and this yields $c_2 = 1$, $c_1 = 0$, $c_0 = 1$. Therefore the minimal polynomial of β over $\mathbb{F}_2$ is $g(x) = x^3 + x^2 + 1$. □

Another method of determining minimal polynomials is as follows. If we wish to find the minimal polynomial g of $\beta \in \mathbb{F}_{q^n}$ over $\mathbb{F}_q$, we calculate the powers $\beta, \beta^q, \beta^{q^2}, \ldots$ until we find the least positive integer d for which $\beta^{q^d} = \beta$. This integer d is the degree of g, and g itself is given by $g(x) = (x - \beta)(x - \beta^q) \ldots (x - \beta^{q^{d-1}})$. The elements $\beta, \beta^q, \ldots, \beta^{q^{d-1}}$ are called the conjugates of β with respect to $\mathbb{F}_q$, they are distinct and g is the minimal polynomial over $\mathbb{F}_q$ of all these elements.

3.5 Example. We compute the minimal polynomials over $\mathbb{F}_2$ of all elements of $\mathbb{F}_{16}$. Let $\zeta \in \mathbb{F}_{16}$ be a root of the primitive polynomial $x^4 + x + 1$ over $\mathbb{F}_2$, so that every nonzero element of $\mathbb{F}_{16}$ can be written as a power of ζ. We have the following index table for $\mathbb{F}_{16}$:

i	ζ^i		i	ζ^i
0	1		8	$1 + \zeta^2$
1	ζ		9	$\zeta + \zeta^3$
2	ζ^2		10	$1 + \zeta + \zeta^2$
3	ζ^3		11	$\zeta + \zeta^2 + \zeta^3$
4	$1 + \zeta$		12	$1 + \zeta + \zeta^2 + \zeta^3$
5	$\zeta + \zeta^2$		13	$1 + \zeta^2 + \zeta^3$
6	$\zeta^2 + \zeta^3$		14	$1 + \zeta^3$
7	$1 + \zeta + \zeta^3$			

The minimal polynomials of the elements β of $\mathbb{F}_{16}$ over $\mathbb{F}_2$ are:

$\beta = 0$: $g_1(x) = x$.

$\beta = 1$: $g_2(x) = x - 1$.

$\beta = \zeta$:   The distinct conjugates of ζ with respect to $\mathbb{F}_2$ are $\zeta, \zeta^2, \zeta^4, \zeta^8$, and the minimal polynomial is

$$g_3(x) = (x - \zeta)(x - \zeta^2)(x - \zeta^4)(x - \zeta^8) = x^4 + x + 1.$$

$\beta = \zeta^3$:   The distinct conjugates of ζ^3 with respect to $\mathbb{F}_2$ are $\zeta^3, \zeta^6, \zeta^{12}, \zeta^{24} = \zeta^9$, and the minimal polynomial is

$$g_4(x) = (x - \zeta^3)(x - \zeta^6)(x - \zeta^9)(x - \zeta^{12}) = x^4 + x^3 + x^2 + x + 1.$$

$\beta = \zeta^5$:   Since $\beta^4 = \beta$, the distinct conjugates of this element with respect to $\mathbb{F}_2$ are ζ^5, ζ^{10}, and the minimal polynomial is

$$g_5(x) = (x - \zeta^5)(x - \zeta^{10}) = x^2 + x + 1.$$

$\beta = \zeta^7$:   The distinct conjugates of ζ^7 with respect to $\mathbb{F}_2$ are $\zeta^7, \zeta^{14}, \zeta^{28} = \zeta^{13}$, $\zeta^{56} = \zeta^{11}$, and the minimal polynomial is

$$g_6(x) = (x - \zeta^7)(x - \zeta^{11})(x - \zeta^{13})(x - \zeta^{14}) = x^4 + x^3 + 1.$$

These elements, together with their conjugates with respect to $\mathbb{F}_2$, exhaust $\mathbb{F}_{16}$.   □

3.6 Theorem. *The number of all monic, irreducible polynomials of degree k over $\mathbb{F}_q$ is given by*

$$I_q(k) = \frac{1}{k} \sum_{d \mid k} \mu\left(\frac{k}{d}\right) q^d.$$

PROOF. Theorem 3.2 implies that the degree of the product of all monic, irreducible polynomials over $\mathbb{F}_q$ whose degree divides k is equal to $\sum_{d|k} I_q(d)d = q^k$. The additive form of the Möbius inversion formula, Theorem 2.19(i), gives the desired result. □

The formula in Theorem 3.6 shows how to check $I_q(k) \geq 1$ for any prime power q and any positive integer k. Here we see the difference between irreducible polynomials over $\mathbb{F}_q$ and irreducible polynomials over $\mathbb{R}$ and $\mathbb{C}$ (see Corollary 2.6).

3.7 Example.

q	k	$I_q(k)$
2	1	2
2	2	1
2	3	2
2	4	3
2	5	6

There is an interesting connection between minimal polynomials and primitive elements of a finite field. We introduce the order (also called the exponent or the period) of a nonzero polynomial over a finite field. The following result motivates the definition of an order.

3.8 Lemma. *Let $f \in \mathbb{F}_q[x]$ be a polynomial of degree $m \geq 1$ with $f(0) \neq 0$. Then there exists a positive integer $e \leq q^m - 1$ such that f divides $x^e - 1$.*

PROOF. $\mathbb{F}_q[x]/(f)$ has q^{m-1} nonzero residue classes. Since the q^m residue classes $x^j + (f), 0 \leq j \leq q^m - 1$, are all nonzero, there exists integers s and t with $0 \leq s < t \leq q^m - 1$ such that $x^t \equiv x^s \bmod f$. Since $(x, f) = 1$, we have $x^{t-s} \equiv 1 \bmod f$, i.e. $f|(x^{t-s} - 1)$ and $0 < t - s \leq q^m - 1$. □

3.9 Definition. Let $0 \neq f \in \mathbb{F}_q[x]$. If $f(0) \neq 0$ then the smallest natural number e with the property that $f|(x^e - 1)$ is called the *order* of f. If $f(0) = 0$ and f is of the form $x^h g$ with $h \in \mathbb{N}$ and $g \in \mathbb{F}_q[x]$, $g(0) \neq 0$, for a unique polynomial g, then the order of f is defined as the order of g, in symbols ord f.

The order of an irreducible polynomial can be characterized by its roots.

3.10 Theorem. *Let $f \in \mathbb{F}_q[x]$ be an irreducible polynomial over $\mathbb{F}_q$ of degree m with $f(0) \neq 0$. Then ord f is equal to the order of any root of f in $\mathbb{F}_{q^m}^*$.*

PROOF. $\mathbb{F}_{q^m}$ is the splitting field of f over $\mathbb{F}_q$. The roots of f have the same order in $\mathbb{F}_{q^m}^*$; let $\alpha \in \mathbb{F}_{q^m}^*$ be a root of f. Then $\alpha^e = 1$ if and only if $f|x^e - 1$.

The result follows from the definition of ord f and the order of α in the group $\mathbb{F}_{q^m}^*$. $\qquad\square$

3.11 Corollary. *If $f \in \mathbb{F}_q[x]$ is irreducible over $\mathbb{F}_q$ of degree m, then* ord f *divides $q^m - 1$.*

PROOF. If $f = cx$ with $c \in \mathbb{F}_q^*$, then ord $f = 1$. Otherwise the result follows from Theorem 3.10 and the fact that the order of $\mathbb{F}_{q^m}^*$ is $q^m - 1$. $\qquad\square$

From the above it follows that the order of a polynomial of degree $m \geq 1$ over $\mathbb{F}_q$ is at most $q^m - 1$. This upper bound is attained for an important class of polynomials, which are based on primitive elements of a finite field.

3.12 Definition. A *primitive element* of $\mathbb{F}_q$ is a generating element of $\mathbb{F}_q^*$. A monic irreducible polynomial of degree m over $\mathbb{F}_q$ is called *primitive* if it is the minimal polynomial of a primitive element of $\mathbb{F}_{q^m}$.

We note the following characterization of primitive polynomials.

3.13 Theorem. *A polynomial $f \in \mathbb{F}_q[x]$ of degree k is primitive over $\mathbb{F}_q$ if and only if f is monic, $f(0) \neq 0$ and the order of f is equal to $q^m - 1$.*

PROOF. If f is primitive over $\mathbb{F}_q$ then f is monic and $f(0) \neq 0$. Now f is irreducible and has as root a primitive element over $\mathbb{F}_{q^m}$, so by Theorem 3.10 ord $f = q^m - 1$. Conversely, it suffices to show that f is irreducible. Suppose on the contrary that f is reducible. We have two cases to consider: Either $f = g_1 g_2$ where g_1 and g_2 are relatively prime polynomials of positive degrees k_1 and k_2, or $f = g$ where $g \in \mathbb{F}_q[x]$, and $g(0) \neq 0$ is irreducible.

In the first case let $e_i = \text{ord } g_i$; then by Lemma 2.8, ord $f \leq e_1 e_2$. By Theorem 3.1, $g | x^{q^{k_i-1}} - 1$ so $e_i \leq q^{k_i} - 1$. Hence ord $f \leq e_1 e_2 \leq (q^{k_1} - 1)(q^{k_2} - 1) < q^{k_1+k_2} - 1 = q^m - 1$, a contradiction.

In the second case, let $e = \text{ord } g$. By Theorem 3.1 and the fact that $g | f$ we have $e | q^m - 1$ so $p | e$ where p is the characteristic of $\mathbb{F}_q$. By Exercise 3.4 $g | x^k - 1$ if $e | k$. So if $k = p^l j$ where $p | j$ we have

$$x^k - 1 = x^{p^l j} - 1 = (x^j - 1)^{p^l}.$$

Since $x^j - 1$ has no repeated roots every irreducible factor of $x^k - 1$ has multiplicity p^l. Let t be the unique integer with $p^{t-1} < b \leq p^t$, then ord $f = ep^t$. But $ep^t \leq (q^n - 1)p^t$ where $n = m/b$, the degree of g. Moreover $ep^t < q^{n+t} - 1$. So $t \leq p^{t-1} \leq b - 1 \leq (b - 1)n$. Now combining these inequalities we have ord $f < q^{n+t} - 1 \leq q^{bn} - 1 = q^m - 1$, a contradiction. Therefore f is irreducible and by Theorem 3.10 the roots of f have order $q^m - 1$, so f is primitive. $\qquad\square$

An important problem is that of the determination of primitive polynomials. One approach is based on the fact that the product of all primitive polynomials over $\mathbb{F}_q$ of degree m is equal to the cyclotomic polynomial Q_e with $e = q^m - 1$. Therefore, all primitive polynomials over $\mathbb{F}_q$ of degree m can be determined by applying one of the factorization algorithms, which we describe below, to the cyclotomic polynomial Q_e.

Another method depends on constructing a primitive element of $\mathbb{F}_{q^m}$ and then determining the minimal polynomial of this element over $\mathbb{F}_q$. To find a primitive element of $\mathbb{F}_{q^m}$, one starts from the order $q^m - 1$ of such an element in the group $\mathbb{F}_{q^m}^*$ and one factorizes it in the form $q^m - 1 = h_1 \ldots h_k$, where the positive integers $h_1, \ldots, h_k$ are pairwise relatively prime. If for each $i, 1 \le i \le k$, one can find an element $\alpha_i \in \mathbb{F}_{q^m}^*$ of order h_i, then the product $\alpha_1 \ldots \alpha_k$ has order $q^m - 1$ and is thus a primitive element of $\mathbb{F}_{q^m}$.

3.14 Example. We determine a primitive polynomial over $\mathbb{F}_3$ of degree 4. Since $3^4 - 1 = 16 \cdot 5$, we first construct two elements of $\mathbb{F}_{81}^*$ of order 16 and 5, respectively. The elements of order 16 are the roots of the cyclotomic polynomial $Q_{16}(x) = x^8 + 1 \in \mathbb{F}_3[x]$. Since the multiplicative order of 3 modulo 16 is 4, Q_{16} factors into two monic irreducible polynomials in $\mathbb{F}_3[x]$ of degree 4. Now

$$x^8 + 1 = (x^4 - 1)^2 - x^4 = (x^4 - 1 + x^2)(x^4 - 1 - x^2),$$

and so $f(x) = x^4 - x^2 - 1$ is irreducible over $\mathbb{F}_3$ and with a root θ of f we have $\mathbb{F}_{81} = \mathbb{F}_3(\theta)$. Furthermore, θ is an element of $\mathbb{F}_{81}^*$ of order 16. It can be verified that $\alpha = \theta + \theta^2$ has order 5. Therefore

$$\zeta = \theta\alpha = \theta^2 + \theta^3$$

has order 80 and is thus a primitive element of $\mathbb{F}_{81}$. The minimal polynomial g of ζ over $\mathbb{F}_3$ is

$$
\begin{aligned}
g(x) &= (x - \zeta)(x - \zeta^3)(x - \zeta^9)(x - \zeta^{27}) = (x - \theta^2 - \theta^3) \\
&\quad \times (x - 1 + \theta + \theta^2)(x - \theta^2 + \theta^3)(x - 1 - \theta + \theta^2) \\
&= x^4 + x^3 + x^2 - x - 1,
\end{aligned}
$$

and we have thus obtained a primitive polynomial over $\mathbb{F}_3$ of degree 4. $\square$

Using the notation of Theorem 3.3 we observe the additional properties of minimal polynomials

(viii) *If $\alpha \ne 0$, then ord M is equal to the order of α in $\mathbb{F}_{q^n}^*$.*

(ix) *M is primitive over $\mathbb{F}_q$ if and only if α is of order $q^d - 1$ in $\mathbb{F}_{q^n}^*$.* $\square$

Next we list some primitive polynomials of degree ≤ 20 over $\mathbb{F}_2$.

Degree	Polynomial
1	$x + 1$
2	$x^2 + x + 1$
3	$x^3 + x + 1$
4	$x^4 + x + 1$
5	$x^5 + x^2 + 1$
6	$x^6 + x + 1$
7	$x^7 + x + 1$
8	$x^8 + x^6 + x^5 + x + 1$
9	$x^9 + x^4 + 1$
10	$x^{10} + x^3 + 1$
11	$x^{11} + x^2 + 1$
12	$x^{12} + x^7 + x^4 + x^3 + 1$
13	$x^{13} + x^4 + x^3 + x + 1$
14	$y^{14} + x^{12} + x^{11} + x + 1$
15	$x^{15} + x + 1$
16	$x^{16} + x^5 + x^3 + x^2 + 1$
17	$x^{17} + x^3 + 1$
18	$x^{18} + x^7 + 1$
19	$x^{19} + x^6 + x^5 + x + 1$
20	$x^{20} + x^3 + 1$

We shall now give a list of some irreducible polynomials $f = a_n x^n + a_{n-1} x^{n-1} + \ldots + a_0$ of degree n over $\mathbb{F}_p$. We abbreviate the polynomials by writing the coefficient vector as $a_n a_{n-1} \ldots a_0$. The column e indicates the order of f.

For $p = 2$:

$n = 1$	e
10	
11	1

$n = 2$	e
111	3

$n = 3$	e
1011	7
1101	7

$n = 4$	e
10011	15
11001	15
11111	5

$n = 5$	e
100101	31
101001	31
101111	31
110111	31
111011	31
111101	31

$n = 6$	e
1000011	63
1001001	9
1010111	21
1011011	63
1100001	63
1100111	63
1101101	63
1110011	63
1110101	21

$n = 7$	e
10000011	127
10001001	127
10001111	127
10010001	127
10011101	127
10100111	127
10101011	127
10111001	127
10111111	127
11000001	127
11001011	127
11010011	127
11010101	127
11100101	127
11101111	127
11110001	127
11110111	127
11111101	127

$n = 8$	e
100011011	51
100011101	255
100101011	255
100101101	255
100111001	17
100111111	85
101001101	255
101011111	255
101100011	255
101100101	255
101101001	255
101110001	255
101110111	85
101111011	85
110000111	255
110001011	85
110001101	255
110011111	51
110100011	85
110101001	255
110110001	51
110111101	85
111000011	255

$n = 8$	e
111001111	255
111010111	17
111011101	85
111100111	255
111110011	51
111110101	255
111111001	85

$n = 9$	e
1000000011	73
1000010001	511
1000010111	73
1000011011	511
1000100001	511
1000101101	511
1000110011	511
1001001011	73
1001011001	511
1001011111	511
1001100101	73

$n = 9$	e
1001101001	511
1001101111	511
1001110111	511
1001111101	511
1010000111	511
1010010101	511
1010011001	73
1010100011	511
1010100101	511
1010101111	511
1010110111	511
1010111101	511
1011001111	511
1011010001	511
1011011011	511
1011110101	511
1011111001	511
1100000001	73

$n = 9$	e
1100010011	511
1100010101	511
1100011111	511
1100100011	511
1100110001	511
1100111011	511
1101001001	73
1101001111	511
1101011011	511
1101100001	511
1101101011	511
1101101101	511
1101110011	511
1101111111	511
1110000101	511
1110001111	511
1110100001	73
1110110101	511
1110111001	511
1111000111	511
1111001011	511
1111001101	511
1111010101	511
1111011001	511
1111100011	511
1111101001	511
1111111011	511

For $p = 3$:

$n = 1$	e
10	
11	2
12	1

$n = 2$	e
101	4
112	8
122	8

$n = 3$	e
1021	26
1022	13
1102	13
1112	13
1121	26
1201	26
1211	26
1222	13

$n = 4$	e
10012	80
10022	80
10102	16
10111	40
10121	40
10202	16
11002	80
11021	20
11101	40
11111	5
11122	80
11222	80
12002	80
12011	20
12101	40
12112	80
12121	10
12212	80

$n = 5$	e
100021	242
100022	121
100112	121
100211	242
101011	242
101012	121
101102	121
101122	121
101201	242
101221	242

$n = 5$	e
102101	242
102112	121
102122	11
102202	121
102211	242
102221	22
110002	121
110012	121
110021	242
110101	242
110111	242
110122	121
111011	242
111121	242
111211	242
111212	121
112001	242
112022	121
112102	11
112111	242
112201	242
112202	121
120001	242
120011	242
120022	121
120202	121
120212	121
120221	242
121012	121
121111	242
121112	121
121222	121
122002	121
122021	242
122101	242
122102	121
122201	22
122212	121

For $p = 5$:

$n = 1$	e
10	
11	2
12	4
13	4
14	1

$n = 2$	e
102	8
103	8
111	3
112	24
123	24
124	12
133	24
134	12
141	6
142	24

$n = 3$	e
1011	62
1014	31
1021	62
1024	31
1032	124
1033	124
1042	124
1043	124
1101	62
1102	124
1113	124
1114	31
1131	62
1134	31
1141	62
1143	124
1201	62
1203	124
1213	124

For $p = 7$:

$n = 3$	e
1214	31
1222	124
1223	124
1242	124
1244	31
1302	124
1304	31
1311	62
1312	124
1322	124
1323	124
1341	62
1343	124
1403	124
1404	31
1411	62
1412	124
1431	62
1434	31
1442	124
1444	31

$n = 1$	e
10	
11	2
12	6
13	8
14	6
15	3
16	1

$n = 2$	e
101	4
102	12
104	12
113	48
114	24
116	16
122	24

$n = 2$	e
123	48
125	48
131	8
135	48
136	16
141	8
145	48
146	16
152	24
153	48
155	48
163	48
164	24
166	16

3.15 Theorem. *There are exactly $\phi(e)/m$ irreducible polynomials of degree m and order e over $\mathbb{F}_q$.*

PROOF. We have $e|(q^m - 1)$ and $e \nmid (q^k - 1)$, for $k < m$, since $\mathbb{F}_{q^m}$ is the smallest extension field of $\mathbb{F}_q$ which contains all orders of an irreducible polynomial of degree m. Since $(x^e - 1)|(x^{q^{m-1}} - 1)$, $\mathbb{F}_{q^m}$ contains all eth roots of unity, and thus contains all $\phi(e)$ primitive eth roots of unity. Let α be such a root; then $\alpha, \alpha^q, \ldots, \alpha^{q^{m-1}}$ are distinct and the minimal polynomial of a primitive eth root of unity has degree m; see 3.3. □

3.16 Theorem. $f = (x^e - 1)/(x - 1) = x^{e-1} + x^{e-2} + \ldots + x + 1$ *is irreducible over $\mathbb{F}_q$ if and only if e is a prime and q is a primitive $(e - 1)$th root of unity modulo e.* □

The proof is left as an exercise.
The following theorem gives a formula for the product of all monic, irreducible polynomials over $\mathbb{F}_q$.

3.17 Theorem. *The product $I(q, n; x)$ of all monic, irreducible polynomials in $\mathbb{F}_q[x]$ of degree n is given by*

$$I(q, n; x) = \prod_{d|n} (x^{q^d} - x)^{\mu(n/d)} = \prod_{d|n} (x^{q^{n/d}} - x)^{\mu(d)}.$$

PROOF. Theorem 3.2 implies $x^{q^n} - x = \prod_{d|n} I(q, d; x)$. Let $f(n) = I(q, n; x)$ and $g(n) = x^{q^n} - x$ for all $n \in \mathbb{N}$ in Theorem 2.19. The multiplicative Möbius inversion formula then yields the result. □

3.18 Example. Let $q = 2$ and $n = 4$.

$$I(2, 4; x) = \frac{x^{16} - x}{x^4 - x} = x^{12} + x^9 + x^6 + x^3 + 1$$

and

$$x^{16} - x = x^{2^4} - x = I(2, 1; x)I(2, 2; x)I(2, 4; x)$$
$$= (x^2 - x)(x^2 + x + 1)(x^{12} + x^9 + x^6 + x^3 + 1). \quad □$$

It can be shown that the following partial factorization of $I(q, n; x)$ holds.

3.19 Theorem. *For $n > 1$ we have $I(q, n; x) = \prod_m Q_m$ where Q_m is the mth cyclotomic polynomial over $\mathbb{F}_q$. The product is extended over all positive divisors m of $q^n - 1$ such that n is the multiplicative order of q modulo m.* □

PROBLEMS

1. Construct $\mathbb{F}_{2^4}$ by using $f(x) = x^4 + x^3 + x^2 + x + 1$, which is irreducible over $\mathbb{F}_2$. Let $f(\alpha) = 0$ and show that α is not a primitive element, but that $\alpha + 1$ is a primitive element. Find the minimal polynomial of $\alpha + 1$.

2. Construct the finite field $\mathbb{F}_{2^{20}}$ by using the primitive polynomial $f(x) = x^{20} + x^3 + 1$ over $\mathbb{F}_2$. Determine all subfields of $\mathbb{F}_{2^{20}}$.

*3. Prove that $x^{p^n} - x + na$ is always divisible by $x^p - x + a$ over $\mathbb{F}_p$, where $a \in \mathbb{F}_p$, p a prime.

*4. Determine primitive polynomials of degrees 21, 22, 23 and 24 over $\mathbb{F}_2$.

5. Show that for every finite field $\mathbb{F}_q$ and positive integer m there exists at least one primitive polynomial over $\mathbb{F}_q$ of degree m.

6. Determine the order of $x^8 + x^7 + x^3 + x + 1$ over $\mathbb{F}_2$.

*7. Let $Q_e(x)$ be the cyclotomic polynomial. Prove that $\text{ord}(Q_e(x)) = e$ for all those e for which $Q_e \in \mathbb{F}_q[x]$ is defined.

8. Calculate $I(3, 4; x)$.

9. Factor $g \in \mathbb{F}_3[x]$ of Example 3.11 in $\mathbb{F}_9[x]$ to obtain primitive polynomials over $\mathbb{F}_9$.

10. Is $x^{28} + x^3 + 1$ a primitive polynomial over $\mathbb{F}_2$?

*11. Determine the order of $x^{27} + x^5 + x^2 + x + 1$ over $\mathbb{F}_2$.

EXERCISES (Solutions in Chapter 8, p. 455)

1. Show that the three irreducible polynomials of degree 4 over $\mathbb{F}_2$ are $x^4 + x + 1$, $x^4 + x^3 + 1$, $x^4 + x^3 + x^2 + x + 1$.

*2. Prove Theorem 3.3.

3. Find the first prime p such that $(x^{19} - 1)(x - 1)^{-1}$ is irreducible over $\mathbb{F}_p$.

4. Prove that the polynomial $f \in \mathbb{F}_q[x]$, with $\bar{f}(0) \neq 0$, divides $x^m - 1$ if an only if the order of f divides the positive integer m.

*5. Prove Theorem 3.16.

*6. Prove Theorem 3.19.

7. Calculate $I(2, 6; x)$ from Theorem 3.17.

8. Calculate $I(2, 6; x)$ from Theorem 3.19.

*9. Let $g_1, \ldots, g_k$ be pairwise relatively prime nonzero polynomials over $\mathbb{F}_q$, and let $f = g_1 \cdot \ldots \cdot g_k$. Prove that ord f is equal to the least common multiple of ord $g_1, \ldots,$ ord g_k. (Use Exercise 4.)

10. Determine the order of $(x^2 + x + 1)^3(x^3 + x + 1)$ over $\mathbb{F}_2$.

11. Given that $f = x^4 + x^3 + x^2 + 2x + 2$ in $\mathbb{F}_3[x]$ is irreducible over $\mathbb{F}_3$ and ord $f = 80$, show that f is primitive over $\mathbb{F}_3$.

§4. Factorization of Polynomials over Finite Fields

A method for the determination of the complete factorization of a polynomial over $\mathbb{F}_q$ into a product of irreducible factors will be given. Such factorizations are useful within mathematics in the factorization of polynomials over $\mathbb{Z}$, in the determination of the Galois group of an equation over $\mathbb{Z}$, and in factorizing elements of finite algebraic number fields into prime elements. Some applications of factorizations of polynomials will be considered in Chapter 4.

The problem of factorizing a given polynomial into a product of irreducible polynomials over the integers is rather old. L. Kronecker gave an algorithm for the explicit determination of such a factorization. However Kronecker's approach is not very economical, since the calculating time for this algorithm increases exponentially with the degree of the given polynomial. More recently so-called *homomorphism methods* have been developed to find the factorization of a given polynomial over $\mathbb{F}_q$. We shall describe an algorithm due to BERLEKAMP. We want to express a given monic polynomial $f \in \mathbb{F}_q[x]$ in the form

$$f = f_1^{e_1} \ldots f_r^{e_r},$$

where f_i are distinct, monic, irreducible polynomials over $\mathbb{F}_q$.

First we want to find out if f has multiple factors, i.e. if $e_i > 1$ for some or all of the r exponents. Suppose $f = f_1^2 f_2$; then the (formal) derivative in $\mathbb{F}_q$ is

$$f' = 2 f_1 f_1' f_2 + f_1^2 f_2',$$

and thus f' is a multiple of f_1. Therefore we form

$$\gcd(f, f') = d$$

before starting with the factorization itself. If $d = 1$ then f is squarefree. If $d \neq 1$ and $d \neq f$ then we factorize d as well as f/d. If $d = f$, then f is of the form $f = g \circ x^p = g^p$, for a suitable $g \in \mathbb{F}_q[x]$ where $p = \operatorname{char} \mathbb{F}_q$. In this case we have to factorize g.

Therefore we consider the factorization of a monic, squarefree polynomial $f \in \mathbb{F}_q[x]$. Berlekamp's algorithm is based on some important properties: First we need a generalization of the Chinese remainder theorem for integers to the case of polynomials over $\mathbb{F}_q$. The Chinese remainder theorem is one of the oldest results in number theory, which has recently found applications in computer programming with fast adders. The old result was used by the ancient Chinese to predict the common period of several astronomical cycles.

4.1 Theorem ("Chinese Remainder Theorem for Polynomials"). *Let $f_1, \ldots, f_r$ be distinct, irreducible polynomials over $\mathbb{F}_q$ and let $g_1, \ldots, g_r$ be arbitrary polynomials over $\mathbb{F}_q$. Then the system of congruences $h \equiv g_i \pmod{f_i}$, $i = 1, 2, \ldots, r$ has a unique solution h modulo $f_1 f_1 \ldots f_r$.*

PROOF. We use the Euclidean algorithm to determine polynomials m_i such that $m_i \prod_{j \neq i} f_j \equiv 1 \pmod{f_i}$. Put $h := \sum_i (g_i m_i \prod_{j \neq i} f_j)$; then $h \equiv g_i \pmod{f_i}$, for all i. If H were also a solution of the system of congruences, then $H - h$ would be divisible by each f_i and therefore $H \equiv h \pmod{\prod_i f_i}$. $\qquad\square$

This theorem is equivalent to the statement:

$$\mathbb{F}_q[x]/(f_1 f_2 \ldots f_r) \cong \bigoplus_{i=1}^r \mathbb{F}_q[x]/(f_i),$$

where $[g]_{f_1 \ldots f_r} \mapsto ([g]_{f_1}, \ldots, [g]_{f_r})$ establishes the isomorphism. In trying to factorize f into the distinct, monic, irreducible divisors f_i over $\mathbb{F}_q$, Theorem 4.1 says that we can find the divisors of f if we can find the polynomial h and the remainders g_i, because

$$f_i = \gcd(h - g_i, f) \quad \text{and} \quad \deg h < \deg \prod f_i = \deg f.$$

If $h \in \mathbb{F}_q[x]$ is such that $h \equiv s_i \pmod{f_i}$, $s_i \in E_q$, then

$$h^q \equiv s_i^q = s_i \equiv h \pmod{f_i}.$$

The multinomial theorem in $\mathbb{F}_q[x]$ yields $h^q = h \circ x^q$ and thus

$$h \circ x^q - h \equiv 0 \pmod{f_i}.$$

Here $\circ$ denotes the substitution of polynomials; so $h \circ x^q = h(x^q)$. Since f_i divides f, it is sufficient to find polynomials h with the property

$$h \circ x^q - h \equiv 0 \pmod{f}. \qquad (*)$$

In order to find such polynomials we construct the $n \times n$ matrix ($n = \deg f$)

$$\mathbf{Q} = \begin{pmatrix} q_{00} & q_{01} & \cdots & q_{0n-1} \\ \vdots & \vdots & & \vdots \\ q_{n-10} & q_{n-11} & \cdots & q_{n-1\,n-1} \end{pmatrix}$$

such that the kth row is given as the coefficient vector of the congruence $x^{qk} \equiv q_{k,n-1}x^{n-1} + \ldots + q_{k1}x + q_{k0} \pmod{f}$.

A polynomial $h = \sum_{i=0}^{n-1} v_i x^i$ is a solution of $(*)$ if and only if

$$(v_0, v_1, \ldots, v_{n-1})\mathbf{Q} = (v_0, v_1, \ldots, v_{n-1}).$$

This follows because

$$h = \sum_i v_i x^i = \sum_i \sum_k v_k q_{ki} x^i = \sum_k v_k x^{qk} = h \circ x^q$$
$$\equiv h^q \pmod{f}.$$

The determination of a polynomial h satisfying $(*)$ can be regarded as solving the system of linear equations $\mathbf{v}(\mathbf{Q} - \mathbf{I}) = \mathbf{0}$, where $\mathbf{v}$ is the coefficient vector of h, $\mathbf{Q} = (q_{ki})$, $\mathbf{I}$ is the $n \times n$ identity matrix and $\mathbf{0}$ is the n-dimensional zero vector. Finding a suitable h is therefore equivalent to determining the null space of $\mathbf{v}(\mathbf{Q} - \mathbf{I}) = \mathbf{0}$. This can be done by using one of the *null space algorithms* (see §5). We use the notation above to prove

4.2 Theorem. (i) $f = \prod_{s \in \mathbb{F}_q} \gcd(f, h - s)$ *is a factorization of f.*

(ii) *The number of distinct irreducible factors f_i of f is equal to the dimension of the null space of the matrix $\mathbf{Q} - \mathbf{I}$.*

PROOF. (i) Since $h^q - h \equiv 0 \pmod{f}$, f is a divisor of $h^q - h = \prod_{s \in \mathbb{F}_q} (h - s)$. Therefore f divides $\prod_{s \in \mathbb{F}_q} \gcd(f, h - s)$. On the other hand, $\gcd(f, h - s)$ divides f. If $s \neq t \in \mathbb{F}_q$, then $h - s$ and $h - t$ are relatively prime, and so are $\gcd(f, h - s)$ and $\gcd(f, h - t)$. Thus $\prod_{s \in \mathbb{F}_q} \gcd(f, h - s)$ divides f. Since both polynomials are monic, they must be equal.

(ii) f divides $\prod_{s \in \mathbb{F}_q} (h - s)$ if and only if each f_i divides $h - s_i$ for some $s_i \in \mathbb{F}_q$. Given $s_1, \ldots, s_r \in \mathbb{F}_q$, then Theorem 4.1 implies the existence of a unique polynomial $h \pmod{f}$, such that $h \equiv s_i \pmod{f_i}$. We have the choice of q^r elements s_i, therefore we have exactly q^r solutions of $h^q - h \equiv 0 \pmod{f}$. We noted above that h is a solution of $(*)$ if and only if

$$(v_0, v_1, \ldots, v_{n-1})\mathbf{Q} = (v_0, v_1, \ldots, v_{n-1}),$$

or equivalently

$$(v_0, v_1, \ldots, v_{n-1})(\mathbf{Q} - \mathbf{I}) = (0, 0, \ldots, 0).$$

This system has q^r solutions. Thus, the dimension of the null space of the matrix $\mathbf{Q} - \mathbf{I}$ is r, the number of distinct monic irreducible factors of f, and the rank of $\mathbf{Q} - \mathbf{I}$ is $n - r$. $\qquad\square$

Let k be the rank of the matrix $\mathbf{Q} - \mathbf{I}$. We have $k = n - r$, so that once the rank k is found, we know that the number of distinct monic irreducible factors of f is given by $n - k$. On the basis of this information we can then decide when the factorization procedure can be stopped. The rank of $\mathbf{Q} - \mathbf{I}$ can be determined by using row and column operations to reduce the matrix to echelon form. However, it is advisable to use only column operations because they leave the null space invariant. Thus, we are allowed to multiply any column of the matrix $\mathbf{Q} - \mathbf{I}$ by a nonzero element of $\mathbb{F}_q$ and to add any multiple of one of its columns to a different column. The rank k is the number of nonzero columns in the column echelon form.

After finding k, we form $r = n - 1$. If $r = 1$, we know that f is irreducible over $\mathbb{F}_q$ and the procedure terminates. In this case the only solutions of (∗) are the constant polynomials and the null space of $\mathbf{Q} - \mathbf{I}$ contains only the vectors of the form $(c, 0, \ldots, 0)$ with $c \in \mathbb{F}_q$. If $r \geq 2$, we take the polynomial h_2 and calculate $\gcd(f, h_2 - s)$ for all $s \in \mathbb{F}_q$. The result will be a nontrivial factorization of $f(x)$ obtained by 4.2(i). If the use of h_2 does not succeed in splitting f into r factors we calculate $\gcd(g, h_3 - s)$ for all $s \in \mathbb{F}_q$ and all nontrivial factors g found so far. This procedure is continued until r factors of f are obtained.

The process described above must eventually yield all the factors. For if we consider two distinct monic irreducible factors of f, say f_1 and f_2, then by the argument above, there exist elements $s_{j1}, s_{j2} \in \mathbb{F}_q$ such that $h_j \equiv s_{j1} \pmod{f_1}$, $h_j \equiv s_{j2} \pmod{f_2}$ for $1 \leq j \leq r$. Suppose we had $s_{j1} = s_{j2}$ for $1 \leq j \leq r$. Then, since any solution h of (∗) is a linear combination of $h_1, \ldots, h_r$ with coefficients in $\mathbb{F}_q$, there would exist for any such h an element $s \in \mathbb{F}_q$ with $h \equiv s \pmod{f_1}$, $h \equiv s \pmod{f_2}$. But there is a solution $h^q \equiv h \pmod{f}$ with $h \equiv 0 \pmod{f_1}$, $h \equiv 1 \pmod{f_2}$. This contradiction proves that $s_{j1} \neq s_{j2}$ for some j with $1 \leq j \leq r$ (in fact, since $h_1 = 1$, we will have $j \geq 2$). Therefore $h_j - s_{j1}$ will be divisible by f_1, but not by f_2. Hence any two distinct monic irreducible factors of f will be separated by h_j.

This approach represents also an irreducibility test, since we see that f is irreducible over $\mathbb{F}_q$ if and only if the dimension of the null space of the matrix $\mathbf{Q} - \mathbf{I}$ is 1. In this case the only solution of $h^q - h \equiv 0 \pmod{f}$ is given by the elements in $\mathbb{F}_q$ and the null space by the vectors $(s, 0, \ldots, 0)$. If the null space has dimension r, then there exists a basis with r monic polynomials $h^{(1)}, \ldots, h^{(r)}$. We summarize these results.

4.3 Theorem (Berlekamp's Algorithm). *Let $f \in \mathbb{F}_q[x]$ be monic of degree n.*
Step 1. Check that f is squarefree, i.e. $\gcd(f, f') = 1$.
Step 2. Form the $n \times n$ matrix $\mathbf{Q} = (q_{ki})$, defined by

$$x^{qk} \equiv \sum_{i=0}^{n-1} q_{ki}x^i \pmod{f}, \qquad 0 \le k \le n-1$$

*Step 3. Find the null space of $\mathbf{Q} - \mathbf{I}$, determine the rank $n - r$ and find r
linearly independent vectors $\mathbf{v}^{(1)}, \ldots, \mathbf{v}^{(r)}$ such that $\mathbf{v}^{(i)}(\mathbf{Q} - \mathbf{I}) = \mathbf{0}$ for $i =
1, 2, \ldots, r$. The integer r is the number of irreducible factors of f. If $r = 1$,
stop. If $r > 1$ go to Step 4.*
*Step 4. Calculate $\gcd(f, h^{(2)} - s)$ for all $s \in \mathbb{F}_q$, where $h^{(2)} := \sum_i v_i^{(2)} x^i$. This
yields a nontrivial decomposition of f into a product of (not necessarily
irreducible) factors. If $\mathbf{v}^{(2)}$, that is $h^{(2)}$ does not give all r factors of f then we
can obtain further factors by computing $\gcd(h^{(k)} - s, w)$, for all $s \in \mathbb{F}_q$, for all
divisors w of f and for $k = 2, 3, 4, \ldots, n - 1$. Here $h^{(k)} := \sum v_i^{(k)} x^i$. The
algorithm ends when all r irreducible factors of f are found.* ☐

4.4 Example. We want to determine the complete factorization of

$$f_1 = x^8 + x^6 + 2x^4 + 2x^3 + 3x^2 + 2x$$

over $\mathbb{F}_5$. We put $f_1 = xf$ and verify that $\gcd(f, f') = 1$. Next we evaluate the
matrix $\mathbf{Q}$, in this case a 7×7 matrix. We have to find $x^{5k} \pmod{f}$, for
$k = 0, 1, 2, \ldots, 6$. The first row of $\mathbf{Q}$ is $(1, 0, 0, 0, 0, 0, 0)$, since $x^0 \equiv
1 \pmod{f}$. We give a systematic procedure for calculating the rows of $\mathbf{Q}$.
In general, let

$$f = x^n + a_{n-1}x^{n-1} + \ldots + a_0$$

and let

$$x^m \equiv a_{m0} + a_{m1}x + \ldots + a_{mn-1}x^{n-1} \pmod{f}.$$

Then

$$x^{m+1} \equiv a_{m0}x + a_{m1}x^2 + \ldots + a_{mn-1}x^n$$
$$\equiv a_{m0}x + a_{m1}x^2 + \ldots + a_{mn-2}x^{n-1}$$
$$+ a_{mn-1}(-a_{n-1}x^{n-1} - \ldots - a_1x - a_0)$$
$$\equiv a_{m+10} + a_{m+11}x + \ldots + a_{m+1\,n-1}x^{n-1} \pmod{f}$$

where

$$a_{m+1,j} = a_{mj-1} - a_{mn-1}a_j,$$
$$a_{m,-1} = 0.$$

Thus we tabulate

m	a_{m0}	a_{m1}	a_{m2}	a_{m3}	a_{m4}	a_{m5}	a_{m6}
0	1	0	0	0	0	0	0
1	0	1	0	0	0	0	0
2	0	0	1	0	0	0	0
3	0	0	0	1	0	0	0
4	0	0	0	0	1	0	0
5	0	0	0	0	0	1	0
6	0	0	0	0	0	0	1
7	3	2	3	3	0	4	0
8	0	3	2	3	3	0	4
9	2	3	0	4	3	4	0
10	0	2	3	0	4	3	4
11	2	3	4	0	0	0	3
12	4	3	2	3	0	2	0
13	0	4	3	2	3	0	2
14	1	4	0	4	2	1	0
15	0	1	4	0	4	2	1
16	3	2	4	2	0	3	2
17	1	2	3	0	2	3	3
18	4	2	1	2	0	4	3
19	4	0	1	0	2	2	4
20	2	2	2	3	0	3	2
21	1	1	3	3	3	3	3
22	4	2	0	2	3	0	3
23	4	0	1	4	2	0	0
24	0	4	0	1	4	2	0
25	0	0	4	0	1	4	2
26	1	4	1	0	0	4	4
27	2	4	1	3	0	1	4
28	2	0	1	3	3	1	1
29	3	4	3	4	3	2	1
30	3	0	2	1	4	2	2

Therefore the matrix $\mathbf{Q}$ is given as

$$\mathbf{Q} = \begin{pmatrix} 1 & 0 & 0 & 0 & 0 & 0 & 0 \\ 0 & 0 & 0 & 0 & 0 & 1 & 0 \\ 0 & 2 & 3 & 0 & 4 & 3 & 4 \\ 0 & 1 & 4 & 0 & 4 & 2 & 1 \\ 2 & 2 & 2 & 3 & 0 & 3 & 2 \\ 0 & 0 & 4 & 0 & 1 & 4 & 2 \\ 3 & 0 & 2 & 1 & 4 & 2 & 2 \end{pmatrix},$$

so that

$$\mathbf{Q} - \mathbf{I} = \begin{pmatrix} 0 & 0 & 0 & 0 & 0 & 0 & 0 \\ 0 & 4 & 0 & 0 & 0 & 1 & 0 \\ 0 & 2 & 2 & 0 & 4 & 3 & 4 \\ 0 & 1 & 4 & 4 & 4 & 2 & 1 \\ 2 & 2 & 2 & 3 & 4 & 3 & 2 \\ 0 & 0 & 4 & 0 & 1 & 3 & 2 \\ 3 & 0 & 2 & 1 & 4 & 2 & 1 \end{pmatrix}.$$

Step 3 requires that we find the null space of $\mathbf{Q} - \mathbf{I}$. Example 5.2 in §5 deals with the matrix $\mathbf{A} = \mathbf{Q} - \mathbf{I}$ and gives as a result of the null space algorithm the two linearly independent vectors

$$\mathbf{v}^{(1)} = (1, 0, 0, 0, 0, 0, 0) \quad \text{and} \quad \mathbf{v}^{(2)} = (0, 3, 1, 4, 1, 2, 1).$$

This means that f has 2 irreducible factors.

In step 4 we calculate $\gcd(f, h^{(2)} - s)$ for all $s \in \mathbb{F}_5$, where

$$h^{(2)} = \mathbf{v}^{(2)} = x^6 + 2x^5 + x^4 + 4x^3 + x^2 + 3x.$$

The result is

$$g = x^5 + 2x^4 + x^3 + 4x^2 + x + 3 \quad \text{for } s = 0.$$

The second factor of f is obtained by division, and turns out to be $x^2 + 3x + 4$. Thus the complete factorization of $f_1 = xf$ over $\mathbb{F}_5$ is of the form $f_1 = x(x^2 + 3x + 4)(x^5 + 2x^4 + x^3 + 4x^2 + x + 3)$. $\qquad\square$

4.5 Example. We want to find the factorization of

$$f = x^8 + x^6 + x^4 + x^3 + 1 \quad \text{over } \mathbb{F}_2.$$

We verify that $\gcd(f, f') = 1$, thus f does not have any multiple factors. The 8×8 matrix $\mathbf{Q}$ can be obtained by means of a recursion. In general, for $\deg f = n$, we have

$$a_{k+1, j} = a_{k, j-1} - a_{k, n-1} f_j,$$

for the coefficients of $x^{2r} \pmod{f}$, where $f = \sum f_j x^j$ and $a_{k,-1} = 0$. Thus in our example

$$f_0 = f_3 = f_4 = f_6 = f_8 = 1, \qquad f_1 = f_2 = f_5 = f_7 = 0.$$

We obtain

k	a_{k0}	a_{k1}	a_{k2}	a_{k3}	a_{k4}	a_{k5}	a_{k6}	a_{k7}
0	1	0	0	0	0	0	0	0
1	0	1	0	0	0	0	0	0
2	0	0	1	0	0	0	0	0
3	0	0	0	1	0	0	0	0
4	0	0	0	0	1	0	0	0
5	0	0	0	0	0	1	0	0
6	0	0	0	0	0	0	1	0
7	0	0	0	0	0	0	0	1
8	1	0	0	1	1	0	1	0
9	0	1	0	0	1	1	0	1
10	1	0	1	1	1	1	0	0
11	0	1	0	1	1	1	1	0
12	0	0	1	0	1	1	1	1
13	1	0	0	0	1	1	0	1
14	1	1	0	1	1	1	0	0

Therefore the matrix $\mathbf{Q}$ consists of the rows for $k = 0, 2, 4, 6, 8, 10, 12, 14$ and is of the form

$$\mathbf{Q} = \begin{pmatrix} 1 & 0 & 0 & 0 & 0 & 0 & 0 & 0 \\ 0 & 0 & 1 & 0 & 0 & 0 & 0 & 0 \\ 0 & 0 & 0 & 0 & 1 & 0 & 0 & 0 \\ 0 & 0 & 0 & 0 & 0 & 0 & 1 & 0 \\ 1 & 0 & 0 & 1 & 1 & 0 & 1 & 0 \\ 1 & 0 & 1 & 1 & 1 & 1 & 0 & 0 \\ 0 & 0 & 1 & 0 & 1 & 1 & 1 & 1 \\ 1 & 1 & 0 & 1 & 1 & 1 & 0 & 0 \end{pmatrix};$$

$$\mathbf{Q} - \mathbf{I} = \begin{pmatrix} 0 & 0 & 0 & 0 & 0 & 0 & 0 & 0 \\ 0 & 1 & 1 & 0 & 0 & 0 & 0 & 0 \\ 0 & 0 & 1 & 0 & 1 & 0 & 0 & 0 \\ 0 & 0 & 0 & 1 & 0 & 0 & 1 & 0 \\ 1 & 0 & 0 & 1 & 0 & 0 & 1 & 0 \\ 1 & 0 & 1 & 1 & 1 & 0 & 0 & 0 \\ 0 & 0 & 1 & 0 & 1 & 1 & 0 & 1 \\ 1 & 1 & 0 & 1 & 1 & 1 & 0 & 1 \end{pmatrix}.$$

The null space of $\mathbf{Q} - \mathbf{I}$ will be calculated in 5.9. From there we have $r = 2$ and

$$\mathbf{v}^{(2)} = x^7 + x^6 + x^5 + x^2 + x.$$

Now we can perform Step 4 in 4.3 and calculate $\gcd(f, \mathbf{v}^{(2)} - 1)$ which yields $x^2 + x + 1$ as an irreducible factor of f. The second factor is $x^6 + x^5 + x^4 + x + 1$.   □

Some additional remarks concerning Berlekamp's algorithm: It may be advantageous to find simple factors of the given polynomial by trial-and-error methods before applying more complicated methods such as 4.3.

If the order of the underlying finite field is large, then the following approach can be used to determine the matrix $\mathbf{Q}$ in Step 2 of 4.3. First we determine an auxiliary table consisting of all coefficients of $x^m \pmod{f}$, for $m = n, n + 1, \ldots, 2n - 2$, where $n = \deg f$. If

$$x^k \equiv c_{n-1} x^{n-1} + \ldots + c_1 x + c_0 \pmod{f},$$

then

$$x^{2k} \equiv c_{n-1}^2 x^{2n-2} + \ldots + (c_1 c_0 + c_0 c_1)x + c_0^2 \pmod{f},$$

where the powers $x^{2n-2}, \ldots, x^n$ can be replaced by combinations of powers of lower degrees from the auxiliary table. In this way $x^q \pmod{f}$ and thus the second row of $\mathbf{Q}$ can be obtained. The remaining rows of $\mathbf{Q}$ are calculated by repeated multiplication of $x^q \pmod{f}$ by itself modulo f.

Step 4 in Berlekamp's algorithm clearly shows that practical calculations are impossible if q is large, since this step requires the computation of $\gcd(f, h - s)$ for all $s \in \mathbb{F}_q$. Berlekamp and Knuth have shown that some of the computations can be performed for all $s \in \mathbb{F}_q$ at once, but only for polynomials of degrees up to $n/2$. In this method the element s remains in the calculations as a parameter. Factorization methods for "large" finite fields are based on work by Berlekamp; see, e.g. LIDL and NIEDERREITER, Chapter 4.

In several applications the factorization of $f = x^n - 1$ over $\mathbb{F}_q$ where $(n, q) = 1$ is of particular importance.

Often it is advantageous to start with the cyclotomic decomposition

$$x^n - 1 = \prod_{d|n} Q_d(x)$$

and then to use the polynomials $h^{(i)}$ to factorize the cyclotomic polynomials $Q_d(x)$.

Alternatively, we note that Step 2 of 4.3 is simpler in this case, namely

$$q_{ij} = 1 \quad \text{if } qi \equiv j \,(\text{mod } n),$$

$$q_{ij} = 0 \quad \text{if } qi \not\equiv j \,(\text{mod } n).$$

A basis for the null space of $Q - I$ can be obtained by choosing $h^{(k)}$ as the coefficient vector of $h = \sum_{k \in K} x^k$ in Step 4 of 4.3, where K is a set of integers closed with respect to multiplication by $q \,(\text{mod } n)$. Any such polynomial h has a nontrivial common factor with $x^n - 1$, which can be determined by Euclid's algorithm. Repetition of this procedure for different polynomials h eventually gives the factorization of f into irreducible factors.

This method of factorizing $x^n - 1$ is based on a partition of the integers $0, 1, \ldots, n - 1$ and uses so-called cyclotomic cosets. We follow PLESS.

4.6 Lemma. *Let f be a polynomial over $\mathbb{F}_p$ and let α be a root of f of order n in a suitable extension field. Let r be the smallest integer such that $p^r \equiv 1 \,(\text{mod } n)$. Then $\alpha, \alpha^p, \ldots, \alpha^{p^{r-1}}$ are all distinct roots of f.*

PROOF. All $\alpha^{p^i}, 0 \le i \le r - 1$, are roots of f. Suppose $i > j$ and that $\alpha^{p^i} = \alpha^{p^j}$ for some i, j, then $\alpha^{p^i - p^j} = 1$. This holds if and only if $p^i \equiv p^j \,(\text{mod } n)$; that is, $p^{i-j} \equiv 1 \,(\text{mod } n)$, which is true if and only if $i - j$ is a multiple of r. $\qquad\square$

Let r be the smallest integer such that $p^r s \equiv s \,(\text{mod } p^m - 1)$, for any s so that $0 \le s < p^m - 1$. Then we define the *cyclotomic coset* C_s containing s by

$$C_s := \{s, ps, p^2 s, \ldots, p^{r-1} s\},$$

where each $p^i s$ is reduced modulo $p^m - 1$. If $\gcd(s, p^m - 1) = 1$, then $r = m$, but if $\gcd(s, p^m - 1) \ne 1$, then r varies with s.

4.7 Example. To demonstrate Lemma 4.6 and the definition of a cyclotomic coset, we consider the polynomial $f = x^4 + x + 1$ over $\mathbb{F}_2$. f has roots α, α^2, $\alpha^{2^2}, \alpha^{2^3}$ and $\alpha^{2^4} = \alpha$ in $\mathbb{F}_{2^4}$. The order of α is 15 and, obviously, 4 is the smallest integer such that $2^4 \equiv 1 \,(\text{mod } 15)$. We consider the set of numbers that appear as powers of α for the minimal polynomial $f = x^4 + x + 1$. Here $C_1 = \{1, 2, 4, 8\}$. We find for the minimal polynomials of $\mathbb{F}_{2^4}$ the following sets of numbers that appear as powers of α:

$$x^4 + x^3 + 1, \qquad\qquad C_7 = \{7, 17, 13, 11\},$$
$$x^4 + x^3 + x^2 + x + 1, \qquad C_3 = \{3, 6, 12, 9\},$$
$$x^2 + x + 1, \qquad\qquad C_5 = \{5, 10\}.$$

$C_0 = \{0\}$, C_1, C_3, C_5, C_7 are all the cyclotomic cosets for $2^4 - 1 = 15$. $\quad\square$

The cyclotomic cosets partition the integers $(\bmod\ p^m - 1)$, that is,

$$\{0, 1, \ldots, p^m - 2\} = \bigcup_s C_s,$$

where s runs through a system of representatives for the cosets modulo $p^m - 1$. It is convenient to denote a coset by C_s, if s is its smallest element.

4.8 Theorem. *Let α be an element in $\mathbb{F}_{p^m}$ and let $m(x)$ be its minimal polynomial. Let ζ be a primitive element in $\mathbb{F}_{p^m}$ and let $\alpha = \zeta^i$. If s is the smallest element in the cyclotomic coset containing i, then*

$$m(x) = \prod_{j \in C_s} (x - \zeta^j).$$

PROOF. If α is primitive, its order is $p^m - 1$, and we can set $i = 1$. Let $m_1 = \prod_{j \in C_1} (x - \alpha^j)$. According to Lemma 4.6, $\alpha, \alpha^p, \ldots, \alpha^{p^{r-1}}$ are all distinct roots of m, and so m_1 divides m. We note that the coefficients of m are elementary symmetric functions of the form $\alpha^s + \alpha^{ps} + \ldots + \alpha^{p^{r-1}s}, \ldots, \alpha^{s+ps+\ldots+p^{r-1}s}$, where each of these coefficients is in $\mathbb{F}_p$. Hence $m = m_1$. If α is not a primitive element, then the order of α in $\mathbb{F}_{p^m}$ is $n = (p^m - 1)/i$, for some i. Let s be the smallest element in the cyclotomic coset containing i and let $m_1 = \prod_{j \in C_s} (x - \beta^j)$. Now $p^r i \equiv i \pmod{p^m - 1}$ if and only if $p^r \equiv 1 \pmod n$, so that $\alpha = \beta^i$, $\alpha^p = \beta^{pi}, \ldots, \alpha^{p^{r-1}} = \beta^{p^{r-1}i}$ are all distinct roots of m, so m_1 divides m. Since all coefficients of m_1 are in $\mathbb{F}_p$, $m_1 = m$ follows. $\quad\square$

4.7 Example (continued). The minimal polynomial of the primitive element α is $(x - \alpha)(x - \alpha^2)(x - \alpha^4)(x - \alpha^8) = x^4 + x + 1$, and the minimal polynomial of the primitive element α^7 is $(x - \alpha^7)(x - \alpha^{14})(x - \alpha^{12})(x - \alpha^{11}) = x^4 + x^3 + 1$. α^3 is not primitive; its order in $\mathbb{F}_{2^4}$ is $5 = (2^4 - 1)/3$ and its minimal polynomial is $(x - \alpha^3)(x - \alpha^6)(x - \alpha^{12})(x - \alpha^9) = x^4 + x^7 + x^2 + x + 1$. The element α^5 is also not primitive; its order in $\mathbb{F}_{2^4}$ is $3 = (2^4 - 1)/5$, and its minimal polynomial is $(x - \alpha^5)(x - \alpha^{10}) = x^2 + x + 1$. $\quad\square$

Cyclotomic cosets are useful in the factorization of polynomials of the form $x^n - 1$. If $n = p^m - 1$ we can find the numbers and degrees of the irreducible factors of $x^{p^{m-1}} - 1$ by computing the cyclotomic cosets for $p^m - 1$. From Example 4.7 we can conclude immediately from calculating C_1, C_3, C_5, C_7 and 4.6 that $x^{15} - 1$ has three factors of degree 4 and one factor of degree 2.

4.9 Example. Let $p^m = 1 = 63$. The cyclotomic cosets (mod 63) are

$$C_0 = \{0\}, \qquad C_1 = \{1, 2, 4, 8, 16, 32\}, \qquad C_3 = \{3, 6, 12, 24, 48, 33\},$$

$$C_5 = \{5, 10, 20, 40, 17, 34\}, \qquad C_7 = \{7, 14, 28, 56, 49, 35\},$$

$$C_9 = \{9, 18, 36\}, \qquad C_{11}\{11, 22, 44, 25, 50, 37\},$$

$$C_{13} = \{13, 26, 52, 41, 19, 38\}, \qquad C_{15} = \{15, 30, 60, 57, 51, 39\},$$

$$C_{21} = \{21, 42\}, \qquad C_{23} = \{23, 46, 29, 58, 53, 43\}, \qquad C_{27} = \{27, 54, 45\},$$

$$C_{31} = \{31, 62, 61, 59, 55, 47\}.$$

Therefore we can conclude that there are nine irreducible factors of $x^{63} - 1$ of degree 6, two irreducible factors of degree 3, and one of each of degrees 2 and 1. To find the factors of degree 6 is not easy.   □

It is easier to find all irreducible factors of $x^n - 1$, $n = p^m - 1$, if we factor $x^{p^m} - x$, since this is a product of all irreducible polynomials whose degrees divide m. We now consider the case $n \neq p^m - 1$. We assume that $\gcd(n, p) = 1$, so that p has an inverse modulo n and generalize the concept of a cyclotomic coset as follows.

4.10 Definition. The *cyclotomic coset modulo n* containing s is

$$C_s = \{s, ps, p^2 s, \ldots, p^{r_s - 1}s\},$$

where each $p^i s$ is reduced modulo n. Here r_s is the smallest integer such that $p^{r_s}s \equiv s \pmod{n}$.

Given n and p, we know that there is a smallest integer m such that n divides $p^m - 1$. This happens if and only if $x^n - 1$ divides $x^{p^{m-1}} - 1$, so that $\mathbb{F}_{p^m}$ is the smallest field of characteristic p that contains all the roots of $x^n - 1$. The roots of $x^n - 1$ are powers of a primitive root of unity. Now Theorem 4.3 can be easily generalized.

4.11 Theorem. *Let α be a root of $x^n - 1 = 0$ in $\mathbb{F}_{p^m}$ and let m be its minimal polynomial. Let ζ be a primitive root of unity in $\mathbb{F}_{p^m}$ and let $\alpha = \zeta^i$. If s is the smallest element in the cyclotomic coset mod n containing i, then*

$$m = \prod_{j \in C_s} (x - \zeta^j). \qquad □$$

Since $x^{n-1} = \prod_{i=0}^{n-1} (x - \alpha_i)$ for n distinct elements $\alpha_0, \ldots, \alpha_{n-1}$ in $\mathbb{F}_{p^m}$— the nth roots of unity—we also have $x^n - 1 = \prod_{i=0}^{n-1} (x - \zeta^i)$ for a primitive nth root of unity ζ. Let $m^{(s)}$ denote the minimal polynomial of ζ^s then Theorem 4.11 implies

$$x^n - 1 = \prod_s m^{(s)},$$

where s runs through a set of coset representatives (mod n). This is the factorization of $x^n - 1$ into irreducible polynomials over $\mathbb{F}_p$. Similar results can be obtained by considering a general finite field $\mathbb{F}_q$ of prime power order.

4.12 Example. Let $p = 2$, $n = 9$. Then $r = 6$ and $9|2^r - 1$, and $x^9 - 1$ splits into linear factors over $\mathbb{F}_{2^4}$. The cyclotomic cosets modulo n are $C_0 = \{0\}$, $C_1 = \{1, 2, 4, 8, 7, 5\}$, $C_3 = \{3, 6\}$. We choose $\zeta \in \mathbb{F}_{2^6}$, where $\zeta^6 + \zeta^3 + 1 = 0$. Then the minimal polynomials according to Theorem 4.11 are

$$m^{(0)} = x + 1, \qquad m^{(1)} = x^6 + x^3 + 1, \qquad m^{(3)} = x^2 + x + 1.$$

Therefore

$$x^9 - 1 = m^{(0)} m^{(1)} m^{(3)}. \qquad \square$$

4.13 Example. To get some information about the factorization of $x^{23} - 1$ over $\mathbb{F}_2$ we compute the cyclotomic cosets modulo 23:

$$C_0 = \{0\}, \qquad C_1 = \{1, 2, 4, 8, 16, 9, 18, 13, 3, 6, 12\},$$

$$C_5 = \{5, 10, 20, 17, 11, 22, 21, 19, 15, 7, 14\}.$$

This implies that $x^{23} - 1$ has two factors of degree 11 and one factor of degree 1, namely $x - 1$. Each is a minimal polynomial of a primitive twenty-third root of unity, which is contained in $\mathbb{F}_{2^{11}}$, since $r = 11$ is the smallest integer such that $23|2^r - 1$. $\qquad \square$

PROBLEMS

1. Let $f_1 = x^2 + x + 1, f_2 = x^3 + x + 1, f_3 = x^4 + x + 1, g_1 = x + 1, g_2 = x^2 + 1$ and $g_3 = x^3 + 1$ be polynomials over $\mathbb{F}_2$. Find a polynomial h such that the simultaneous congruences $h \equiv g_i \pmod{f_i}$, $i = 1, 2, 3$, are satisfied.

2. Show that $x^9 + x^4 + 1$ is irreducible over $\mathbb{F}_2$.

3. Prove the extended Euclidean algorithm for finding polynomials a and b over $\mathbb{F}_q$ such that $\gcd(f, g) = af + bg$, for given polynomials f and g over $\mathbb{F}_q$.

4. Find x^{12} modulo $x^5 + 2x^3 + x + 1$ over $\mathbb{F}_3$.

*5. Prove Theorem 4.11.

*6. Let f be an irreducible polynomial in $\mathbb{F}_q[x]$ of degree n and define the matrix $\mathbf{Q} = (q_{ki})$ as in Step 2 of Theorem 4.3. Prove that the characteristic polynomial $\det(x\mathbf{I} - \mathbf{Q})$ of $\mathbf{Q}$ is equal to $x^n - 1$.

7. Factorize $\quad x^{17} + x^{14} + x^{13} + x^{12} + x^{11} + x^{10} + x^9 + x^8 + x^7 + x^5 + x^4 + x + 1$ over $\mathbb{F}_2$.

8. Show "quickly" that $Q_{45}(x)$ is the product of two irreducible polynomials.

9. Factorize $x^8 + x^6 - 3x^4 - 3x^3 + 8x^2 + 2x - 5$ over $\mathbb{F}_{13}$ and also over $\mathbb{F}_2$.

10. Determine the order of $x^{16} + x^3 + 1$ over $\mathbb{F}_2$.

11. Determine the cyclotomic cosets mod 31 over $\mathbb{F}_2$. What can you say about the factorization of $x^{31} - 1$ over $\mathbb{F}_2$?

12. What can you say about the factorization of $x^{19} - 1$ over $\mathbb{F}_2$?

13. Factorize $x^{13} - x$ over $\mathbb{F}_2$.

14. Determine the degrees of the factors of $x^{127} - 1$ over $\mathbb{F}_2$.

*15. Let $\mathbb{F}_4 = \mathbb{F}_2(\alpha)$. Factorize $x^5 + \alpha x^4 + x^3 + (1 + \alpha)x + \alpha$ over $\mathbb{F}_4$.

EXERCISES (Solutions in Chapter 8, p. 457)

1. How many distinct irreducible factors divide

$$f = x^5 + 2x^4 + x^3 + x^2 + 2 \quad \text{in } \mathbb{F}_3[x]?$$

2. Determine the factorization of $f = x^{12} + x^8 + x^7 + x^6 + x^2 + x + 1$ over $\mathbb{F}_2$.

3. Determine the factorization of $f = x^9 + x + 1$ over $\mathbb{F}_2$.

*4. Determine the number of factors of $f = x^4 + 1 \in \mathbb{F}_p[x]$ over $\mathbb{F}_p$ for all primes p, using Berlekamp's algorithm.

5. Kronecker's algorithm can be used to determine the divisors of degree $\leq s$ of a nonconstant polynomial f in $\mathbb{Q}[x]$; it works as follows:
   (i) Without loss of generality suppose $f \in \mathbb{Z}[x]$.
   (ii) Choose distinct integers $a_0, \ldots, a_s$ which are not zeros of f and then determine all divisors of $\bar{f}(a_i), 0 \leq i \leq s$.
   (iii) For each $(s + 1)$-tuple $(b_0, \ldots, b_s)$ for which $b_i | f(a_i)$, determine polynomials $g \in \mathbb{Q}[x]$ of degree $\leq s$ such that $\bar{g}(a_i) = b_i, 0 \leq i \leq s$.
   (iv) Determine which of these polynomials g divides f. If $\deg f = n \geq 1, [n/2] = s$ and the only divisors of f are constants, then f is irreducible. In the case of nonconstant divisors g of f repeated application of (ii) and (iii) gives the complete factorization of f over $\mathbb{Q}$. Factorize

$$f = \tfrac{1}{3}x^4 - x^3 + \tfrac{1}{2}x^2 - \tfrac{1}{2}x + \tfrac{1}{6} \in \mathbb{Q}[x],$$

6. Determine the factorization of $x^7 + x^6 + x^5 - x^3 + x^2 - x - 1$ over $\mathbb{F}_3$.

*7. Determine the cyclotomic cosets mod 21 over $\mathbb{F}_2$ and find the factorization of $x^{21} - 1$ over $\mathbb{F}_2$.

8. What are the degree of the irreducible factors of $x^{17} - 1$ over $\mathbb{F}_2$? What is the smallest field of characteristic 2 in which $x^{17} - 1$ splits into linear factors?

§5. The Nullspace of a Matrix (Appendix to §4)

We saw in §4 that we have to solve a system of linear equations over $\mathbb{F}_q$ in order to obtain the factorization of a polynomial over $\mathbb{F}_q$. For the algorithm of Berlekamp it was necessary to find the nullspace of the equation $v(Q - I) = 0$, where Q is a given matrix, I is the identity matrix and v is the unknown vector. We now describe an algorithm in the formulation of KNUTH and a special algorithm in the binary case due to BERLEKAMP.

5.1 Theorem (Nullspace Algorithm). *Let K be a field. Let $\mathbf{A} = (a_{ij}) \in K_n^n$, with rank $A = r$. The following steps give the basis vectors $\mathbf{v}^{(i)}$ of the solution space of the equation $\mathbf{v}\mathbf{A} = \mathbf{0}$, where $i = 1, \ldots, n - r$.*

  Step 1. (Loop for k). *Perform step 2 for $k = 0, 1, \ldots, n - 1$ and stop.*
  Step 2. (a) *If all $a_{kj}(0 \le j \le n - 1)$ are zero, then*

$$\mathbf{v}^{(k)} := (v_0^{(k)}, v_1^{(k)}, \ldots, v_{n-1}^{(k)}), \quad where \quad v_j^{(k)} := \begin{cases} 1 & if \, j = k, \\ 0 & otherwise. \end{cases}$$

  (b) *If there is some $j \in \{0, 1, \ldots, n - 1\}$ with $a_{kj} \ne 0$ and $a_{sj} = 0$ for $s < k$, multiply the jth column of $\mathbf{A}$ by $-1/a_{kj}$. This gives the element -1 in the place (k, j), which is circled. Suitable column operations transform all remaining elements of the kth row into 0.*
  (c) *If for all $a_{kj} \ne 0$ in the jth column exactly one a_{sj} is -1, for $s = 0, 1, \ldots, k - 1$, and all other elements in the column are 0 for $s < k$, then the output is $\mathbf{v}^{(k)} := (v_0^{(k)}, v_1^{(k)}, \ldots, v_{n-1}^{(k)})$, defined by*

$$v_j^{(k)} = \begin{cases} a_{kt} & if \, a_{jt} = -1, \\ 1 & if \, j = k. \\ 0 & otherwise. \end{cases} \qquad \square$$

5.2 Example. Determine the nullspace of

$$(a_{ij}) = \mathbf{Q} - \mathbf{I} = \begin{pmatrix} 0 & 0 & 0 & 0 & 0 & 0 & 0 \\ 0 & 4 & 0 & 0 & 0 & 1 & 0 \\ 0 & 2 & 2 & 0 & 4 & 3 & 4 \\ 0 & 1 & 4 & 4 & 4 & 2 & 1 \\ 2 & 2 & 2 & 3 & 4 & 3 & 2 \\ 0 & 0 & 4 & 0 & 1 & 3 & 2 \\ 3 & 0 & 2 & 1 & 4 & 2 & 1 \end{pmatrix}$$

over $\mathbb{Z}_5$. (This is the matrix $\mathbf{Q} - \mathbf{I}$ of Example 4.4.) We have $k = 0, 1, \ldots, 6$ in the notation of 5.1; for $k = 0$ all elements of $\mathbf{Q} - \mathbf{I}$ are zero, thus Step 2(a) yields the vector

$$\mathbf{v}^{(0)} = (1, 0, 0, 0, 0, 0, 0).$$

For $k = 1, j = 1$ or 5 satisfy the conditions of Step 2(b). We choose $j = 1$, and then $a_{11} = -1$ in $\mathbb{Z}_5$ and suitable column operations transform the matrix into

$$\begin{pmatrix} 0 & 0 & 0 & 0 & 0 & 0 & 0 \\ 0 & ④ & 0 & 0 & 0 & 0 & 0 \\ 0 & 2 & 2 & 0 & 4 & 0 & 4 \\ 0 & 1 & 4 & 4 & 4 & 3 & 1 \\ 2 & 2 & 2 & 3 & 4 & 0 & 2 \\ 0 & 0 & 4 & 0 & 1 & 3 & 2 \\ 3 & 0 & 2 & 1 & 4 & 2 & 1 \end{pmatrix}.$$

The circled entry indicates that the element a_{11} is $4 = -1$ in $\mathbb{Z}_5$. For $k = 2$ we choose $j = 4$ amongst the possible $j = 2, 4, 6$ and obtain

$$\begin{pmatrix} 0 & 0 & 0 & 0 & 0 & 0 & 0 \\ 0 & ④ & 0 & 0 & 0 & 0 & 0 \\ 0 & 0 & 0 & 0 & ④ & 0 & 0 \\ 0 & 4 & 2 & 4 & 4 & 3 & 2 \\ 2 & 0 & 0 & 3 & 4 & 0 & 3 \\ 0 & 2 & 1 & 0 & 1 & 3 & 1 \\ 3 & 3 & 0 & 1 & 4 & 2 & 2 \end{pmatrix}.$$

For $k = 3, j = 3$ we have

$$\begin{pmatrix} 0 & 0 & 0 & 0 & 0 & 0 & 0 \\ 0 & ④ & 0 & 0 & 0 & 0 & 0 \\ 0 & 0 & 0 & 0 & ④ & 0 & 0 \\ 0 & 0 & 0 & ④ & 0 & 0 & 0 \\ 2 & 2 & 1 & 3 & 1 & 4 & 4 \\ 0 & 2 & 1 & 0 & 1 & 3 & 1 \\ 3 & 2 & 2 & 1 & 3 & 0 & 4 \end{pmatrix}.$$

For $k = 4, j = 5$:

$$\begin{pmatrix} 0 & 0 & 0 & 0 & 0 & 0 & 0 \\ 0 & ④ & 0 & 0 & 0 & 0 & 0 \\ 0 & 0 & 0 & 0 & ④ & 0 & 0 \\ 0 & 0 & 0 & ④ & 0 & 0 & 0 \\ 0 & 0 & 0 & 0 & 0 & ④ & 0 \\ 1 & 3 & 4 & 4 & 4 & 3 & 3 \\ 3 & 2 & 2 & 1 & 3 & 0 & 4 \end{pmatrix}.$$

Finally, for $k = 5$ and $j = 2$

$$\begin{pmatrix} 0 & 0 & 0 & 0 & 0 & 0 & 0 \\ 0 & ④ & 0 & 0 & 0 & 0 & 0 \\ 0 & 0 & 0 & 0 & ④ & 0 & 0 \\ 0 & 0 & 0 & ④ & 0 & 0 & 0 \\ 0 & 0 & 0 & 0 & 0 & ④ & 0 \\ 0 & 0 & ④ & 0 & 0 & 0 & 0 \\ 0 & 3 & 2 & 4 & 1 & 1 & 0 \end{pmatrix}.$$

In the case $k = 6$ we see that there is no suitable j with property 2(b), therefore the output according to Step 2(c) is $v^{(6)} = (0, 3, 1, 4, 1, 2, 1)$. Thus the nullspace algorithm outputs two linearly independent vectors $v^{(0)}$ and $v^{(6)}$. $\qquad\square$

BERLEKAMP describes a modification of this algorithm based on the so-called *reduced triangular idempotent form* of A (in short RTI form). A is in RTI form if:

> each element below the main diagonal is zero,
> each element in the main diagonal is zero or one,
> each element in a column with main diagonal element zero
> or in a row with main diagonal element one is zero.

5.3 Example. The matrix

$$\begin{pmatrix} 0 & 2 & 4 & 0 & 3 & 1 \\ 0 & 1 & 0 & 0 & 0 & 0 \\ 0 & 0 & 1 & 0 & 0 & 0 \\ 0 & 0 & 0 & 0 & 2 & 3 \\ 0 & 0 & 0 & 0 & 1 & 0 \\ 0 & 0 & 0 & 0 & 0 & 1 \end{pmatrix}$$

over $\mathbb{Z}_5$ is in RTI form.   □

The definition of the RTI form of a matrix $\mathbf{A} = (a_{ij})$, $0 \le i, j < n$, can also be given in the following form: $\mathbf{A}$ is in RTI form if

$$a_{ij} = 0 \quad \text{for } i > j,$$
$$a_{ij} = 0 \quad \text{or } 1 \quad \text{for } i = j,$$
$$a_{ij} = 0 \quad \text{if } a_{ii} = 1 \text{ or } a_{jj} = 0.$$

It can be verified directly that $\mathbf{A}^2 = \mathbf{A}$ if $\mathbf{A}$ is in RTI form. The following theorem gives a connection between RTI matrices $\mathbf{A}$ and the solution of equations $\mathbf{vA} = \mathbf{0}$.

5.4 Theorem. *Let $\mathbf{A}$ be in* RTI *form. The vector $\mathbf{v}$ is a solution of $\mathbf{vA} = \mathbf{0}$ if and only if $\mathbf{v}$ is a linear combination of rows of $\mathbf{A} - \mathbf{I}$.*

PROOF. $\mathbf{A}$ is idempotent, therefore $\mathbf{A}^2 = \mathbf{A}$, and thus $\mathbf{A}(\mathbf{A} - \mathbf{I}) = (\mathbf{A} - \mathbf{I})\mathbf{A} = \mathbf{0}$. Therefore, if $\mathbf{v}$ is any linear combination of rows of $\mathbf{A} - \mathbf{I}$, then $\mathbf{vA} = \mathbf{0}$. The only nonzero rows of $\mathbf{A} - \mathbf{I}$ have -1 on the main diagonal, so rank $(\mathbf{A} - \mathbf{I})$ = dimension of nullspace of $\mathbf{A}$. So linear combinations of rows of $\mathbf{A} - \mathbf{I}$ are the only solutions of the equation $\mathbf{vA} = \mathbf{0}$.   □

Next we give an example of how to transform a square matrix in RTI form. The reduction algorithm consists of the following steps, which are repeated n times.

5.5 Theorem (RTI Algorithm). *Given a matrix (a_{ij}), repeat the following loop n times:*

Step 1. If $a_{00} \ne 0$, go to 2. If $a_{00} = 0$, exchange the first column with the jth column nearest the first column for which $a_{0j} \ne 0$ and $a_{jj} = 0$. If there is no such column, exchange the first column with the jth column nearest the first column, for which $a_{0j} \neq 0$ and $a_{jj} \neq 0$. If there is no such column, stop.

Step 2. Normalize the first column, i.e. divide by its top element.

Step 3. Suitable column operations transform all other entries of the first row into 0.

Step 4. Replace the first row by the second row, the second row by the third row, etc.... the nth row by the first row. Similarly for the columns.   □

If this algorithm is performed k times, the resulting $n - k$ columns consist of zeros only in the last k rows and the lower right $k \times k$ submatrix is of triangular idempotent form.

5.6 Example. Applying 5.5, $k = 4$ times we obtain the matrix over $\mathbb{Z}_5$ of the form

$$\begin{pmatrix} 0 & 0 & 3 & 1 & 4 & 0 & 2 \\ 1 & 2 & 2 & 3 & 4 & 0 & 0 \\ 4 & 3 & 2 & 4 & 3 & 4 & 1 \\ 0 & 0 & 0 & 1 & 0 & 0 & 0 \\ 0 & 0 & 0 & 0 & 0 & 3 & 2 \\ 0 & 0 & 0 & 0 & 0 & 1 & 0 \\ 0 & 0 & 0 & 0 & 0 & 0 & 1 \end{pmatrix}.$$

Performing Steps 1 to 4 we obtain

$$\begin{pmatrix} 2 & 4 & 2 & 1 & 0 & 3 & 1 \\ 3 & 1 & 2 & 4 & 4 & 2 & 2 \\ 0 & 0 & 1 & 0 & 0 & 0 & 0 \\ 0 & 0 & 0 & 0 & 3 & 2 & 0 \\ 0 & 0 & 0 & 0 & 1 & 0 & 0 \\ 0 & 0 & 0 & 0 & 0 & 1 & 0 \\ 0 & 0 & 0 & 0 & 0 & 0 & 1 \end{pmatrix}.$$

Thus we see that step $k = 5$ yields the 5×5 submatrix in RTI form. Two more steps give the 7×7 matrix in RTI form.   □

5.7 Theorem. Let $\bar{A}$ be the RTI matrix corresponding to the matrix A. Then the nullspace of A is equal to the nullspace of $\bar{A}$.

PROOF. The algorithm 5.5 uses only admissible column operations.   □

5.8 Corollary. If $\bar{A}$ is the RTI matrix A then those row vectors of the matrix $(b_{ij}) = B = \bar{A} - I$, for which $b_{ii} = -1$, form a basis of the nullspace of A.

PROOF. Only the rows with $b_{ii} = -1$ in the matrix B are nonzero rows.
    □

BERLEKAMP gives a special, faster algorithm in case of sparse matrices, i.e. if "many" of the n^2 elements of the $n \times n$ matrix $\mathbf{A}$ are zero.

5.9 Example. Solve $v\mathbf{A} = \mathbf{0}$ over $\mathbb{Z}_2$, where $\mathbf{A}$ over $\mathbb{Z}_2$ is

$$
\begin{pmatrix}
0 & 0 & 0 & 0 & 0 & 0 & 0 & 0 \\
0 & 1 & 1 & 0 & 0 & 0 & 0 & 0 \\
0 & 0 & 1 & 0 & 1 & 0 & 0 & 0 \\
0 & 0 & 0 & 1 & 0 & 0 & 1 & 0 \\
1 & 0 & 0 & 1 & 0 & 0 & 1 & 0 \\
1 & 0 & 1 & 1 & 1 & 0 & 0 & 0 \\
0 & 0 & 1 & 0 & 1 & 1 & 0 & 1 \\
1 & 1 & 0 & 1 & 1 & 1 & 0 & 1
\end{pmatrix}.
$$

(This is the matrix $\mathbf{Q} - \mathbf{I}$ of Example 4.5.) The nonzero elements in each column of $\mathbf{A}$ correspond to the indices of coordinates v_i, the sum of which must be zero. We tabulate these indices; the rows of the table correspond to the rows of $v\mathbf{A} = \mathbf{0}$.

Equation No.	Index of variables
0	4 5 7
1	1 7
2	1 2 5 6
3	3 4 5 7
4	2 5 6 7
5	6 7
6	3 4
7	6 7

Firstly we eliminate all equations with not more than two variables v_i, e.g. equation 1, since $v_1 + v_7 = 0$ implies $v_1 = v_7$ in $\mathbb{Z}_2$. In general, the unknowns v_k with higher index are expressed in terms of those with lower index, and all v_k are deleted from the remaining equations. The index k is circled to indicate that this variable is defined by the equation. For instance, we have $k = 7$ in equation 1, so in the remaining equations we replace 7 by 1. In equation 6 the index 4 is circled and replaced by 3 in the remaining equations. This is continued until all v_k are expressed in terms of smaller indices. This yields

0	4̶ 5̶ 7̶ 4̶ 2̶ ③
1	1 ⑦
2	1̶ 2̶ ⑤ 6̶ 1
3	3̶ 4̶ 5̶ 7̶ 1 ②
4	2̶ 5̶ 6̶ 7̶ 4̶
5	⑥ 7̶ 1
6	3 ④
7	6̶ 7̶ 1

From this we see: $v_3 = 0$ from equation 0. $v_3 + v_4 = 0 \Rightarrow v_4 = 0$ from equation 6, etc., and finally $v_1 = v_2 = v_5 = v_6 = v_7$. For $v_0 = 1$ and $v_1 = 0$ we obtain,

$$\mathbf{v}^{(1)} = (1, 0, 0, 0, 0, 0, 0, 0).$$

For $v_0 = 0$ and $v_1 = 1$ we have

$$\mathbf{v}^{(2)} = (0, 1, 1, 0, 0, 1, 1, 1).$$

For $v_0 = v_1 = 0$ and $v_0 = v_1 = 1$:

$$\mathbf{v}^{(3)} = (0, 0, 0, 0, 0, 0, 0, 0) \quad \text{and} \quad \mathbf{v}^{(4)} = (1, 1, 1, 0, 0, 1, 1, 1).$$

We see immediately that $\mathbf{v}^{(3)}$ and $\mathbf{v}^{(4)}$ are linearly dependent on $\mathbf{v}^{(1)}$ and $\mathbf{v}^{(2)}$. Therefore we have two linearly independent vectors $\mathbf{v}^{(1)}$ and $\mathbf{v}^{(2)}$ and thus the dimension of the nullspace of $\mathbf{A}$ is 2.   □

NOTES

Most of the definitions and theorems in section 1 can be found in nearly any of the introductory books on modern algebra, e.g. FRALEIGH, HERSTEIN, LANG, MACLANE and BIRKHOFF, VAN DER WAERDEN, and more detailed accounts of field theory are given in ALBERT, NAGATA, WINTER, to mention a few. The discovery of so-called p-adic fields by Hensel led Steinitz to the abstract formulation of the theory of fields which was developed in the fundamental paper of STEINITZ, where he introduced prime fields, separable elements, etc.; earlier contributions in the area of field theory were made by KRONECKER and WEBER, who designed the first axiomatic field theory. One of the most fundamental theorems in the theory of fields is Theorem 1.32, due to KRONECKER. Kronecker used the idea of Cauchy that the complex numbers can be defined as residue classes in $\mathbb{R}[x]/(x^2 + 1)$. If we apply Kronecker's approach to $\mathbb{Z}/(p)$ we obtain "Galois imaginaries" due to Serret and Dedekind.

Richard Dedekind introduced the term "field" (in German "Körper") for the first time, although the concept and some properties of special types of fields were known and studied before him, e.g. by Gauss, Galois, Abel.

Leibniz created the name "transcendentals" when he proved in 1682 that $\sin x$ cannot be an algebraic function of x. The existence of a transcendental number was verified by Liouville in 1844; in 1873 Hermite proved the transcendence of e and in 1882 Lindemann showed that π is transcendental, which concluded the "squaring of the circle" problem of the ancient Greeks. The concepts (but not the modern terminology) of constructing an extension field by adjoining an element to a given field, and of irreducible polynomials and conjugate roots were studied by Galois. Abel and Galois defined the elements of the underlying field as those quantities which can be expressed as rational functions of given quantities. In 1744, A. Legendre recognized that any root of an algebraic equation with rational coefficients is an algebraic number, and that numbers that are not algebraic are transcendental. E. E. Kummer, a pupil of Gauss and Dirichlet, introduced algebraic numbers of the form $a_0 + a_1\alpha + \ldots + a_{p-2}\alpha^{p-2}$, $a_i \in \mathbb{Z}$, α an imaginary pth root of unity, and then studied divisibility, primality and other properties of these numbers. In order to obtain unique factorization for his numbers he created the theory of "ideal numbers". For instance, consider $\{a + b\sqrt{-5}\,|\,a, b \in \mathbb{Z}\}$; then $6 = 2 \cdot 3 = (1 + \sqrt{-5})(1 - \sqrt{-5})$, where all factors are prime. Using the ideal numbers

$$\alpha = \sqrt{2}, \qquad \rho_1 = \frac{1 + \sqrt{-5}}{\sqrt{2}}, \qquad \rho_2 = \frac{1 - \sqrt{-5}}{\sqrt{2}}$$

we have $6 = \alpha^2 \rho_1 \rho_2$ uniquely expressed as a product of four factors. Beginning in 1837, and following work of C. F. Gauss, Kummer studied the arithmetic of cyclotomic fields for some 25 years in an effort to solve Fermat's Last Theorem (namely the conjecture that $x^n + y^n = z^n$ is unsolvable in nonzero integers x, y, z for $n > 2$) in the case of n being an odd prime. (Kummer succeeded in proving the conjecture for primes $p < 100$.)

Throughout the long history of mathematics over several millennia, the problem of solving equations has been an important one. We know from the clay tablets of the ancient Babylonians, from papyri of the ancient Egyptians and from old Chinese and Hindu writings that the ancient civilizations dealt with the solution of equations of the first and second degree. Although the interest in such problems diminished during and after the classical and Alexandrian Greek period because of the predominance of geometry, this interest greatly increased early in the sixteenth century due to the Italian mathematicians, who solved equations of degrees 3 and 4 by radicals. From then on until the eighteenth century the solution of algebraic equations by radicals, i.e. "expressions involving roots", was a natural problem to consider. Since the middle of the eighteenth century the search was on for an *a priori* proof of the fundamental theorem of algebra, because all attempts at finding solutions by radicals failed for general equations of degree ≥ 5. After contributions by Tschirnhaus, Leibniz, de Moivre, Euler, d'Alembert and others, Lagrange attempted a proof of the fundamental theorem in the early 1770's. C. F. Gauss gave several complete proofs, his first one in 1799. Today there are more than 100 different proofs known. Lagrange studied the relationship between the roots $\alpha_1, \ldots, \alpha_n$ and the coefficients $a_0, \ldots, a_{n-1}$ of a polynomial equation, namely

$$x^n + a_{n-1}x^{n-1} + \ldots + a_1 x + a_0 = (x - \alpha_1) \ldots (x - \alpha_n),$$

where

$$-a_{n-1} = \alpha_1 + \ldots + \alpha_n, \; a_{n-2} = \alpha_1\alpha_2 + \ldots + \alpha_{n-1}\alpha_n, \ldots, (-1)^n a_0 = \alpha_1 \ldots \alpha_n.$$

Using these relationships Lagrange devised a technique for solving third- and fourth-degree equations by finding a related equation whose degree was smaller than the degree of the original equation. When he applied this approach to the general fifth-degree equation he obtained a related equation which was of degree six. From this Lagrange deduced that the solution of the general fifth-degree equation by radicals might be impossible.

Work by Ruffini and Gauss led Abel and Galois to the complete solution of the problem. In 1824–1826, N. H. Abel showed that the solution by radicals of a general equation of fifth degree is impossible (nowadays called the Abel–Ruffini theorem). E. Galois studied and extended work by Lagrange, Gauss and Abel, and he gave the complete answer that the general equation of degree n is not solvable by radicals if $n \geq 5$. Galois' work (summarized under the name Galois theory) also supplied a criterion for constructability that dispensed with some of the famous Greek construction problems. The solutions to the problems of the solvability of equations by radicals in the nineteenth century made way for a wider development of algebraic concepts in the decades following Galois' death in 1832.

The Greek school of sophists, which was the first school of philosophers in ancient Athens, singled out constructions with ruler and compass alone, mostly for philosopical rather than mathematical reasons. Therefore Euclid in his *Elements* describes only construction problems of this type, now referred to as the classical Greek construction problems. The problems such as constructing a square equal in area to a given circle, trisecting any angle, constructing the side of a cube whose volume is double that of a cube with given edge, had to wait over 2000 years for their solution, namely a demonstration of the impossibility of certain such constructions. One version of the origin of the problem of doubling the cube, found in a work by Erathosthenes (approx. 284–192 B.C.), relates that the Delians, suffering from pestilence, consulted an oracle who advised constructing an altar double the size of the existing one. The famous Plato was asked to help; he answered that the God of the oracle did not really need an altar of twice the size, but he censured the Greeks for their indifference to mathematics and their lack of respect for geometry.

Cotes, de Moivre and Vandermonde studied solutions of $x^n - 1 = 0$ and divided the circle into n equal parts. Gauss in his fundamental work *Disquisitiones arithmeticae* also studied this cyclotomic equation for n an odd prime. He arrived at the famous result that a regular n-gon is constructible, using ruler and compass only, if n is a prime of the form $2^{2^k} + 1$. This was later generalized to the result that the regular n-gon is constructible with ruler and compass if and only if all the odd primes dividing n are primes of the form $2^{2^k} + 1$ whose squares do not divide n. Thus, for instance, the regular 60-gon is constructible, since $60 = 2^2 \cdot 3 \cdot 5$, where $3 = 2 + 1$ and $5 = 2^2 + 1$.

In most textbooks on abstract or applied algebra, the theory of finite fields is covered in a few pages. More extensive treatments can be found, for instance, in ALBERT, BERLEKAMP, DORNHOFF and HOHN, MACWILLIAMS and SLOANE, MCDONALD, IRELAND and ROSEN. A comprehensive book on finite fields (with an extensive bibliography) is LIDL and NIEDERREITER.

The history of finite fields up to 1915 is presented in DICKSON (Ch. 8); Some of the early results of Fermat, Euler, Lagrange, Legendre and Gauss were concerned with many properties of the special finite field $\mathbb{Z}_p$. The concept of a finite field in general (that is, not only considered as the prime field $\mathbb{Z}_p$) first occurred in a paper of GALOIS in 1830. However, Gauss initiated a project of developing a theory of "higher congruences", as equations over finite fields were called. R. Dedekind observed, after following Dirichlet's lectures, that Gauss's project could be carried out by establishing an analogy with the theory of ordinary congruences and then emulating Dirichlet's approach to the latter topic. This resulted in an important paper by Dedekind (1857), which put the theory of higher congruences on a sound basis and also played a role in the conceptual development of algebraic number theory and abstract algebra. A posthumous paper by Gauss investigated the arithmetic in the polynomial rings $\mathbb{F}_p[x]$; in particular, he realized that in these rings the Euclidean algorithm and unique factorization are valid. The construction of extension fields of $\mathbb{F}_p$ was described by Galois in 1830, who used an "imaginary" root i of an irreducible polynomial over $\mathbb{F}_p$ of degree n and showed that the expressions $a_0 + a_1 i + \ldots + a_{n-1} i^{n-1}$ with $a_j \in \mathbb{F}_p$ form a field with p^n elements. During the nineteenth century further research on "higher congruences", was carried out by Dedekind, Schöneman, Kummer.

LIDL and NIEDERREITER (in their Notes to Chapters 2–9) refer to a large number of varied contributions to the theory of finite fields. This book also contains tables of primitive roots modulo p due to ALANEN and KNUTH and has further references to tables representing elements of finite fields.

Properties of polynomials over finite fields are described in ALBERT (Ch. 5), and BERLEKAMP; see also MACWILLIAMS and SLOANE, McDONALD, BLAKE and MULLIN, to mention just a few more recently published books. ALBERT calls the order of a polynomial its exponent, BERLEKAMP calls it the period. In the older literature the word primitive was used for a polynomial whose coefficients had gcd equal to 1. Our primitive polynomials are sometimes also called indexing polynomials.

The term "irreducible polynomial" over $\mathbb{Q}$ goes back to the seventeenth century, when Newton and Leibniz gave methods for determining irreducible factors of a given polynomial. Gauss showed that $(x^n - 1)/(x - 1)$ is irreducible over $\mathbb{Q}$ if n is an odd prime and he also verified that the roots of $(x^n - 1)/(x - 1) = 0$ are cyclic. He also found the formula for the number of irreducible polynomials over $\mathbb{F}_p$ of fixed degree; our proof of this formula and the Möbius multiplicative inversion formula is due to Dedekind. Theorem 3.19 is still known as Dedekind's formula.

The factorization algorithm based on the matrix $\mathbf{Q}$ (Berlekamp's algorithm) was first developed by BERLEKAMP and it also occurs in CHILDS, KNUTH, LIDL and NIEDERREITER, the latter book also contains other factorization algorithms for small and large finite fields. CANTOR and ZASSENHAUS proposed a new probabilistic algorithm for factorizing univariate polynomials over finite fields. RABIN redesigned Berlekamp's deterministic algorithm into a probabilistic method for finding factorizations of polynomials over large finite fields. LAZARD gives several modifications of such algorithms. All these algorithms are probabilistic in the sense that the time of computation depends on random choices, but the validity of the results does not depend on them. For computer implementations of factorization algorithms, see, e.g. MUSSER (*JACM* 22 (1975) 291–308) or MOORE and NORMAN (Proceedings SYMSAC 1981, New York, *ACM* (1981) 109–116).

Cyclotomic cosets and applications to the factorization of $x^n - 1$ are studied in MACWILLIAMS and SLOANE, PLESS; BLAKE and MULLIN call them q-chains. Factorization tables for the binomials $x^n - 1$ can be found in MCELIECE. The nullspace algorithms given in the last section can be found in the given form in KNUTH and BERLEKAMP.

We indicated that applications of finite fields and polynomials over finite fields to areas outside algebra will be treated in Chapters 4 and 5. There are also other uses of finite fields which we cannot describe here due to space limitations. For instance, the analysis of switching circuits (see Chapter 2) can be based on arithmetic in finite fields. The calculation of switching functions or general logic functions can also be described in terms of finite fields. Other applications comprise finite-state algorithms and linear systems theory.

CHAPTER 4
Coding Theory

In this and the following sections we shall describe an area of applications of modern algebra which has become increasingly important during the last two decades. We consider the problem of safe transmission of a message over a channel, which can be affected by "noise". This is the problem of coding and decoding of information. Examples of this situation, in communication systems, are: radio, television, telegraph, telephone, and data storage systems, such as records, tapes, films, etc.

The theory of finite fields and the theory of polynomials over finite fields have both influenced coding theory considerably. The origin of coding theory comes from a famous theorem of C. Shannon which guarantees the existence of codes that can transmit information at rates close to capacity with an arbitrarily small probability of error. Here we shall not be concerned with any aspects of information theory. We refer the reader interested in such aspects to VAN LINT. We shall describe some of the methods for the construction of error-correcting and error-detecting codes, which form part of algebraic coding theory. In §1 we introduce linear codes, §2 contains basic properties of cyclic codes and §3 gives special types of cyclic codes.

§1. Linear Codes

We begin by describing a simple model of a communication transmission system (see Figure 4.1). Messages go through the system starting from the sender (or source). We shall only consider senders with a finite number of discrete signals (e.g. telegraph) in contrast to continuous sources (e.g. radio). In most systems the signals emanating from the source cannot be transmitted

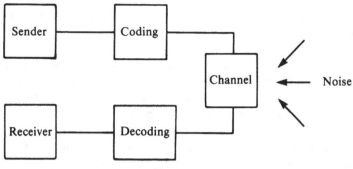

Figure 4.1

directly by the channel. For instance, a binary channel cannot transmit words in the usual latin alphabet. Therefore an encoder performs the important task of data reduction and suitably transforms the message into usable form. Accordingly one distinguishes between source encoding and channel encoding. The former reduces the message to its essential (recognizable) parts, the latter adds redundant information to enable detection and correction of possible errors in the transmission. Similarly, on the receiving end one distinguishes between channel decoding and source decoding, which invert the corresponding channel and source encoding besides detecting and correcting errors.

One of the main aims of coding theory is to design methods for transmitting messages error free, cheap and as fast as possible. There is of course the possibility of repeating the message. However, this is time consuming (see 1.5 and the following comments), inefficient and crude. We also note that the possibility of errors increases with an increase in the length of messages. We want to find efficient algebraic methods (codes) to improve the reliability of the transmission of messages. There are many types of algebraic codes; we shall consider only a few of them. These codes use algebraic methods to give simple coding and decoding algorithms which can be easily implemented.

1.1 Example. We describe a simple model for a transmission channel, called the *binary symmetric channel*. Let p be the probability that a binary signal is received correctly and then $q = 1 - p$ is the probability of incorrect reception. We assume that errors in transmission of successive signals are independent. Then the transmission probabilities for this channel are as follows

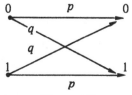

Figure 4.2

The probability that in the transmission of n symbols we have k errors is given by $\binom{n}{k}p^{n-k}q^k$. $\square$

In the following we assume that the elements of a finite field represent the underlying alphabet for coding. Coding consists of transforming a block of k *message symbols* $a_1 a_2 \ldots a_k$, $a_i \in \mathbb{F}_q$, into a *codeword* $\mathbf{x} = x_1 x_2 \ldots x_n$, $x_i \in \mathbb{F}_q$, where $n \geq k$. Here the first k symbols x_i are the message symbols, i.e. $x_i = a_i$, $1 \leq i \leq k$; the remaining $n - k$ elements $x_{k+1}, \ldots, x_n$ are *check symbols* (or *control symbols*). Codewords will be written in one of the forms $\mathbf{x}$, $x_1 x_2 \ldots, x_n$, $(x_1, x_2, \ldots, x_n)$, or $x_1, x_2, \ldots, x_n$. The check symbols can be obtained from the message symbols in such a way that the codewords $\mathbf{x}$ satisfy the system of linear equations

$$\mathbf{H}\mathbf{x}^T = \mathbf{0},$$

where $\mathbf{H}$ is a given $(n - k) \times n$ matrix with elements in $\mathbb{F}_q$. A "*standard form*" for $\mathbf{H}$ is $(\mathbf{A}, \mathbf{I}_{n-k})$, with $\mathbf{A}$ an $(n - k) \times k$ matrix and $\mathbf{I}_{n-k}$ the $(n - k) \times (n - k)$ identity matrix.

1.2 Example. Let $q = 2$, $n = 6$, $k = 3$. The message $a_1 a_2 a_3$ is encoded as the codeword $\mathbf{x} := a_1 a_2 a_3 x_4 x_5 x_6$. Here the check symbols x_4, x_5, x_6 are such that for this given matrix

$$\mathbf{H} = \begin{pmatrix} 0 & 1 & 1 & 1 & 0 & 0 \\ 1 & 0 & 1 & 0 & 1 & 0 \\ 1 & 1 & 0 & 0 & 0 & 1 \end{pmatrix} =: (\mathbf{A}, \mathbf{I}_3)$$

we have $\mathbf{H}\mathbf{x}^T = \mathbf{0}$, i.e.

$$a_2 + a_3 + x_4 = 0,$$
$$a_1 + a_3 + x_5 = 0,$$
$$a_1 + a_2 + x_6 = 0.$$

Therefore

$$x_4 = a_2 + a_3, \qquad x_5 = a_1 + a_3, \qquad x_6 = a_1 + a_2.$$

Thus the check symbols x_4, x_5, x_6 are determined by a_1, a_2, a_3. The equations in the system $\mathbf{H}\mathbf{x}^T = \mathbf{0}$ are also called *check equations*. If the message $\mathbf{a} = 011$ is transmitted, then the corresponding codeword is $\mathbf{x} = 011011$. Altogether there are 2^3 codewords:

$$000000, 011011, 110110, 001110,$$
$$100011, 111000, 010101, 101101.$$ $\square$

In general we define

1.3 Definition. Let $\mathbf{H}$ be an $(n - k) \times n$ matrix with elements in $\mathbb{F}_q$. The set of all n-dimensional vectors $\mathbf{x}$ satisfying $\mathbf{H}\mathbf{x}^T = \mathbf{0}$ over $\mathbb{F}_q$ is called a *linear* (*block*) *code* C over $\mathbb{F}_q$ of *block length* n. The matrix $\mathbf{H}$ is called the *parity-check matrix* of the code C. C is also called a *linear* (n, k) *code*.

If $\mathbf{H}$ is of the form $(\mathbf{A}, \mathbf{I}_{n-k})$ then the first k symbols of the codeword $\mathbf{x}$ are called *message* (or *information*) symbols, and the last $n - k$ symbols in $\mathbf{x}$ are the *check symbols*. C is then also called a *systematic* linear (n, k) code. If $q = 2$, then C is a *binary code*. k/n is called *transmission* (or *information*) *rate*.

1.4 Remark. The set C of solutions $\mathbf{x}$ of $\mathbf{H}\mathbf{x}^T = \mathbf{0}$, i.e. the solution space of this system of equations, forms a subspace of $\mathbb{F}_q^n$ of dimension k. Since the codewords form an additive group, C is also called a *group code*. C can also be regarded as the nullspace (see Chapter 4, §4, §5) of the matrix $\mathbf{H}$.

1.5 Example ("Repetition Code"). If each codeword of a code consists of only one message symbol $a_1 \in \mathbb{F}_2$ and the $n - 1$ check symbols $x_2 = \ldots = x_n$ are all equal to a_1 (a_1 is "repeated" $n - 1$ times), then we obtain a binary $(n, 1)$ code with parity-check matrix

$$\mathbf{H} = \begin{pmatrix} 1 & 1 & 0 & \ldots & 0 \\ 1 & 0 & 1 & \ldots & 0 \\ \vdots & \vdots & \vdots & & \vdots \\ 1 & 0 & 0 & \ldots & 1 \end{pmatrix}.$$

There are only two codewords in this code, namely $00\ldots0$ and $11\ldots1$. □

It is often impracticable, impossible or too expensive to send the original message more than once. Especially in the transmission of information from satellites or other spacecraft, it is impossible to repeat such messages owing to severe time limitations. In repetition codes we can, of course, also consider codewords with more than one message symbol. If we transmit a message of length k three times and compare corresponding "coordinates" x_i, x_{k+i}, x_{2k+i} of the codeword $x_1 \ldots x_i \ldots x_k x_{k+1} \ldots x_{k+i} \ldots x_{2k} x_{2k+1} \ldots x_{2k+i} \ldots x_{3k}$, then a "majority decision" decides which word has been sent, e.g. if $x_i = x_{k+i} \neq x_{2k+i}$, then probably x_i has been transmitted.

1.6 Example ("Parity-Check Code"). This is a binary $(n, n - 1)$ code with parity-check matrix $\mathbf{H} = (11\ldots1)$. Each codeword has one check symbol and all codewords are given by all binary vectors of length n with an even number of 1's. If the sum of ones of the received word is odd then at least one error must have occurred in the transmission. Such codes are used in banking. The last digit of the account number usually is a control digit. □

We have seen that a message $\mathbf{a} = a_1 \ldots a_k$ can be encoded as codeword $\mathbf{x} = x_1 \ldots x_k x_{k+1} \ldots x_n$, with $x_1 = a_1, \ldots, x_k = a_k$. The *check equations*

$(\mathbf{A}, \mathbf{I}_{n-k})\mathbf{x}^T = \mathbf{0}$ yield

$$\begin{pmatrix} x_{k+1} \\ \vdots \\ x_n \end{pmatrix} = -\mathbf{A}\begin{pmatrix} x_1 \\ \vdots \\ x_k \end{pmatrix} = -\mathbf{A}\begin{pmatrix} a_1 \\ \vdots \\ a_k \end{pmatrix}.$$

Thus we obtain

$$\begin{pmatrix} x_1 \\ \vdots \\ x_n \end{pmatrix} = \begin{pmatrix} \mathbf{I}_k \\ -\mathbf{A} \end{pmatrix}\begin{pmatrix} a_1 \\ \vdots \\ a_k \end{pmatrix}.$$

We transpose and denote this equation as

$$(x_1, \ldots, x_n) = (a_1, \ldots, a_k)(\mathbf{I}_k, -\mathbf{A}^T).$$

1.7 Definition. The matrix $\mathbf{G} = (\mathbf{I}_k, -\mathbf{A}^T)$ is called a *canonical generator matrix* (or *canonical basic matrix* or *encoding matrix*) of a linear (n, k) code with parity-check matrix $\mathbf{H} = (\mathbf{A}, \mathbf{I}_{n-k})$.

In this case we have $\mathbf{GH}^T = \mathbf{0}$.

1.8 Example.

$$\mathbf{G} = \begin{pmatrix} 1 & 0 & 0 & 0 & 1 & 1 \\ 0 & 1 & 0 & 1 & 0 & 1 \\ 0 & 0 & 1 & 1 & 1 & 0 \end{pmatrix}$$

is a canonical generator matrix for the code of Example 1.2. The 2^3 codewords $\mathbf{x}$ of this binary code can be obtained from $\mathbf{x} = \mathbf{aG}$ with $\mathbf{a} = a_1a_2a_3$, $a_i \in \mathbb{F}_2$.   □

A code can have several parity-check matrices and generator matrices. It can be easily shown that a basis of a code C can be formed from the rows of a generator matrix of C. In fact, the code C defined by $\mathbf{G}$ of Definition 1.7 is equal to the row space of $\mathbf{G}$, i.e. the space generated by the row vectors of $\mathbf{G}$. More generally, any $k \times n$ matrix whose row space is equal to C is called a generator matrix of C.

In the following we shall also consider codes C with parity-check matrix $\mathbf{H}$ and generator matrix $\mathbf{G}$, which are not of the form $(\mathbf{A}, \mathbf{I})$ and $(\mathbf{I}, -\mathbf{A}^T)$, respectively. We can show that every linear code is equivalent to a linear systematic code in the following sense. Let $\mathbf{G}$ be an $m \times n$ generator matrix over $\mathbb{F}_q$ for a linear code C. Then $\mathbf{G}$ is of rank m. Let $\mathbf{M}$ be an invertible $m \times m$ matrix over $\mathbb{F}_q$; then $\mathbf{MG}$ is an $m \times n$ matrix of rank m, and its rows are linear combinations of the rows of $\mathbf{G}$, so $\mathbf{MG}$ is also a generator matrix for the code C. If $\mathbf{P}$ is any $n \times n$ permutation matrix (with one 1 in each row and each column, other entries 0) then $\mathbf{MGP}$ is a generator matrix for an m-dimensional code C'. The codes C and C' are said to be *equivalent*. Now, since **G** has rank m, some m columns of **G** are linearly independent. Therefore there is a permutation matrix **P** such that **GP** has its first m columns linearly independent. If **GP** = (**BD**), with **B** an invertible matrix, and **D** an $m \times (n - m)$ matrix then we have **MGP** = ($\mathbf{I}_m\mathbf{E}$), with **E** = **MD**.

If a message $\mathbf{a} = a_1 \ldots a_k$ is encoded as $\mathbf{x} = x_1 \ldots x_n$ and is transmitted through a "noisy" channel, the codeword **x** may be altered to a vector $\mathbf{y} = y_1 \ldots y_n$, where **y** may be different to **x**. Alternatively, **G** can be transformed into its now reduced echelon form.

1.9 Definition. If **x** is a transmitted codeword and **y** is received then $\mathbf{e} = \mathbf{y} - \mathbf{x} = e_1 \ldots e_n$ is called an *error word* (or *error vector*, or *error*).

The decoder, on receiving **y**, has to decide which codeword has been transmitted. To do this he has to decide the most likely error. If all codewords are equally likely then this procedure makes the probability for incorrect decoding minimal. Thus it is called "maximum likelihood decoding". It works such that if **v** is received, then **v** is decoded to **w**, where **e** is the vector of least weight with $\mathbf{v} = \mathbf{w} + \mathbf{e}$ for $\mathbf{w} \in C$.

1.10 Definition. (i) The *Hamming distance* $d(\mathbf{x}, \mathbf{y})$ between two vectors $\mathbf{x} = x_1 \ldots x_n$ and $\mathbf{y} = y_1 \ldots y_n$ in $\mathbb{F}_q^n$ is the number of coordinates in which **x** and **y** differ.

(ii) The *Hamming weight* $w(\mathbf{x})$ of a vector $\mathbf{x} = x_1 \ldots x_n$ in $\mathbb{F}_q^n$ is the number of nonzero coordinates x_i. In short $w(\mathbf{x}) = d(\mathbf{x}, \mathbf{0})$.

We leave it as an exercise to show

1.11 Theorem. *The Hamming distance d is a metric on $\mathbb{F}_q^n$, and the Hamming weight w is a norm on $\mathbb{F}_2^n$.* $\square$

One of the most important properties of linear codes is based on the following definition.

1.12 Definition. The *minimum distance* $d_{\min}$ of a linear code C is given as

$$d_{\min} := \min_{\substack{\mathbf{u}, \mathbf{v} \in C \\ \mathbf{u} \neq \mathbf{v}}} d(\mathbf{u}, \mathbf{v}).$$

For linear codes we have $d(\mathbf{u}, \mathbf{v}) = d(\mathbf{u} - \mathbf{v}, \mathbf{0}) = w(\mathbf{u} - \mathbf{v})$. Therefore

1.13 Theorem. *The minimum distance of C is equal to the least weight of all nonzero codewords.* $\square$

For instance, for the code in 1.2 we have $d_{\min} = 3$. Two arbitrary codewords differ in at least $d_{\min}$ coordinates. A linear code of length n, dimension k and minimum distance d is called an (n, k, d) *code*.

A simple decoding rule is to decode a received vector **y** as the codeword closest to **x** with respect to Hamming distance, i.e. one chooses an error vector **e** of least weight. This decoding method is called "*nearest neighbor decoding*" and amounts to comparing **y** with all q^k codewords and choosing the closest amongst them. The nearest neighbor decoding is the maximum likelihood decoding if the probability p for correct transmission is $> \frac{1}{2}$. We will assume this from now on. Obviously this procedure is impossible for large k and one of the aims of coding theory is to find codes with faster decoding algorithms.

1.14 Definition. The set $S_r(\mathbf{x}) := \{\mathbf{y} \in \mathbb{F}_q^n \mid d(\mathbf{x}, \mathbf{y}) \le r\}$ is called the *sphere of radius r* about $\mathbf{x} \in \mathbb{F}_q^n$.

In decoding we distinguish between the detection and the correction of errors. We say a code can *correct* t errors and can *detect* $t + s$, $s \ge 0$, errors, if the structure of the code makes it possible to correct up to t errors and to detect (but not necessarily correct) $t + j$, $0 < j \le s$, errors, which occurred during transmission over a channel.

1.15 Theorem. *A linear code C with minimum distance $d_{\min}$ can correct up to t errors and can detect $t + j$, $0 < j \le s$, errors, iff $2t + s \le d_{\min}$.*

PROOF. Let C be a subset of $\mathbb{F}_q^n$, then "nearest neighbor decoding" shows that any received word with t or fewer errors must be in a sphere of radius t about the transmitted codeword. In order to correct t errors, the spheres $S_t(\mathbf{x})$ must be disjoint. Suppose $\mathbf{u} \in S_t(\mathbf{x})$, and $\mathbf{u} \in S_t(\mathbf{y})$ with $\mathbf{x}, \mathbf{y} \in C$, $\mathbf{x} \ne \mathbf{y}$. Then $d(\mathbf{x}, \mathbf{y}) \le d(\mathbf{x}, \mathbf{u}) + d(\mathbf{u}, \mathbf{y}) \le 2t$, in contradiction to $d_{\min} \ge 2t + s$. In order to detect $t + j$ errors, $0 < j \le s$, it suffices to show that $t + s$ errors can be detected. We leave this as an exercise. $\qquad\square$

Thus the code in 1.2 can detect up to two errors and correct one error. Because of this theorem one tries to construct codes with distance as large as possible.

The theorem is often stated in the following special form: A linear code C with minimum distance d can correct t errors if and only if $t = [(d - 1)/2]$. (Here the square brackets denote the greatest integer less than $(d - 1)/2$.)

The real problem of coding theory is not merely to minimize errors, but to do so without reducing the transmission rate unnecessarily. Errors can be corrected by lengthening the code blocks, but this reduces the number of message symbols that can be sent per second. To maximize the transmission rate we want code blocks which are numerous enough to encode a given message alphabet, but at the same time no longer than is necessary to achieve a given Hamming distance. In other words one of the main problems in coding theory is:

Given block length n and Hamming distance d, find the maximum number, $A(n, d)$, of binary blocks of length n which are at distances $\ge d$ from each other.

The following table gives the first few values of $A(n, d)$ for $d = 3$, i.e. the maximum sizes of single-error-correcting codes:

n	3	4	5	6	7	8	9
$A(n, 3)$	2	2	4	8	16	20	Not known, but 38, 39 or 40

Let C be a code over $\mathbb{F}_q$ of length n with M codewords. Suppose C can correct t errors, i.e. C is t-error-correcting. There are $(q - 1)^m \binom{n}{m}$ vectors of length n and weight m over $\mathbb{F}_q$. The spheres of radius t about the codewords of C are disjoint and each of the M spheres contains $1 + (q - 1)\binom{n}{1} + \ldots + (q - 1)^t \binom{n}{t}$ vectors. The total number of vectors in $\mathbb{F}_q^n$ is q^n. Thus we have

1.16 Theorem ("Hamming Bound"). *The parameters q, n, t, M of a t-error-correcting code C over $\mathbb{F}_q$ of length n with M codewords satisfy the inequality*

$$M\left(1 + (q - 1)\binom{n}{1} + \ldots + (q - 1)^t \binom{n}{t}\right) \leq q^n. \qquad \square$$

If all vectors of $\mathbb{F}_q^n$ are within or on spheres of radius t about codewords of a linear (n, k) code, then we obtain a special class of codes.

1.17 Definition. A t-error-correcting code over $\mathbb{F}_q$ is called *perfect* if equality holds in Theorem 1.16.

We shall mention perfect codes later on (see Theorem 3.17). Let $\mathbf{u} = u_1 \ldots u_n$ and $\mathbf{v} = v_1 \ldots v_n$ be vectors in $\mathbb{F}_q^n$ and let $\mathbf{u} \cdot \mathbf{v} = u_1 v_1 + \ldots + u_n v_n$ denote the dot product of $\mathbf{u}$ and $\mathbf{v}$ over $\mathbb{F}_q$. If $\mathbf{u} \cdot \mathbf{v} = 0$ then $\mathbf{u}$ and $\mathbf{v}$ are called *orthogonal.*

1.18 Definition. Let C be a linear (n, k) code over $\mathbb{F}_q$. The *dual* (or *orthogonal*) *code* $C^\perp$ of C is defined by

$$C^\perp = \{\mathbf{u} \,|\, \mathbf{u} \cdot \mathbf{v} = 0, \forall \, \mathbf{v} \in C\}.$$

Since C is a k-dimensional subspace of the n-dimensional vector space $\mathbb{F}_q^n$, the orthogonal complement is of dimension $n - k$ and an $(n, n - k)$ code. It can be shown that if the code C has a generator matrix $\mathbf{G}$ and parity-check matrix $\mathbf{H}$ then $C^\perp$ has generator matrix $\mathbf{H}$ and parity-check matrix $\mathbf{G}$. Orthogonality of the two codes can be expressed by $\mathbf{G}\mathbf{H}^T = \mathbf{H}\mathbf{G}^T = \mathbf{0}$. We summarize some of the simple properties of linear codes, the proofs of which are left as exercises.

1.19 Theorem. *Let* **H** *be a parity-check matrix of a* (n, k, d) *code C. Then the following hold.*

(i) dim $C = n - $ rank **H**;
(ii) $d = $ rank **H** $+ 1$ *iff every* $n - k$ *columns of* **H** *are linearly independent*;
(iii) $n - k \geq d - 1$;
(iv) d *is the minimum distance of C if and only if any* $s \leq d - 1$ *columns of* **H** *are linearly independent.*   □

To verify the existence of linear (n, k) codes with minimum distance d over $\mathbb{F}_q$, it suffices to show that there exists an $(n - k) \times n$ matrix **H** with condition 1.19(iv) satisfied. The construction of such a matrix proves

1.20 Theorem ("Gilbert–Varshamov Bound"). *If*

$$q^{n-k} > \sum_{i=0}^{d-2} \binom{n-1}{i}(q-1)^i,$$

then we can construct a linear (n, k) *code over* $\mathbb{F}_q$ *with minimum distance* $\geq d$.

PROOF. We construct an $(n - k) \times n$ parity check matrix **H** for such a code. Let the first column of **H** be any nonzero $(n - k)$-tuple over $\mathbb{F}_q$. The second column is any $(n - k)$-tuple over $\mathbb{F}_q$, which is not a scalar multiple of the first column. Suppose $j - 1$ columns have been chosen so that any $d - 1$ of them are linearly independent. There are at most

$$\sum_{i=0}^{d-2} \binom{j-1}{i}(q-1)^i$$

vectors obtained by taking linear combinations of $d - 2$ or fewer of these $j - 1$ columns. If the inequality of the theorem holds it will be possible to choose a jth column which is linearly independent of any $d - 2$ of the first $j - 1$ columns. This construction can be carried out in such a way that **H** has rank $n - k$. To show that the resulting code has minimum distance $d_{\min} \geq d$ follows from the fact that no $d - 1$ columns of **H** are linearly dependent.   □

Let C be a linear (n, k) code over $\mathbb{F}_q$ and let $1 \leq i \leq n$ be such that C contains a codeword with nonzero ith component. Let D be the subspace of C containing all codewords with ith component zero. In C/D there are q elements which correspond to q choices for the ith component of a codeword. If $|C| = M$ denotes the number of elements in C, then $M/|D| = |C/D|$, i.e. $|D| = q^{k-1}$. The sum of the weights of the codewords in C is $\leq nq^{k-1}(q-1)$. Since the total number of codewords of nonzero weight is $q^k - 1$, the minimum distance d must satisfy the following inequality.

1.21 Theorem ("Plotkin Bound"). *If there is a code of length n with M codewords and minimum distance d over $\mathbb{F}_q$, then*

$$d \le \frac{nM(q-1)}{(M-1)q}.$$
□

In what follows, we shall describe a simple decoding algorithm for linear codes. Let C be a linear (n, k) code over $\mathbb{F}_q$, regarded as a subspace of $\mathbb{F}_q^n$. The factor space $\mathbb{F}_q^n/C$ consists of all cosets $\mathbf{a} + C = \{\mathbf{a} + \mathbf{x} \mid \mathbf{x} \in C\}$ for arbitrary $\mathbf{a} \in \mathbb{F}_q^n$. Each coset contains q^k vectors. There is a partition of $\mathbb{F}_q^n$ of the form

$$\mathbb{F}_q^n = C \cup \{\mathbf{a}^{(1)} + C\} \cup \ldots \cup \{\mathbf{a}^{(t)} + C\} \quad \text{for } t = q^{n-k} - 1.$$

If a vector $\mathbf{y}$ is received then $\mathbf{y}$ must be an element of one of these cosets, say $\mathbf{a}^{(i)} + C$. If the codeword $\mathbf{x}^{(1)}$ has been transmitted then the error vector $\mathbf{e}$ is given as $\mathbf{e} = \mathbf{y} - \mathbf{x}^{(1)} \in \mathbf{a}^{(i)} + C - \mathbf{x}^{(1)} = \mathbf{a}^{(i)} + C$. Thus we have the following decoding rule.

1.22 Theorem. *If a vector $\mathbf{y}$ is received then the possible error vectors $\mathbf{e}$ are the vectors in the coset containing $\mathbf{y}$. The most likely error is the vector $\bar{\mathbf{e}}$ with minimum weight in the coset of $\mathbf{y}$. Thus $\mathbf{y}$ is decoded as $\bar{\mathbf{x}} = \mathbf{y} - \bar{\mathbf{e}}$.*
□

Next we show how to find the coset of $\mathbf{y}$ and apply Theorem 1.22. The vector of minimum weight in a coset is called the *coset leader*. If there are several such vectors then we arbitrarily choose one of them as coset leader. Let $\mathbf{a}^{(1)}, \ldots, \mathbf{a}^{(t)}$ be the coset leaders. We first establish the following table (due to Slepian).

$$
\begin{array}{cccc}
\mathbf{x}^{(1)} = \mathbf{0} & \mathbf{x}^{(2)} & \cdots & \mathbf{x}^{(q^k)} \\
\mathbf{a}^{(1)} + \mathbf{x}^{(1)} & \mathbf{a}^{(1)} + \mathbf{x}^{(2)} & \cdots & \mathbf{a}^{(1)} + \mathbf{x}^{(q^k)} \\
\cdots\cdots & \cdots\cdots & \cdots & \cdots\cdots \\
\mathbf{a}^{(t)} + \mathbf{x}^{(1)} & \mathbf{a}^{(t)} + \mathbf{x}^{(2)} & \cdots & \mathbf{a}^{(t)} + \mathbf{x}^{(q^k)}
\end{array}
$$

$\left.\right\}$ codewords in C

$\left.\right\}$ other cosets

$\underbrace{\quad\quad}$
coset leaders

If a vector $\mathbf{y}$ is received then we have to find $\mathbf{y}$ in the table. Let $\mathbf{y} = \mathbf{a}^{(i)} + \mathbf{x}^{(j)}$; then the decoder decides that the error $\bar{\mathbf{e}}$ is the coset leader $\mathbf{a}^{(i)}$. Thus $\mathbf{y}$ is decoded as the codeword $\bar{\mathbf{x}} = \mathbf{y} - \bar{\mathbf{e}} = \mathbf{x}^{(j)}$. The codeword $\bar{\mathbf{x}}$ occurs as the first element in the column of $\mathbf{y}$. The coset of $\mathbf{y}$ can be found by evaluating the so-called syndrome.

1.23 Definition. Let H be the parity-check matrix of a linear (n, k) code. Then the vector $S(\mathbf{y}) := H\mathbf{y}^T$ of length $n - k$ is called the *syndrome* of $\mathbf{y}$.

1.24 Theorem. (i) $S(\mathbf{y}) = \mathbf{0} \Leftrightarrow \mathbf{y} \in C$.
(ii) $S(\mathbf{y}^{(1)}) = S(\mathbf{y}^{(2)}) \Leftrightarrow \mathbf{y}^{(1)} + C = \mathbf{y}^{(2)} + C$.

PROOF. (i) This follows from the definition of C with the parity-check matrix $\mathbf{H}$.

(ii)

$$S(\mathbf{y}^{(1)}) = S(\mathbf{y}^{(2)}) \Leftrightarrow \mathbf{H}\mathbf{y}^{(1)T} = \mathbf{H}\mathbf{y}^{(2)T}$$
$$\Leftrightarrow \mathbf{H}(\mathbf{y}^{(1)T} - \mathbf{y}^{(2)T}) = 0$$
$$\Leftrightarrow \mathbf{y}^{(1)} - \mathbf{y}^{(2)} \in C$$
$$\Leftrightarrow \mathbf{y}^{(1)} + C = \mathbf{y}^{(2)} + C. \qquad \square$$

Thus we can define the cosets via syndromes. Let $\mathbf{e} = \mathbf{y} - \mathbf{x}$, $\mathbf{x} \in C, \mathbf{y} \in \mathbb{F}_q^n$; then

$$S(\mathbf{y}) = S(\mathbf{x} + \mathbf{e}) = S(\mathbf{x}) + S(\mathbf{e}) = S(\mathbf{e}),$$

i.e. $\mathbf{y}$ and $\mathbf{e}$ are in the same coset. We may now state a different formulation of the decoding algorithm 1.22.

1.22' Theorem (Decoding Algorithm). *If* $\mathbf{y} \in \mathbb{F}_q^n$ *is a received vector, find* $S(\mathbf{y})$, *and the coset leader* $\bar{\mathbf{e}}$ *with syndrome* $S(\mathbf{y})$. *Then the most likely transmitted codeword is* $\bar{\mathbf{x}} = \mathbf{y} - \bar{\mathbf{e}}$ *and we have* $d(\bar{\mathbf{x}}, \mathbf{y}) = \min\{d(\mathbf{x}, \mathbf{y}) \mid \mathbf{x} \in C\}$. $\square$

1.25 Example. Let C be a binary linear $(4, 2)$ code with generator matrix $\mathbf{G}$ and parity-check matrix $\mathbf{H}$, where

$$\mathbf{G} = \begin{pmatrix} 1 & 0 & 1 & 1 \\ 0 & 1 & 0 & 1 \end{pmatrix}, \qquad \mathbf{H} = \begin{pmatrix} 1 & 0 & 1 & 0 \\ 1 & 1 & 0 & 1 \end{pmatrix}.$$

The corresponding coset table is

message	0 0	1 0	0 1	1 1	
codewords	0 0 0 0	1 0 1 1	0 1 0 1	1 1 1 0	$(0 \ 0)^T$
	1 0 0 0	0 0 1 1	1 1 0 1	0 1 1 0	$(1 \ 1)^T$
other	0 1 0 0	1 1 1 1	0 0 0 1	1 0 1 0	$(0 \ 1)^T$
cosets	0 0 1 0	1 0 0 1	0 1 1 1	1 1 0 0	$(1 \ 0)^T$

coset   syndromes
leaders

If $\mathbf{y} = 1111$ is received, then $S(\mathbf{y}) = \mathbf{H}\mathbf{y}^T = (01)^T$. Thus the error $\bar{\mathbf{e}} = 0100$ and $\mathbf{y}$ is decoded as $\bar{\mathbf{x}} = \mathbf{y} - \bar{\mathbf{e}} = 1011$ and the corresponding message is 10. $\square$

In large linear codes, finding explicitly the coset leaders with least weights is practically impossible. (A $(50, 20)$ code over $\mathbb{F}_2$ has approximately 10^9 cosets.) Therefore one constructs codes which are more systematic, such as Hamming codes. In the binary case we see that $S(\mathbf{y}) = \mathbf{H}\mathbf{e}^T$ for $\mathbf{y} = \mathbf{x} + \mathbf{e}$, $\mathbf{y} \in \mathbb{F}_2^n$, $\mathbf{x} \in C$ so that we have

1.26 Theorem. *In a binary code the syndrome is equal to the sum of those columns of the parity-check matrix* **H** *in which errors occurred.* ☐

In a single-error-correcting code the columns of **H** must be nonzero, otherwise we could not detect an error in the ith place of the message if the ith column is zero. All columns must be pairwise distinct, otherwise we could not distinguish between the errors.

1.27 Definition. A binary code C_m of length $n = 2^m - 1$, $m \geq 2$, with an $m \times (2^m - 1)$ parity-check matrix **H** whose columns consist of all nonzero binary vectors of length m, is called a *binary Hamming code*.

With the aid of 1.19(ii) it is easily verified that C_m is a $(2^m - 1, 2^m - 1 - m, 3)$ code. The rank of **H** is m. Decoding in Hamming codes is particularly simple. Theorem 1.15 tells us that C_m can correct errors **e** with weight $w(\mathbf{e}) = 1$ and C_m can detect errors with $w(\mathbf{e}) \leq 2$. We choose the lexicographical order for the columns of **H**, i.e. the ith column is the binary representation of i. If an error occurs in the ith column then $S(\mathbf{y}) = \mathbf{H}\mathbf{y}^T = \mathbf{H}\mathbf{e}^T$ is the binary representation of i.

1.28 Example. Let C_3 be the $(7, 4, 3)$ Hamming code with parity-check matrix

$$\mathbf{H} = \begin{pmatrix} 0 & 0 & 0 & 1 & 1 & 1 & 1 \\ 0 & 1 & 1 & 0 & 0 & 1 & 1 \\ 1 & 0 & 1 & 0 & 1 & 0 & 1 \end{pmatrix}.$$

The first column is the binary representation of $1 = (001)^T$, the second column is $2 = (010)^T$, etc. If $S(\mathbf{y}) = (101)^T$, then we know that an error occurred in the fifth column, since $(101)^T$ is the binary representation of 5. ☐

Interchanging columns of **H** gives the parity-check matrix in the form $\mathbf{H}' = (\mathbf{A}, \mathbf{I}_m)$, where **A** consists of all columns with at least two 1's. We know that C_m is up to "equivalence" (in the sense of the remark after 1.8) uniquely determined, i.e. any linear $(2^m - 1, 2^m - 1 - m, 3)$ code is equivalent to C_m.

We summarize some further properties and remarks for Hamming codes.

1.29 Remarks. (i) If we choose

$$\mathbf{H}'' = \begin{pmatrix} 1 & 1 & 1 & 0 & 1 & 0 & 0 \\ 0 & 1 & 1 & 1 & 0 & 1 & 0 \\ 0 & 0 & 1 & 1 & 1 & 0 & 1 \end{pmatrix}$$

as the parity-check matrix for the $(7, 4, 3)$ Hamming code in 1.28 then we see that the code is cyclic, i.e. a cyclic shift of a codeword is also a codeword. We shall study this property in §2.

(ii) The Hamming code C_m is a perfect 1-error-correcting linear code.

(iii) A generalized Hamming code over $\mathbb{F}_q$ can be defined by an $m \times (q^m - 1)/(q - 1)$ parity-check matrix $\mathbf{H}$ such that no two columns of $\mathbf{H}$ are multiples of each other. This gives us a $((q^m - 1)/(q - 1), (q^m - 1)/(q - 1) - m, 3)$ code over $\mathbb{F}_q$.

(iv) Hamming codes cannot correct any errors $\mathbf{e}$ with $w(\mathbf{e}) \geq 2$. A generalization of Hamming codes are the BCH codes, which will be studied in the next sections.

(v) We obtain a so-called "*extended code*", if we add a new element, the negative sum of the first n symbols, in an (n, k, d) code. The parity-check matrix $\bar{\mathbf{H}}$ of an extended code can be obtained from the parity-check matrix $\mathbf{H}$ of the original code by the addition of a column of zeros and then a row of ones. The code C_3 of Example 1.28 can be extended to an $(8, 4, 4)$ extended Hamming code with parity-check matrix

$$\bar{\mathbf{H}} = \begin{pmatrix} 1 & 1 & 1 & 1 & 1 & 1 & 1 & 1 \\ 0 & 0 & 0 & 1 & 1 & 1 & 1 & 0 \\ 0 & 1 & 1 & 0 & 0 & 1 & 1 & 0 \\ 1 & 0 & 1 & 0 & 1 & 0 & 1 & 0 \end{pmatrix}.$$

Any nonzero codeword in this extended code is of weight at least 4. By Theorem 1.15 this code can correct all errors of weight 1 and detect all errors of weight ≤ 3.

(vi) The dual code of the binary $(2^m - 1, 2^m - 1 - m, 3)$ Hamming code C_m is the *binary simplex code* with parameters $(2^m - 1, m, 2^{m-1})$.

(vii) The dual of an extended $(2^m, 2^m - 1 - m, 4)$ Hamming code is called a *first-order Reed–Muller code*. This code can also be obtained by "lengthening" a simplex code.

A code is *lengthened* by the addition of the codeword $11 \ldots 1$ and then is extended by adding an overall parity check. (This has the effect of adding one more message symbol.) The first-order Reed–Muller code has parameters $(2^m, m + 1, 2^{m-1})$. The decoding algorithm for these codes is based on the fast Fourier transform and is not inherently binary, and so the first-order Reed–Muller codes can be used well on so-called Gaussian channels. This is the main reason that all of NASA's Mariner deep-space probes launched between 1969 and 1977 were equipped with a $(32, 6, 15)$ Reed–Muller code for communication purposes. Other deep-space probes have been equipped with convolutional codes (see McELIECE, MASSEY).

The *weight distribution* of a code is the number of codewords of any weight in the code. It is useful for the practical computation of the probability of correct decoding. We can describe this by the list of numbers A_i, which gives the number of codewords of weight i in the code. A_0 is always 1. As an example we can calculate the weight distribution of the $(7, 4, 3)$ Hamming code. The sum of the rows of the generator matrix of this code is $(1, 1, 1, 1, 1, 1, 1) = \mathbf{a}$, say, so $A_7 = 1$. The only other possible nonzero weights are 3 and 4, since with any $\mathbf{x}$ in the code, $\mathbf{a} + \mathbf{x}$ is also in the code. Therefore the weight distribution is $A_0 = A_7 = 1$, $A_3 = A_4 = 7$. All other A_i are 0.

1.30 Definition. Let A_i denote the number of codewords $\mathbf{x} \in C$ of weight i, $0 \le i \le n$. Then the polynomial

$$A(X, Y) = \sum_{i=0}^{n} A_i X^i Y^{n-i}$$

in $\mathbb{C}[X, Y]$ is called the *weight enumerator* of C.

There is an elementary way to establish the following Theorem 1.34 on weight distribution. However, we use characters to describe the results more elegantly.

A homomorphism χ from the additive group $(\mathbb{F}_q, +)$, $\mathbb{F}_q$ for short, into the multiplicative group $(\mathbb{C}, \cdot)$ is called a *character* of $(\mathbb{F}_q, +)$, and $\chi_0 \colon \mathbb{F}_q \to 1$ is called the *trivial character*.

1.31 Lemma. (i) $\sum_{a \in \mathbb{F}_q} \chi(a) = 0$ *for a nontrivial character of* $\mathbb{F}_q$.
   (ii) *The characters of* $\mathbb{F}_q$ *form a group which is isomorphic to the group* $(\mathbb{F}_q, +)$.
   (iii) *For any* $a, b \in \mathbb{F}_q$, $a \ne -b$, *we have* $\sum \chi(a)\chi(b) = 0$ *where the summation runs through all characters of* $\mathbb{F}_q$.

PROOF. (i) Since χ is nontrivial, there is some $b \in \mathbb{F}_q$ with $\chi(b) \ne 1$. Then

$$\chi(b) \sum_{a \in \mathbb{F}_q} \chi(a) = \sum_{a \in \mathbb{F}_q} \chi(b)\chi(a) = \sum_{a \in \mathbb{F}_q} \chi(a + b) = \sum_{a \in \mathbb{F}_q} \chi(a).$$

Thus $(\chi(b) - 1) \sum \chi(a) = 0$, and the result follows.

(ii) It is obvious that the characters of $\mathbb{F}_q$ form a group with respect to pointwise multiplication. If $q = p^m$, then $\mathbb{F}_q$ is isomorphic to $(\mathbb{Z}_p)^m$. Then χ is of the form $(a_1, \dots, a_m) \to (\xi_1^{a_1}, \dots, \xi_m^{a_m})$, where ξ_i are pth roots of unity in $\mathbb{C}$. The set of these m-tuples of pth roots of unity forms a group, which is isomorphic to $(\mathbb{Z}_p)^m$ and therefore to $(\mathbb{F}_q, +)$.

(iii) Since we have $a + b \ne 0$, there exists a nontrivial χ_1 such that

$$\chi_1(a)\chi_1(b) = \chi_1(a + b) \ne 1.$$

Then

$$\chi_1(a)\chi_1(b) \sum_\chi \chi(a)\chi(b) = \sum_\chi \chi(a)\chi(b)$$

and the result follows as in (i). $\square$

1.32 Definition. Let χ be a nontrivial additive character of $\mathbb{F}_q$ and let $\mathbf{v} \cdot \mathbf{u}$ denote the dot product of $\mathbf{v}, \mathbf{u} \in \mathbb{F}_q^n$. For fixed $\mathbf{v} \in \mathbb{F}_q^n$ we define the mapping

$$\chi_\mathbf{v} \colon \mathbb{F}_q^n \to \mathbb{C}, \mathbf{u} \mapsto \chi(\mathbf{v} \cdot \mathbf{u}) \quad \text{for } \mathbf{u} \in \mathbb{F}_q^n.$$

If V is a vector space over $\mathbb{C}$ and f a mapping from $\mathbb{F}_q^n$ into V, then we define

$$g_f \colon \mathbb{F}_q^n \to V, \mathbf{u} \to \sum_{\mathbf{v} \in \mathbb{F}_q^n} \chi_\mathbf{v}(\mathbf{u}) f(\mathbf{v}) \quad \text{for } \mathbf{u} \in \mathbb{F}_q^n.$$

1.33 Lemma. *Let W be a subspace of $\mathbb{F}_q^n$, $W^\perp$ its orthogonal complement, $f \colon \mathbb{F}_q^n \to V$ a mapping from $\mathbb{F}_q^n$ into a vector space V over $\mathbb{C}$ and χ a nontrivial additive character of $\mathbb{F}_q$. Then*

$$\sum_{\mathbf{v} \in W} g_f(\mathbf{u}) = |W| \sum_{\mathbf{v} \in W^\perp} f(\mathbf{v}).$$

PROOF. We have

$$\sum_{\mathbf{u} \in W} g_f(\mathbf{u}) = \sum_{\mathbf{u} \in W} \sum_{\mathbf{v} \in \mathbb{F}_q^n} \chi_\mathbf{v}(\mathbf{u}) f(\mathbf{v}) = \sum_{\mathbf{v} \in \mathbb{F}_q^n} \sum_{\mathbf{u} \in W} \chi(\mathbf{v} \cdot \mathbf{u}) f(\mathbf{v})$$

$$= |W| \sum_{\mathbf{v} \in W^\perp} f(\mathbf{v}) + \sum_{\mathbf{v} \notin W^\perp} \sum_{a \in \mathbb{F}_q} \sum_{\substack{\mathbf{u} \in W \\ \mathbf{v} \cdot \mathbf{u} = a}} \chi(a) f(\mathbf{v}).$$

We use 1.31(i) and the fact that the number of vectors $\mathbf{u} \in W$ with $\mathbf{v} \cdot \mathbf{u} = a$ is a constant, for a $\mathbf{v} \notin W^\perp$. Then

$$\sum_{\mathbf{u} \in W} g_f(\mathbf{u}) = |W| \sum_{\mathbf{v} \in W^\perp} f(\mathbf{v}) + \frac{|W|}{q} \sum_{\mathbf{v} \notin W^\perp} f(\mathbf{v}) \sum_{a \in \mathbb{F}_q} \chi(a) = |W| \sum_{\mathbf{v} \in W^\perp} f(\mathbf{v}). \qquad \square$$

Now let V be the space of polynomials in two indeterminates X and Y over $\mathbb{C}$ and let f be defined by $f(\mathbf{v}) = X^{w(\mathbf{v})} Y^{n - w(\mathbf{v})}$, where $w(\mathbf{v})$ denotes the weight of $\mathbf{v} \in \mathbb{F}_q^n$.

1.34 Theorem (MacWilliams Identity). *Let C be a linear (n, k) code over $\mathbb{F}_q$ and $C^\perp$ its dual code. If $A(X, Y)$ is the weight enumerator of C and $A^\perp(X, Y)$ is the weight enumerator of $C^\perp$, then*

$$A^\perp(X, Y) = \frac{1}{q^k} A(Y - X, Y + (q - 1)X).$$

PROOF. Let $f \colon \mathbb{F}_q^n \to \mathbb{C}[X, Y]$ be as given above. Then

$$A^\perp(X, Y) = \sum_{\mathbf{v} \in W^\perp} f(\mathbf{v}).$$

Let g_f be as in 1.32 and for $v \in \mathbb{F}_q$ let

$$|v| := \begin{cases} 1 & \text{if } v \neq 0, \\ 0 & \text{if } v = 0. \end{cases}$$

For $\mathbf{u} = (u_1, \ldots, u_n) \in \mathbb{F}_q^n$ we have

$$g_f(\mathbf{u}) = \sum_{\mathbf{v} \in \mathbb{F}_q^n} \chi(\mathbf{v} \cdot \mathbf{u}) X^{w(\mathbf{v})} Y^{n-w(\mathbf{v})}$$

$$= \sum_{v_1, \ldots, v_n \in \mathbb{F}_q} \chi(u_1 v_1 + \ldots + u_n v_n) X^{|v_1| + \ldots + |v_n|} Y^{(1-|v_1|) + \ldots + (1-|v_n|)}$$

$$= \sum_{v_1, \ldots, v_n \in \mathbb{F}_q} \prod_{i=1}^{n} (\chi(u_i v_i) X^{|v_i|} Y^{1-|v_i|})$$

$$= \prod_{i=1}^{n} \sum_{v \in \mathbb{F}_q} (\chi(u_i v) X^{|v|} Y^{1-|v|}).$$

For $u_i = 0$ we have $\chi(u_i v) = \chi(0) = 1$, so the corresponding factor in the product is $(q-1)X + Y$. For $u_i \neq 0$ the factor is $Y + X \sum_{v \in \mathbb{F}_q^*} \chi(v) = Y - X$. Therefore

$$g_f(\mathbf{u}) = (Y - X)^{w(\mathbf{u})} (Y + (q-1)X)^{n-w(\mathbf{u})}.$$

Lemma 1.33 implies

$$|C| A^\perp(X, Y) = |C| \sum_{\mathbf{v} \in C^\perp} f(\mathbf{v}) = \sum_{\mathbf{u} \in C} g_f(\mathbf{u}) = A(Y - X, Y + (q-1)X).$$

The required identity follows, since $|C| = q^k$.   □

1.35 Corollary. *Let $X = Z$ and $Y = 1$ in the weight enumerators $A(X, Y)$ and $A^\perp(X, Y)$ and let the resulting polynomials be denoted by $A(Z)$ and $A^\perp(Z)$, respectively. Then the MacWilliams identity can be written in the form*

$$A^\perp(Z) = \frac{1}{q^k} (1 + (q-1)Z)^n A\left(\frac{1 - Z}{1 + (q-1)Z}\right).$$

1.36 Example. Let C be the binary $(3, 2)$ code with codewords $\{000, 001, 101, 110\}$. Then $C^\perp = \{000, 111\}$. It is easy to calculate the weights of the code words, namely $A_0 = 1$, $A_2 = 3$, $A_0^\perp = 1$, $A_3^\perp = 1$. Therefore

$$A(X, Y) = Y^3 + 3X^2 Y \quad \text{and} \quad A^\perp(X, Y) = Y^3 + X^3.$$

We can verify the MacWilliams identity in this case.

$$\frac{1}{q^k} A(Y - X, Y + (q-1)X) = \tfrac{1}{4}((Y + X)^3 + 3(Y - X)^2(Y + X))$$

$$= Y^3 + X^3 = A^\perp(X, Y).$$ □

1.37 Example. Let C_m be the binary Hamming code of length $n = 2^m - 1$ and dimension $n - m$. The dual code $C_m^\perp$ has as its generator matrix the parity-check matrix $\mathbf{H}$ of C_m, which consists of all nonzero column vectors of length m over $\mathbb{F}_2$. $C_m^\perp$ consists of the zero vector and $2^m - 1$ vectors of weight 2^{m-1}. Thus

$$A^{\perp}(X, Y) = Y^n + (2^m - 1)X^{2^{m-1}} Y^{2^{m-1}-1}.$$

By Theorem 1.34 the weight enumerator for C_m is

$$A(X, Y) = \frac{1}{n+1}((Y + X)^n + n(Y - X)^{(n+1)/2}(Y + X)^{(n-1)/2}).$$

In the form of Corollary 1.35 let $A(X, 1) = A(Z) = \sum_{i=0}^{n} A_i X^i$. Then it can be verified that $A(Z)$ satisfies the differential equation

$$(1 - Z)^2 \frac{dA(Z)}{dZ} + (1 + nZ)A(Z) = (1 + Z)^n,$$

with initial condition $A(0) = A_0 = 1$. This is equivalent to

$$iA_i = \binom{n}{i-1} - A_{i-1} - (n - i + 2)A_{i-2} \quad \text{for } i = 2, 3, \ldots, n,$$

with initial conditions $A_0 = 1$, $A_1 = 0$, which enables us to calculate the A_i recursively.   □

We note that the MacWilliams identity can also be expressed in the following equivalent form. The notation is as in 1.30. Let C be an (n, k) code over $\mathbb{F}_q$. Then

$$\sum_{r=0}^{n} \binom{i}{r} A_i = q^{k-r} \sum_{i=0}^{n} (-1)^i \binom{n - i}{n - r} A_i^{\perp} \quad \text{for } 0 \le r \le n.$$

PROBLEMS

1. Let G be generator matrix of a binary $(5, 2)$ code

$$G = \begin{pmatrix} 0 & 1 & 1 & 1 & 1 \\ 1 & 0 & 0 & 1 & 0 \end{pmatrix}.$$

   Determine a parity-check matrix, all syndromes and coset leaders for this code.

*2. Let C be a linear (n, k) code over $\mathbb{F}_q$ with minimum distance $d \ge 2t + 1$. Let N be the cardinality of the set $\{u \mid u \in \mathbb{F}_q^n, d(u, v) \le t\}$. Prove that $N \le q^{n-k}$.

3. A linear $(5, 3)$ code over $\mathbb{F}_4$ is defined by the generator matrix

$$G = \begin{pmatrix} 1 & 0 & 0 & 1 & 1 \\ 0 & 1 & 0 & 1 & \alpha \\ 0 & 0 & 1 & 1 & \beta \end{pmatrix}.$$

   Determine its parity-check matrix, its codewords, its dual code. Prove that it is a single-error-correcting perfect code.

*4. Design a circuit for encoding and decoding with a $(7, 4)$ Hamming code.

5. Show that in a binary code either all vectors have even weight or half have even weight and half have odd weight.

6. Let

$$G = \begin{pmatrix} 1 & 0 & 1 & 1 \\ 0 & 1 & 2 & 1 \end{pmatrix}$$

be the generator matrix of a ternary $(4, 2)$ code C. Find a parity-check matrix for this code. Also find the weight distribution of C.

*7. If C is a ternary self-orthogonal code (see Exercise 11) show that $A_j = 0$ unless 3 divides j.

8. Let H be the parity-check matrix of a $(7, 4)$ Hamming code. Suppose two words of the code were sent and the words $1\ 0\ 0\ 1\ 0\ 1\ 0$ and $1\ 1\ 0\ 1\ 0\ 1\ 1$ were received from a noisy channel. Decode these words.

$$H = \begin{pmatrix} 0 & 0 & 0 & 1 & 1 & 1 & 1 \\ 0 & 1 & 1 & 0 & 0 & 1 & 1 \\ 1 & 0 & 1 & 0 & 1 & 0 & 1 \end{pmatrix}.$$

9. For the $(15, 11)$ Hamming code with columns of the parity-check matrix in natural order using maximum likelihood decoding,
    (i) find the parity-check matrix;
    (ii) determine the information rate;
    (iii) decode the received word $1\ 1\ 1\ 1\ 0\ 0\ 1\ 0\ 1\ 1\ 0\ 0\ 0\ 1\ 0$;
    (iv) find parity check equations (similar to Example 1.2).

10. Determine the probability of making no error when using a $(7, 4)$ Hamming code with maximum likelihood decoder over a binary symmetric channel.

11. Let C be a $(5, 3)$ linear code over $\mathbb{F}_2$ with generator matrix

$$G = \begin{pmatrix} 1 & 1 & 1 & 0 & 0 \\ 0 & 0 & 1 & 1 & 0 \\ 1 & 1 & 1 & 1 & 1 \end{pmatrix}.$$

Determine its dual code and the weight enumerators for C and its dual code.

EXERCISES (Solutions in Chapter 8, p. 462)

1. Prove Theorem 1.11.

2. Show that a code can at the same time correct $\leq r$ errors and detect $r + t$ errors if its minimum distance is $\geq r + t + 1$.

3. Let G_1 and G_2 be generator matrices of a linear (n_1, k, d_1) code and (n_2, k, d_2) code, respectively. Show that the codes with generator matrices

$$\begin{pmatrix} G_1 & 0 \\ 0 & G_2 \end{pmatrix} \quad \text{and} \quad (G_1 \quad G_2)$$

are $(n_1 + n_2, 2k, \min\{d_1, d_2\})$ and $(n_1 + n_2, k, d)$ codes, respectively, where $d \geq d_1 + d_2$.

4. A linear code $C \subseteq \mathbb{F}_2^5$ is defined by the generator matrix

$$G = \begin{pmatrix} 0 & 1 & 0 & 0 & 1 \\ 0 & 0 & 1 & 0 & 1 \\ 1 & 0 & 0 & 1 & 1 \end{pmatrix}.$$

Determine the rank of G, the minimum distance of C, a parity-check matrix for C and all the codewords.

5. Determine the dual code $C^\perp$ to the code C given in Exercise 4. Find the table of all cosets of $\mathbb{F}_2^5$ mod $C^\perp$, and determine the coset leaders and syndromes. If $y = 0\ 1\ 0\ 0\ 1$ is received, which word is most likely to have been transmitted?

*6. Prove that all binary Hamming codes and the repetition codes of odd block length are perfect. Also prove that the general Hamming codes over $\mathbb{F}_q$ defined in 1.29(iii) are perfect.

*7. The *covering radius* r for a binary code of length n is defined as the smallest number s so that spheres of radius s about codewords cover the vector space $\mathbb{F}_2^n$. Show that the covering radius is the weight of the coset of largest weight.

*8. Use the definition of equivalent codes to show that the binary linear codes with generator matrices

$$G_1 = \begin{pmatrix} 1 & 1 & 1 & 0 \\ 0 & 1 & 1 & 0 \\ 0 & 0 & 1 & 1 \end{pmatrix} \quad \text{and} \quad G_2 = \begin{pmatrix} 1 & 0 & 1 & 1 \\ 0 & 1 & 1 & 1 \\ 1 & 0 & 0 & 1 \end{pmatrix},$$

respectively, are equivalent.

9. Determine the generator matrix, all codewords and the dual code of the $(4, 2)$ code over $\mathbb{F}_3$ with parity-check matrix

$$H = \begin{pmatrix} 1 & 1 & 1 & 0 \\ 1 & 2 & 0 & 1 \end{pmatrix}.$$

*10. Let $C_i^\perp$ be the dual code of a code C_i, $i = 1, 2$. Prove:
   (i) $(C_i^\perp)^\perp = C_i$.
   (ii) $(C_1 + C_2)^\perp = C_1^\perp \cap C_2^\perp$.
   (iii) If C_1 is the $(n, 1, n)$ binary repetition code, then $C_1^\perp$ is the $(n, n-1, 2)$ parity-check code.

*11. A code C is *self-orthogonal* if $C \subseteq C^\perp$. Prove:
   (i) If the rows of a generator matrix G for a binary (n, k) code C have even weight and are orthogonal to each other, then C is self-orthogonal, and conversely.
   (ii) If the rows of a generator matrix G for a binary (n, k) code C have weights divisible by 4 and are orthogonal to each other, then C is self-orthogonal and all weights in C are divisible by 4.

12. Let C be a binary $(7, 3)$ code with generator matrix

$$G = \begin{pmatrix} 0 & 0 & 0 & 1 & 1 & 1 & 1 \\ 0 & 1 & 1 & 0 & 0 & 1 & 1 \\ 1 & 0 & 1 & 0 & 1 & 0 & 1 \end{pmatrix}.$$

Show that C is self-orthogonal. Find its dual code.

13. Let

$$H = \begin{pmatrix} 1 & 0 & 0 & 1 & 1 & 0 & 1 \\ 0 & 1 & 0 & 1 & 0 & 1 & 1 \\ 0 & 0 & 1 & 0 & 1 & 1 & 1 \end{pmatrix}$$

be a parity-check matrix of the $(7, 4)$ Hamming code. If $y = 1110011$ is received, determine the codeword which was most likely sent.

14. Use the Hamming $(7, 4)$ code to encode the messages 0110 and 1011. Also encode these messages using the extended modified Hamming $(8, 4)$ code, which is obtained from the $(7, 4)$ Hamming parity-check matrix by adding a column of 0's and filling the new row with 1's. Decode 11001101 by this $(8, 4)$ code.

*15. Find the parity-check matrix and generator matrix of the ternary Hamming code of length 13. Decode 0110100000011 using this code.

16. Verify the MacWilliams identity for:
   (i) the self-dual code $\{00, 11\}$;
   (ii) the $(7, 4, 3)$ Hamming code.

17. Find the weight distribution of the extended Hamming $(8, 4)$ code.

*18. Let C be a binary linear (n, k) code with weight enumerator $A(x, y) = \sum_{j=0}^{n} A_i x^i y^{n-i}$ and let $A^{\perp}(x, y) = \sum_{i=0}^{n} A_i^{\perp} x^i y^{n-i}$ be the weight enumerator of the dual code $C^{\perp}$. Show the following identity for $r = 0, 1, \ldots$:

$$\sum_{i=0}^{n} i^r A_i = \sum_{i=0}^{n} (-1)^i A_i^{\perp} \sum_{t=0}^{r} t!\, S(r, t) 2^{k-t} \binom{n-i}{n-t},$$

where

$$S(r, t) = \frac{1}{t!} \sum_{j=0}^{t} (-1)^{t-j} \binom{t}{j} j^r$$

is a Stirling number of the second kind and the binomial coefficient $\binom{m}{h}$ is defined to be 0 whenever $h > m$ or $h < 0$. Write down the identity for $r = 0$, 1 and 2.

§2. Cyclic Codes

Cyclic codes are codes which have been studied extensively in the literature. They have relatively simple coding and decoding algorithms, which can easily be implemented. Moreover, cyclic codes also have important algebraic properties. Many of the linear codes are also cyclic.

Let $(\mathbb{F}_q^n, +, \cdot)$ be the n-dimensional vector space of n-tuples $(a_0, a_1, \ldots, a_{n-1}) \in \mathbb{F}_q^n$, with the usual operations of addition of n-tuples and scalar multiplication of n-tuples by elements in $\mathbb{F}_q$. The mapping

$$Z: \mathbb{F}_q^n \to \mathbb{F}_q^n, (a_0, a_1, \ldots, a_{n-1}) \mapsto (a_{n-1}, a_0, \ldots, a_{n-2})$$

is a linear mapping, called a *"cyclic shift"*.

We shall consider the linear algebra $A = (\mathbb{F}_q[x], +, \cdot, .)$ with the operation $+$ of polynomial addition, polynomial multiplication $\cdot$ (not composition) and multiplication of polynomials by elements of $\mathbb{F}_q$. Then $(\mathbb{F}_q[x], +, \cdot)$ is a ring and $(\mathbb{F}_q[x], +, \cdot)$ is a vector space over $\mathbb{F}_q$. We define a subspace V_n of this vector space by

$$V_n := \{v \in \mathbb{F}_q[x] | \text{degree } v < n\}$$

$$= \{v_0 + v_1 x + \ldots + v_{n-1}x^{n-1} | v_i \in \mathbb{F}_q, 0 \le i \le n - 1\}.$$

We can identify the two spaces V_n and $\mathbb{F}_q^n$ by the isomorphism

$$\tau: (\mathbb{F}_q^n, +, .) \to V_n,$$

$$\tau: (v_0, v_1, \ldots, v_{n-1}) \mapsto v_0 + v_1 x + \ldots + v_{n-1}x^{n-1}.$$

From now on we shall use both the word vector and polynomial for

$$v = (v_0, v_1, \ldots, v_{n-1}) = v_0 + v_1 x + \ldots + v_{n-1}x^{n-1}.$$

Consider polynomial ring $R = (\mathbb{F}_q[x], +, \cdot)$ and its factor ring $R/(x^n - 1)$, modulo the principal ideal generated by $x^n - 1$ in R. Then the mapping

$$\omega: \mathbb{F}_q^n \to R/(x^n - 1), \qquad \omega(v_0, v_1, \ldots, v_{n-1}) = v_0 + \ldots + v_{n-1}x^{n-1}$$

is an isomorphism of the additive group $\mathbb{F}_q^n$ onto the factor group of all polynomials of degree $< n$ over $\mathbb{F}_q$, denoted by V_n or $(V_n, +)$. V_n is also an algebra over $\mathbb{F}_q$ with respect to the operations of addition, scalar multiplication by elements in $\mathbb{F}_q$ and multiplication $*$ of polynomials modulo $x^n - 1$, defined by

$$v_1 * v_2 = v :\Leftrightarrow v_1 v_2 \equiv v \pmod{x^n - 1},$$

for all $v_1, v_2 \in \mathbb{F}_q[x]$, $v \in V_n$. If $\deg v_1 v_2 < n$, then $v_1 v_2 = v_1 * v_2$.

2.1 Example. Let $n = 5$, $q = 7$. Then

$$(x^4 + x^3 + x^2 + x + 1) * (x^3 + x^2 + 1) = 3(x^4 + x^3 + x^2 + x + 1).$$

This is because

$$(x^4 + x^3 + x^2 + x + 1)(x^3 + x^2 + 1)$$

$$= x^7 + 2x^6 + 2x^5 + 3x^4 + 3x^3 + 2x^2 + x + 1$$

$$\equiv x^2 + 2x + 2 + 3x^4 + 3x^3 + 2x^2 + x + 1$$

$$\equiv 3x^4 + 3x^3 + 3x^2 + 3x + 3 \pmod{x^5 - 1}. \qquad \square$$

For the linear map Z we have $Z(v) = x * v$ and more generally $Z^i(v) = x^i * v$ for $v \in \mathbb{F}_q[x]$, $i \in \mathbb{N}$. Here $Z^i = Z \circ Z \circ \ldots \circ Z$ (i times). Cyclic shifts can be implemented by using *linear shift registers*. The basic building blocks of a shift register are *delay* (or storage) *elements* and *adders*. The delay element (flip-flop) has one input and one output and is regulated by an external synchronous clock so that its input at a particular time appears as its output one unit of time later. The adder has two inputs and one output, the output being the sum in $\mathbb{F}_q$ of the two inputs. The delay element and adder can be represented as follows.

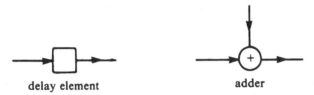

delay element   adder

Figure 4.3

In addition a *constant multiplier* and a *constant adder* are used, which multiply or add constant elements of $\mathbb{F}_q$ to the input. Their representation is as follows.

constant multiplier   constant adder

Figure 4.4

A shift register is built by interconnecting a finite number of these devices in a suitable way along closed loops. As an example, we give a binary shift register with four delay elements a_0, a_1, a_2 and a_3 and two binary adders. At time 0, four binary elements, say 1, 1, 0 and 1 are placed in a_0, a_1, a_2 and a_3, respectively. These positions can be interpreted as message positions. After one time interval $a_0 = 1$ is output, $a_1 = 1$ is shifted into a_0, $a_2 = 0$ into a_1, $a_3 = 1$ into a_2 and the new element is entered into a_3. If the shift register is of the form

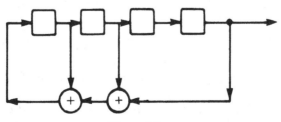

Figure 4.5

then this new element is $a_0 + a_2 + a_3$. To summarize the outputs for seven time intervals, we obtain

Outputs	a_0 a_1 a_2 a_3	Time
	1 1 0 1	0
1	1 0 1 0	1
1 1	0 1 0 0	2
1 1 0	1 0 0 0	3
1 1 0 1	0 0 0 1	4
1 1 0 1 0	0 0 1 1	5
1 1 0 1 0 0	0 1 1 0	6
1 1 0 1 0 0 0	1 1 0 1	7

Continuing this process for time $8, 9, \ldots$ we see that the output vector 1101000 generated from the four initial entries 1101 will be repeated. The process can be interpreted as encoding the message 1101 into a code word 1101000. The code generated in this way is a linear code, with the property that whenever $(v_0, v_1, \ldots, v_{n-1})$ is in the code then the cyclic shift $(v_{n-1}, v_0, v_1, \ldots, v_{n-2})$ is also in the code.

2.2 Definition. A k-dimensional subspace C of $\mathbb{F}_q^n$ is called a *cyclic code* if $Z(v) \in C$ for all $v \in C$, that is

$$v = (v_0, v_1, \ldots, v_{n-1}) \in C \Rightarrow (v_{n-1}, v_0, \ldots, v_{n-2}) \in C$$

for $v \in \mathbb{F}_q^n$.

If C is a cyclic, linear (n, k) code, for brevity, we call C a cyclic (n, k) code.

2.3 Example. Let $C \subseteq \mathbb{F}_2^7$ be defined by the generator matrix

$$\begin{pmatrix} 1 & 1 & 1 & 0 & 1 & 0 & 0 \\ 0 & 1 & 1 & 1 & 0 & 1 & 0 \\ 0 & 0 & 1 & 1 & 1 & 0 & 1 \end{pmatrix} = \begin{pmatrix} g^{(1)} \\ g^{(2)} \\ g^{(3)} \end{pmatrix}.$$

We show that C is cyclic. Each codeword of C is a linear combination of the linearly independent vectors $g^{(1)}$, $g^{(2)}$, $g^{(3)}$. C is cyclic if and only if $Z(g^{(i)}) \in C$ for $i = 1, 2, 3$. We have

$$Z(g^{(1)}) = g^{(2)}, \qquad Z(g^{(2)}) = g^{(3)}, \qquad Z(g^{(3)}) = g^{(1)} + g^{(2)}. \qquad \square$$

We now prove an elegant algebraic characterization of cyclic codes.

2.4 Theorem. *A linear code* $C \subseteq V_n$ *is cyclic if and only if* C *is a principal ideal in* V_n, *generated by* $g \in C$.

PROOF. Let C be cyclic and let $f = \sum a_i x^i \in F_q[x]$. Then for $v \in C$ we have $f * v = \sum a_i(x^i * v) = \sum a_i Z^i(v) \in C$, and moreover $C \leq V_n$. Now C is a principal ideal (g) with $g \in C$ where g is an element in C of least nonzero degree.

Conversely, let $C = (g)$ and $v \in C$. Since $Z(v) = x * v \in C$, we see that C is cyclic.   □

2.5 Remark and Definition. The polynomial $g \in C$ in Theorem 2.4 can be assumed to be monic. If we suppose in addition that $g \mid x^n - 1$ (this can be derived from 2.4) then g is uniquely determined and is called the *generator polynomial* of C. The elements of C are called *codewords*, *code polynomials* or *code vectors*.

2.6 Remark. Let $g = g_0 + g_1 x + \ldots + g_m x^m \in V_n$, $g \mid (x^n - 1)$ and $\deg g = m < n$. Let C be a linear (n, k) code, with $k = n - m$, defined by the generator matrix

$$G = \begin{pmatrix} g_0 & g_1 & \cdots & g_m & 0 & \cdots & 0 \\ 0 & g_0 & \cdots & g_{m-1} & g_m & \cdots & 0 \\ \multicolumn{7}{c}{\dotfill} \\ 0 & 0 & \cdots & g_0 g_1 & & \cdots & g_m \end{pmatrix} = \begin{pmatrix} g \\ xg \\ \vdots \\ x^{k-1}g \end{pmatrix}.$$

Then C is cyclic. The rows of G are linearly independent and rank $G = k$, the dimension of C.

A message $a_0 a_1 \ldots a_{k-1}$ can be encoded by using a code $C = (g)$ in such a way that we calculate $a_0 a_1 \ldots a_{k-1} * g$.

2.7 Example. Suppose messages for transmission are elements of F_2^3. Let $g = 1 + x^3$ be the generator polynomial of a cyclic code. Then all possible messages can be encoded as codewords in the following way: We compute $a * g$ for each message polynomial a.

$$\begin{array}{ll} 000 \mapsto 000000 & 100 \mapsto 100100 \\ 001 \mapsto 001001 & 101 \mapsto 101101 \\ 010 \mapsto 010010 & 110 \mapsto 110110 \\ 011 \mapsto 011011 & 111 \mapsto 111111 \end{array}$$

The corresponding generator matrix is of the form

$$\begin{pmatrix} 1 & 0 & 0 & 1 & 0 & 0 \\ 0 & 1 & 0 & 0 & 1 & 0 \\ 0 & 0 & 1 & 0 & 0 & 1 \end{pmatrix}.$$ □

If $x^n - 1 = g_1 \ldots g_t$ is the complete factorization of $x^n - 1$ into irreducible polynomials over F_q, then the cyclic codes (g_i) generated by polynomials

g_i are called *maximal cyclic codes*. A maximal cyclic code is a maximal ideal in $\mathbb{F}_q[x]/(x^n - 1)$. All cyclic codes over $\mathbb{F}_q$ can be obtained by choosing any of the 2^t divisors of $x^n - 1$ as generator polynomials of the principal ideals. Many of these codes will be equivalent. Since a cyclic code is a principal ideal in $\mathbb{F}_q[x]/(x^n - 1)$ we shall use some of the properties of ideals and of principal ideal rings for the construction of codes.

If g is a generator polynomial of a cyclic code C then $g \mid (x^n - 1)$. Therefore $h = (x^n - 1)/g$ is also a generator polynomial of a cyclic code. Let $\deg g = m = n - k$, so that $\deg h = k$.

2.8 Definition. Let g be a generator polynomial of a cyclic code C. Then $h = (x^n - 1)/g$ is called a *check polynomial* of C. For the sake of brevity we denote $h = \sum_{i=0}^{k} h_i x^i$, $h_k \neq 0$, by h.

2.9 Theorem. *Let h be a check polynomial of a cyclic code $C \subseteq V_n$ with generator polynomial g. Let $v \in V_n$. Then*

$$v \in C \Leftrightarrow v * h = 0.$$

PROOF. Let $v \in C$; then there is a polynomial a such that $v = ag = a * g$. Since $g * h = 0$, we have $v * h = a * g * h = 0$. Conversely, if $v * h = 0$, then $vh \equiv 0 \ (\mathrm{mod}(x^n - 1))$. Moreover, $x^n - 1 = gh$ implies $v = ag$ for a suitable polynomial a, i.e. $v \in C$.   □

In Theorem 2.9, let $v = \sum v_i x^i$. Then the coefficient of x^i in the product $v * h$ is given by

$$\sum_{i=0}^{n-1} v_i h_{j-i} = 0, \qquad j = 0, 1, \ldots, n - 1,$$

where the indices are calculated mod n. Thus we see

$$v \in C \Leftrightarrow Hv^T = 0,$$

where

$$H = \begin{pmatrix} 0 & \cdots & 0 & h_k & & \cdots & h_1 & h_0 \\ 0 & \cdots & 0 & h_k & h_{k-1} & \cdots & h_0 & 0 \\ \cdots\cdots\cdots\cdots\cdots\cdots\cdots\cdots\cdots\cdots \\ h_k & \cdots & & h_1 & h_0 & 0 & \cdots & 0 & 0 \end{pmatrix}$$

is a parity-check matrix of the given code $C = (g)$. Hence we have

2.10 Theorem. *Let C be a cyclic (n, k) code with generator polynomial g. Then the dual code $C^\perp$ of C is a cyclic $(n, n - k)$ code with generator polynomial*

$$g^\perp = x^{\deg h}(h \circ x^{-1}), \quad \text{where} \quad h = (x^n - 1)/g.$$

We note that a polynomial of the form $x^{\deg h}(h \circ x^{-1})$ is called the *reciprocal polynomial* of h. The code generated by h is equivalent to the dual code $C^\perp$ of $C = (g)$; therefore $C^\perp = (h)$ is also called the dual code of C.

In Chapter 5 we shall consider linear recurrence relations of the form $\sum_{j=0}^{k} f_j a_{i+j} = 0$, $i = 0, 1, \ldots$, which are periodic of period n. Here $f_j \in F_q$ are coefficients of $f(x) = f_0 + f_1 x + \ldots + f_k x^k$, $f_0 \neq 0$, $f_k = 1$. The set of the n-tuples of the first n terms of each possible solution, considered as polynomials modulo $x^n - 1$, is the ideal generated by $g \in V_n$, where g is the reciprocal polynomial of $(x^n - 1)/f(x)$ of degree $n - k$. Thus there is a close relationship between cyclic codes and linear recurrence relations. Moreover, this relationship facilitates the implementation of cyclic codes on feedback shift registers.

2.11 Example. Let $f(x) = x^3 + x + 1$; f divides $x^7 - 1$ over F_2. Associated with f is the linear recurrence relation $a_{i+3} + a_{i+1} + a_i = 0$, which defines a $(7, 3)$ cyclic code. This code encodes 111, say, as 1110010. The generator polynomial is the reciprocal polynomial of $(x^7 - 1)/f(x)$, namely $g(x) = x^4 + x^3 + x^2 + 1$. □

The following summarizes some results on principal ideals. Let J_1 and J_2 be ideals in $F_q[x]/(x^n - 1) = V_n$. Then the intersection $J_1 \cap J_2$ is an ideal while the (set-theoretic) union of J_1 and J_2 is not an ideal in general. The ideal $J_1 + J_2$ is the smallest ideal containing J_1 and J_2. The product $J_1 J_2 := \{\sum a_{i_1} a_{i_2} \mid a_{i_1} \in J_1, a_{i_2} \in J_2\}$ is an ideal of V_n, too. Let J_i, $i = 1, 2$, be two ideals in V_n with generating elements g_i. Then

(i) $J_1 + J_2$ is generated by $\gcd(g_1, g_2)$;
(ii) $J_1 \cap J_2$ is generated by $\text{lcm}(g_1, g_2)$;
(iii) $J_1 J_2$ is generated by $\gcd(g_1 g_2, x^n - 1)$.

Let $x^n - 1 = g_1 \ldots g_t$ be the factorization of $x^n - 1$ into irreducible polynomials over F_q. The relation between a code C and its dual $C^\perp$ in Theorem 2.10 means that $(x^n - 1)/g_i$ is a generator polynomial of a so-called "*minimal code*" and g_i is a generator polynomial of the corresponding maximal code. The codes generated by $(x^n - 1)/g_i$ are also called *irreducible cyclic codes*.

2.12 Example. For $n = 7$ and $q = 2$ we have $x^7 - 1 = (x - 1)(x^3 + x^2 + 1)(x^3 + x + 1)$ as the factorization into irreducibles. Let $g = x^3 + x + 1$; then G is a generator matrix of a cyclic code and since $h = (x^7 - 1)/g$, H is check matrix for this code and at the same time the generator matrix of a minimal code:

$$G = \begin{pmatrix} 1 & 1 & 0 & 1 & 0 & 0 & 0 \\ 0 & 1 & 1 & 0 & 1 & 0 & 0 \\ 0 & 0 & 1 & 1 & 0 & 1 & 0 \\ 0 & 0 & 0 & 1 & 1 & 0 & 1 \end{pmatrix}, \quad H = \begin{pmatrix} 0 & 0 & 1 & 0 & 1 & 1 & 1 \\ 0 & 1 & 0 & 1 & 1 & 1 & 0 \\ 1 & 0 & 1 & 1 & 1 & 0 & 0 \end{pmatrix}. \quad \square$$

For a better understanding of the following remarks we refer the reader to the relevant parts of Chapter 3, §1. Let $M_i = ((x^n - 1)/g_i)$ be principal ideals, then $M_i \cap M_j = \{0\}$ for $i \neq j$ and $M_i + M_j = \{m_i + m_j \mid m_i \in M_i, m_j \in M_j\}$, where $\gcd(h_i, h_j) = (x^n - 1)/g_i g_j$ is a generating element of $M_i + M_j$, because of (i) after 2.11.

For any ideal $J = (g)$ in V_n with $g = g_{i_1} \ldots g_{i_s}$ we have

$$(x^n - 1)/g_{j_1} \ldots g_{j_r} = g_{i_1} \ldots g_{i_s},$$

if $\{j_1, \ldots, j_r\} \cup \{i_1, \ldots, i_s\} = \{1, \ldots, k\}$ and J is the direct sum of the ideals $M_{j_1}, \ldots, M_{j_r}$.

Next we summarize simple coding and decoding rules:

Coding. Let $a = a_0 \ldots a_{k-1} \in \mathbb{F}_q^k$ be a message with $k = n - m$ symbols. Then a, if regarded as a polynomial in $\mathbb{F}_q[x]$, is encoded as $a * g$. The map $\mu \colon V_k \to V_n$, $a \mapsto a * g$ is linear and injective and we have $a * g \in C = (g)$.

Decoding. A received word w will be divided by g. If a remainder occurs in the division then an error must have occurred in the transmission. To recognize the error e we determine $w * h$, where h is the check polynomial of C. Here we have $w = v + e$; see 2.9 and 2.15.

2.13 Example. (i) Let $g = 1 + x^2 + x^3$ be a generator polynomial of a binary $(7, 4)$ code. The message $a = 1010$ is encoded as $a * g = 1001110$.

(ii) Let $w = 1100001$ be a received word; then w/g is $0111 = x + x^2 + x^3$ with remainder $1 + x^2$. This shows that an error must have occurred. To correct it, we find $(x^7 - 1)/g = h = 10111 = 1 + x^2 + x^3 + x^4$ and $w * h = 1001011$. This equals $e * h$, since $w * h = (v + e) * h = 0 + e * h$ (see 2.9). Division by h gives us the error $e = 1010000$. Therefore $v = w - e = 0110001$ is the transmitted codeword. The original message a is obtained by division by g, namely $a = 0111$. □

A canonical generator matrix of the form $\mathbf{G} = (\mathbf{I}_k, -\mathbf{A}_{k \times m})$ can be obtained by using the division algorithm for polynomials. Let $\deg g = m = n - k$; then there are unique polynomials $a^{(j)}$ and $r^{(j)}$ with $\deg r^{(j)} < m$, such that

$$x^j = a^{(j)} g + r^{(j)}, \qquad j \in \mathbb{N}.$$

Therefore

$$x^j - r^{(j)} \in C.$$

The k polynomials $g^{(j)} = x^k(x^j - r^{(j)})$ considered modulo $x^n - 1$, $m \leq j \leq n - 1$, are linearly independent and form a matrix of the required form, where $\mathbf{A}$ is the $k \times m$ matrix whose ith row is the vector of coefficients of $r^{(n-k-1+i)}$.

2.14 Example. Let $n = 7$, $q = 2$ and $x^7 - 1 = (x + 1)(x^3 + x + 1)(x^3 + x^2 + 1)$. The polynomial $g = x^3 + x^2 + 1$ generates a cyclic $(7, 4)$ code with check polynomial $h = x^4 + x^3 + x^2 + 1$. The canonical generator matrix has the rows $g^{(j)}$ for $j = 3, 4, 5, 6$. Thus

$$G = \begin{pmatrix} 1 & 0 & 0 & 0 & 1 & 0 & 1 \\ 0 & 1 & 0 & 0 & 1 & 1 & 1 \\ 0 & 0 & 1 & 0 & 1 & 1 & 0 \\ 0 & 0 & 0 & 1 & 0 & 1 & 1 \end{pmatrix} \quad \text{and} \quad H = \begin{pmatrix} 1 & 1 & 1 & 0 & 1 & 0 & 0 \\ 0 & 1 & 1 & 1 & 0 & 1 & 0 \\ 1 & 1 & 0 & 1 & 0 & 0 & 1 \end{pmatrix}. \quad \square$$

We mention also that the syndrome $S(w)$ with respect to a parity-check matrix H of the form (A, I_m) can be easily determined.

2.15 Theorem. *Let C be a cyclic code with generator polynomial g and parity-check matrix H. Then the syndrome $S(w)$ of a received vector satisfies $S(w) \equiv w \pmod{g}$; i.e. it is the remainder when dividing w by g.*

Cyclic codes can be described in various ways by using roots of the generator polynomial. The simple proofs are left as exercises.

2.16 Theorem. (i) *Let g be a generator polynomial of a cyclic code C over $\mathbb{F}_q$ and let $\alpha_1, \ldots, \alpha_m$ be roots of g in an extension field of $\mathbb{F}_q$. Then*

$$v \in C \Leftrightarrow \bar{v}(\alpha_1) = \bar{v}(\alpha_2) = \ldots = \bar{v}(\alpha_m) = 0.$$

(ii) *Let $g = g^{(1)} \ldots g^{(t)}$ be a generator polynomial of a cyclic code C over $\mathbb{F}_q$ and let α_i be roots of $g^{(i)}$ in an extension of $\mathbb{F}_q$. Then*

$$v \in C \Leftrightarrow \bar{v}(\alpha_1) = \bar{v}(\alpha_2) = \ldots = \bar{v}(\alpha_t) = 0.$$

As a special form of (i) *we state:*

(iii) *Let $g = \prod_{i \in K} (x - \zeta^i)$ be a generator polynomial of a cyclic (n, k) code over $\mathbb{F}_q$, let ζ be a primitive nth root of unity and let K be a union of cyclotomic cosets mod n (see Chapter 3, §4). Then*

$$v(x) \in C \Leftrightarrow v(\zeta^i) = 0 \quad \text{for all } i \in K. \qquad \square$$

As an example of the concurrence of the description of a cyclic code by a generator polynomial or by roots of code polynomials we prove the following results.

2.17 Theorem. *The binary cyclic code of length $n = 2^m - 1$ for which the generator polynomial is the minimal polynomial over $\mathbb{F}_2$ of a primitive element of $\mathbb{F}_{2^m}$ is equivalent to the binary $(n, n - m)$ Hamming code.*

PROOF. Let ζ be a primitive element of $\mathbb{F}_{2^m}$ and let $g = (x - \zeta)(x - \zeta^2) \ldots (x - \zeta^{2^{m-1}})$ be the minimal polynomial of ζ over $\mathbb{F}_2$. For the cyclic code C generated by g we construct an $m \times (2^m - 1)$ matrix H for which the jth column is $(c_0, c_1, \ldots, c_{m-1})^T$ if $\zeta^{j-1} = \sum_{i=0}^{m-1} c_i \zeta^i$, $c_i \in \mathbb{F}_2$. If $v = (v_0, v_1, \ldots, v_{n-1})$ and $v(x) = v_0 + v_1 x + \ldots + v_{n-1} x^{n-1} \in \mathbb{F}_2[x]$, then Hv^T

corresponds to $v(\zeta)$ expressed in the basis $\{1, \zeta, \ldots, \zeta^{m-1}\}$. So $\mathbf{H}v^T = 0$ holds exactly when g divides v, so $\mathbf{H}$ is a parity-check matrix of C. The proof is complete, since the columns of $\mathbf{H}$ are a permutation of the binary representations of the numbers $1, 2, \ldots, 2^m - 1$. $\qquad\square$

2.18 Example. $x^4 + x + 1$ is primitive over $\mathbb{F}_2$ and thus has a primitive element $\zeta \in \mathbb{F}_{2^4}$ as a root. If we use vector notation for the 15 elements $\zeta^j \in \mathbb{F}_2^*$, $j = 0, 1, \ldots, 14$ expressed in terms of the basis $\{1, \zeta, \zeta^2, \zeta^3\}$ and if we form a 4×15 matrix with these vectors as columns, then we get the parity-check matrix $\mathbf{H}$ of a code equivalent to the $(15, 11)$ Hamming code. A message $v = (v_0, v_1, \ldots, v_{10}) = \sum v_i x^i$ is encoded into a code word $w = v * (x^4 + x + 1)$. Suppose the received polynomial contains one error, i.e. $w + x^{e-1}$ is received when w is transmitted. Then the syndrome is $w(\zeta) + \zeta^{e-1} = \xi^{e-1}$ and the decoder knows that there is an error in the eth position. $\qquad\square$

2.19 Theorem. *Let $C \subseteq V_k$ be a cyclic code with generator polynomial g and let $\alpha_1, \ldots, \alpha_{n-k}$ be the roots of g. Then $v \in V_n$ is a code polynomial if and only if the coefficient vector $(v_0, \ldots, v_{n-1})$ of v is in the nullspace of the matrix*

$$\mathbf{H} = \begin{pmatrix} 1 & \alpha_1 & \alpha_1^2 & \cdots & \alpha_1^{n-1} \\ \vdots & \vdots & \vdots & & \vdots \\ 1 & \alpha_{n-k} & \alpha_{n-k}^2 & \cdots & \alpha_{n-k}^{n-1} \end{pmatrix}.$$

PROOF. Let $v(x) = v_0 + v_1 x + \ldots + v_{n-1}x^{n-1}$; then $v(\alpha_i) = 0$, i.e. $(1, \alpha_i, \ldots, \alpha_i^{n-1})(v_0, v_1, \ldots, v_{n-1})^T = (0, \ldots, 0)$, $1 \le i \le n - k$, if and only if $\mathbf{H}v^T = 0$. $\qquad\square$

In the case of cyclic codes, the syndrome can often be replaced by a simpler entity. Let α be a primitive nth root of unity in $\mathbb{F}_{q^m}$ and let the generator polynomial g be the minimal polynomial of α over $\mathbb{F}_q$. We know that g divides $v \in V_n$ if and only if $v(\alpha) = 0$. Therefore we can replace the parity-check matrix $\mathbf{H}$ in Theorem 2.19 by $(1 \quad \alpha \quad \alpha^2 \quad \cdots \quad \alpha^{n-1})$. Then the role of the syndrome is played by $S(u) = \mathbf{H}u^T$, and $S(u) = u(\alpha)$ since $u = (u_0, \ldots, u_{n-1})$ can be regarded as a polynomial $u(x)$ with coefficients u_i.

If v denotes the transmitted codeword and w is the received word, suppose $e^{(j)}(x) = x^{j-1}$, $1 \le j \le n$, is an error polynomial with a single error. Let $w = v + e^{(j)}$. Then

$$w(\alpha) = v(\alpha) + e^{(j)}(\alpha) = e^{(j)}(\alpha) = \alpha^{j-1}.$$

Here $e^{(j)}(\alpha)$ is called the *error-location number*. $S(w) = \alpha^{j-1}$ indicates the error uniquely, since $e^{(i)}(\alpha) \ne e^{(j)}(\alpha)$ for $i \ne j$, $1 \le i \le n$.

The nth roots of unity $\{\zeta^i \,|\, i \in K\}$ in 2.16(iii) are called *zeros* of the code C. The remaining nth roots of unity are zeros of $h = (x^n - 1)/g$. The following theorem shows that the generator polynomial does not have to be the monic polynomial of least degree in C. Let $C = (g)$ be a cyclic (n, k) code. We use the notation of Theorem 2.16(iii).

2.20 Theorem. *The polynomials g and fg are generator polynomials of the same code, if for a polynomial $f \in V_n$ we have $f(\zeta^i) \neq 0$ for all $i \in K$, i.e. f does not introduce new zeros of C.* ☐

A cyclic code can be described in terms of a primitive nth root of unity ζ and the set $K = \{[k]_n \mid g(\zeta)^k = 0\}$. The following theorem shows that the code is independent of the choice of ζ.

2.21 Theorem. *Let ζ and μ be arbitrary primitive nth roots of unity in an extension field of $\mathbb{F}_q$ and let K be the union of cyclotomic cosets mod n. Then the polynomials*

$$g_\zeta = \prod_{i \in K} (x - \zeta^i) \quad \text{and} \quad g_\mu = \prod_{i \in K} (x - \mu^i)$$

are generator polynomials of equivalent codes.

PROOF. See e.g. BLAKE and MULLIN. ☐

A different characterization of a cyclic (n, k) code C with generator polynomial g over $\mathbb{F}_q$ follows. Let ζ and K be as above. Then

$$C = \{v \in V_n \mid v(\zeta^i) = 0 \text{ for all } i \in K\}.$$

2.22 Definition. The *Mattson–Solomon polynomial F_v* of a vector $v \in V_n$ is

$$F_v = \frac{1}{n} \sum_{j=1}^{n} f_j x^{n-j}, \quad \text{where} \quad f_j = v(\zeta^j) = \sum_{i=0}^{n-1} v_i \zeta^{ij}, \quad 1 \leq j \leq n.$$

2.23 Theorem. *Let $v = (v_0, \ldots, v_{n-1}) \in C$ and f_j as in 2.22. Then $v_s = F_v(\zeta^s)$, $s = 0, 1, \ldots, n-1$.*

PROOF.

$$F_v(\zeta^s) = \frac{1}{n} \sum_{j=1}^{n} f_j \zeta^{-js} = \frac{1}{n} \sum_{j=1}^{n} \zeta^{-j} \sum_{i=0}^{n-1} v_i \zeta^{ij}$$

$$= \frac{1}{n} \sum_{i=0}^{n-1} v_i \sum_{j=1}^{n} \zeta^{(i-s)j} = v_s, \quad \text{since} \quad \sum_{j=1}^{n} \zeta^{(i-s)j} = 0, \quad i \neq s. \quad ☐$$

The Mattson–Solomon polynomial F_v for arbitrary vectors in $\mathbb{F}_{q^m}^n$ is also called a *discrete Fourier transform* of v. Theorem 2.23 enables us to describe a cyclic code C in the form

$$C = \{(F_v(\zeta^0), F_v(\zeta^1), \ldots, F_v(\zeta^{n-1})) \mid v \in C\}.$$

2.24 Lemma. *Let $g = \prod_{i \in K} (x - \zeta^i)$ be a generator polynomial of a cyclic code C and let $\{1, 2, \ldots, d - 1\}$ be a subset of K, with K as above. Then* $\deg F_v \leq n - d$ *for $v \in C$.*

PROOF. For any $j < d$ we have $v(\zeta^j) = 0$.   □

2.25 Theorem. *If there are exactly r nth roots of unity which are zeros of F_v, then the weight of v is $n - r$.*

PROOF. The weight of v is the number of nonzero coordinates of v_s. There are r values s, such that $v_s = 0$.   □

This theorem will be used to prove Theorem 3.4.

The discrete Fourier transforms can be used for the formulation of the subject of coding. In such an approach the engineering aspects of signal processing and the algebraic theory of error control codes are drawn much closer together. We give a brief indication of the introduction of the so-called *spectral setting* for cyclic codes. We modify our definition 2.22 as follows. Let α be an element of $\mathbb{F}_{q^m}$ of order n, where n divides $q^m - 1$ for some $m \in \mathbb{N}$.

The *discrete Fourier transform* (DFT) of $c = (c_0, \ldots, c_{n-1}) \in \mathbb{F}_q^n$ is the vector $C = (C_0, \ldots, C_{n-1})$ given by

$$C_j = \sum_{i=0}^{n-1} \alpha^{ij} c_i, \qquad j = 0, \ldots, n - 1.$$

If $n = q^m - 1$ then α is a primitive element of $\mathbb{F}_{q^m}$. In the engineering context the discrete index i is called *time*, c is the *time domain function* or *signal*, j is the *frequency* and C is the *frequency domain function* or the *spectrum*. The DFT has many interesting properties. Let p be the characteristic of $\mathbb{F}_q$.

2.26 Lemma. (i) *Let C be the DFT of $c \in \mathbb{F}_q^n$. Then the components of c are*

$$c_i = \frac{1}{n} \sum_{j=0}^{n-1} \alpha^{-ij} C_j, \qquad i = 0, \ldots, n - 1,$$

where n is taken mod p.

(ii) *Let $c_i = e_i g_i$, $i = 0, \ldots, n - 1$, be the components of $c \in \mathbb{F}_q^n$. Then the DFT C of c has components*

$$C_j = \frac{1}{n} \sum_{k=0}^{n-1} E_{j-k} G_k,$$

where all subscripts are taken mod n.   □

The proof of (i) is analogous to the proof of 2.23, and (ii) is left as an exercise. The DFT can also transform polynomials $c \in \mathbb{F}_q[x]$ into polynomials C. The roots of polynomials are closely related to the spectrum, since $c(x)$ has a root at α^j if and only if C_j equals zero. $C(x)$ has a root at α^{-i} if and only if c_i equals zero. Thus speaking of roots of polynomials is the same as speaking of spectral components equal to zero.

An alternative description of cyclic codes, similar to the one given before Lemma 2.24, is the following: The set of words of F_q^n whose spectrum is zero in components $j_1, \ldots, j_{n-k}$ is a cyclic code. The transform of each such word is called a *codeword spectrum*. Although each codeword in such a cyclic code is a vector with components in F_q, the codeword spectrum is in $F_{q^m}^n$. Hence, a cyclic code can be interpreted as the inverse discrete Fourier transforms of all spectral vectors which are zero in several prescribed components, provided that the DFT has a value in F_q. We shall briefly return to this description when we consider special cyclic codes. Here we conclude with an application of the following result (see BLAHUT).

2.27 Theorem. *Let n be a divisor of 2^{m-1} and let $C_j \in F_{2^m}$, $j = 0, \ldots, n - 1$. The components of the inverse DFT c are all elements of F_2 if and only if $C_j^2 = C_{2j(\mathrm{mod}\ n)}$, $j = 0, \ldots, n - 1$.*

PROOF. By definition

$$C_j^2 = \left(\sum_{i=0}^{n-1} \alpha^{ij} c_i \right)^2 = \sum_{i=0}^{n-1} \alpha^{2ij} c_i^2 = \sum_{i=0}^{n-1} \alpha^{2ij} c_i = C_{2j},$$

where the subscripts are taken mod n. If $c_i \notin F_2$ for some i, then $c_i^2 \neq c_i$. Therefore

$$C_j^2 = \sum_{i=0}^{n-1} \alpha^{2ij} c_i^2 \neq \sum_{i=0}^{n-1} \alpha^{2ij} c_i = C_{2j}. \qquad \square$$

2.28 Example. We consider again the construction of the binary $(7, 4)$ Hamming code, but now we use DFT. We have $n = 7$, $q = 2$, $m = 6$, $\alpha \in F_{2^6}$, ord $\alpha = 7$, and we find the following codeword spectra and codewords:

Codeword spectra (frequency domain codewords)							Codewords (time domain codewords)						
C_0	C_1	C_2	C_3	C_4	C_5	C_6	c_0	c_1	c_2	c_3	c_4	c_5	c_6
0	0	0	0	0	0	0	0	0	0	0	0	0	0
0	0	0	α^0	0	α^0	α^0	1	1	1	0	1	0	0
0	0	0	α^1	0	α^4	α^2	0	0	1	1	1	0	1
0	0	0	α^2	0	α^1	α^4	0	1	0	0	1	1	1
0	0	0	α^3	0	α^5	α^6	1	1	0	1	0	0	1
0	0	0	α^4	0	α^2	α^1	0	1	1	1	0	1	0
0	0	0	α^5	0	α^6	α^3	1	0	0	1	1	1	0

(continued on next page)

Codeword spectra (frequency domain codewords)							Codewords (time domain codewords)						
C_0	C_1	C_2	C_3	C_4	C_5	C_6	c_0	c_1	c_2	c_3	c_4	c_5	c_6
0	0	0	α^6	0	α^3	α^5	1	0	1	0	0	1	1
1	0	0	0	0	0	0	1	1	1	1	1	1	1
1	0	0	α^0	0	α^0	α^0	0	0	0	1	0	1	1
1	0	0	α^1	0	α^4	α^2	1	1	0	0	0	1	0
1	0	0	α^2	0	α^1	α^4	1	0	1	1	0	0	0
1	0	0	α^3	0	α^5	α^6	0	0	1	0	1	1	0
1	0	0	α^4	0	α^2	α^1	1	0	0	0	1	0	1
1	0	0	α^5	0	α^6	α^3	0	1	1	0	0	0	1
1	0	0	α^6	0	α^3	α^5	0	1	0	1	1	0	0

$\square$

PROBLEMS

1. Let G be a generator matrix of a $(5, 3)$ code over F_3:

$$G = \begin{pmatrix} 2 & 1 & 1 & 0 & 0 \\ 1 & 1 & 0 & 2 & 0 \\ 0 & 0 & 2 & 1 & 1 \end{pmatrix}.$$

   Determine a parity-check matrix for this code. Is this a cyclic code?

2. How many binary cyclic $(8, k)$ codes are there for $k = 1, 2, \ldots, 7$? Answer the same question for ternary cyclic $(10, k)$ codes.

3. The polynomial $g = x^6 + x^5 + x^4 + x^3 + 1$ is the generator polynomial for a cyclic code over F_2 with block length $n = 15$. Find the parity-check polynomial and the generator and parity-check matrix for this code. How many errors can this code correct?

4. Determine the generator polynomial of the $(9, 7)$ Hamming code over F_8.

5. Design circuits for the multiplication and also for the division of polynomials over a finite field.

6. Give the degrees of the following polynomials in an (n, k) code: generator, parity-check, message, codeword, error received and syndrome polynomial.

*7. Let d be the minimum distance of a cyclic code C. Show that every error of weight less than $\frac{1}{2}d$ has a unique syndrome.

8. How many cyclic codes of block length 15 are there?

9. Show that the $(21, 18)$ Hamming code over F_4 is not cyclic.

10. Prove that the Hamming codes of block length $n = (q^m - 1)/(q - 1)$ over F_q are cyclic if m and $q - 1$ are relatively prime.

*11. Let $f(x) = \sum_{i=0}^{k} b_i x^{k-i}$, $b_0 = 1$, be a polynomial over $\mathbb{F}_2$ such that $f | x^n - 1$. Show that

$$C = \left\{ (a_0, a_1, \ldots, a_{n-1}) \,\middle|\, \sum_{j=0}^{k} a_{i+k-j} b_j = 0, i = 0, 1, \ldots, n - k + 1 \right\}$$

is a linear, cyclic (n, k) code over $\mathbb{F}_2$.

12. If e is the idempotent generator of a code C of length n, show that the generator polynomial g of C equals $\gcd(e, x^n - 1)$. (See Exercise 12 for a definition.)

13. If $x^n - 1 = gh$ and the code $C = (g)$ has idempotent generator e, show that (h) has idempotent generator $1 + e$. (See Exercise 12 for a definition.)

*14. Let $n = (q^m - 1)/(q - 1)$ and β be a primitive nth root of unity in $\mathbb{F}_{q^m}$, $m \geq 2$. Prove that the nullspace of the matrix $\mathbf{H} = (1 \beta \beta^2 \ldots \beta^{n-1})$ is a code over $\mathbb{F}_q$ with minimum distance at least 3 if and only if $\gcd(m, q - 1) = 1$.

*15. A code C is called *reversible* if $(a_0, a_1, \ldots, a_{n-1}) \in C$ holds if and only if $(a_{n-1}, \ldots, a_1, a_0) \in C$.
   (i) Prove that a cyclic code $C = (g)$ is reversible if and only if with each root of g also the reciprocal value of that root is a root of g.
   (ii) Prove that any cyclic code over $\mathbb{F}_q$ of length n is reversible if -1 is a power of q modulo n.

EXERCISES (Solutions in Chapter 8, p. 467)

1. Determine all codewords of a code with generator polynomial $g = 1 + x + x^3$ over $\mathbb{F}_2$, if the length k of the messages is 4. Which of the following received words have detectable errors: 1000111, 0110011, 0100011.

2. Show that there are noncyclic Hamming codes.

3. Determine a table of coset leaders and syndromes for the binary $(3, 1)$ code generated by $g = 1 + x + x^2$. Do likewise for the binary $(7, 3)$ code generated by $1 + x^2 + x^3 + x^4$.

*4. Prove properties (i), (ii), (iii) stated immediately after Example 2.11.

5. Show that $g = x^4 + x^2 + x + 1$ generates a binary code equivalent to the $(8, 4)$ extended Hamming code. Find $d_{\min}$ of this code.

*6. Let $x^n - 1 = g_1 \ldots g_t$ be the factorization of $x^n - 1$ into irreducible polynomials over $\mathbb{F}_q$. Prove that (g_i), $1 \leq i \leq t$, is a maximal ideal in $\mathbb{F}_q[x]/(x^n - 1)$, and $(x^n - 1)/g_i$ generates a minimal ideal M_i, such that $M_i \cap M_j = 0$ and $M_i + M_j = \{m_i + m_j \,|\, m_i \in M_i, m_j \in M_j\}$ has generator $\gcd((x^n - 1)/g_i, (x^n - 1)/g_j) = (x^n - 1)/g_i g_j$.

7. Prove: The binary cyclic code with generator polynomial $1 + x$ is the $(n - 1)$-dimensional code C consisting of all even weight vectors of length n. A binary cyclic code $C = (g)$ contains only even weight vectors if and only if $1 + x$ divides g.

8. A code C is called *self-orthogonal* if $C \subseteq C^{\perp}$.

   (i) Let $x^n - 1 = gh$ over $\mathbb{F}_q$. Prove that a cyclic code C with generator polynomial g is self-orthogonal if and only if the reciprocal polynomial of h divides g.

   (ii) Consider a cyclic code of block length $n = 7$, with generator polynomial $1 + x^2 + x^3 + x^4$. Is this code self-orthogonal?

*9. Show that if $\operatorname{ord}(g) = n$ then the (n, k) binary code generated by g has $d_{\min} \geq 3$. If $\operatorname{ord}(g) < 9$ find $d_{\min}$. Is this true for nonbinary codes?

*10. A binary $(9, 3)$ code C is defined by

$$(v_0, v_1, \ldots, v_8) \in C \Leftrightarrow v_0 = v_1 = v_2, \, v_3 = v_4 = v_5, \, v_6 = v_7 = v_8.$$

Show that C is equivalent to a cyclic code and find a generator polynomial for C.

11. A binary cyclic code of length 63 has a generator polynomial $x^5 + x^4 + 1$. Determine the minimum distance of this code.

12. Let n be an odd integer and let $q = 2$. Prove: For every ideal C in V_n there is a unique polynomial $e \in C$, called the *idempotent* of C, with the following properties:

   (i) $e = e^2$;

   (ii) e generates C;

   (iii) e is a unit for C, i.e. $ea = a$ for all $a \in C$. Furthermore, show that a binary polynomial f is an idempotent in V_n if and only if the set K of powers of x that occur with nonzero coefficients in f is a union of cyclotomic cosets for n.

13. Find the generator polynomials, dimensions and idempotent generators, for all binary cyclic codes of length $n = 7$. Identify dual codes and self-orthogonal codes.

14. Let $C = (f)$ over $\mathbb{F}_q$, $L = \{i, 0 \leq i \leq n - 1 \mid f(\zeta^i) = 0\}$, and ζ be a primitive root of unity over $\mathbb{F}_q$. Prove $C = (g)$, with $g = \operatorname{lcm}\{M^{(i)} \mid i \in L\}$ for minimal polynomials $M^{(i)}$ of ζ^i.

*15. Prove that the coordinates v_i of $v = v_0 v_1 \ldots v_{n-1}$ can be obtained from

$$v = v_0 + v_1 x + \ldots + v_{n-1} x^{n-1} \quad \text{by} \quad v_i = \frac{1}{n} \sum_{j=0}^{n-1} v(\zeta^j) \zeta^{-ij},$$

where ζ is a primitive nth root of unity over $\mathbb{F}_q$.

16. Prove the following properties of the Mattson–Solomon polynomials:

   *(i) If v is a binary vector then F_v is an idempotent in the polynomial ring over $\mathbb{F}_{2^m} \bmod (x^n - 1)$.

   (ii) $v = u + w$ in $V_n \Rightarrow F_v = F_u + F_w$.

   *(iii) $v = u * w$ in $V_n \Leftrightarrow F_v = F_u F_w$.

   (iv) $v = uw := (u_0 w_0, u_1 w_1, \ldots, u_{n-1} w_{n-1}) \Leftrightarrow F_v = \frac{1}{n}(F_u * F_w) = \sum_{i=0}^{n-1} u_i w_i x^i$.

17. What is the block length of the shortest cyclic binary code whose generator polynomial is $x^7 + x + 1$? Repeat the question for $x^9 + x + 1$.

*§3. Special Cyclic Codes

In this section we shall consider some important examples of cyclic codes. In Theorem 2.17 we showed that the cyclic code of block length $2^m - 1$ over $\mathbb{F}_{2^m}$ is a Hamming code C_m, if we choose the minimal polynomial of a primitive element ζ of $\mathbb{F}_{2^m}$ as a generating polynomial. The matrix $\mathbf{H} = (1, \zeta, \zeta^2, \ldots, \zeta^{n-1})$ is a parity-check matrix for C_m. This is a special case of Theorem 2.19. In the following example we shall generalize these codes and motivate the theory of these codes.

3.1 Example. Let $\zeta \in \mathbb{F}_{2^4}$ be a root of $x^4 + x + 1 \in \mathbb{F}_2[x]$ and let α_1, α_2 of Theorem 2.9 be ζ, and ζ^3, respectively. Then ζ and ζ^3 have the minimal polynomials $M^{(1)} = x^4 + x + 1$ and $M^{(3)} = x^4 + x^3 + x^2 + x + 1$, over $\mathbb{F}_2$, respectively. They are divisors of $x^{15} - 1$, because $\zeta^{15} = 1$ and $(\zeta^3)^{15} = 1$. Hence a cyclic code C over $\mathbb{F}_2$ is defined by the generator polynomial $g = M^{(1)} M^{(3)}$ and parity-check matrix

$$\mathbf{H} = \begin{pmatrix} 1 & \zeta & \cdots & \zeta^{13} & \zeta^{14} \\ 1 & \zeta^3 & \cdots & \zeta^{39} & \zeta^{42} \end{pmatrix}.$$

We shall see in 3.4 that $d_{\min} \geq 5$ for this code. This means (by 1.15) that C can correct up to two errors and can detect up to four errors. C is a cyclic $(15, 7)$ code since $n = 15$, $k = 15 - \deg g = 7$. We have

$$v \in C \Leftrightarrow S(v) = \mathbf{H}v^T = 0 \Leftrightarrow S_1 = S_3 = 0,$$

where

$$S_1 := \sum_{i=0}^{14} v_i \zeta^i \quad \text{and} \quad S_3 := \sum_{i=0}^{14} v_i \zeta^{3i}$$

are the components of the syndrome $S(v) = (S_1, S_3)^T$ of v with respect to $\mathbf{H}$. If we use binary representation for elements of $\mathbb{F}_{2^4}$ then $\mathbf{H}$ is of the form

$$\mathbf{H} = \begin{pmatrix}
1 & 0 & 0 & 0 & 1 & 0 & 0 & 1 & 1 & 0 & 1 & 0 & 1 & 1 & 1 \\
0 & 1 & 0 & 0 & 1 & 1 & 0 & 1 & 0 & 1 & 1 & 1 & 1 & 0 & 0 \\
0 & 0 & 1 & 0 & 0 & 1 & 1 & 0 & 1 & 0 & 1 & 1 & 1 & 1 & 0 \\
0 & 0 & 0 & 1 & 0 & 0 & 1 & 1 & 0 & 1 & 0 & 1 & 1 & 1 & 1 \\
1 & 0 & 0 & 0 & 1 & 1 & 0 & 0 & 0 & 1 & 1 & 0 & 0 & 0 & 1 \\
0 & 0 & 0 & 1 & 1 & 0 & 0 & 0 & 1 & 1 & 0 & 0 & 0 & 1 & 1 \\
0 & 0 & 1 & 0 & 1 & 0 & 0 & 1 & 0 & 1 & 0 & 0 & 1 & 0 & 1 \\
0 & 1 & 1 & 1 & 1 & 0 & 1 & 1 & 1 & 1 & 0 & 1 & 1 & 1 & 1
\end{pmatrix}.$$

Here we use the fact that $\zeta^4 + \zeta + 1 = 0$ in $\mathbb{F}_{2^4}$. The columns of $\mathbf{H}$ can be obtained as follows: The upper half of the first column is the coefficient vector $\zeta^0 + 0\zeta^1 + 0\zeta^2 + 0\zeta^3$, i.e. $(1000)^T$. The upper half of the second column is the coefficient vector of $0\zeta^0 + 1\zeta^1 + 0\zeta^2 + 0\zeta^3$, i.e. $(0100)^T$, etc. The lower half of the second column is the coefficient vector of $0\zeta^0 + 0\zeta^1 + 0\zeta^2 + 1\zeta^3$, etc. Suppose the received vector $v = (v_0, v_1, \ldots, v_{14})$ is a vector with at most two errors. Let $e = x^{a_1} + x^{a_2}$, $0 \le a_1, a_2 \le 14$, $a_1 \ne a_2$, be the error vector. We have

$$S_1 = \zeta^{a_1} + \zeta^{a_2}, \qquad S_3 = \zeta^{3a_1} + \zeta^{3a_2}.$$

We let

$$X_1 := \zeta^{a_1}, \qquad X_2 := \zeta^{a_2}$$

be the error-location numbers, then

$$S_1 = X_1 + X_2, \qquad S_3 = X_1^3 + X_2^3,$$

or

$$X_2 = S_1 + X_1, \qquad X_2^3 = S_1^3 + S_1^2 X_1 + S_1 X_1^2 + X_1^3.$$

Hence

$$S_3 = S_1^3 + S_1^2 X_1 + S_1 X_1^2$$

and

$$1 + \frac{S_1}{X_1} + \frac{S_1^2 + S_3/S_1}{X_1^2} = 0.$$

If two errors occurred then $1/X_1$ and $1/X_2$ are two roots in $\mathbb{F}_{2^4}$ of the polynomial

$$\sigma = 1 + S_1 x + \left(S_1^2 + \frac{S_3}{S_1} \right) x^2, \tag{1}$$

called the *error-locator polynomial*. If only one error occurred, then $S_1 = X_1$, $S_3 = X_1^3$ and $S_1^3 + S_3 = 0$. Hence

$$\sigma = 1 + S_1 x. \tag{2}$$

If no error occurred then

$$S_1 = S_3 = 0. \tag{3}$$

To summarize, first we have to find the syndrome of the received vector v, then we check to see if a maximum of one error occurred, determine σ, check if one or no error has occurred and find the error location and roots of σ. Equation (2) is always solvable in $\mathbb{F}_{2^4}$. If σ in (1) is irreducible, then we know that we have a detectable error e with more than two error locations, which is not correctable. □

Suppose the vector $v = 100111000000000$ is the received word. Then the syndrome $S(v) = (S_1, S_3)^T$ is given by

$$S_1 = 1 + \zeta^3 + \zeta^4 + \zeta^5,$$
$$S_3 = 1 + \zeta^9 + \zeta^{12} + \zeta^{15}.$$

Here ζ is a primitive element of $\mathbb{F}_{2^4}$ with $\zeta^4 + \zeta + 1 = 0$. We use the following powers of ζ in our calculations:

$$\begin{array}{ll}
\zeta^4 = 1 + \zeta, & \zeta^{10} = 1 + \zeta + \zeta^2, \\
\zeta^5 = \zeta + \zeta^2, & \zeta^{11} = \zeta + \zeta^2 + \zeta^3, \\
\zeta^6 = \zeta^2 + \zeta^3, & \zeta^{12} = 1 + \zeta + \zeta^2 + \zeta^3, \\
\zeta^7 = 1 + \zeta + \zeta^3, & \zeta^{13} = 1 + \zeta^2 + \zeta^3, \\
\zeta^8 = 1 + \zeta^2, & \zeta^{14} = 1 + \zeta^3, \\
\zeta^9 = \zeta + \zeta^3, & \zeta^{15} = 1.
\end{array}$$

Then

$$S_1 = \zeta^2 + \zeta^3 \quad \text{and} \quad S_3 = 1 + \zeta^2.$$

The polynomial σ is, according to (1),

$$\sigma = 1 + (\zeta^2 + \zeta^3)x + \left(1 + \zeta + \zeta^2 + \zeta^3 + \frac{1 + \zeta^2}{\zeta^2 + \zeta^3}\right)x^2.$$

We determine the roots of σ by trial-and-error and find that ζ and ζ^7 are roots. Hence

$$\frac{1}{X_1} = \zeta = \frac{1}{\zeta^{14}} \quad \text{and} \quad \frac{1}{X_2} = \zeta^7 = \frac{1}{\zeta^8}$$

in $\mathbb{F}_{2^4}$. Thus we know that errors must have occurred in the positions corresponding to x^8 and x^{14}, i.e. in the ninth and fifteenth components of the received word v. Therefore the corrected transmitted codeword is $w = 100111001000001$. Then w is decoded by dividing it by the generator polynomial $g = 1 + x^4 + x^6 + x^7 + x^8$. This gives $w/g = x^6 + x^5 + x^3 + 1$ with remainder 0. Therefore the original message was $a = 1001011$. □

We describe a class of codes which has been introduced by BOSE, CHAUDHURI and HOCQUENGHEM and which are therefore all called BCH codes.

3.2 Definition and Theorem. Let $c, m \in \mathbb{N}$, let ζ be a primitive nth root of unity in $\mathbb{F}_{q^m}$, where m is the multiplicative order of q modulo n, and let $d \in \mathbb{N}$ with $2 - c \le d \le n$ and $I = \{c, c+1, \dots, c+d-2\}$. Then the *BCH code* $C \subseteq V_n$ *of designed distance* d is a cyclic code over $\mathbb{F}_q$ of length n defined by the following equivalent conditions:

(i) $v \in C \Leftrightarrow v(\zeta^i) = 0$ for all $i \in I$.
(ii) The polynomial $g = \text{lcm}\{M^{(i)} | i \in I, M^{(i)}$ is the minimal polynomial of $\zeta^i\}$ is a generator polynomial of C, and ζ^i for $i \in I$ are the roots of g.

(iii) A parity-check matrix of C is the matrix

$$
\mathbf{H} = \begin{pmatrix}
1 & \zeta^c & \cdots & \zeta^{c(n-1)} \\
1 & \zeta^{c+1} & \cdots & \zeta^{(c+1)(n-1)} \\
 & & & \\
1 & \zeta^{c+d-2} & \cdots & \zeta^{(c+d-2)(n-1)}
\end{pmatrix}.
$$

Here C is the nullspace of $\mathbf{H}$.   □

3.3 Remark. If $c = 1$ then the code defined in 3.2 is called a *narrow-sense BCH code*. If $n = q^m - 1$ the BCH code is called *primitive*. The dual of a BCH code is, in general, not a BCH code. If $c = 1$ and $d \geq 2t + 1$ the code is called a *t-error correcting code*. The dimension of C is $\geq n - m(d - 1)$.   □

A practical example for the use of BCH codes is the European and trans-Atlantic information communication system, which has been using such codes for many years. The message symbols are of length 231 and the generator polynomial is of degree 24 such that $231 + 24 = 255 = 2^8 - 1$ is the length of the codewords. The code detects at least six errors, and its failure probability is 1 in 16 million.

3.4 Theorem. *A BCH code of designed distance d defined by 3.2 has minimum distance $d_{\min} \geq d$.*

PROOF. We show that no $d - 1$ or fewer columns of $\mathbf{H}$ are linearly dependent over $\mathbb{F}_{q^m}$. We choose a set of $d - 1$ columns with first elements $\zeta^{ci_1}, \zeta^{ci_2}, \ldots, \zeta^{ci_{d-1}}$ and form the $(d - 1) \times (d - 1)$ determinant. We can factorize the divisors ζ^{ci_k}, $k = 1, 2, \ldots, d - 1$, and obtain

$$
\zeta^{c(i_1 + \ldots + i_{d-1})} \begin{vmatrix}
1 & 1 & \cdots & 1 \\
\zeta^{i_1} & \zeta^{i_2} & \cdots & \zeta^{i_{d-1}} \\
\cdots\cdots\cdots\cdots\cdots\cdots\cdots\cdots\cdots \\
\zeta^{(d-2)i_1} & \zeta^{(d-2)i_2} & \cdots & \zeta^{(d-2)i_{d-1}}
\end{vmatrix}
$$

$$
= \zeta^{c(i_1 + \ldots + i_{d-1})} \prod_{1 \leq k < j \leq d-1} (\zeta^{i_j} - \zeta^{i_k}) \neq 0.
$$

The determinant is the Vandermonde determinant and is nonzero, which proves that the minimum distance of the code is at least d.   □

We note that there is a simple proof of Theorem 3.4 in the case $c = 1$, by using Lemma 2.24. Let v be a codeword in the BCH code of 3.4, so that $\deg F_v \leq n - d$. Theorem 2.25 shows that the weight of v must be $m - n$.

In the spectral setting introduced at the end of §2, a primitive t-error-correcting BCH code of length $n = q^m - 1$ is the set of all vectors with n

components in $\mathbb{F}_q$ whose spectrum is zero in a specified block of $2t$ consecutive components.

For $n = q - 1$ we obtain the Reed–Solomon codes. Encoding of these codes is very simple. Since some set of $2t$ consecutive frequencies has been given zeros as components, the message symbols are loaded into the remaining $n - 2t$ symbols, and the result is inverse Fourier transformed to produce the (time-domain) codeword.

The following definition introduces this special class of BCH codes.

3.5 Definition. A *Reed–Solomon code* (in short: RS code) is a BCH code of designed distance d and of length $n = q - 1$ over $\mathbb{F}_q$.

If we let $c = m = 1$ in Definition 3.2 then the generator polynomial of an RS code is given by

$$g = \prod_{i=1}^{d-1} (x - \zeta^i),$$

where ζ is a primitive element of $\mathbb{F}_q$.

3.6 Theorem. *The minimum distance of an RS code with generator polynomial* $g = \prod_{i=1}^{d-1} (x - \zeta^i)$ *is* d.

PROOF. Theorem 3.4 shows that the minimum distance of the RS code is at least d. If we consider a linear code of length n and dimension k and if we collect all codewords with $k - 1$ zeros, then we obtain a subcode of dimension at least 1 and minimum distance at most $n - k + 1$. In particular, this is true for the given code, where the dimension is $k = n + 1 - d$. $\quad\square$

3.7 Examples of BCH codes. (i) Let $\zeta \in \mathbb{F}_{2^m}$ be a primitive element with minimal polynomial $M \in \mathbb{F}_2[x]$. Let $n = 2^m - 1$, $\deg M = m$, $I = \{1, 2\}$, and then according to 3.2, $C = (M)$ is a BCH code and is also a binary Hamming code with m check symbols.

(ii) Let $\zeta \in \mathbb{F}_{2^4}$ be a primitive element with minimal polynomial $M^{(1)}(x) = x^4 + x + 1$ over $\mathbb{F}_2$. The powers ζ^i, $0 \le i \le 14$, can be written as linear combinations of 1, ζ, ζ^2, ζ^3. Thus we obtain a parity-check matrix $\mathbf{H}$ of a code equivalent to the $(15, 11)$ Hamming code,

$$\mathbf{H} = \begin{pmatrix} 1 & 0 & 0 & 0 & 1 & 0 & 0 & 1 & 1 & 0 & 1 & 0 & 1 & 1 & 1 \\ 0 & 1 & 0 & 0 & 1 & 1 & 0 & 1 & 0 & 1 & 1 & 1 & 1 & 0 & 0 \\ 0 & 0 & 1 & 0 & 0 & 1 & 1 & 0 & 1 & 0 & 1 & 1 & 1 & 1 & 0 \\ 0 & 0 & 0 & 1 & 0 & 0 & 1 & 1 & 0 & 1 & 0 & 1 & 1 & 1 & 1 \end{pmatrix}$$

$$= (1 \quad \zeta \quad \zeta^2 \quad \zeta^3 \quad \zeta^4 \quad \zeta^5 \quad \zeta^6 \quad \zeta^7 \quad \zeta^8 \quad \zeta^9 \quad \zeta^{10} \quad \zeta^{11} \quad \zeta^{12} \quad \zeta^{13} \quad \zeta^{14}).$$

This code can be regarded as a narrow-sense BCH code of designed distance $d = 3$ over $\mathbb{F}_2$. Its minimum distance is also 3 and therefore this code can correct one error. For decoding a received word $w \in \mathbb{F}_2^{15}$, we have to find the syndrome $\mathbf{H}w^T$. In this case the syndrome is given as $w(\zeta)$ in the basis $\{1, \zeta, \zeta^2, \zeta^3\}$. We obtain it by dividing $w(x)$ by $M^{(1)}(x)$; say $w(x) = a(x)M^{(1)}(x) + r(x)$, where $\deg r < 4$. Then $w(\zeta) = v(\zeta)$, i.e. the components of the syndrome are equal to the coefficients of $r(\zeta)$. For instance, if $w = 010110001011101$ is the received word, then $r(x) = 1 + x$, and hence $\mathbf{H}w^T = (1100)^T = 1 + \zeta$. Next we have to find the error e with weight ≤ 1 and having the same syndrome as w. Thus we must find the exponent j, $0 \leq j \leq 14$, such that $\zeta^j = \mathbf{H}w^T$. In this numerical example $j = 4$, and thus the received word w has an error in the fifth position. The transmitted codeword was $v = 010100001011101$.

(iii) Let $q = 3$, $n = 15$ and $d = 4$. Then $x^4 + x + 1$ is irreducible over $\mathbb{F}_2$ and roots are primitive elements of $\mathbb{F}_{2^4}$. If ζ is such a root, then ζ^2 is a root, and ζ^3 is then a root of $x^4 + x^3 + x^2 + x + 1$. Thus

$$g(x) = (x^4 + x + 1)(x^4 + x^3 + x^2 + x + 1)$$

is a generator polynomial of a narrow-sense BCH code with $d = 4$. This is also a generator for a BCH code with designed distance $d = 5$, since ζ^4 is a root of $x^4 + x + 1$. The dimension of this code is $15 - \deg g = 7$. This code was considered in detail in Example 3.1.

(iv) Let $q = 3$, $n = 8$ and $d = 4$ and let ζ be a primitive element of $\mathbb{F}_{3^2}$ with minimal polynomial $x^2 - x - 1$ over $\mathbb{F}_3$. Then ζ^2 has minimal polynomial $x^2 + 1$, and ζ^3 has $x^2 - x - 1$ as its minimal polynomial. The polynomial $g(x) = (x^2 - x - 1)(x^2 + 1) = x^4 - x^3 - x - 1$ is the generating polynomial for a BCH code of length 8, dimension 4 over $\mathbb{F}_3$ and has minimum distance 4, since g is a polynomial of weight 4.

(v) Let $q = 2$, $n = 23$ and $d = 5$. The polynomial $x^{23} - 1$ has the following factorization into irreducibles over $\mathbb{F}_2$

$$x^{23} - 1 = (x - 1)g_0(x)g_1(x) = (x - 1)(x^{11} + x^9 + x^7 + x^6 + x^5 + x + 1)$$
$$\times (x^{11} + x^{10} + x^6 + x^5 + x^4 + x^2 + 1).$$

The roots of these polynomials are in $\mathbb{F}_{2^{11}}$ and they are the primitive twenty-third roots of unity over $\mathbb{F}_2$, since $2^{11} - 1 = 23.89$. If ζ is such a root then ζ^j, for $j = 1, 2, 4, 8, 16, 9, 18, 13, 3, 6, 12$, are its conjugates. Each of the cyclic codes in $\mathbb{F}_2^{23}$ generated by the irreducible factors g_0, g_1 of $x^{23} - 1$ of degree 11 is a BCH code of dimension 12 and of designed distance $d = 5$. These codes are equivalent versions of the so-called *binary Golay code*. We state that its minimum distance is 7 and note that again $d_{\min} \geq d$.   □

BCH codes are very powerful since for any positive integer d we can construct a BCH code of minimum distance $\geq d$. To find a BCH code for a larger minimum distance, we have to increase the length n and hence increase the number m, i.e. the degree of $\mathbb{F}_{q^m}$ over $\mathbb{F}_q$. A BCH code of designed distance $d \geq 2t + 1$ will correct t or fewer errors, but at the same time, in order to achieve the desired minimum distance, we must use code words of great length.

We describe a general decoding algorithm for BCH codes and follow LIDL and NIEDERREITER, Chapter 9. This will generalize the approach taken in Example 3.1. Let us denote by $v(x)$, $w(x)$ and $e(x)$ the transmitted code polynomial, the received polynomial, and the error polynomial, respectively, so that $w(x) = v(x) + e(x)$. Also w, v, e denote the corresponding coefficient vectors. The syndrome of w is calculated as

$$S(w) = Hw^T = (S_c, S_{c+1}, \ldots, S_{c+d-2})^T,$$

where

$$S_j = w(\zeta^j) = v(\zeta^j) + e(\zeta^j) \quad \text{for } c \leq j \leq c + d - 2.$$

The error polynomial is $e(x) = \sum_{i=1}^r c_i x^{a_i}$, where $r \leq t$ errors occur and where $a_1, \ldots, a_r$ are distinct elements in $\{0, 1, \ldots, n-1\}$. The elements $X_i = \zeta^{a_i}$ in $\mathbb{F}_{q^m}$ are called *error-location numbers*, the elements $c_i \in \mathbb{F}_q^*$ are the *error values*. Since $v(\zeta^j) = 0$, we obtain

$$S_j = e(\zeta^j) = \sum_{i=1}^r c_i X_i^j \quad \text{for } c \leq j \leq c + d - 2.$$

Because of the computational rules in $\mathbb{F}_{q^m}$ we have

$$S_j^q = \left(\sum_{i=1}^r c_i X_i^j \right)^q = \sum_{i=1}^r c_i^q X_i^{jq} = \sum_{i=1}^r c_i X_i^{jq} = S_{jq}. \qquad (*)$$

The unknowns here are the elements X_i and c_i, $i = 1, \ldots, r$, which we pair into (X_i, c_i). The S_j are known quantities by calculation from the received vector w. In the binary case $q = 2$ we have $c_i = 1$ for all i, so that any error is completely characterized by the X_i alone.

Next we determine the coefficients σ_i defined by the identity

$$\prod_{i=1}^r (X_i - x) = \sum_{i=0}^r (-1)^i \sigma_{r-i} x^i = \sigma_r - \sigma_{r-1} x + \ldots + (-1)^r \sigma_0 x^r.$$

Therefore $\sigma_0 = 1$, and $\sigma_1, \ldots, \sigma_r$ are the elementary symmetric polynomials in $X_1, \ldots, X_r$. We substitute X_i for x and have

$$(-1)^r \sigma_r + (-1)^{r-1} \sigma_{r-1} X_i + \ldots + (-1) \sigma_1 X_i^{r-1} + X_i^r = 0 \quad \text{for } i = 1, \ldots, r.$$

Multiplying by $c_i X_i^j$ and summing these equations for $i = 1, \ldots, r$ yields

$$(-1)^r \sigma_r S_j + (-1)^{r-1} \sigma_{r-1} S_{j+1} + \ldots + (-1) \sigma_1 S_{j+r-1} + S_{j+r} = 0$$

$$\text{for } j = c, c+1, \ldots, c + r - 1.$$

Now we consider the determinant in the X_i of the system below, observe that it is nonzero and thus obtain

3.8 Lemma. *The system of equations*

$$\sum_{i=1}^{r} c_i X_i^j = S_j, \qquad j = c, c+1, \ldots, c+r-1$$

in the unknowns c_i is solvable if the X_i are distinct elements of $\mathbf{F}_{q^m}^*$. □

3.9 Lemma. *The system of equations*

$$(-1)^r \sigma_r S_j + (-1)^{r-1} \sigma_{r-1} S_{j+1} + \ldots + (-1)\sigma_1 S_{j+r-1} + S_{j+r} = 0,$$

$$j = c, c+1, \ldots, c+r-1$$

in the unknowns $(-1)^i \sigma_i$, $i = 1, 2, \ldots, r$, is solvable uniquely if and only if r errors occur.

PROOF. The matrix of the system with entries S_j can be decomposed as the product $\mathbf{VDV}^T$ of matrices of the form

$$\mathbf{V} = \begin{pmatrix} 1 & 1 & \ldots & 1 \\ X_1 & X_2 & \ldots & X_r \\ \vdots & \vdots & & \vdots \\ X_1^{r-1} & X_2^{r-1} & \ldots & X_r^{r-1} \end{pmatrix} \quad \text{and} \quad \mathbf{D} = \begin{pmatrix} c_1 X_1 & 0 & \ldots & 0 \\ 0 & c_2 X_2^c & \ldots & 0 \\ \vdots & \vdots & & \vdots \\ 0 & 0 & & c_r X_r^c \end{pmatrix}.$$

The matrix of the given system of equations is nonsingular if and only if $\mathbf{V}$ and $\mathbf{D}$ are nonsingular. We observe that $\mathbf{V}$ is a Vandermonde matrix and it can be shown that it is nonsingular if and only if the X_i, $i = 1, \ldots, r$, are distinct. The matrix $\mathbf{D}$ is nonsingular if and only if all the X_i and c_i are nonzero. Both of these conditions hold if and only if r errors occur. □

We generalize the polynomial σ of (1) in Example 3.1 and introduce the general error-locator polynomial:

$$s(x) = \prod_{i=1}^{r} (1 - X_i x) = \sum_{i=0}^{r} (-1)^i \sigma_i x^i,$$

where the σ_i are as above. The roots of $s(x)$ are $X_1^{-1}, X_2^{-1}, \ldots, X_r^{-1}$. In order to find these roots, one can use a search method which is due to Chien. First we want to know if ζ^{n-1} is an error-location number, that is if $\zeta = \zeta^{-(n-1)}$ is a root of $s(x)$. This can be tested by forming

$$-\sigma_1 \zeta + \sigma_2 \zeta^2 + \ldots + (-1)^r \sigma_r \zeta^r.$$

If this expression is equal to -1, then ζ^{n-1} is an error-location number since then $s(\zeta) = 0$. More generally, ζ^{n-m} is tested for $m = 1, 2, \ldots, n$ in the same way. In the binary case, the finding of error locations is equivalent to correcting errors. We summarize the BCH decoding algorithm, now writing τ_i for $(-1)^i \sigma_i$.

3.10 BCH Decoding. *Let v be a codeword and suppose that at most t errors occur in transmitting it by using a BCH code of designed distance $d \geq 2t+1$.*

For decoding a received word w, the following steps should be executed:
  Step 1. Determine the syndrome of w, where

$$S(\zeta) = (S_c, S_{c+1}, \ldots, S_{c+d-2})^T.$$

Let

$$S_j = \sum_{i=1}^{r} c_i X_i^j, \qquad c \le j \le c + d - 2.$$

Step 2. Determine the maximum number $r \le t$ of equations of the form

$$S_{j+r} + S_{j+r-1}\tau_1 + \ldots + S_j\tau_r = 0, \qquad c \le j \le c + d - r - 2,$$

such that the coefficient matrix of the τ_i is nonsingular and thus obtain the number r of errors that have occurred. Then form the error-locator polynomial

$$s(x) = \prod_{i=1}^{r} (1 - X_i x) = \sum_{i=0}^{r} \tau_i x^i.$$

Determine the coefficients τ_i from the S_j.
  Step 3. Solve $s(x) = 0$ by substituting the powers of ζ into $s(x)$. Thus find the error-location numbers X_i (by performing a Chien search).
  Step 4. Introduce the X_i into the first r equations of Step 1 to determine the error values c_i. Then find the transmitted word v from $v(x) = w(x) - e(x)$.

3.11 Remark. The most difficult step in this algorithm is Step 2. One way of facilitating the required calculations is to use an algorithm due to Berlekamp and Massey (see BERLEKAMP or LIDL and NIEDERREITER, Chapter 8) to determine the unknown coefficients τ_i in the system of Step 2. This system represents a linear recurrence relation for the S_j (see Chapter 5, §3).

3.12 Example (Example 3.1 revisited). For $c = 1$, $n = 15$, $q = 2$ as parameters we consider the BCH code with designed distance $d = 5$, which can correct up to two errors and was introduced in 3.1. Again, let $M^{(i)}$ denote the minimal polynomial of ζ^i over $\mathbb{F}_2$, where ζ^i is a primitive element of $\mathbb{F}_{2^4}$, and a root of $x^4 + x + 1$. Then

$$M^{(1)} = M^{(2)} = M^{(4)} = 1 + x + x^4,$$

$$M^{(3)} = M^{(6)} = M^{(12)} = M^{(9)} = 1 + x + x^2 + x^3 + x^4.$$

Therefore a generator polynomial of the BCH code will be

$$g = M^{(1)}M^{(3)} = 1 + x^4 + x^6 + x^7 + x^8.$$

The code is a $(15, 7)$ code, with parity-check polynomial

$$h = (x^{15} - 1)/g = 1 + x^4 + x^6 + x^7.$$

A generator matrix of the $(15, 7)$ BCH code is obtained by taking the vectors corresponding to $g, xg, x^2g, x^3g, x^4g, x^5g, x^6g$ as the basis of the code. Then

$$G = \begin{pmatrix}
1 & 0 & 0 & 0 & 1 & 0 & 1 & 1 & 1 & 0 & 0 & 0 & 0 & 0 & 0 \\
0 & 1 & 0 & 0 & 0 & 1 & 0 & 1 & 1 & 1 & 0 & 0 & 0 & 0 & 0 \\
0 & 0 & 1 & 0 & 0 & 0 & 1 & 0 & 1 & 1 & 1 & 0 & 0 & 0 & 0 \\
0 & 0 & 0 & 1 & 0 & 0 & 0 & 1 & 0 & 1 & 1 & 1 & 0 & 0 & 0 \\
0 & 0 & 0 & 0 & 1 & 0 & 0 & 0 & 1 & 0 & 1 & 1 & 1 & 0 & 0 \\
0 & 0 & 0 & 0 & 0 & 1 & 0 & 0 & 0 & 1 & 0 & 1 & 1 & 1 & 0 \\
0 & 0 & 0 & 0 & 0 & 0 & 1 & 0 & 0 & 0 & 1 & 0 & 1 & 1 & 1
\end{pmatrix}.$$

Suppose now that the received word w is given by

$$1 \quad 0 \quad 0 \quad 1 \quad 0 \quad 0 \quad 1 \quad 1 \quad 0 \quad 0 \quad 0 \quad 0 \quad 1 \quad 0 \quad 0,$$

or written as a polynomial,

$$w(x) = 1 + x^3 + x^6 + x^7 + x^{12}.$$

We calculate the syndrome according to Step 1, using equation $(*)$ before Lemma 3.8 to simplify the work:

$$S_1 = e(\zeta) = v(\zeta) = 1,$$
$$S_2 = e(\zeta^2) = v(\zeta^2) = 1,$$
$$S_3 = e(\zeta^3) = v(\zeta^3) = \zeta,$$
$$S_4 = e(\zeta^4) = v(\zeta^4) = 1.$$

Then Step 2 yields the system of linear equations

$$S_2\tau_1 + S_1\tau_2 = S_3,$$
$$S_3\tau_1 + S_2\tau_2 = S_4,$$

or, after substituting for S_j,

$$\tau_1 + \tau_2 = \zeta^4$$
$$\zeta^4\tau_1 + \tau_2 = 1.$$

Since these two equations are linearly independent, two errors must have occurred, i.e. $r = 2$. We solve the system and find $\tau_1 = 1, \tau_2 = \zeta$. Substituting into $s(x)$ and noting $\tau_0 = 1$, we find $s(x) = 1 + x + \zeta x^2$. As roots in $\mathbb{F}_{16}$ we find $X_1^{-1} = \zeta^8$, $X_2^{-1} = \zeta^6$, and hence $X_1 = \zeta^7$, $X_2 = \zeta^9$. Therefore we know that errors must have occurred in positions 8 and 10 of the code word. In order to correct these errors in the received polynomial we form

$$w(x) - e(x) = (1 + x^3 + x^6 + x^7 + x^{12}) - (x^7 + x^9)$$
$$= 1 + x^3 + x^6 + x^9 + x^{12},$$

and thus obtain the transmitted code polynomial $v(x)$. The corresponding codeword is

$$1 \quad 0 \quad 0 \quad 1 \quad 0 \quad 0 \quad 1 \quad 0 \quad 0 \quad 1 \quad 0 \quad 0 \quad 1 \quad 0 \quad 0.$$

The initial message can be recovered by dividing the corrected polynomial $v(x)$ by $g(x)$. This gives

$$v(x)/g(x) = 1 + x^3 + x^4,$$

which yields the corresponding message word $1\ 0\ 0\ 1\ 1\ 0\ 0$. □

3.13 Remark. The decoding algorithm of 3.10 can be presented in a different way by using a suitable notation for the error vector. Let v be a codeword, w the received word after using a BCH code C of designed distance $d \geq 2t + 1$, which can correct $r \leq t$ errors. Let $e = w - v$ be the error vector, with $e = \sum_{i \in E} e_i x^i$, $E = \{n_1, \ldots, n_r\}$.

Step 1. Determine the syndrome $S(w) = \mathbf{H}w^T = (S_c, S_{c+1}, \ldots, S_{c+d-2})$.

Step 2. Determine the error-locator polynomial σ from the S_j's:

$$\sigma = \sum_{k=0}^{r} \sigma_k x^k, \quad \text{where } \sigma_0 = 1.$$

Step 3. Solve the equation $\sigma = 0$. Since $\sigma = \prod_{i \in E} (1 - \zeta^i x)$, we have for each $i \in \{1, 2, \ldots, n - 1\}$,

$$\frac{1}{\zeta^i} \text{ is a root of } \sigma \Rightarrow i \in E.$$

Step 4. Determine e_i with $S_j = \sum_{i \in E} e_i \zeta^{ij}$, $c \leq j \leq c + d - 2$. □

The calculations of the syndrome yield $S(w) = w(\zeta^i) = e(\zeta^i) = f_j$, where f_j is a coefficient of the Mattson–Solomon polynomial. If r errors occurred, then r coordinates e_i are nonzero, and they are error-locations.

For the determination of the error-locating polynomial σ in Step 2 we have

$$1 + \sigma_1 \zeta^{-i} + \ldots + \sigma_r \zeta^{-ir} = 0 \quad \text{for each } i \in E.$$

Using S_j from Step 1 we obtain

$$S_{j+r} + S_{j+r-1} \sigma_1 + \ldots + S_j \sigma_r = 0, \qquad c \leq j \leq c + d - r - 2.$$

Since the S_i are linear combinations of r elements in $\mathbb{F}_{q^m}$, at most r of these equations can be linearly independent. If the first r equations are linearly independent then

$$S_j = \sum_{i=1}^{r} e_{n_i} \zeta^{jn_i}, \qquad c \leq j \leq c + d - 2.$$

There are special methods available to determine σ (see specialist books on coding theory).

Step 3 requires us to find the roots of σ. We have to check whether $\sigma(\zeta^{-i}) = 0$, $0 \le i \le n - 1$, i.e. whether zeros of σ are the reciprocals of the error locations ζ^i.

We conclude this chapter with some remarks on important types of codes and recall that a t-error-correcting code over $\mathbb{F}_q$ is perfect if equality holds in the Hamming bound (Theorem 1.16). In the binary case, if $t = 3$, a sphere about a codeword contains $1 + n + \binom{n}{2} + \binom{n}{3}$ points and this is a power of 2 for $n = 23$. Similarly, $1 + 2n + 4\binom{n}{2}$ is a power of 3 for $n = 11$, so for these values the Hamming bound is also satisfied as an equation. Probably this observation led Marcel Golay to find the corresponding codes and give the difficult existence proof for the famous and very important binary $(23, 12, 7)$ code and the ternary $(11, 6, 5)$ code. Both are perfect codes and are called the *binary* and *ternary Golay codes*, respectively. There are various ways of defining Golay codes, and one possibility for the binary Golay code was given in Example 3.7(v).

Golay codes are a special type of code, called quadratic residue codes. In the following, let n be an odd prime and q a prime power such that $(n, q) = 1$, q is assumed to be a quadratic residue modulo n. U_0 denotes the set of quadratic residues mod n, that is the set of squares in $\mathbb{F}_n^*$, and U_1 denotes the set of nonsquares in $\mathbb{F}_n$. If ζ is a primitive nth root of unity in $\mathbb{F}_{q^{n-1}}$, then

$$g_0 = \prod_{i \in U_0} (x - \zeta^i), \qquad g_1 = \prod_{j \in U_1} (x - \zeta^j)$$

are polynomials over $\mathbb{F}_q$ and we have $x^n - 1 = (x - 1)g_0 g_1$.

3.14 Definition. The *quadratic residue codes* (*QR codes*) C_0^+, C_0, C_1^+, C_1 are the cyclic codes with generator polynomials g_0, $(x - 1)g_0$, g_1 and $(x - 1)g$, respectively.

It can be easily verified that the codes C_0 and C_1 (and also C_0^+ and C_1^+) are equivalent and that $C_0^\perp = C_0^\perp$, $C_1^\perp = C_1^\perp$, if the prime n satisfies $n \equiv -1 \pmod 4$. If $n \equiv 1 \pmod 4$ then $C_0^\perp = C_1^+$, $C_1^\perp = C_0^+$. Extended quadratic residue codes of length $n + 1$ are obtained from QR codes by adding an overall parity check component.

3.15 Definition. The *extended QR codes* C_∞^+ are defined by adding an overall parity-check component c_∞ to C_0^+.

It can be shown that the binary QR code C_0 consists of all code words of even weight of the code C_0^+. If G_0 is an $r \times n$ generator matrix for the binary code C_0 then $\begin{pmatrix} G_0 \\ 1 \end{pmatrix}$ is a generator matrix of C_0^+. A well-known result from number theory is that $q = 2$ is a quadratic residue mod n if and only if $n \equiv \pm 1 \pmod 8$. A generator matrix for extended QR codes can be obtained by using idempotent generators (see Exercises 12, 13 in §2). Let $q = 2$ and n be a prime $\equiv \pm 1 \pmod 8$ and $e = \sum_{i \in U_0} x^i$. Then it can be shown that e is an idempotent generator of C_0 if $n \equiv 1 \pmod 8$, and e is an idempotent generator of C_0^+ if $n \equiv -1 \pmod 8$. Now let $\mathbf{G}$ be a circulant $n \times n$ matrix over $\mathbb{F}_2$ with $e = (e_0, e_1, \ldots, e_{n-1})$ as its first row, i.e.

$$
\mathbf{G} = \begin{pmatrix}
e_0 & e_1 & \cdots & e_{n-2} & e_{n-1} \\
e_1 & e_2 & \cdots & e_{n-1} & e_0 \\
\multicolumn{5}{c}{\dotfill} \\
e_{n-1} & e_0 & \cdots & e_{n-3} & e_{n-2}
\end{pmatrix}.
$$

Let $\mathbf{a} = (0, \ldots, 0)$ for $n \equiv 1 \pmod 8$ and $\mathbf{a} = (1, \ldots, 1)$ for $n \equiv -1 \pmod 8$. Then

$$
\mathbf{G}_\infty = \left(\begin{array}{c|ccc} 1 & 1 & \cdots & 1 \\ \hline \mathbf{a}^T & & \mathbf{G} & \end{array} \right)
$$

is a generator matrix for the extended QR code C_∞^+. An important result on the minimum distance of a QR code is the following (see VAN LINT).

3.16 Theorem. *The minimum distance of the binary* QR *code* C_0^+ *of length* n *is the odd number* d, *for which*

(i) $d^2 > n$ *if* $n \equiv 1 \pmod 8$,
(ii) $d^2 - d + 1 \geq n$ *if* $n \equiv -1 \pmod 8$. $\square$

Now we observe that the binary (23, 12) Golay code is a binary QR code of length $n = 23$. Its minimal distance is 7, because of Theorem 3.16(ii). If we extend this binary (23, 12, 7) Golay code, we obtain the so-called *binary* (24, 12, 8) *Golay code*. More generally, given an (n, k) code C for n even, then we obtain a *punctured code* by removing a column of a generator matrix of C. Thus the binary (23, 12, 7) Golay code can be obtained by puncturing the (24, 12, 8) Golay code.

The *ternary* (11, 6, 5) *Golay code* can be defined as follows. Let $\mathbf{S}_5$ be the 5×5 circulant matrix over $\mathbb{F}_3$, whose first row is $(0, 1, -1, -1, 1)$; thus

$$
\mathbf{S}_5 = \begin{pmatrix}
0 & 1 & -1 & -1 & 1 \\
1 & 0 & 1 & -1 & -1 \\
-1 & 1 & 0 & 1 & -1 \\
-1 & -1 & 1 & 0 & 1 \\
1 & -1 & -1 & 1 & 0
\end{pmatrix}.
$$

Then the matrix

$$
\begin{pmatrix}
1 & \mathbf{0} & 1 \\
\mathbf{0} & \mathbf{I}_5 & \mathbf{S}_5
\end{pmatrix}
$$

is a generator matrix for the ternary $(11, 6, 5)$ Golay code. This code is also a QR code. We consider the factorization

$$x^{11} - 1 = (x - 1)(x^5 - x^3 + x^2 - x - 1)(x^5 + x^4 - x^3 + x^2 - 1)$$

over $\mathbb{F}_3$. This factorization must be the factorization $x^{11} - 1 = (x - 1)g_0g_1$ in the sense of QR codes. Again the codes (g_0) and (g_1) are equivalent.

We say that an (n, k, d) code with certain properties over $\mathbb{F}_q$ is *unique* if any two (n, k, d) codes over $\mathbb{F}_q$ with these properties are equivalent. It has been shown (see VAN LINT, PLESS) that the Hamming codes, the ternary Golay codes, the binary $(24, 12, 8)$ code and the binary $(23, 12, 7)$ code are unique.

We conclude with a remarkable result on perfect codes, which involves the Golay codes and classifies all perfect codes. Besides the binary repetition code of odd length with two codewords $111 \ldots 1$ and $00 \ldots 0$, the only trivial perfect codes are codes which consist of only one codeword or which contain the whole vector space; (see Example 1.5). The 1-error-correcting Hamming codes over $\mathbb{F}_q$ of length $n = (q^m - 1)/(q - 1)$, dimension $k = (q^m - 1)/(q - 1) - m$ and minimum distance $d = 3$ are perfect. The binary $(23, 12, 7)$ Golay code and the $(11, 6, 5)$ Golay code over $\mathbb{F}_3$ are perfect. It has been shown in the mid-1970's that there are no other perfect codes.

3.17 Theorem. *The only nontrivial multiple-error-correcting perfect codes are equivalent to either the binary $(23, 12, 7)$ Golay code or the ternary $(11, 6, 5)$ Golay code. The only nontrivial single-error-correcting perfect codes have the parameters of the Hamming codes.* $\square$

PROBLEMS

1. A BCH code C can be defined in the following ways: Let $a_0, a_1, \ldots, a_{n-1}$ be an arbitrary but fixed ordering of the nonzero elements of $\mathbb{F}_{2^m}$, $n = 2^m - 1$.

    (i) $v = (v_0, v_1, \ldots, v_{n-1})$ is a codeword if and only if

    $$\sum_{i=0}^{n-1} v_i a_i^j = 0 \quad \text{for } j = 1, 2, 3, \ldots, 2t.$$

    (ii) if $a_i = a^i$, a a primitive root of $\mathbb{F}_{2^m}$, then

    $$v \in C \Leftrightarrow \sum_{i=0}^{n-1} v_i a^{ij}, \quad j = 1, 2, \ldots, 2t.$$

    (iii) $v \in C \Leftrightarrow v(a^j) = 0, j = 1, 3, \ldots, 2t - 1$ and $\deg v(x) \le n - 1$.

    (iv) $$v \in C \Leftrightarrow \sum_{i=0}^{n-1} \frac{v_i}{x - a_i} \equiv 0 \pmod{x^{2t}}.$$

    Verify the equivalence of these definitions. Then use (iii) to prove that there is a unique polynomial $g(x) \in \mathbb{F}_2[x]$ such that $v(x) \in C \Leftrightarrow v(x) \equiv 0 \pmod{g(x)}$ and $\deg v(x) \le n - 1$. Moreover, prove $\dim C = k = n - \deg g(x)$. Also, define RS codes according to (i), (ii), (iii) or (iv).

*2. Let A_i denote the number of Golay codewords of weight i. Prove that $A_i = A_{23-i}$, $0 \le i \le 23$, for the $(23, 12)$ Golay code.

3. Describe a RS $(15, 13)$ code over $\mathbb{F}_{16}$ by determining its generator polynomial and the number of errors it will correct.

4. Find the generating polynomial of the $(7, 3)$ RS code over $\mathbb{F}_{2^3}$.

5. The polynomial $g(x) = x^6 + 3x^5 + x^4 + x^3 + 2x^2 + 2x + 1$ over $\mathbb{F}_4$ is a generator polynomial for a $(15, 9)$ code over $\mathbb{F}_4$. Encode $x^8 + 2x^2 + x + 1$. Find the parity-check polynomial. Is $x^{11} + 3x^2 + 2x + 1$ a codeword? Determine the syndrome of $x^{10} + 3x^2 + x + 2$.

*6. Prove: The minimum distance d of any quadratic residue code of block length p satisfies $d \ge \sqrt{p}$.

7. Find the generator polynomial of the binary cyclic $(17, 9, 5)$ code.

*8. Let $G(x)$ be a polynomial of degree s over $\mathbb{F}_{q^m}$ and let $a_0, a_1, \dots, a_{n-1}$ be any subset of $\mathbb{F}_{q^m}$ such that $G(a_i) \ne 0$, $i = 0, 1, \dots, n-1$. Then a *Goppa code* C is defined as follows: $v = (v_0, v_1, \dots, v_{n-1}) \in \mathbb{F}_q^n$ is a codeword if and only if

$$\sum_{i=0}^{n-1} \frac{v_i}{x - a_i} \equiv 0 \, (\mathrm{mod}\, G(x)).$$

Show that the Goppa code defined in this way is a linear (n, k) code over $\mathbb{F}_q$ satisfying the inequalities $d_{\min} \ge s + 1$ and $k \ge n - ms$.

*9. Let P be the set of all polynomials of degree $\le d$ in m variables over $\mathbb{F}_2$. Let $v_0, \dots, v_{2^m - 1}$ be a list of all 2^m binary vectors $(x_1, \dots, x_m)$. For each $f \in P$ define a binary vector of length 2^m by evaluating $(f(v_0), \dots, f(v_{2^m - 1}))$. The set of all vectors obtained in this way from polynomials in P is called the dth *order Reed–Muller* code of length 2^m, abbreviated as $\mathrm{RM}(m, d)$.
  (i) Show that $\mathrm{RM}(m, d)$ is a binary (n, k) linear code with $n = 2^m$ and

$$k = 1 + \binom{m}{1} + \dots + \binom{m}{d}.$$

  (ii) Prove that the minimum distance of $\mathrm{RM}(m, d)$ is 2^{m-d}.
  (iii) Show that the dual of $\mathrm{RM}(m, d)$ is $\mathrm{RM}(m, m - d - 1)$.

*10. A generator matrix for $\mathrm{RM}(m, d)$ can be defined as an array $\mathbf{G} = (\mathbf{G}_0, \mathbf{G}_1, \dots, \mathbf{G}_d)^T$, where $\mathbf{G}_0$ is the vector of length $n = 2^m$ containing all ones; $\mathbf{G}_1$ is an m by 2^m matrix that has each binary m-tuple appearing once as a column; the rows of $\mathbf{G}_r$, $2 \le r \le d$, are constructed from $\mathbf{G}_1$ by calculating them as all possible products of r rows of $\mathbf{G}_1$. Usually the leftmost column of $\mathbf{G}_1$ is the all zeros column, the rightmost the all ones, the others are binary m-tuples in increasing order. Determine $\mathbf{G}$ for $\mathrm{RM}(16, 3)$, and for $\mathrm{RM}(16, 2)$.

11. A *burst* of length b is a vector whose nonzero digits occur in b (cyclically) consecutive components, the first and the last of which are nonzero. Let C be a linear (n, k) code. Prove:
  (i) If C has every nonzero codeword a burst of length $> b$, then $n - k \ge b$.
  (ii) If C corrects all bursts of length $\le b$, then $n - k \ge 2b$ (*Rieger bound*).

12. A *Fire code* is a cyclic code over $\mathbb{F}_q$ with generator polynomial $g(x) = (x^{2b-1} - 1)r(x)$, where $r(x)$ is an irreducible polynomial of degree $m \geq b$ and of order e. $r(x)$ does not divide $x^{2b-1} - 1$. The block length of the Fire code is the order of $g(x)$. Show that this block length is $n = e(2b - 1)$. If $g(x)$ is primitive, prove that the Fire code is a

$$((q^m - 1)(2b - 1), (q^m - 1)(2b - 1) - m - 2b + 1) \text{ code.}$$

   (We note without proof that a Fire code can correct all burst errors of length $\leq b$.)

13. Let $g(x) = x^6 + x^3 + x^2 + x + 1$ be the generator polynomial of a binary code of length 15. Enumerate the burst errors of length ≤ 3 and show that this code can correct burst errors of length 3.

14. The inequality $k/n \leq 1 - 2b/n$, derived from the Rieger bound, shows that the best burst-error-correcting codes have rate very close to $1 - 2b/n$, that is $n - k - 2b$ very close to 0. Compare the Rieger bounds for a Fire code, a BCH code, and a RS code, all considered with respect to burst-error-correcting.

15. Construct an (n, k) Fire code which corrects all bursts of ≤ 5 errors.

16. What is the length of an error burst that can be corrected by a $(19437, 19408)$ Fire code?

17. Show that the binary $(23, 12)$ Golay code has no nonzero codeword of weight 4 or less.

*18. A ternary $(12, 6)$ Golay code is defined by its generator matrix $\mathbf{G} = (\mathbf{I} | \mathbf{A})$, where A is of the form

$$\mathbf{A} = \begin{pmatrix} 0 & 1 & 1 & 1 & 1 & 1 \\ 1 & 0 & 1 & -1 & -1 & 1 \\ 1 & 1 & 0 & 1 & -1 & -1 \\ 1 & -1 & 1 & 0 & 1 & -1 \\ 1 & -1 & -1 & 1 & 0 & 1 \\ 1 & 1 & -1 & -1 & 1 & 0 \end{pmatrix}.$$

   Show that this code is self-orthogonal. Verify that it has minimum distance 6.

EXERCISES (Solutions in Chapter 8, p. 473)

1. Determine the errors in Example 3.1, if the syndrome of a received word is given as $(10010110)^T$. Find a generator matrix of this code.

2. Determine the dimension of a 5-error-correcting BCH code over $\mathbb{F}_3$ of length 80.

3. A binary 2-error-correcting BCH code of length 31 is defined by the zero ζ of $x^5 + x^2 + 1 \in \mathbb{F}_{2^5}[x]$. If a received word has syndrome $(1110011101)^T$, find the errors.

4. Find a generator polynomial of the 3-error-correcting BCH code of length 15 by using the primitive element ζ of $\mathbb{F}_{2^4}$, where $\zeta^4 = \zeta^3 + 1$.

5. Determine a generator polynomial g for a BCH code of length 31 with minimum distance $d = 9$.

*6. Let ζ be a primitive element of $\mathbb{F}_{2^4}$ with $\zeta^4 + \zeta + 1 = 0$, and let $g = x^{10} + x^8 + x^5 + x^4 + x^2 + x + 1$ be a generator polynomial of a $(15, 5)$ BCH code. Suppose the word $v = 110001001101000$ is received. Then determine the corrected received word and decode it.

7. Suppose $n = ad$, and show that the binary BCH code of length n and designed distance d has $d_{\min} = d$.

*8. Find all narrow sense primitive binary BCH codes of length $n = 15$.

*9. Find the generator polynomial of the double error correcting narrow sense ternary code of length 8 over $\mathbb{F}_3$. ($\zeta \in \mathbb{F}_{3^2}$ satisfying $\zeta^2 + \zeta + 2 = 0$ is primitive.) Decode $v = 22001001$.

10. Determine whether the dual of an arbitrary BCH code is a BCH code. Is the dual of an arbitrary Reed–Solomon code a RS code?

11. Find a generator polynomial of the RS code of length 4 over $\mathbb{F}_5$ with designed distance 3.

12. Let C be the $(n = q^m - 1, k)$ RS code of minimum distance d. Prove that the extended code, which is obtained by adding to each codeword an overall parity check, is a (q^m, k) code of minimum distance $d + 1$.

13. For any positive integer m and $t \leq 2^{m-1} - 1$, prove the existence of a binary BCH code of length $n = 2^m - 1$ that is t-error-correcting and has dimension $\geq n - mt$.

*14. Verify that the QR codes C_0^+ and C_1^+ are equivalent.

*15. Show that the weight of all codewords of the ternary $(11, 6, 5)$ Golay code is not congruent to 1 (mod 3).

*16. Show that the ternary Golay code has minimum distance 5 and that the code is perfect.

NOTES

SHANNON's paper marks the beginning of information theory as a mathematical discipline. The first nontrivial example of an error-correcting code over a finite field is contained in this paper; nowadays it would be referred to as the $(7, 4)$ Hamming code. A short history of the developments in algebraic coding theory is given in BLAKE; this is a collection of important papers on the subject. There is quite a number of monographs and textbooks on algebraic coding theory; we refer the beginner, who is interested in knowing somewhat more than we present, to the introductory texts by SLOANE, PLESS or VAN LINT. Some of the books on applied algebra or combinatorics also contain material on algebraic coding theory, which is easily accessible; see, e.g. BIRKHOFF and BARTEE, DORNHOFF and HOHN,

STREET and WALLIS, LIDL and NIEDERREITER. Amongst the more advanced texts and monographs we mention the very comprehensive volume MACWILLIAMS and SLOANE (with an extensive bibliography on coding), PETERSON and WELDON, BERLEKAMP, BLAHUT, and BLAKE and MULLIN. The following books contain sections on coding in the general context of information theory: McELIECE, GUIASU, GALLAGER.

Cyclic codes were introduced by Prange, most of the books mentioned above contain the basics of cyclic codes (and some books much more). For connections between cyclic codes, shift registers and linear recurring sequences (Chapter 5, §3) see ABRAMSON, BERLEKAMP (Chapter 5), ZIERLER. The engineering approach to coding was described by R. E. BLAHUT in order to bring coding closer to the subject of signal processing.

Fourier transforms over a finite field were discussed by POLLARD, but MATTSON and SOLOMON used similar relationships already a decade earlier.

BCH codes were introduced by Hocquenghem and Bose and Ray-Chaudhuri in the binary case and GORENSTEIN and ZIERLER in the non-binary case. Any standard book on coding includes a description of BCH codes and their properties, in particular a presentation of decoding procedures.

The interrelation between coding and combinatorics is of advantage to both disciplines and the books by BLAKE and MULLIN, PLESS, MACWILLIAMS and SLOANE, WALLIS and STREET and VAN LINT pay particular attention to this relationship.

The vast material on algebraic coding makes it impossible to give details on references for other types of codes. It should suffice to refer the interested reader to MACWILLIAMS and SLOANE, who treat some thirty kinds of codes, and give encoding and decoding procedures and nearly 1500 references.

In conclusion we mention a few applications of codes. One of the important areas where codes are used is digital communication by satellite. The $(32, 6)$ Reed–Muller codes were used in all of NASA's Mariner class deep-space probes from 1969 to 1977. Block codes are used for satellite communication with mobile airborne terminals. The $(7, 2)$ Reed–Solomon (RS) code was used in the U.S. Defence Department Tactical Satellite whereas the $(31, 15)$ RS code has been adopted for tactical military communication links in the United States (see BERLEKAMP). As a code for high-speed data transmission with the INTELSAT V satellite the double-error-correcting, triple-error-detecting $(128, 112)$ BCH code has been selected. This is an extended version of the $(127, 112)$ BCH code. The RS code of length 63 over $\mathbb{F}_{2^6}$ was used for correcting errors in binary data stored in a photodigital mass memory. Many thousands of lines of 6.63 binary bits of data are stored on a memory chip, each line representing one code word of the RS code. DORNHOFF and HOHN describe a method based on cyclic codes, called the cyclic redundancy check, to correct errors in recorded data on magnetic tape.

BHARGAVA *et al.* refer to some other examples of applications: Digital communication via satellite has been considered for the U.S. Postal Service Electronic Message System, where binary error probabilities as low as 10^{-12} have been stated as goals to be achieved with appropriate error-control techniques. Long cyclic codes have been suggested for a UHF satellite system, where erasure bursts are due to radio-frequency interference with pulse lengths significantly longer than the bit transmission time. It was thought that codes of block length no greater than 128 should suffice for the U.S. Navy's UHF demand-assignment multiple access system.

In contrast to block codes, convolutional codes (which are not discussed in this text) are suitable when the message symbols to be transmitted are given serially in long sequences rather than in blocks, when block codes would be more appropriate. Several deep-space probes, such as Voyager, NASA's Pioneer missions, the INTELSAT SPADE system and West Germany's Helios have used convolutional codes. The INTELSAT SPADE system, for example, uses threshold decoding. Another decoding scheme was successfully used on a two-way digital video link with the Canadian experimental satellite Hermes, where the use of a Viterbi decoding scheme could be translated into a substantial reduction of antenna site and a smaller overall Earth station cost.

CHAPTER 5

Further Applications of Fields and Groups

This chapter contains some topics from combinatorics, cryptology and recurring sequences. The material is mainly selected to give a brief indication of some further applications of algebraic concepts, mainly groups and fields, to areas such as the design of experiments, transmission of secret information or communication with satellites. The selected areas and examples of further applications form only a sample. We refer the interested reader to the special literature on these topics, and also to the notes at the end of the chapter. Some additional references on other topics of applications of abstract algebra are given at the end of the bibliography.

§1. Combinatorial Applications

A. Hadamard Matrices

We begin this section with the definition of so-called Hadamard matrices. Because of the use of Hadamard transforms, these matrices are important in algebraic coding theory, information science and physics. They are also useful in various problems for determining: weights; voltage or resistance; concentration of chemicals; or frequencies of spectra.

1.1 Definition. A *Hadamard matrix* $\mathbf{H}_n$ of order n is an $n \times n$ matrix with elements in $\{1, -1\}$, such that

$$\mathbf{H}_n^T \mathbf{H}_n = \mathbf{H}_n \mathbf{H}_n^T = n\mathbf{I}.$$

This implies that the scalar product of any two distinct rows (or columns) of H_n is equal to 0, i.e. any two distinct rows (or columns) are orthogonal.

1.2 Examples.

$$H_1 = (1), \quad H_2 = \begin{pmatrix} 1 & 1 \\ 1 & -1 \end{pmatrix}, \quad H_4 = \begin{pmatrix} 1 & 1 & 1 & 1 \\ 1 & -1 & 1 & -1 \\ 1 & 1 & -1 & -1 \\ 1 & -1 & -1 & 1 \end{pmatrix},$$

$$H_8 = \begin{pmatrix} 1 & 1 & 1 & 1 & 1 & 1 & 1 & 1 \\ 1 & -1 & 1 & -1 & 1 & -1 & 1 & -1 \\ 1 & 1 & -1 & -1 & 1 & 1 & -1 & -1 \\ 1 & -1 & -1 & 1 & 1 & -1 & -1 & 1 \\ 1 & 1 & 1 & 1 & -1 & -1 & -1 & -1 \\ 1 & -1 & 1 & -1 & -1 & 1 & -1 & 1 \\ 1 & 1 & -1 & -1 & -1 & -1 & 1 & 1 \\ 1 & -1 & -1 & 1 & -1 & 1 & 1 & -1 \end{pmatrix}.$$ □

H_n is called a Hadamard matrix, since its determinant attains a bound originally given by Hadamard, namely $|\det H_n| = n^{n/2}$. Since transition from H_n to $-H_n$ and the interchanging of signs of rows or columns of H_n does not alter the defining property of H_n, we may assume that the first row and first column consist of ones only. In this case H_n is called "*normalized*".

1.3 Theorem. *If a Hadamard matrix of order n exists, then n is equal to 1 or 2 or is a multiple of* 4.

PROOF. For $n = 1$ and 2 Hadamard matrices exist. Let $n \geq 3$ and suppose the first three rows of H_n are of the form

$$\begin{array}{cccc cccc cccc cccc}
1 & 1 & \ldots & 1 & 1 & 1 & \ldots & 1 & 1 & 1 & \ldots & 1 & 1 & 1 & \ldots & 1 \\
1 & 1 & \ldots & 1 & 1 & 1 & \ldots & 1 & -1 & -1 & \ldots & -1 & -1 & -1 & \ldots & -1 \\
1 & 1 & \ldots & 1 & -1 & -1 & \ldots & -1 & 1 & 1 & \ldots & 1 & -1 & -1 & \ldots & -1
\end{array}$$

$$\underbrace{}_{i\text{ times}} \quad \underbrace{}_{j\text{ times}} \quad \underbrace{}_{k\text{ times}} \quad \underbrace{}_{l\text{ times}}$$

Since the rows are orthogonal, we have

$$i + j - k - l = 0,$$
$$i - j + k - l = 0,$$
$$i - j - k + l = 0.$$

This implies that $i = j = k = l$, and thus $n = 4i$. □

It is an unsolved problem whether or not Hadamard matrices of order $n = 4k$ exist for all $k \in \mathbb{N}$. Hadamard matrices can be constructed in the following way. If $\mathbf{H}_n$ is a Hadamard matrix of order n, then

$$\mathbf{H}_{2n} = \begin{pmatrix} \mathbf{H}_n & \mathbf{H}_n \\ \mathbf{H}_n & -\mathbf{H}_n \end{pmatrix}$$

is a Hadamard matrix of order $2n$. More generally, let χ denote the *Legendre symbol*, defined by

$$\chi(a) = \begin{cases} 1 & \text{if } a \neq 0 \text{ is a square,} \\ -1 & \text{if } a \text{ is not a square,} \\ 0 & \text{if } a = 0. \end{cases}$$

for any $a \in \mathbb{F}_q$.

1.4 Theorem. *If* $\mathbb{F}_q = \{a_0, a_1, \ldots, a_{q-1}\}$, $q = p^e = 4t - 1$, *then the following* $(q + 1) \times (q + 1)$ *matrix is a Hadamard matrix:*

$$\begin{pmatrix} 1 & 1 & 1 & \ldots & 1 & 1 \\ 1 & -1 & \chi(a_1) & \ldots & \chi(a_{q-2}) & \chi(a_{q-1}) \\ 1 & \chi(a_{q-1}) & -1 & \ldots & \chi(a_{q-3}) & \chi(a_{q-2}) \\ \multicolumn{6}{c}{\dotfill} \\ 1 & \chi(a_1) & \chi(a_2) & \ldots & \chi(a_{q-1}) & -1 \end{pmatrix}. \qquad \square$$

Since all entries are ± 1 the theorem is proved by showing that the inner product of any two distinct rows is 0.

Hadamard matrices of orders $2^k(q + 1)$ with $k \geq 0$ and prime powers q of the form $q = 4t - 1$ can be obtained by using 1.4 and the construction mentioned before this theorem. By starting from the Hadamard matrix $\mathbf{H}_1$ of Example 1.2, one can construct Hadamard matrices of orders 2^k, $k \geq 0$. At present the smallest order of a Hadamard matrix that has not been constructed is 268.

Some Hadamard matrices (namely those obtained by the so-called *Paley construction* (see Theorem 1.4)) give rise to an important class of nonlinear codes. Let $\mathbf{H}_n$ be a normalized Hadamard matrix of order n. We replace all $+1$'s in $\mathbf{H}_n$ by 0's and all -1's by 1's, and then $\mathbf{H}_n$ is changed into a binary Hadamard matrix $\mathbf{A}_n$. The rows of $\mathbf{H}_n$ are orthogonal, therefore any two rows of $\mathbf{A}_n$ agree in $\frac{1}{2}n$ places and differ in $\frac{1}{2}n$ places and so are Hamming distance $\frac{1}{2}n$ apart. A *nonlinear code* is defined as a set of M vectors of length n (with components from some field F) such that any two vectors differ in at least d places and d is the smallest number with this property. We call such a code an $\langle n, M, d \rangle$ code. The following nonlinear codes can be obtained from the matrix $\mathbf{A}_n$.

(i) an $\langle n - 1, n, \frac{1}{2}n \rangle$ code $\mathcal{A}_n$ consists of the rows of $\mathbf{A}_n$ with the first column deleted;

(ii) an $\langle n - 1, 2n, \frac{1}{2}n - 1 \rangle$ code $\mathcal{B}_n$ consists of $\mathcal{A}_n$ together with the complements of all its codewords;

(iii) an $\langle n, 2n, \frac{1}{2}n \rangle$ code $\mathcal{C}_n$ consists of the rows of $\mathbf{A}_n$ and their complements.

For instance, the codewords of a $\langle 7, 8, 4 \rangle$ code are

$$\mathbf{A}_8 = \begin{pmatrix} 0 & 0 & 0 & 0 & 0 & 0 & 0 \\ 1 & 0 & 0 & 1 & 0 & 1 & 1 \\ 1 & 1 & 0 & 0 & 1 & 0 & 1 \\ 1 & 1 & 1 & 0 & 0 & 1 & 0 \\ 0 & 1 & 1 & 1 & 0 & 0 & 1 \\ 1 & 0 & 1 & 1 & 1 & 0 & 0 \\ 0 & 1 & 0 & 1 & 1 & 1 & 0 \\ 0 & 0 & 1 & 0 & 1 & 1 & 1 \end{pmatrix}.$$

For further properties and examples see MACWILLIAMS and SLOANE.

1.5 Example. The problem is to determine the weight of four objects by using a chemical balance with two pans. Suppose we know that the scales have error ε at each weighing. This ε is a random variable with mean 0 and variance σ^2, independent of the weight of the objects. Let x_i denote the exact weights of the objects, let y_i be the weights found by reading off the scales, and let ε_i be the (unknown) errors, i.e.

$$x_i = y_i + \varepsilon_i \quad \text{for } i = 1, 2, 3, 4.$$

Estimates for the unknown weights x_i are

$$\hat{x}_i = y_i = x_i - \varepsilon_i, \qquad i = 1, 2, 3, 4,$$

where each has variance σ^2.

Alternatively, we can use the following weighing design. We make the measurements z_i,

$$z_1 = x_1 + x_2 + x_3 + x_4 + e_1,$$
$$z_2 = x_1 - x_2 + x_3 - x_4 + e_2,$$
$$z_3 = x_1 + x_2 - x_3 - x_4 + e_3,$$
$$z_4 = x_1 - x_2 - x_3 + x_4 + e_4,$$

where again the x_i denote the true weights and e_i denote errors. These equations mean that in the first weighing all four objects are placed in the left pan, in the second weighing objects 1 and 3 are in the left pan and 2 and 4 in the right, and so on. The coefficient matrix of the x_i is a Hadamard matrix $\mathbf{H}_4$. The estimates for the weights are

$$\hat{x}_1 = \tfrac{1}{4}(z_1 + z_2 + z_3 + z_4),$$
$$\hat{x}_2 = \tfrac{1}{4}(z_1 - z_2 + z_3 - z_4),$$
$$\hat{x}_3 = \tfrac{1}{4}(z_1 + z_2 - z_3 - z_4),$$
$$\hat{x}_4 = \tfrac{1}{4}(z_1 - z_2 - z_3 + z_4),$$

which implies that

$$\hat{x}_1 = x_1 + \tfrac{1}{4}(e_1 + e_2 + e_3 + e_4), \quad \text{etc.}$$

The variances or the mean squared errors in the estimates are $\varepsilon_1 =$ mean$(\hat{x}_1 - x_1)^2 = \sigma^2/4 = \varepsilon_2 = \varepsilon_3 = \varepsilon_4$, which is an improvement by a factor of 4 over weighing the objects one at a time. This holds under the assumption that the errors e_i in the ith measurement are independent of the quantity being measured, i.e. that the weight of the objects should be light in comparison with the mass of the balance. Also we assumed mean$(e_i) = 0$, mean$(e_i^2) = \sigma^2$ and mean$(e_i e_j) = 0$.   □

Hotelling proved that Hadamard matrices make the best chemical balance weighing designs, i.e. they are the best matrices to be used to design the weighing of objects on chemical balances as described above. If there are n objects and a Hadamard matrix of order n is used, then the mean squared error in each unknown x_i is reduced by a factor of n.

SLOANE describes an application of so-called S-matrices in measuring the spectrum of a beam of light. Instead of n objects whose weights are to be determined, a beam of light has to be divided into n components of different wavelengths and their intensities have to be found. This is done by using a multiplexing spectrometer with several exit slits allowing several components to pass freely through a mask and be measured. The S-matrices used for such a device are defined by beginning with a normalized Hadamard matrix $\mathbf{H}_n$ of order n. Then an S-*matrix* of order $n - 1$, $\mathbf{S}_{n-1}$, is the $(n - 1) \times (n - 1)$ matrix of 0's and 1's obtained by omitting the first row and column of $\mathbf{H}_n$ and then changing 1's to 0's and -1's to 1's. The S-matrices $\mathbf{S}_1$, $\mathbf{S}_3$ and $\mathbf{S}_4$ are obtained from Example 1.2.

$$\mathbf{S}_1 = (1), \qquad \mathbf{S}_3 = \begin{pmatrix} 1 & 0 & 1 \\ 0 & 1 & 1 \\ 1 & 1 & 0 \end{pmatrix}, \qquad \mathbf{S}_7 = \begin{pmatrix} 1 & 0 & 1 & 0 & 1 & 0 & 1 \\ 0 & 1 & 1 & 0 & 0 & 1 & 1 \\ 1 & 1 & 0 & 0 & 1 & 1 & 0 \\ 0 & 0 & 0 & 1 & 1 & 1 & 1 \\ 1 & 0 & 1 & 1 & 0 & 1 & 0 \\ 0 & 1 & 1 & 1 & 1 & 0 & 0 \\ 1 & 1 & 0 & 1 & 0 & 0 & 1 \end{pmatrix}.$$

We leave it as an exercise to show

1.6 Lemma. *Let* $\mathbf{S}_n$ *be an S-matrix of order n. Then* $\mathbf{S}_n$ *satisfies*

(i) $\mathbf{S}_n \mathbf{S}_n^T = \tfrac{1}{4}(n + 1)(\mathbf{I}_n + \mathbf{J}_n)$;

(ii) $\mathbf{S}_n \mathbf{J}_n = \mathbf{J}_n \mathbf{S}_n = \tfrac{1}{2}(n + 1)\mathbf{J}_n$;

(iii) $\mathbf{S}_n^{-1} = \dfrac{2}{n+1}(2\mathbf{S}_n^T - \mathbf{J}_n),$

where $\mathbf{J}_n$ is an $n \times n$ matrix of 1's. (Properties (i) and (ii) for an arbitrary $n \times n$ matrix of 0's and 1's imply that $\mathbf{S}_n$ is an S-matrix.) □

The most important examples for practical purposes are cyclic S-matrices, which have the property that each row is a cyclic shift to the left or to the right of the previous row, e.g.

$$\begin{pmatrix} 1 & 1 & 1 & 0 & 1 & 0 & 0 \\ 0 & 1 & 1 & 1 & 0 & 1 & 0 \\ 0 & 0 & 1 & 1 & 1 & 0 & 1 \\ 1 & 0 & 0 & 1 & 1 & 1 & 0 \\ 0 & 1 & 0 & 0 & 1 & 1 & 1 \\ 1 & 0 & 1 & 0 & 0 & 1 & 1 \\ 1 & 1 & 0 & 1 & 0 & 0 & 1 \end{pmatrix}.$$

Cyclic S-matrices can be constructed by using maximal length shift-register sequences (see §3 of this chapter). In this construction the first row of the $n \times n$ matrix is taken to be one period's worth of the output from a shift-register whose feedback polynomial is a primitive polynomial (see 3.3.12) of degree m. In §3 we shall see that the period of such a shift-register is $n = 2^m - 1$.

1.7 Example. The polynomial $x^4 + x + 1$ over $\mathbb{F}_2$ determines the following shift-register

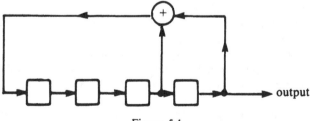

Figure 5.1

The output from this shift-register has period $n = 2^m - 1 = 15$. If it initially contains 1 0 0 0 in the delay elements one period's worth of the output is

0 0 0 1 0 0 1 1 0 1 0 1 1 1 1 1.

An S-matrix of order 15 having this sequence as its first row is

$$
S_{15} = \begin{pmatrix}
0 & 0 & 0 & 1 & 0 & 0 & 1 & 1 & 0 & 1 & 0 & 1 & 1 & 1 & 1 \\
0 & 0 & 1 & 0 & 0 & 1 & 1 & 0 & 1 & 0 & 1 & 1 & 1 & 1 & 0 \\
0 & 1 & 0 & 0 & 1 & 1 & 0 & 1 & 0 & 1 & 1 & 1 & 1 & 0 & 0 \\
1 & 0 & 0 & 1 & 1 & 0 & 1 & 0 & 1 & 1 & 1 & 1 & 0 & 0 & 0 \\
0 & 0 & 1 & 1 & 0 & 1 & 0 & 1 & 1 & 1 & 1 & 0 & 0 & 0 & 1 \\
0 & 1 & 1 & 0 & 1 & 0 & 1 & 1 & 1 & 1 & 0 & 0 & 0 & 1 & 0 \\
1 & 1 & 0 & 1 & 0 & 1 & 1 & 1 & 1 & 0 & 0 & 0 & 1 & 0 & 0 \\
1 & 0 & 1 & 0 & 1 & 1 & 1 & 1 & 0 & 0 & 0 & 1 & 0 & 0 & 1 \\
0 & 1 & 0 & 1 & 1 & 1 & 1 & 0 & 0 & 0 & 1 & 0 & 0 & 1 & 1 \\
1 & 0 & 1 & 1 & 1 & 1 & 0 & 0 & 0 & 1 & 0 & 0 & 1 & 1 & 0 \\
0 & 1 & 1 & 1 & 1 & 0 & 0 & 0 & 1 & 0 & 0 & 1 & 1 & 0 & 1 \\
1 & 1 & 1 & 1 & 0 & 0 & 0 & 1 & 0 & 0 & 1 & 1 & 0 & 1 & 0 \\
1 & 1 & 1 & 0 & 0 & 0 & 1 & 0 & 0 & 1 & 1 & 0 & 1 & 0 & 1 \\
1 & 1 & 0 & 0 & 0 & 1 & 0 & 0 & 1 & 1 & 0 & 1 & 0 & 1 & 1 \\
1 & 0 & 0 & 0 & 1 & 0 & 0 & 1 & 1 & 0 & 1 & 0 & 1 & 1 & 1
\end{pmatrix} . \qquad \square
$$

Problems

1. Let $a_1, \ldots, a_n$ denote the $n = 2^m$ distinct binary m-tuples. Show that the matrix $H = (h_{ij})$ where $h_{ij} = (-1)^{a_i \cdot a_j}$ is a Hadamard matrix of order n.

*2. Let $\otimes$ denote the Kronecker product of matrices, e.g. $H_4 = H_2 \otimes H_2$ for Hadamard matrices. Hence

$$
H_4 = \begin{pmatrix} H_2 & H_2 \\ H_2 & -H_2 \end{pmatrix}.
$$

In general we define a Hadamard matrix recursively by $H_{2^{k+1}} = H_2 \otimes H_{2^k}$. Let H_n be obtained in this way, where $n = 2^m$. If we replace 1 and -1 by 0 and 1, respectively, we obtain a matrix C_n of 0's and1's. Show that the 2^{m+1} rows of C_n and $J - C_n$ form a linear subspace of dimension $m + 1$ in $(\mathbb{F}_2)^n$. This defines a linear code of length n and dimension $m + 1$, (see Chapter 4, §1). For NASA's Mariner 9 space probe such a code was used with $m = 5$ and $n = 32$.

3. Let I_k denote the $k \times k$ unit matrix and let H_n be a Hadamard matrix where $n = 2^m$. Define

$$
M_n^{(i)} = I_{2^{m-i}} \otimes H_2 \otimes I_{2^{i-1}} \quad \text{for } 1 \leq i \leq m.
$$

Prove that $H_n = \prod_{i=1}^{m} M_n^{(i)}$.

4. If A is a matrix of zeros and ones and if B is the matrix obtained from A by replacing every 0 by -1, show that $B = 2A - J$. Again J denotes the matrix whose entries are all ones.

5. A *conference matrix* $\mathbf{C}_n$ of order n is an $n \times n$ matrix with diagonal entries 0 and others $+1$ or -1 which satisfies

$$\mathbf{C}_n \mathbf{C}_n^T = (n-1)\mathbf{I}_n.$$

(The name arises from the use of such matrices in the design of networks having the same attenuation between every pair of terminals.) Show that n must be even for $\mathbf{C}_n$ to exist. Then let $n = 4t + 2 = p + 1$, p an odd prime. Let $\mathbf{C}_n$ be of the form

$$\mathbf{C}_n = \begin{pmatrix} 0 & 1 & 1 & \cdots & 1 \\ 1 & & & & \\ \vdots & & \mathbf{B}_{n-1} & & \\ 1 & & & & \end{pmatrix},$$

where $\mathbf{B}_{n-1} = (b_{ij})$ is a symmetric matrix.

A conference matrix $\mathbf{C}_n$ can be constructed by setting $b_{ij} = 0$ if $i = j$, $b_{ij} = 1$ if $j - 1$ is a square (mod p), and $b_{ij} = -1$ if $j - 1$ is not a square (mod p). Verify that $\mathbf{B}_{n-1}$ is cyclic. Construct conference matrices for $n = 6$ and for $n = 14$. Generalize this construction by replacing p by any odd prime power q. The rows and columns of $\mathbf{B}_{n-1}$ are labeled with the elements of the finite field $\mathbb{F}_q$. What conditions do you have to impose on b_{ij} in this generalized situation? Construct $\mathbf{C}_{10}$.

*6. Design the weighing of eight objects similar to Example 1.5.

7. Construct two more S-matrices other than those given in the text.

*8. Prove Lemma 1.6.

B. Balanced Incomplete Block Designs

R. A. Fisher in his book *The Design of Experiments* was the first to indicate how experiments can be organized systematically so that statistical analysis can be applied. In the planning of experiments it often occurs that results are influenced by phenomena outside the control of the experimenter. The first applications of experimental design methods were in the design of agricultural and biological experiments. Suppose v types of fertilizers have to be tested on b different plants. In ideal circumstances we would test each plant with respect to each fertilizer, i.e. we would have b plots (blocks) each of size v. Such an array is called 2-(v, v, b) design. However, this is often not economical, and so one tries to design experiments such that each plant is tested with k fertilizers and any two fertilizers together are applied on the same plant λ times. Such an experiment is called "balanced". It is a 2-(v, k, λ) design with b blocks, and is called "incomplete" if $k < v$.

If we have to compare m types of wheat with respect to their average yield on a given plot, then we subdivide the plot into m^2 subplots and plant each type of wheat exactly once (in one the plots) in each row of m

plots. This arrangement is fair in avoiding undesirable influences on the experiment by quality variations in different subplots. It is called a Latin square. In general, if we have to test the effect of r different conditions with m possibilities for each condition, this leads to a set of r orthogonal Latin squares. Sections B and C contain the formal definitions.

1.8 Definition. A *balanced incomplete block design* (BIBD) with parameters (v, b, r, k, λ) is a pair (P, B) with the following properties:

  (i) P is a set with v elements;
 (ii) $B = \{B_1, \ldots, B_b\}$ is a subset of $\mathcal{P}(P)$ with b elements;
(iii) each B_i has exactly k elements where $k < v$;
 (iv) each unordered pair (p, q) with $p, q \in P, p \neq q$, occurs in exactly λ elements of B.

The sets $B_1, \ldots, B_b$ are called the *blocks* of the BIBD. Each $a \in P$ occurs in exactly r sets of B.

Such a BIBD is also called a (v, b, r, k, λ) *configuration* or $2\text{-}(v, k, \lambda)$ *tactical configuration* or *design*. The term "balance" indicates that each pair of elements occurs in exactly the same number of blocks; the term "incomplete" means that each block contains less than v elements. A BIBD is *symmetric* if $v = b$.

1.9 Definition. The *incidence matrix* of a (v, b, r, k, λ) configuration is the $v \times b$ matrix $\mathbf{A} = (a_{ij})$, where

$$a_{ij} = \begin{cases} 1 & \text{if } i \in B_j, \\ 0 & \text{otherwise,} \end{cases}$$

here i denotes the ith element of the configuration.

We summarize some properties of incidence matrices.

1.10 Theorem. *Let $\mathbf{A}$ be the incidence matrix of a (v, b, r, k, λ) configuration. Then*

  (i) $\mathbf{A}\mathbf{A}^T = (r - \lambda)\mathbf{I}_v + \lambda \mathbf{J}_{vv}$;
 (ii) $\det(\mathbf{A}\mathbf{A}^T) = [r + (v - 1)\lambda](r - \lambda)^{v-1}$;
(iii) $\mathbf{A}\mathbf{J}_{bb} = r\mathbf{J}_{vb}$;
 (iv) $\mathbf{J}_{vv}\mathbf{A} = k\mathbf{J}_{vb}$;

where $\mathbf{J}_{mn}$ is the $m \times n$ matrix with all entries equal to 1 and $\mathbf{I}_n$ is the $n \times n$ identity matrix. $\qquad\square$

We prove (ii) and leave the other parts to the reader. If we subtract the first column from all other columns, and then add the second, third, $\ldots$, vth row to the first, then all elements above the main diagonal are zero. On the main diagonal the first entry is $r + (v - 1)\lambda$, and the others are $r - \lambda$.

1.11 Theorem. *The following conditions are necessary for the existence of a BIBD with parameters* v, b, r, k, λ:

(i) $bk = rv$;
(ii) $r(k - 1) = \lambda(v - 1)$;
(iii) $b \geq v$.

PROOF. (i) We count the number of ordered pairs consisting of a block and an element of that block as bk. Each element of P must occur in r blocks B_i, therefore there are vr ordered pairs of an element and a block containing it. The numbers bk and vr count the number of elements of the same set, so $bk = vr$.

(ii) follows by a similar argument.

(iii) If $r > \lambda$ then $\mathbf{A}^T$ has rank v, by 1.10(ii). On the other hand, since the rank of $\mathbf{A}$ can be at most b, and since rank $\mathbf{A} \geq$ rank $\mathbf{A}\mathbf{A}^T = v$, we have $b \geq v$. $\qquad\qquad\square$

1.12 Example. Let $\{1, 2, 3, 4, 5, 6, 7\}$ be the set of elements (varieties) and $\{\{1, 2, 3\}, \{3, 4, 5\}, \{1, 5, 6\}, \{1, 4, 7\}, \{3, 6, 7\}, \{2, 5, 7\}, \{2, 4, 6\}\}$ be the set of blocks. Then we obtain a $(7, 7, 3, 3, 1)$ configuration. This is the so-called *Fano geometry* (Fano plane), the simplest example of a finite projective plane over $\mathbb{F}_2$. The blocks are the sets of points on lines, the elements are the seven points of the plane.

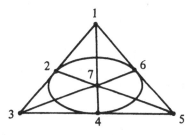

Figure 5.2 $\qquad\qquad\square$

There is a special type of algebraic structure, near-rings, which can be used in constructing block designs.

1.13 Definition. A set N together with two operations $+$ and $\circ$ is a *near-ring* provided that:

(i) $(N, +)$ is a group (not necessarily abelian);
(ii) $(N, \circ)$ is a semigroup;
(iii) $\forall\, n, n', n'' \in N: (n + n') \circ n'' = n \circ n'' + n' \circ n''$.

1.14 Examples. (i) Let $(\Gamma, +)$ be a (not necessarily abelian) group with zero
   0. Near-rings are, e.g. $M(\Gamma) := (\Gamma^\Gamma, +, \circ)$, $M_0(\Gamma) := \{f \in M(\Gamma) | f(0) = 0\}$
   and $M_c(\Gamma) := \{f \in M(\Gamma) | f \text{ constant}\}$.

(ii) Let $(\Gamma, +)$ be a group and $S \subseteq \text{End}(\Gamma)$. Then $M_S(\Gamma) :=$ $\{f \in M_0(\Gamma)| \forall\, s \in S: f \circ s = s \circ f\}$ is a near-ring (which is going to play an important role in the sequel). If, for instance, $(\Gamma, +) = (\mathbb{R}, +)$ and $S = \{s_\lambda | \lambda \in \mathbb{R}\}$ with $s_\lambda : \mathbb{R} \to \mathbb{R}, x \to \lambda x$ then $M_S(\Gamma)$ consists of the "homogeneous functions":

$$M_S(\Gamma) = \{f \in \mathbb{R}^{\mathbb{R}}| \forall\, x, \lambda \in \mathbb{R}: f(\lambda x) = \lambda f(x)\}.$$

(iii) Let V be a vector space. $M_{\text{aff}}(V) := \{f \in V^V | f$ is an affine map, i.e. the sum of a linear and a constant map$\}$ is another important near-ring.   □

Of course, every ring is a near-ring. Hence near-rings are generalized rings. Two ring axioms are missing: the commutativity of addition and (much more important) the second distributive law.

1.15 Definition. Let N be a near-ring and $n_1, n_2 \in N$.

$$n_1 \equiv n_2 :\Leftrightarrow \forall\, n \in N: n \circ n_1 = n \circ n_2.$$

Then $\equiv$ turns out to be an equivalence relation.

1.16 Definition. The near-ring N is called *planar*, if $|N/\!\!\equiv| \geq 3$ and if all equations $x \circ a = x \circ b + c$ $(a, b, c \in N, a \not\equiv b)$ have exactly one solution $x \in N$.

This condition is motivated by geometry. Basically it says that two "nonparallel lines have exactly one point of intersection". From the general theory (see, e.g. PILZ) we get the following example which will be of considerable use in the sequel.

1.17 Example. Consider $N := (\mathbb{Z}_5, +, *)$ with $+$ as usual and $n * 0 = 0$, $n * 1 = n * 2 = n$, $n * 3 = n * 4 = 4n$ for all $n \in N$. Then $1 \equiv 2$ and $3 \equiv 4$. N is planar; for instance, the equation $x * 2 = x * 3 + 1$ with $2 \not\equiv 3$ has the unique solution $x = 3$.   □

Planar near-rings produce BIBD's with parameters v, b, r, k, λ and excellent "*efficiency*" $E := \lambda v/rk$. This E is a number between 0 and 1 and it estimates, for the statistical analysis of experiments, the "quality" of the design; BIBD's with efficiency $E \geq 0.75$ are usually considered to be "good" ones.

One gets BIBD's from planar near-rings in the following way. We define for $a \in N, N$ a planar near-ring, $g_a: N \to N, n \mapsto n \circ a$ and form $G :=$ $\{g_a | a \in N\}$. Call $a \in N$ "group-forming" if $a \circ N$ is a subgroup of $(N, +)$. Let us call all sets $a \circ N + b$ $(a \in N^*, b \in N)$ "blocks". Then these blocks, together with N as the set of "points", form a tactical configuration with parameters

$$(v, b, r, k, \lambda) = \left(v, \frac{\alpha_1 v}{|G|} + \alpha_2 v, \alpha_1 + \alpha_2 |G|, |G|, \lambda \right),$$

where $v = |N|$ and $\alpha_1 (\alpha_2)$ denote the number of orbits of F under the group $G\backslash\{0\}$ which consists entirely of group-forming (nongroup-forming, respectively) elements. This tactical configuration is a BIBD if and only if either all elements are group-forming (in this case we get $\lambda = 1$) or just 0 is group-forming (in which case we arrive at $\lambda = |G|$).

1.18 Example. We consider the problem of testing combinations of six out of ten fertilizers, each on three fields. Definition 1.8 tells us that we have to look for a BIBD with parameters $(v, 10, 6, 3, \lambda)$. By 1.11(i) we get $10 \cdot 3 = v \cdot 6$, whence $v = 5$. Similarly, 1.11(ii) forces $6 \cdot 2 = \lambda \cdot (5 - 1)$, whence $\lambda = 3$ (all under the assumption that there actually exists a BIBD with these said parameters). Hence we are searching for a planar near-ring of order 5 and the near-ring N of 1.17 comes to our mind.

We construct the blocks $a * N + b \, (a \neq 0)$:

$$1 * N + 0 = 4 * N + 0 = \{0, 1, 4\} =: B_1,$$
$$1 * N + 1 = 4 * N + 1 = \{1, 2, 0\} =: B_2,$$
$$1 * N + 2 = 4 * N + 2 = \{2, 3, 1\} =: B_3,$$
$$1 * N + 3 = 4 * N + 3 = \{3, 4, 2\} =: B_4,$$
$$1 * N + 4 = 4 * N + 4 = \{4, 0, 3\} =: B_5,$$
$$2 * N + 0 = 3 * N + 0 = \{0, 2, 3\} =: B_6,$$
$$2 * N + 1 = 3 * N + 1 = \{1, 3, 4\} =: B_7,$$
$$2 * N + 2 = 3 * N + 2 = \{2, 4, 0\} =: B_8,$$
$$2 * N + 3 = 3 * N + 3 = \{3, 0, 1\} =: B_9,$$
$$2 * N + 4 = 3 * N + 4 = \{4, 1, 2\} =: B_{10}.$$

Obviously, only 0 is group-forming. By our theorem above, we get a BIBD. The parameters do not have to be computed by the formula given (we end up with $|G| = 3$, $v = 5$, $\alpha_1 = 0$, $\alpha_2 = 2$); they can be obtained directly from this list:

  (i) $v =$ five points (namely $0, 1, 2, 3, 4$);
  (ii) $b =$ ten blocks;
  (iii) Each point lies in exactly $r =$ six blocks;
  (iv) Each block contains precisely $k =$ three elements;
  (v) Every pair of different points appears in $\lambda =$ three blocks.

In order to solve our fertilizer problem, we divide the whole field into five

parts which we call 0, 1, 2, 3, 4. Then we use the fertilizer F_i on every field
of the block B_i $(i = 1, 2, \ldots, 10)$:

Field 0	Field 1	Field 2	Field 3	Field 4
F_1	F_1			F_1
F_2	F_2	F_2		
	F_3	F_3	F_3	
		F_4	F_4	F_4
F_5			F_5	F_5
F_6		F_6	F_6	
	F_7		F_7	F_7
F_8		F_8		F_8
F_9	F_9		F_9	
	F_1	F_1		F_1

Then we get: every field contains exactly six fertilizers, and every fertilizer
is applied on three fields. Finally, every pair of different fields have three
fertilizers in common. The efficiency E of this BIBD is already $E = \lambda v / rk = 15/18 = 0.8\dot{3}$. ☐

This BIBD is a special case of designs found by J. R. Clay and G. Ferrero:
Let $F := \mathbb{F}_{p^n}$ (see §2 of Chapter 3) and choose a nontrivial divisor s of
$p^n - 1$. Then 3.2.2 gives us the information that $(F^*, \cdot)$ is cyclic with a
subgroup S of order s. Next we choose representatives $f_1 = 1, f_2, \ldots, f_m$ of
the cosets $Sf_1, \ldots, Sf_m$ of S in $(F^*, \cdot)$ and define $*_s$ in F by

$$a *_s b := \begin{cases} 0 & \text{if } b \neq 0 \\ a \cdot s_b & \text{if } b \neq 0, b \in Sf_i, b = s_b f_i. \end{cases}$$

Then $N = (F, +, *_s)$ is a planar near-ring and the sets $aN + b$
$(a \in N^*, b \in N)$ form a BIBD with parameters

(i) $\left(p^n, \dfrac{p^n(p^n - 1)}{p^m(p^m - 1)}, \dfrac{p^n - 1}{p^m - 1}, p^m, 1 \right)$ if s is of the form $s = p^k - 1$,

(ii) $\left(p^n, \dfrac{p^n(p^n - 1)}{s}, \dfrac{(s + 1)(p^n - 1)}{s}, s + 1, s + 1 \right)$ otherwise.

In our example above we had $p = 5$, $n = 1$, $s = 2$, $S = \{1, 3\}$, $f_1 = 1, f_2 = 2$
(hence this is case (ii)).

PROBLEMS

1. Prove that in a symmetric BIBD with $k \leq v - 1$ the following are equivalent:
   (i) $b = v$; (ii) $r = k$; (iii) any two blocks have λ common points.

*2. Prove parts (i), (iii) and (iv) of Theorem 1.10.

3. If **A** is the incidence matrix of a (v, k, λ) configuration, show that **A** is a Hadamard matrix if and only if $v = 4(k - \lambda)$.

4. Show that there exists no $(8, 12, 3, 2, 1)$ configuration.

5. Which designs can be used to test nine types of washing powder with the help of twelve households, or with the assistance of three households on four consecutive days.

6. The complement of a design D is the design obtained by changing 0 to 1 and 1 to 0 throughout the incidence matrix of D. If D is a (v, b, r, k, λ) configuration, show that its complement is a $(v, b, b - r, v - k, b - 2r + \lambda)$ configuration. Find a $(9, 12, 8, 6, 5)$ configuration.

7. Consider a $(4n - 1, 2n - 1, n - 1)$ configuration and take the rows of its incidence matrix as codewords of a code. How many errors will the code correct? Derive a code from the Fano plane which corrects one error.

8. Design an experiment which tests 22 brands of breakfast cereal by distributing them to households in such a way that each household tests seven brands over a 7-week period and each pair of brands is tested by two different households.

9. Similarly for 16 brands to be tested such that each household tests six brands over 6 weeks such that each pair of brands is tested by two different households.

10. A (v, b, r, k, λ) configuration is *resolvable* if the blocks may be grouped into s sets each of which is a partition of the set P. Prove that if a configuration is resolvable, then $b/r = v/k$ are integers.

11. Design an experiment to test nine brands of an item on test persons in the following way over several days: each test can test exactly three items at the same time; the experimenter wishes to test all nine brands each day the experiment runs; each brand should be tested against each other by the same test person exactly once.

12. A college tennis club wants to organize a night-time tennis competition for a period of about 12 weeks. It is assumed that there are either 14 or maybe 13 or 12 players available. The tennis courts can only be used for three matches at night and the club would like to have three players come each time and play three matches, where one is doing the umpiring, whilst the other two players play, etc. Can the matches be scheduled so that each player plays each other player twice and each player comes to the club just once a week? Should the club persuade one or two of the 14 players not to participate?

13. The testing of toothpaste consists of giving 13 subjects 13 brands of toothpaste in differently colored tubes so that no toothpaste has an advantage over any other toothpase. It is required that each subject tests the same number of tubes, each brand is used the same number of times, each pair of brands is compared by a subject the same number of times. Which symmetric BIBD should be used to design this experiment? Can this BIBD satisfy the requirement that each brand is given each color the same number of times so that no brand has the advantage of being contained in a favorably colored tube?

*14. Establish a BIBD with parameters $(9, 18, 10, 5, 5)$ by using the approach of Example 1.18. Determine the efficiency of this BIBD.

C. Steiner Systems, Difference Sets and Latin Squares

Example 1.12 can be generalized in the following way. A configuration consisting of k blocks in a set A of v elements is called a *t-design*, if any subset of *t elements in A* is contained in eactly λ blocks, in symbols: *t*-(v, k, λ) design. The Fano geometry in 1.12 is a 2-$(7, 3, 1)$ design. A 2-(v, k, λ) design is a BIBD.

1.19 Definition. A *Steiner system* is a *t*-design with $\lambda = 1$; a *t*-$(v, k, 1)$ design is denoted by $S(t, k, v)$.

A *projective geometry of order n* is a Steiner system $S(2, n + 1, n^2 + n + 1)$ with $n \geq 2$.

An *affine geometry of order n* is a Steiner system $S(2, n, n^2)$ with $n \geq 2$.

1.20 Example. We consider the Fano geometry as $S(2, 3, 7)$. For $v = 1 + q + q^2$ and $k = 1 + q$ we obtain a projective geometry, denoted by $PG(2, q)$. This is a finite projective geometry of order q, q a prime power. □

At this point we mention an interesting, but still unsolved problem: Is there a finite projective plane of order 10? In 1.12 we saw an example of a projective plane with seven "points" and seven "lines". If we transfer this notation to the problem of the existence of a finite projective plane, of order 10, we would have to find a plane with 111 points and 111 lines, where each line contains 11 points and each point lies on 11 lines. Even extensive use of large computers has not produced a solution to this problem.

If we denote the points of the Fano geometry by the integers (mod 7), i.e. 0, 1, 2, 3, 4, 5, 6, and if the lines are defined in such a way that we successively add the integers (mod 7) to the triple (0, 1, 3), then we obtain the Fano geometry of 1.12. This approach can be generalized.

1.21 Definition. A set of residue classes $\{d_1, d_2, \ldots, d_k\}$ is called a *difference set modulo v*, or a (v, k, λ) difference set, if the set of all differences $d_i - d_j, i \neq j$, represents all the nonzero residues (mod v) with each difference occurring exactly λ times.

We have $k(k - 1) = \lambda(v - 1)$. There is a simple connection between designs and difference sets.

1.22 Theorem. *Let* $\{d_1, \ldots, d_k\}$ *be a* (v, k, λ) *difference set. Then* $(\mathbb{Z}_v, \{B_0, B_1, \ldots, B_{v-1}\})$ *is a* (v, k, λ) *design, where*

$$B_j := \{[j + d_i]_v | i = 1, 2, \ldots, k\}, \qquad j = 0, 1, \ldots, v - 1.$$

Thus the elements of B_j are the residue classes (mod v).

PROOF. A residue a modulo v occurs precisely in those blocks with subscripts $a - d_1, \ldots, a - d_k$ modulo v, and thus each of the v points is incident with the same number k of blocks. For a pair of distinct residues a, c modulo v, we have $a, c \in B_j$ if and only if $a \equiv d_r + j \pmod{v}$ and $c \equiv d_s + j \pmod{v}$, for some d_r, d_s. Therefore, $a - c \equiv d_r - d_s \pmod{v}$, and conversely, for every solution (d_r, d_s) of this congruence, both a and c occur in the block with subscript $a - d_r$ modulo v. By hypothesis, there are exactly λ solutions (d_r, d_s) of this congruence and so all the conditions for a symmetric (v, k, λ) block design are satisfied. $\square$

1.23 Example. $\{0, 1, 2, 4, 5, 8, 10\} \in \mathbb{Z}_{15}$ form a $(15, 7, 3)$ difference set. The blocks

$$B_j = \{j, j + 1, j + 2, j + 4, j + 5, j + 8, j + 10\}, \qquad 0 \le j \le 14$$

form a $(15, 7, 3)$ design, which can be interpreted as a three-dimensional projective geometry $PG(3, 2)$. It consists of 15 planes and 15 points, the residue classes $(\bmod\ 15)$. Each of these planes is a Fano geometry $PG(2, 2)$. $\square$

Theorem 1.22 can be generalized in the following way.

1.24 Theorem. Let $\{d_{i1}, \ldots, d_{ik}\}$, $i = 1, 2, \ldots, t$, be a system of difference sets $(\bmod\ v)$, where v is a prime power. Then the following form the blocks of a BIBD:

$$\{d_{i1} + d, \ldots, d_{ik} + d\}, d \in \mathbb{F}_v.$$

The elements of the difference sets are the elements of the BIBD. We have $b = tv$ and $r = tk$. $\square$

The simple proof is left as an exercise.

1.25 Example. Let $v = 6t + 1$ be a prime; then a BIBD with $b = t(6t + 1)$, $k = 3$, and $r = 3t$ can be obtained from the system of difference sets

$$\{0, \zeta^s, \zeta^{s+t}\},$$

where ζ is a primitive element of $\mathbb{F}_v$ and $s = 0, 1, \ldots, t - 1$.

A BIBD with $v = 6t + 1$ a prime power, $b = t(6t + 1)$, $k = 3$ and $r = 3t$ can also be obtained from the system of difference sets

$$\{\zeta^i, \zeta^{i+2t}, \zeta^{i+4t}\}, \qquad i = 0, 1, \ldots, t - 1.$$ $\square$

Finally we formulate the concept of a Latin square.

1.26 Definition. An $n \times n$ matrix $\mathbf{L} = (a_{ij})$ over $A = \{a_1, \ldots, a_n\}$ is called a *Latin square* of order n if each row and column contains each element of A exactly once. Two Latin squares (a_{ij}) and (b_{ij}), usually over A, of order n are called *orthogonal* if all n^2 ordered pairs (a_{ij}, b_{ij}) are distinct.

1.27 Example. The two Latin squares

$$L_1 = \begin{pmatrix} a & b & c \\ c & a & b \\ b & c & a \end{pmatrix} \quad \text{and} \quad L_2 = \begin{pmatrix} a & b & c \\ b & c & a \\ c & a & b \end{pmatrix}$$

superimposed give the square

$$\begin{pmatrix} aa & bb & cc \\ cb & ac & ba \\ bc & ca & ab \end{pmatrix}.$$

Thus L_1 and L_2 are orthogonal. $\qquad\qquad\qquad\qquad\qquad\qquad\qquad\square$

Latin squares first appeared in parlor games, e.g. in the problem of arranging the Jacks, Queens, Kings and Aces of a card deck into a 4×4 matrix, such that in each row and column each suit and each value of a card occurs exactly once.

The famous *Euler's officers problem* goes back to Leonard Euler, who posed the following question (in 1779 and more generally in 1782): Is it possible to arrange 36 officers of six regiments with six ranks into a 6×6 square in a parade, such that each "row" and each "column" of the square has exactly one officer of each regiment and each rank? In the formulation of Definition 1.26, this means, do there exist two orthogonal Latin squares of order 6. Euler conjectured that there are no two orthogonal Latin squares of order 6. More generally, he conjectured: There are no two orthogonal Latin squares of order n, where $n \equiv 2 \pmod 4$.

In 1899, G. Tarry proved that Euler's officers problem cannot be done. But Euler's more general conjecture was disproved by Bose, Shrikhande and Parker in 1959. They showed that for any $n > 6$ there are at least two orthogonal Latin squares of order n.

1.28 Theorem. *For any positive integer n there is a Latin square of order n.* $\qquad\qquad\qquad\qquad\qquad\qquad\qquad\qquad\qquad\qquad\qquad\qquad\square$

This follows immediately, since Latin squares are essentially the operation tables of (not necessarily associative) finite "groups" (called *finite loops*). We might take $\mathbb{Z}_n$, for instance.

Next we show that for $n = p$, a prime, there are $n - 1$ pairwise orthogonal Latin squares of order n. We form

$$L_j := \begin{pmatrix} 0 & 1 & \cdots & p-1 \\ j & 1+j & \cdots & p-1+j \\ 2j & 1+2j & \cdots & p-1+2j \\ \cdots\cdots\cdots\cdots\cdots\cdots\cdots\cdots\cdots\cdots\cdots\cdots\cdots \\ (p-1)j & 1+(p-1)j & \cdots & p-1+(p-1)j \end{pmatrix}$$

for $j = 1, 2, \ldots, p - 1$, where all elements in L_j are reduced (mod p). The L_j's are Latin squares and L_i and L_j are orthogonal, if $i \neq j$. For suppose (a, b) were a pair which appeared twice, say in row α and column β as well as in row γ and column δ, then

$$\beta + \alpha j \equiv \delta + \gamma j \equiv a \ (\text{mod } p),$$

$$\beta + \alpha i \equiv \delta + \gamma i \equiv b \ (\text{mod } p).$$

Thus $\alpha(i - j) \equiv \gamma(i - j) \ (\text{mod } p)$, i.e. $\alpha \equiv \gamma, \beta \equiv \delta \ (\text{mod } p)$.

This construction can be generalized to Latin squares over $\mathbb{F}_q, q = p^e$.

1.29 Theorem. *Let* $0, 1, a_2, \ldots, a_{q-1}$ *be the elements of* $\mathbb{F}_q$. *Then the Latin squares* $L_i, 1 \leq i \leq q - 1$, *form a set of* $q - 1$ *pairwise orthogonal Latin squares, where*

$$L_i := \begin{pmatrix} 0 & 1 & \cdots & a_{q-1} \\ a_i & a_i + 1 & \cdots & a_i + a_{q-1} \\ a_i a_2 & a_i a_2 + 1 & \cdots & a_i a_2 + a_{q-1} \\ \cdots\cdots\cdots\cdots\cdots\cdots\cdots\cdots\cdots\cdots\cdots\cdots \\ a_i a_{q-1} & a_i a_{q-1} + 1 & \cdots & a_i a_{q-1} + a_{q-1} \end{pmatrix}.$$

If ζ *is a primitive element of* $\mathbb{F}_q$, *then the following* $q - 1$ *Latin squares* L_i' *are pairwise orthogonal,* $i = 0, 1, \ldots, q - 2$:

$$L_i' = \begin{pmatrix} 0 & 1 & \cdots & a_{q-1} \\ \zeta^{0+i} & 1 + \zeta^{0+i} & \cdots & a_{q-1} + \zeta^{0+i} \\ \zeta^{1+i} & 1 + \zeta^{1+i} & \cdots & a_{q-1} + \zeta^{1+i} \\ \cdots\cdots\cdots\cdots\cdots\cdots\cdots\cdots\cdots\cdots\cdots\cdots \\ \zeta^{q-2+i} & 1 + \zeta^{q-2+i} & \cdots & a_{q-1} + \zeta^{q-2+i} \end{pmatrix}.$$

L_{i+1}' *can be obtained from* L_i' *by a cyclic exchange of the last* $q - 1$ *rows.*

PROOF. We prove the first part of the theorem. Each L_i is a Latin square. Let $a_{mn}^{(i)} = a_i a_{m-1} + a_{n-1}$ be the (m, n) entry of L_i. For $i \neq j$, suppose $(a_{mn}^{(i)}, a_{mn}^{(j)}) = (a_{rs}^{(i)}, a_{rs}^{(j)})$ for some $1 \leq m, n, r, s \leq q$. Then

$$(a_i a_{m-1} + a_{n-1}, a_j a_{m-1} + a_{n-1}) = (a_i a_{r-1} + a_{s-1}, a_j a_{r-1} + a_{s-1}),$$

so

$$a_i(a_{m-1} - a_{r-1}) = a_{s-1} - a_{n-1}$$

and

$$a_j(a_{m-1} - a_{r-1}) = a_{s-1} - a_{n-1}.$$

Since $a_i \neq a_j$, it follows that $a_{m-1} = a_{r-1}, a_{s-1} = a_{n-1}$ hence $m = r, n = s$. Thus the ordered pairs of corresponding entries from L_i and L_j are all different and so L_i and L_j are orthogonal. $\square$

1.30 Example. Three pairwise orthogonal Latin squares of order 4 are given by

$$
L_1 = \begin{pmatrix} 0 & 1 & \zeta & \zeta^2 \\ 1 & 0 & \zeta^2 & \zeta \\ \zeta & \zeta^2 & 0 & 1 \\ \zeta^2 & \zeta & 1 & 0 \end{pmatrix}, \quad
L_2 = \begin{pmatrix} 0 & 1 & \zeta & \zeta^2 \\ \zeta & \zeta^2 & 0 & 1 \\ \zeta^2 & \zeta & 1 & 0 \\ 1 & 0 & \zeta^2 & \zeta \end{pmatrix},
$$

$$
L_3 = \begin{pmatrix} 0 & 1 & \zeta & \zeta^2 \\ \zeta^2 & \zeta & 1 & 0 \\ 1 & 0 & \zeta^2 & \zeta \\ \zeta & \zeta^2 & 0 & 1 \end{pmatrix}.
$$

Here ζ is a primitive element of $\mathbb{F}_4$ and a zero of $f = x^2 + x + 1$. L_1 is the addition table of $\mathbb{F}_4$. □

1.31 Example. As an example of the use of Latin squares in experimental design, suppose it is required to test the effect of five different treatments on wool. Five processes P_i are to be used to treat the wool and both the process and the operator affect the final texture of the wool.

Ideally we would test the effect of every treatment with every operator and every process. However, using the technique of analysis of variance, it suffices that each type of treated wool is treated once by each operator and once using each process.

If we have five operators O_i, $1 \le i \le 5$, then the required set of experiments can be specifid by a 5×5 Latin square as follows

	O_1	O_2	O_3	O_4	O_5
P_1	W_0	W_1	W_2	W_3	W_4
P_2	W_1	W_2	W_3	W_4	W_0
P_3	W_2	W_3	W_4	W_0	W_1
P_4	W_3	W_4	W_0	W_1	W_2
P_5	W_4	W_0	W_1	W_2	W_3

Here the columns specify the operator and the rows specify the process to be used. The entries in the square indicate the type of wool to be used.

Now suppose that the performance of the operators varied with the day of the week. This could be accounted for by using a Latin square which was orthogonal to the first to specify the day of the week on which the experiment was to be carried out. Such a suitable Latin square is shown below. We then have each process, each operator and each type of wool associated with each day of the week just once.

$$\begin{pmatrix} 0 & 1 & 2 & 3 & 4 \\ 2 & 3 & 4 & 0 & 1 \\ 4 & 0 & 1 & 2 & 3 \\ 1 & 2 & 3 & 4 & 0 \\ 3 & 4 & 0 & 1 & 2 \end{pmatrix}.$$

It may happen that only four operators are available. In this case, the first Latin square can still be used but with its last column (say) deleted. Each of the types of wool would still be treated by each of the four operators, but now each process would work on only four of the different wool types. To cope with these more general problems, one uses balanced incomplete block designs. □

1.32 Example. The following method is due to Olderogge and is based on orthogonal Latin squares for synthesizing binary codes. This technique allows the construction of code words of length $n^2 + 4n + 1$, where n^2 of the symbols are the appropriate message symbols. It can be shown that such a code will correct all single and double error patterns, and even a majority of patterns of three or four errors. We illustrate this procedure by example (following DENES and KEEDWELL).

Consider the case $n = 5$ and suppose the message of binary digits is: 11100, 00110, 11011, 01101, 01010. To obtain the remaining $4 \cdot 5 + 1 = 21$ additional parity-check digits which are to be transmitted, we partition the message into five blocks, each block containing five symbols as shown. Each block is taken to be a row of a 5×5 matrix $K = (k_{ij})$

$$K = \begin{array}{c c} \begin{array}{ccccc} 0 & 0 & 1 & 1 & 0 \end{array} & \\ \hline \begin{pmatrix} 1 & 1 & 1 & 0 & 0 \\ 0 & 0 & 1 & 1 & 0 \\ 1 & 1 & 0 & 1 & 1 \\ 0 & 1 & 1 & 0 & 1 \\ 0 & 1 & 0 & 1 & 0 \end{pmatrix} & \begin{array}{c} 1 \\ 0 \\ 0 \\ 1 \\ 0 \end{array} \end{array}$$

We border the matrix K by parity-check digits, one for each row and column. If the sum of the digits of the rth column is $i \pmod 2$ we head that column by i, and similarly for other columns. The rows are labeled in a corresponding way.

Next we construct a 5×5 matrix $M = ((a_{ij}, b_{ij}))$ whose entries are ordered pairs of elements from two orthogonal Latin squares $A = (a_{ij})$, $B = (b_{ij})$, each of order 5. Let

$$M = \begin{pmatrix} 11 & 22 & 33 & 44 & 55 \\ 34 & 45 & 51 & 12 & 23 \\ 52 & 13 & 24 & 35 & 41 \\ 25 & 31 & 42 & 53 & 14 \\ 43 & 54 & 15 & 21 & 32 \end{pmatrix}.$$

Using **M**, we construct from **K** a further 5×5 matrix $\mathbf{L} = (c_{ij})$ by defining $c_{ij} = k_{a_{ij}b_{ij}}$, for all $1 \le i, j \le 5$, For instance, $c_{23} = k_{51} = 0$. We border the matrix **L** with parity-check digits as we did for the matrix **K**. Then we have

$$
\mathbf{L} = \begin{array}{c} \begin{matrix} 0 & 0 & 0 & 0 & 0 \end{matrix} \\ \hline \begin{pmatrix} 1 & 0 & 0 & 0 & 0 \\ 1 & 1 & 0 & 1 & 1 \\ 0 & 1 & 1 & 0 & 0 \\ 1 & 1 & 1 & 1 & 0 \\ 1 & 1 & 0 & 1 & 1 \end{pmatrix} \begin{matrix} 1 \\ 0 \\ 0 \\ 0 \\ 1 \end{matrix} \end{array} .
$$

To get the codeword to be transmitted we now adjoin in turn the five parity-check digits for the rows of **K**, the five parity-check digits for the columns of **K**, the five parity-check digits for the rows of **L**, the five parity-check digits for the columns of **L**, and finally a parity-check digit taken over the whole preceding set of $5^2 + 4 \cdot 5$ components giving a total word length of $5^2 + 4 \cdot 5 + 1$ components, or of $n^2 + 4n + 1$ components in the general case.

Let 1 1 1 0 0, 0 0 1 1 0, 1 1 0 1 1, 0 1 1 0 1, 0 1 0 1 0 be the message $a_1 a_2 \ldots a_{25}$, regarded as a 25-tupole over $\mathbb{Z}_2$. The parity check digits to be added to the message to obtain a codeword are: 1 0 0 1 0 from the row, 0 0 1 1 0 from the column parity-checks of **K**, respectively, 1 0 0 0 1 from the row, 0 0 0 0 0 from the column parity-checks of **L**, and 0 is the overall parity-check symbol.

This code is successful in correcting all double errors. We observe firstly that, if the two errors occur in different rows and columns of **K** then two of the five parity checks on the rows of **K** will fail, and two will fail on the columns. Thus there will be four possible cells whose entries are among the two in error. These four cells occur two in each of two rows and two in each of two columns of **K**. In **L**, the contents k_{rs}, k_{rt} of the two cells of the same row of **K** necessarily occur in different rows and a similar result holds for columns. Therefore the two cells whose entries are incorrect can be distinguished by the examination of the parity check digits associated with **L**.

If the two errors occur in the same row (or column) of **K** then just two of the five parity-checks on the columns (rows) of **K** will be in error. To determine which is the row (column) of **K** which contains the two errors, we again use the parity-checks on **L**. Finally, cases when one or both of the errors occur among the parity-checks themselves are easily detected and corrected. □

PROBLEMS

1. Show that Steiner systems $S(5, 7, 13)$ and $S(2, 7, 43)$ cannot exist.

2. Prove that if a Steiner system $S(3, 4, v)$ exists, then v must be of the form $v = 6r + 2$ or $6r + 4$.

3. Prove that if a Steiner system $S(t, k, v)$ exists, so does a system $S(t - s, k - s, v - s)$ for each $s < t$.

*4. One of the most interesting Steiner systems is $S(5, 8, 24)$, with important applications in coding (see PLESS). Determine the number of its eight-element sets, show that each of the 24 elements in a set A are in 253 eight-element sets and each pair of elements in A lies in 77 eight-element sets. What can you say about triples, quadruples and quintuples of elements of A?

*5. Prove that the codewords of weight 3 in a $(2^m - 1, 2^m - 1 - m, 3)$ Hamming code C form a Steiner system $S(2, 3, 2^m - 1)$ and the codewords of the code, which are obtained from C by adding an overall parity-check, form a Steiner system $S(3, 4, 2^m)$. Determine the number of codewords of weight 3 in C.

*6. Prove Theorem 1.24.

7. Construct a $(31, 6, 1)$ difference set.

8. A (v, k, λ) group difference set is a subset $S = \{x_i | i \in I\}$ of an additive abelian group G of order v if among the set of elements $\{x_i - x_j | i \neq j, 1 \leq i, j \leq k\}$ the nonzero elements of G each occur λ times. Prove $\lambda(v - 1) = k(k - 1)$. Construct a $(27, 13, 6)$ group difference set in $\mathbb{F}_{3^3}$ where $x^3 - x - 2 = 0$.

*9. Establish a one-to-one connection between a complete set of mutually orthogonal Latin squares $L_1, L_2, \ldots, L_{s-1}$ and PG(2, s). Consider the special case $s = 4$ first.

10. Let B be a BIBD (v, b, r, k, λ). Prove: if $v = b$ is even, then $r - \lambda = k - \lambda$. Is it possible to have a $(46, 46, 10, 10, 2)$ BIBD?

11. An *orthogonal array* (N, c, n, t) of size N, c constraints, level n and strength t is a rectangular arrangement of n different elements in N columns and c rows, such that in any t rows the N columns contain all the n^t possible ordered sets of n elements the same number of times. If this number is the index λ of the array, then $N = \lambda n^t$. For $\lambda = 1$, $t = 2$ the array is an orthogonal set of Latin squares.

Let $f_j(x) = \sum_{i=0}^{t-1} a_i x^i$, $j = 1, 2, \ldots, q^t$, denote the s^t different polynomials over $\mathbb{F}_q = \{b_1, \ldots, b_q\}$, let $t \leq q$. Prove that the matrix with entry $f_j(b_i)$ in the ith row and jth column is an orthogonal array (q^t, q, q, t). Construct $(16, 5, 4, 2)$ by expanding $(16, 4, 4, 2)$ by the addition of a row which contains, in each column, the leading coefficient of the polynomial to which that column corresponds. Prove in general that such an extended array is still orthogonal.

EXERCISES (Solutions in Chapter 8, p. 480)

1. Determine a Hadamard matrix of order 12.

2. Prove that if Hadamard matrices $\mathbf{H}_m$ and $\mathbf{H}_n$ exist, then $\mathbf{H}_{mn}$ is a Hadamard matrix.

3. Let $\mathbf{H}$ be a Hadamard matrix of order h. Then prove:
   (i) $|\det \mathbf{H}| = h^{1/2h}$;
   (ii) Hadamard matrices may be changed into other Hadamard matrices by permuting rows and columns and by multiplying rows and columns by -1.
   (iii) if H is a normalized Hadamard matrix of order $4n$, then every row (column) except the first contains $2n$ minus ones and $2n$ plus ones, and

moreover n minus ones in any row (column) overlap with n minus ones in each other row (column).

4. An $r \times n$ *Latin rectangle* is an $r \times n$ array made out of the integers $1, 2, \ldots, n$ such that no integer is repeated in any row or in any column. Give an example of a 3×5 Latin rectangle, and extend it to a Latin square of order 5. Can a Latin rectangle always be extended to a Latin square?

*5. Let $L^{(k)} = (a_{ij}^k)$, where $a_{ij}^k = i + jk \pmod 9$. Which of the $L^{(k)}, 1 \le k \le 8$, are Latin squares? Are $L^{(2)}$ and $L^{(5)}$ orthogonal?

*6. How many normalized Latin squares of order $n \le 5$ are there?

7. Show that for $n \ge 2$ there can be at most $n - 1$ mutually orthogonal Latin squares of order n.

8. List the points and lines of PG(2, 3).

9. Show that $\{0, 4, 5, 7\}$ is a difference set of residues modulo 13 which yields a projective geometry PG(2, 3).

*10. Verify that $\{0, 1, 2, 3, 5, 7, 12, 13, 16\}$ is a difference set of residues modulo 19. Determine the parameters v, k and λ.

11. Decide whether $D = \{0, 1, 2, 32, 33, 12, 24, 29, 5, 26, 27, 22, 18\}$ is a difference set (mod 40). Determine the parameters v, k, λ.

12. Show that the following system of blocks from elements $1, 2, \ldots, 9$ is a design and determine the parameters v, b, r, k and λ:

$$
\begin{array}{cccc}
123 & 147 & 159 & 168 \\
456 & 258 & 267 & 249 \\
789 & 369 & 348 & 357.
\end{array}
$$

*13. Let $P_1, \ldots, P_t$ be distinct points in a t-(v, k, λ) design, let λ_i be the number of blocks containing $P_1, \ldots, P_i$, $1 \le i \le t$, and let $\lambda_0 = b$ be the total number of blocks. Prove that λ_i is independent of the choice of $P_1, \ldots, P_i$ and that

$$
\lambda_i = \frac{\lambda \dbinom{v - i}{t - i}}{\dbinom{k - i}{t - i}} = \lambda \frac{(v - i)(v - i - 1) \ldots (v - t + 1)}{(k - i)(k - i - 1) \ldots (k - t + 1)}
$$

for $0 \le i \le t$. This implies that a t-(v, k, λ) design is an i-(v, k, λ_i) design for $1 \le i \le t$.

14. Set up an "incidence table" for PG(2, 2) assuming that the point P_r and the line l_s are incident if

$$
r + s \equiv 0, 1 \text{ or } 3 \pmod 7.
$$

Does this geometry satisfy the following axioms?
Axiom 1. Any two distinct points are incident with just one line.
Axiom 2. Any two lines meet in at least one point.
Axiom 3. There exist four points no three of which are collinear.
Axiom 4. The three diagonal points of a quadrangle are never collinear.

15. A magic square of order n consists of the integers 1 to n^2 arranged in an $n \times n$ array such that the sums of entries in rows, column and diagonals are all the same. Let $A = (a_{ij})$ and $B = (b_{ij})$ be two orthogonal Latin squares of order n with entries in $\{0, 1, \ldots, n - 1\}$ such that the sum of entries in each of the diagonals of A and B is $n(n - 1)/2$. Show that $M = (na_{ij} + b_{ij} + 1)$ is a magic square of order n. Construct a magic square of order 4 from two orthogonal Latin squares.

16. De la Loubère's method for constructing magic squares is as follows: Place a 1 in the middle square of the top row. The successive integers are then placed in their natural order in a diagonal line which slopes upwards to the right with the following modifications.
    (i) When the top row is reached, the next integer is placed in the square in the bottom row as if it came immediately above the top row.
    (ii) When the right-hand column is reached, the next integer is placed in the left-hand column as if it immediately succeeded the right-hand column.
    (iii) When a square already contains a number or when the top right-hand square is reached, the next integer is placed vertically below it in the next row.
    Illustate this construction for a 5×5 array.

*17. Show that from a normalized Hadamard matrix of order $4t$, $t \geq 2$, one can construct a symmetric $(4t - 1, 2t - 1, t - 1)$ block design.

18. Room squares were described by T. G. Room in an article entitled "A New Type of Magic Square" in the *Mathematics Gazette*, 1955. However Room squares were first introduced into duplicate bridge under the name of Howell Movements by E. C. Howell about 1897.
    A Room square of side n is an $n \times n$ array of cells, n odd, in which each cell is either empty or contains an unordered pair of symbols from the set $\{\infty, 0, 1, 2, \ldots, n - 1\}$ with the following properties holding:
    (i) The array is "Latin" by columns (every symbol occurs exactly once in a column).
    (ii) The array is Latin by rows.
    (iii) All lines contain exactly $\frac{1}{2}(n - 1)$ blank cells.
    (iv) All pairs are distinct (each possible pair occurs once and only once).
    Show that there is no Room square of side $n = 3$, and one of side $n = 5$.

19. (i) Solve the following special case of the "Kirkman Schoolgirl Problem": A schoolmistress took her nine girls for a daily walk, the girls arranged in rows of three girls. Plan the walk for four consecutive days so that no girl walks with any of her classmates in any triplet more than once.
    *(ii) Try to solve the same problem for 15 girls and five rows of three girls for seven consecutive days.

20. In a school consisting of b pupils, t athletics teams are formed. The athletics teacher puts k pupils on each team, and arranges for every pupil to play in the same number of teams. He also arranges for each pair of pupils to play together the same number of times. Find how many teams a pupil plays in and how often two pupils play in the same team.

§2. Algebraic Cryptography

Cryptology, or the general abstract encryption and decryption problem can be paralleled with the basic process of encoding and decoding. Again we consider a transmitter who wants to communicate a message to a receiver through an unreliable communication channel. Here the unreliability of the channel can be caused by an unauthorized interceptor, who tries to obtain or alter the original message.

Whereas in coding the main aim is fast and correct transmission of a message over a noisy channel, in cryptology, the aim is safe transmission by disguising the original message in such a way that this unauthorized interceptor of the transmitted message would be unable to understand the message. Usually in this context the words cipher, enciphering and deciphering (or encryption and decryption) are used instead of code, encoding and decoding. Cryptology comprises *cryptography* and *cryptanalysis*. The former deals with the designing of cipher systems, the latter is concerned with methods to decipher a received or intercepted cipher. The nature of such methods can be mathematical, linguistic or technical in the sense of engineering; here we shall concentrate on cryptography.

The following problem areas may arise if a sender A wants to send a message to a receiver B: How can both be certain that nobody else can receive the message? How can A ensure that B receives all of A's message? How can they ensure that B receives only A's message and no others? These questions highlight the main problems in communication and cryptography: secrecy and authenticity of message transfer.

Cryptography has important applications in data security in computing. We indicate a few examples of such applications. Often secret data are stored in computers or are exchanged between computers. Cryptographic devices can prevent that unauthorized personnel can obtain these data. Another example is represented by electronic mail and electronic transfers of money. In both cases secrecy and authenticity are often crucial. It is also essential to design security devices in many other instances, for instance to prevent industrial espionage or sabotage of important installations such as power stations or military installations. Another application of cryptology lies in deciphering ancient secret documents written in ciphers. This branch is often called historical cryptanalysis; it is several hundred years old. Secret ciphers have existed for centuries to protect messages from intelligent opponents. Thus the problem areas of secrecy and authenticity (i.e. protection from forgery) have very old beginnings. However, nowadays the most important aspects of cryptography are in the governmental as well as in the commercial and private domain of data security in the widest sense. Contemporary cryptology has been characterized as a multi-million dollar problem awaiting solution and as an important question in applied mathematics. In this text we shall just describe some algebraic methods in cryptography emphasizing recent developments in public-key cryptography. First we introduce some terminology.

Cryptography, or a cryptographic system (*cryptosystem* or *cipher*) consists of transforming a message (called the *plaintext*) via an *enciphering function* (or encryption) in such a way that only an authorized recipient can invert the transformation and restore the message. A message can be a readable text for a collection of computer data. The process of transforming the plaintext into a ciphertext (or *cryptogram*) is called *enciphering* and each enciphering function is defined by an *enciphering algorithm*, which is common to every enciphering function of a given family, and a *key k*, which is a parameter particular to one transformation. The key is known to the sender and the (authorized) receiver but not to those from whom the plaintext is to be concealed. The key and ciphertext must determine the plaintext uniquely. Recovering the plaintext from a received ciphertext is called *deciphering* and is performed by a *deciphering transformation* (or function) which is defined by a deciphering algorithm and a key *k*. For a given key, the corresponding deciphering function is the inverse of the enciphering function. An important feature of classical forms of cryptosystems or ciphers is that sender and receiver use a single key *k*. We refer to such systems as *single key cryptosystems* in order to distinguish them from systems using two different keys, the newer *public-key cryptosystems.* Such systems are very well suited for protecting information transmitted over a computer network or for enciphering a user's private files. The security of cryptosystems lies in the large range of choice for the key *k*, which is known only to the sender and the receiver. It must be computationally impracticable, virtually impossible, or extremely difficult and time consuming (depending on the specific task) to find the value of the key, even if corresponding values of plaintext and cryptogram are known to an unauthorized interceptor. Thus in secure cryptosystems it should be computationally infeasible for an interceptor to determine the deciphering function, even if the plaintext corresponding to the intercepted ciphertext is known. It should also be computationally infeasible for an interceptor to systematically determine the plaintext from intercepted ciphertext without knowing the deciphering function.

A. Single Key (Symmetric) Cryptosystems

In classical cryptography two basic transformations of plaintext messages exist: *transposition ciphers* rearrange characters according to some rule (e.g. the plaintext is written into a matrix by rows), whereas *substitution ciphers* replace plaintext letters (or characters) by corresponding characters of a ciphertext alphabet. If only one cipher alphabet is used, the cryptosystem is called *monoalphabetic.* Cryptosystems, in which a ciphertext letter may represent more than one plaintext letter are called *polyalphabetic.* Such systems use a sequence of monoalphabetic cryptosystems. In deciphering unknown cryptosystems it was said by cryptanalysts that success is measured by four things in the order named: perseverance, careful methods of analysis, intuition and luck. There are, however, some statistical and other techniques at the cryptanalysts disposal. Often frequency analysis on the characters of a cryptogram, or on pairs or triples of characters lead to success. However, we shall not consider statistical methods.

Let A denote the plaintext alphabet and B be a cipher alphabet. These alphabets may be letters of the English alphabet, or n-tuples of letters, or elements of $\mathbb{Z}_q$ (particularly $q = 2$, $q = 26$ or $q = 96$ as in some commercial systems), or elements of a finite field. We shall replace letters by numbers in $\mathbb{Z}_q$.

2.1 Definition. Any one-to-one (or injective) mapping from A into B is called a *key*.

The given problem is to encipher a text consisting of letters in A, i.e. a word $a_1 a_2 \ldots a_n$, $a_i \in A$. If we use the same mapping to encipher each a_i then we use a "fixed key", otherwise we use a "variable key". In the latter case the key depends on some parameters, e.g. time. Time dependent variable keys are called context free; text dependent variable keys are called context sensitive.

A fixed key $f: A \to B$, can be extended to words with letters in A (i.e. to the free semigroups on A, see Chapter 6) by considering $a_1 a_2 \ldots a_n \to f(a_1)f(a_2) \ldots f(a_n)$. Similarly, a variable key $f_1, f_2, \ldots$ can be extended by $a_1 a_2 \ldots a_n \to f_1(a_1)f_2(a_2) \ldots f_n(a_n)$. Looking for the pre-images is the process of deciphering. We shall describe five methods of enciphering: by using linear functions on $\mathbb{Z}_q$, by using matrices over $\mathbb{Z}_q$, by using permutations on $\mathbb{Z}_q$, by using scalar products and finally by using big prime numbers. First we give an example of a simple substitution cipher. We verify easily that the mapping $a \to an + k$ from $\mathbb{Z}_q$ into itself is injective if and only if $\gcd(n, q) = 1$.

2.2 Definition. The mapping $f: \mathbb{Z}_q \to \mathbb{Z}_q$, $a \to an + k$, with fixed $n, k \in \mathbb{Z}_q$ and $\gcd(n, q) = 1$ is called a *modular encoding*; for $n = 1$ we call this cipher a *Caesar cipher*.

The key with $n = 1$ and $k = 3$, i.e. cyclic shift of the alphabet by three letters, is of historical interest since it was used by the Roman emperor Gaius Julius Caesar. A simple Caesar cipher can be deciphered if only one association of an original letter with a letter in the cipher alphabet is known.

2.3 Example. Let $A = \mathbb{Z}_{26}$. We identify A with 0, more precisely with $[0]_{26}$, B with 1, C with 2, etc.:

A	B	C	D	E	F	G	H	I	J	K	L	M	N	O	P	Q	R	S	T	U	V	W	X	Y	Z
0	1	2	3	4	5	6	7	8	9	10	11	12	13	14	15	16	17	18	19	20	21	22	23	24	25

If we use the modular encoding $a \to 3a + 2$ then the word ALGEBRA i.e. 0 11 6 4 1 17 0 is enciphered as 2 9 20 14 5 1 2, namely CJUOFBC.   □

Simple substitution ciphers use only one function from plaintext characters to ciphertext characters and can thus be easily broken by observing single-letter frequency distribution of the characters (= letters), which is preserved in the ciphertext. Polyalphabetic substitution ciphers conceal letter frequencies by using multiple substitution.

2.4 Definition. A *periodic substitution cipher* with period p consists of p cipher alphabets $B_1, \dots, B_p$ and the keys $f_i: A \to B_i$, $1 \le i \le p$. A plaintext message $m = m_1 \dots m_p m_{p+1} \dots m_{2p} \dots$ is enciphered as $f_1(m_1) \dots f_p(m_p) f_1(m_{p+1}) \dots f_p(m_{2p}) \dots$ by repeating the sequence of functions $f_1, \dots, f_p$ every p characters.

A popular example of a periodic substitution cipher is the *Vigenère cipher,* falsely ascribed to the sixteenth century French cryptologist Claise de Vigenère (see KAHN). In this cipher the alphabets are $\mathbb{Z}_{26}$ and the key is given by a sequence of letters (often a key word $k = k_1 \dots k_p$, where k_i indicates the cyclic shift in the ith alphabet. So $f_i(m) = m + k_i \pmod{26}$. A Vigenère square can be used for enciphering. This is a special Latin square (see Chapter 5, §1) with the letters A, B, …, Z of the alphabet as first row and column.

$$
\begin{array}{cccc}
A & B & \dots & Y \; Z \\
B & C & \dots & Z \; A \\
\multicolumn{4}{c}{\dotfill} \\
Y & Z & \dots & W \; X \\
Z & A & \dots & X \; Y
\end{array}
$$

A periodic column method of period p can be described by a simple key word of length p. Let "GALOIS" be a key word, thus $p = 6$, and then the keys f_i are given by the following simple tableau.

A	B	C	...
G	H	I	... $= f_1(A)f_1(B)f_1(C)\dots,$
A	B	C	... $= f_2(A)f_2(B)f_2(C)\dots,$
L	M	N	... $= f_3(A)f_3(B)f_3(C)\dots,$
O	P	Q	... $= f_4(A)f_4(B)f_4(C)\dots,$
I	J	K	... $= f_5(A)f_5(B)f_5(C)\dots,$
S	T	U	... $= f_6(A)f_6(B)f_6(C)\dots.$

In this case the word "ALGEBRA" is enciphered as GLRSJJG = $f_1(A)f_2(L)f_3(G)f_4(E)f_5(B)f_6(R)f_7(A)$. We note that such a column method is essentially a sequence $f_1, f_2, \dots$ of period 6 of Caesar ciphers on $\mathbb{Z}_{26}$.

Two more examples of substitution ciphers on $\mathbb{Z}_q$ are obtained from the sequence (f_i), $m = 1, 2, \ldots$, where f_i is applied to the ith letter a_i in the plaintext and f_i is given as

$$f_i: \mathbb{Z}_q \to \mathbb{Z}_q, \, a \to b_i := ka + d_i$$

where $\gcd(k, q) = 1$, and such that

$$d_i = ca_{i-1} \quad \text{with } c, a_0 \text{ given,} \tag{1}$$

or

$$d_i = cb_{i-1} \quad \text{with } c, b_0 \text{ given.} \tag{2}$$

2.5 Example. (i) If we number the letters from A to Z as in Example 2.3, then ALGEBRA, i.e. 0 11 6 4 1 17 0 $= a_1\, a_2\, a_3\, a_4\, a_5\, a_6\, a_7$, is enciphered according to (1) by means of $f_i(a) = 3a + a_{i-1}$ and $a_0 := 4$ as follows.

$$f_1(0) = f_1(a_1) = 3a_1 + 4 = 4, \qquad f_5(1) = 7,$$
$$f_2(11) = f_2(a_2) = 3a_2 + 0 = 7, \qquad f_6(17) = 0,$$
$$f_3(6) = f_3(a_3) = 3a_3 + 11 = 3, \qquad f_7(0) = 17.$$
$$f_4(4) = f_4(a_4) = 3a_4 + 6 = 18,$$

Thus ALGEBRA is enciphered as 4 7 3 18 7 0 17, i.e. EHDSHAR.
(ii) If we use (2) with $f_i(a) = 3a + f_{i-1}(a_{i-1})$ and $f_0(a_0) := 4$, we obtain

$$f_1(0) = f_1(a_1) = 3a_1 + f_0(a_0) = 0 + 4 = 4,$$
$$f_2(11) = f_2(a_2) = 3a_2 + f_1(a_1) = 7 + 4 = 11, \qquad f_5(1) = 18,$$
$$f_3(6) = f_3(a_3) = 3a_3 + f_2(a_2) = 18 + 11 = 3, \qquad f_6(17) = 17,$$
$$f_4(4) = f_4(a_4) = 3a_4 + f_3(a_3) = 12 + 3 = 15, \qquad f_7(0) = 17.$$

Thus ALGEBRA is enciphered as 4 11 3 15 18 17 17, i.e. ELDPSRR.
□

The Vigenère cipher was considered unbreakable in its day, but it has been successfully cracked in the nineteenth century due to work by Kasiski. As in the Caesar cipher, a shift is applied to the alphabet, but the length of the shift varies, usually in a periodic way. For example, cf. STILLWELL, our opponent might decide to use shifts of lengths 1, 7, 4, 13, 5 over and over again as a running key. He then writes the sequence

$$1, 7, 4, 13, 5, 1, 7, 4, 13, 5, 1, 7, 4, 13, 5, \ldots$$

(call this the *key* sequence) for as long as necessary and "adds" it to the message, say

Message	S	E	N	D	M	O	R	E	M	E	N	A	N	D	M	O	R	E	A	R	M	S
Key Sequence	1	7	4	13	5	1	7	4	13	5	1	7	4	13	5	1	7	4	13	5	1	7
Enciphered Message	T	L	R	Q	R	P	Y	I	Z	J	O	H	R	Q	R	P	Y	I	N	W	N	Z.

The changing shifts even out the overall letter frequencies, defeating the kind of analysis used to break Caesar ciphers, but the characteristic frequencies are retained in subsequences of the enciphered message corresponding to repetition in the key sequence (every five places in the above example). If we can find the length of the key's period, letters can be identified by frequency analysis.

The period can indeed be discovered, by looking for repeated blocks in the enciphered message. Some of these will be accidental, but a large proportion will result from matches between repeated words or subwords of the message and repeated blocks in the key sequence. When this happens, the distance between repetitions will be a multiple of the period. In our example, the block RQRPYI is undoubtedly a true repeat; the distance between its two occurrences is 10, indicating that the period length is 10 or 5. Examining all the repeats in a longer enciphered message we will find a majority at distances which are multiples of 5, at which time we will know that the period is 5.

The ultimate generalization of the Vigenère cipher was proposed by the American engineer G. S. Vernam. Let the key sequence be arbitrarily long and *random*, and use successive blocks of it for successive messages. If we assume an upper bound N on the length of all possible messages then we take the number of keys to be at least as large as N. All keys are equally likely. A commonly used method is a modulo 26 adder (or a mod 2 adder for implementation in microelectronics). If the message is $\mathbf{m} = m_1 m_2 \ldots m_n$ and each m_i is represented by one of the integers from 0 to 25 and each k_i is a number between 0 and 25 then the resulting cryptogram is $\mathbf{c} = c_1 c_2 \ldots$ where $c_i = m_i + k_i \pmod{26}$. This system is called a *one-time pad* and its name is derived from the fact that for enciphering one utilizes written pads of random numbers to obtain the sequence $k_1 k_2 \ldots k_n$ and one uses each key tape only once. This is a cumbersome method, because both sender and receiver need to keep a copy of the long key sequence, but it is clearly unbreakable—the randomness of the key means that any two message sequences of the same length are equally likely to have produced the message, the cryptogram. A one-time pad is used for the hot line between Washington and Moscow.

The increase in security from Caesar cipher to one-time pad depends on increasing the length of the key. For a Caesar cipher the key is a single number between 1 and 26 (the length of shift), for a periodic Vigenère cipher it is a finite sequence of numbers, and for the one-time pad it is a potentially infinite sequence. The longer the key, the harder the cipher is to break, but for all the classical ciphers it is possible for an opponent to reconstruct the key by an amount of work which does not grow too exhorbitantly, relative to key size.

Next we consider matrix methods based on investigations by HILL. A mapping from the space $(\mathbb{Z}_q)_n$ into itself, defined by

$$\mathbf{b} = \mathbf{K}\mathbf{a} + \mathbf{d} \quad \text{with} \quad \mathbf{K} \in (\mathbb{Z}_q)_n^n, \qquad \mathbf{d} \in (\mathbb{Z}_q)_n$$

is injective if and only if $\gcd(\det \mathbf{K}, q) = 1$. Therefore we can generalize modular enciphering as follows.

2.6 Definition. A *Hill cipher* (*key*) with matrix $\mathbf{K}$ is a mapping from $(\mathbb{Z}_q)_n$ into itself, defined by $\mathbf{a} \mapsto \mathbf{K}\mathbf{a} + \mathbf{d}$, such that $\gcd(\det \mathbf{K}, q) = 1$.

For enciphering we subdivide the plaintext into blocks of n letters each, replace each letter by its corresponding element in $\mathbb{Z}_q$ and write this in columns. Then we apply the given linear transformation to each block $\mathbf{a}$. In this context the so-called *involutory* (or self-inverse) matrices $\mathbf{K}$, defined by $\mathbf{K}^2 = \mathbf{I}$ or $\mathbf{K} = \mathbf{K}^{-1}$, are convenient to use since in deciphering we do not have to evaluate $\mathbf{K}^{-1}$ separately. A cryptogram which is constructed according to 2.6 can be deciphered if a plaintext word of sufficient minimal length (depending on n) is known. For $n = 3$ and $q = 26$ one can show that such a word of length ≥ 4 is sufficient. For $n = 2$ such a cryptogram can be deciphered even if no word of minimal length is known in the plaintext. In this case all possible involutory 2×2 matrices mod q are determined and tried. (There are 736 such matrices.) For larger n, deciphering by this trial and error method is no longer feasible, since, e.g. there are 22 involutory 3×3 matrices mod 2, 66,856 such matrices mod 13, and therefore 1,360,832 involutory 3×3 matrices mod 26.

2.7 Example. We assign to the 26 letters of the alphabet the integers from 0 to 25 as follows

A	B	C	D	E	F	G	H	I	J	K	L	M	N	O	P	Q	R	S	T	U	V	W	X	Y	Z
19	2	21	0	4	7	6	9	17	24	11	15	14	13	12	16	18	1	25	20	3	22	5	8	23	10

Using the involutory matrix

$$\mathbf{K} = \begin{pmatrix} 4 & 7 \\ 9 & 22 \end{pmatrix} \in (\mathbb{Z}_{26})_2^2$$

we form

$$\begin{pmatrix} b_1 \\ b_2 \end{pmatrix} = \mathbf{K} \begin{pmatrix} a_1 \\ a_2 \end{pmatrix}. \tag{3}$$

Here a_1, a_2 are the numbers of letters in the plaintext, which has been divided into blocks of two. If the plaintext is CRYPTOLOGY, we divide this into pairs, CR YP TO LO GY; in case of an odd number of letters we add an arbitrary letter. To each pair of letters corresponds a pair a_1, a_2 in (3), which determines b_1, b_2. Since CR $\leftrightarrow a_1 a_2 = 21\ 1$, we have

$$b_1 = 4 \cdot 21 + 7 \cdot 1 = 13 \leftrightarrow N,$$

$$b_2 = 9 \cdot 21 + 22 \cdot 1 = 3 \leftrightarrow U,$$

i.e. CR is enciphered as NU, etc. □

These methods are fixed transposition methods. Variable methods can be obtained which are similar to (1) and (2) (preceding 2.5) by subdividing the plaintext into blocks, each of which contains n letters, such that by substitution with elements in $\mathbb{Z}_q$, the plain text is of the form

$$\underbrace{a_{11}a_{12}\ldots a_{1n}}_{\mathbf{a}_1^T}\underbrace{a_{21}a_{22}\ldots a_{2n}}_{\mathbf{a}_2^T\ldots}\ldots =: \mathbf{a}_1^T\mathbf{a}_2^T\ldots.$$

A variable matrix method is obtained by letting $f_m(\mathbf{a}) := \mathbf{b}_m := \mathbf{K}\mathbf{a} + \mathbf{d}_m$, where $\gcd(\det \mathbf{K}, q) = 1$, and with

$$\mathbf{d}_m = \mathbf{C}\mathbf{a}_{m-1} \quad \text{with given } \mathbf{C} \in (\mathbb{Z}_q)^n_n, \ \mathbf{a}_0 \text{ given}, \tag{4}$$

or

$$\mathbf{d}_m = \mathbf{C}\mathbf{b}_{m-1} \quad \text{with given } \mathbf{C} \in (\mathbb{Z}_q)^n_n, \ \mathbf{b}_0 \text{ given}. \tag{5}$$

2.8 Example. A variable key according to (4) for enciphering CRYP-TOLOGY is given by

$$\mathbf{K} = \begin{pmatrix} 1 & 2 & 3 \\ 2 & 5 & 6 \\ 1 & 2 & 4 \end{pmatrix}, \quad \mathbf{C} = \begin{pmatrix} 4 & 1 & 1 \\ 2 & 0 & 3 \\ 1 & 2 & 0 \end{pmatrix}, \quad \mathbf{a}_0 = \begin{pmatrix} 1 \\ 2 \\ 3 \end{pmatrix},$$

$$\begin{pmatrix} b_{i1} \\ b_{i2} \\ b_{i3} \end{pmatrix} = \begin{pmatrix} 1 & 2 & 3 \\ 2 & 5 & 6 \\ 1 & 2 & 4 \end{pmatrix} \begin{pmatrix} a_{i1} \\ a_{i2} \\ a_{i3} \end{pmatrix}$$

$$+ \begin{pmatrix} 4 & 1 & 1 \\ 2 & 0 & 3 \\ 1 & 2 & 0 \end{pmatrix} \begin{pmatrix} a_{i-1,1} \\ a_{i-1,2} \\ a_{i-1,3} \end{pmatrix}. \tag{6}$$

We use the simple association $A \leftrightarrow 0$, $B \leftrightarrow 1$, etc. as in 2.3. For the first group of three letters we obtain

$$C \leftrightarrow 2, \ R \leftrightarrow 17, \ Y \leftrightarrow 24,$$

i.e. $\mathrm{CRY} = (a_{i1}, a_{i2}, a_{i3}) = (2, 17, 24)$. Thus

$$\begin{pmatrix} b_{11} \\ b_{12} \\ b_{13} \end{pmatrix} = \begin{pmatrix} 1 & 2 & 3 \\ 2 & 5 & 6 \\ 1 & 2 & 4 \end{pmatrix} \begin{pmatrix} 3 \\ 18 \\ 25 \end{pmatrix} + \begin{pmatrix} 4 & 1 & 1 \\ 2 & 0 & 3 \\ 1 & 2 & 0 \end{pmatrix} \begin{pmatrix} 1 \\ 2 \\ 3 \end{pmatrix} = \begin{pmatrix} 19 \\ 23 \\ 14 \end{pmatrix} = \begin{pmatrix} \mathrm{T} \\ \mathrm{X} \\ \mathrm{O} \end{pmatrix}.$$

Deciphering means determining the inverse transformation of the transformation (6). ☐

Enciphering by using a linear system of equations $\mathbf{C} = \mathbf{KP}$ can be done by replacing the constant matrix $\mathbf{K}$ by a matrix with variable elements, i.e. by a matrix with parameters. We assume that the determinant of $\mathbf{K}$ is independent of the parameters and also that it is relatively prime to 26. A simple form for such matrices is the lower triangular matrix

$$\mathbf{K} = \begin{pmatrix} t_{11} & 0 & \cdots & 0 \\ t_{21} & t_{22} & \cdots & 0 \\ & & \ddots & \\ t_{n1} & t_{n2} & \cdots & t_{nn} \end{pmatrix},$$

where $t_{11}, \ldots, t_{nn}$ are fixed numbers with $\gcd(t_{ii}, 26) = 1$, while the other t_{ij} $(i > j)$ might be variable.

2.9 Example. Let $\mathbf{K}(t)$ be the matrix

$$\mathbf{K}(t) = \begin{pmatrix} 1 & 0 & 0 & 0 \\ t & 3 & 0 & 0 \\ 2t+1 & 2 & 5 & 0 \\ 1 & t & t+1 & 7 \end{pmatrix} \in (\mathbb{Z}_{26})_4^4,$$

with an arbitrary parameter $t \in \mathbb{Z}_{26}$. For each such t we have $\det \mathbf{K}(t) = 1$. ☐

The notation $\mathbf{K}(t)$ indicates that the matrix $\mathbf{K}$ depends on t. Let

$$\mathbf{b} = \mathbf{K}(t)\mathbf{a} + \mathbf{d}$$

and assign a special value to t for each vector $\mathbf{a}_i$ in the plaintext, say k_i; then 2.7 gives

$$\mathbf{b}_i = \mathbf{K}(k_i)\mathbf{a}_i^T.$$

In general $\mathbf{K}^{-1}(t)$ is complicated, but we can make $\mathbf{K}^{-1}(t)$ linear in t by letting $\mathbf{K}(t) = \mathbf{G} + t\mathbf{H}$, with $(\det \mathbf{G}, 26) = 1$, where $\mathbf{G}$, $\mathbf{H}$ are constant matrices satisfying $\mathbf{H} = \mathbf{XG}$, and $\mathbf{X}^2 = \mathbf{0}$. Then

$$\mathbf{K}^{-1}(t) = \mathbf{G}^{-1} - t\mathbf{G}^{-1}\mathbf{X}.$$

A suitable matrix $\mathbf{X}$ with the property $\mathbf{X}^2 = \mathbf{0}$ is, say, $\mathbf{X} = \mathbf{QNQ}^{-1}$, where $\mathbf{Q}$

is an arbitrary invertible matrix and N is a matrix of the form

$$N = \begin{pmatrix} 0 & n_1 & & & & & & & \\ 0 & 0 & & & & & & & \\ & & 0 & n_2 & & & & & \\ & & 0 & 0 & & & & & \\ & & & & \ddots & & & & \\ & & & & & 0 & n_r & & \\ & & & & & 0 & 0 & & \\ & & & & & & & 0 & \\ & & & & & & & & \ddots \\ & & & & & & & & & 0 \end{pmatrix}$$

with $r \le [n/2]$. A suitable "key matrix" $K(t)$ is of the form

$$K(t) = G + tXG = G + tQNQ^{-1}G.$$

It is advantageous to use involutory matrices $K(t)$. Then $K(t)^2 = I$ or $K(t) = K^{-1}(t)$. Such matrices are determined by choosing

$$K(t) = Q(J - tN)Q^{-1},$$

where Q is an arbitrary invertible matrix with constant elements, N is as above, and J is given by

$$J = \begin{pmatrix} 1 & j_1 & & & & & & & \\ 0 & -1 & & & & & & & \\ & & 1 & j_2 & & & & & \\ & & 0 & -1 & \cdot & & & & \\ & & & & \ddots & & & & \\ & & & & & 1 & j_r & & \\ & & & & & 0 & -1 & & \\ & & & & & & & a_1 & \\ & & & & & & & & \ddots \\ & & & & & & & & \cdot & a_s \end{pmatrix}$$

with $j_1, \ldots, j_r$ constant, where $a_1, \ldots, a_s$ are either $+1$ or -1 and $s + 2r = n$.

We now describe permutation keys. Let $\mathbb{F}_q$ be a finite field with $q = p^n$ elements and let ζ be a primitive element of $\mathbb{F}_q$. Each element of $\mathbb{F}_q$ can be expressed in the form

$$a = (c_0, \ldots, c_{n-1}) = c_0 \zeta^{n-1} + c_1 \zeta^{n-2} + \ldots + c_{n-1}, \qquad c_i \in \mathbb{F}_p.$$

$\mathbb{F}_q$ consists of all elements of the form $\{0, \zeta, \ldots, \zeta^{q-1}\}$. The primitive element is a zero of a primitive irreducible polynomial f over $\mathbb{F}_p$. We denote the integer

$$c_0 p^{n-1} + c_1 p^{n-2} + \ldots + c_{n-1}, \qquad c_i \in \mathbb{F}_p$$

as the *field-value* of the element a in $\mathbb{F}_q$.

2.10 Example. Let $f = x^3 + 2x + 1$ be a primitive polynomial with zero ζ in $\mathbb{F}_{3^3}$. The elements of $\mathbb{F}_{3^3}$ are given in the following table. Column (1) will be used in the enciphering of plaintexts. Column (2) represents the field values of the elements in $\mathbb{F}_{3^3}$. In column (3) we describe the elements of $\mathbb{F}_{3^3}$ in vector form $c_0 c_1 \ldots c_{n-1}$. Column (4) gives the exponents of the powers ζ^i of the elements of the cyclic group $\mathbb{F}_{3^3}^*$.

(1)	(2)	(3)	(4)	(1)	(2)	(3)	(4)
A	0	000	0	O	14	112	11
B	1	001	26	P	15	120	4
C	2	002	13	Q	16	121	18
D	3	010	1	R	17	122	7
E	4	011	9	S	18	200	15
F	5	012	3	T	19	201	25
G	6	020	14	U	20	202	8
H	7	021	16	V	21	210	17
I	8	022	22	W	22	211	20
J	9	100	2	X	23	212	5
K	10	101	21	Y	24	220	23
L	11	102	12	Z	25	221	24
M	12	110	10	ω	26	222	19
N	13	111	6				$\square$

2.11 Definition. A polynomial $g \in \mathbb{F}_q[x]$ is called a *permutation polynomial* of $\mathbb{F}_q$, if the polynomial function $\bar{g}$ induced by g is a permutation of $\mathbb{F}_q$.

If A is an alphabet with q letters then a key can be defined as follows. First we determine a one-to-one correspondence between the q letters of A and the q letters of $\mathbb{F}_q$. If $P_1 P_2 P_3 \ldots$ is a plaintext then let $a_1 a_2 a_3 \ldots$ be the corresponding sequence of elements in $\mathbb{F}_q$. This sequence will be mapped into $\bar{g}(a_1)\bar{g}(a_2)\bar{g}(a_3) \ldots$ by using a permutation polynomial g over $\mathbb{F}_q$. The elements $\bar{g}(a_i)$ are represented as letters of A and form a cryptogram. If $f: A \to \mathbb{F}_q$ is bijective and $P \in A$ then

$$P \to f(P) \to g(f(P)) \to f^{-1}(g(f(P)))$$

is a key $f^{-1} \circ g \circ f$. If we use permutation polynomials g_i in enciphering, we obtain a variable key.

2.12 Example (2.10 continued). Columns (1) and (3) give the association between the letters of the natural alphabet and $\mathbb{F}_{3^3}$ and ω denotes the twenty-seventh letter. Thus A = 000, H = 021, etc. Let $g_i = a_i x + b_i$, $a \neq 0$, and let a_i and b_i be determined by

$$a_{i+2} = a_{i+1} a_i \quad \text{and} \quad b_{i+2} = b_{i+1} + b_i, \qquad i = 1, 2, \ldots$$

with initial values

$$a_1 = 021 = \zeta^{16}, \qquad a_2 = 111 = \zeta^6, \qquad b_1 = 002, \qquad b_2 = 110.$$

The enciphering of the word GALOIS is given by

(a)	G	A	L	O	I	S
(b)	020	000	102	112	022	200
(c)	14	0	12	11	22	15
(d)	16	6	22	2	24	1
(e)	120	111	202	002	211	021
(f)	002	110	112	222	001	220
(g)	122	221	011	221	212	211
(h)	R	Z	E	Z	X	W

The rows can be calculated as follows: (a) plaintext; (b) vector form; (c) exponents of the elements $\alpha = \zeta^t$; (d) exponents t of the powers ζ^t of a_i; (e) vector form of the products $a_i\alpha_i$, i.e. (c) + (d) mod 26; (f) vector form of b_i; (g) vector form of $y_i = a_i\alpha_i + b_i$ of (e) + (f); (h) cipher description of (f) and (a). □

PROBLEMS

1. Use ALGEBRA as a key word (sequence) in a Vigenère cipher and encipher CRYPTOGRAPHY.

2. Find a random sequence of 0–1 by flipping a coin and use this sequence in a one-time pad to encipher THEMESSAGEISFAKED.

3. Use the key word ALGEBRA in a Vigenère cipher to encipher the message SECRET CODES.

4. Encipher the message CIPHER according to the method of Example 2.5.

5. Let $\mathbb{Z}_{29}$ be the plaintext and ciphertext alphabet of a Hill cipher. Let

$$\mathbf{K} = \begin{pmatrix} 27 & 1 \\ 16 & 14 \end{pmatrix}$$

be the enciphering matrix. Encipher the message 18 4 2 20 17 8 19 24. Also find the inverse of $\mathbf{K} \pmod{29}$.

6. Decipher the cryptogram TXO of Example 2.8 which was obtained by using the variable matrix key given in 2.8.

7. Let $\mathbf{K}$ be the matrix of a Hill key; let $n = 3$. Use $\mathbf{K}$ to encipher SECURITYISVITAL,

$$\mathbf{K} = \begin{pmatrix} 17 & 17 & 5 \\ 21 & 18 & 21 \\ 2 & 2 & 19 \end{pmatrix}.$$

Also find $\mathbf{K}^{-1} \pmod{26}$. Decipher LNSHDLEWMTRW.

8. We use the notation immediately after 2.9. Let

$$
N = \begin{pmatrix} 0 & 1 & 0 \\ 0 & 0 & 0 \\ 0 & 0 & 0 \end{pmatrix}, \qquad Q = \begin{pmatrix} 17 & 6 & 12 \\ 6 & 13 & 24 \\ 12 & 24 & 23 \end{pmatrix},
$$

$$
X = \begin{pmatrix} 24 & 13 & 18 \\ 10 & 0 & 14 \\ 20 & 0 & 2 \end{pmatrix}, \qquad G = \begin{pmatrix} 25 & 2 & 16 \\ 2 & 25 & 10 \\ 16 & 10 & 3 \end{pmatrix}.
$$

Determine $K(t)$ and $K^{-1}(t)$. Encipher the message KEY.

9. Let N, Q and X be as before and let

$$
G = \begin{pmatrix} 3 & 9 & 10 \\ 6 & 19 & 2 \\ 12 & 14 & 3 \end{pmatrix}.
$$

Construct an involutory matrix $K(t)$.

10. How many involutory 3×3 matrices are there (mod 2)?

11. Use the permutation polynomials $g_i = a_i x^3 + b_i$ of $\mathbb{F}_{27}$ (with a_i and b_i given in Example 2.12) to encipher the message SECRET according to the method of 2.12.

12. A Hill cipher can also be constructed by using blocks of four message characters abcd and forming a 2×2 matrix

$$
M = \begin{pmatrix} a & b \\ c & d \end{pmatrix}
$$

which is enciphered as the matrix product KM with a Hill key K. Suppose letters A through Z are denoted by 0 through 25, a blank by 26, a full stop by 27 and a comma by 28. Suppose the key

$$
K = \begin{pmatrix} 1 & 2 \\ 3 & 4 \end{pmatrix}
$$

was used for enciphering to obtain MA.AFWHEHbFLSVCL. Decipher this cryptogram.

B. Public-Key Cryptosystems

In the usual form of cryptosystems discussed so far the secrecy of the transformation of a message is preserved by making the algorithm depend on a secret key. We use the same key for enciphering and deciphering and call such systems *single-key systems*. A different type of system, the *public-key cryptosystem*, is based on the observation that encipherment procedure and key do not have to be the same as decipherment procedure and key. The purpose of public-key cryptosystems is to remove the difficult requirements that the key must be kept secret. Distribution of keys in a large organization can be expensive and risky and the problem increases as the size of the communicating community increases.

In public-key cryptosystems, the desirable features of the cipher in terms of an enciphering method E and a deciphering method D are:

(i) Deciphering an enciphered message M yields M, i.e.

$$D(E(M)) = M.$$

(ii) E and D can be easily computed.

(iii) E can be made public without revealing D.

(iv) Enciphering a deciphered message M yields M, i.e.

$$E(D(M)) = M.$$

In 1976 Diffie and Hellman called ciphers with such properties public-key cryptosystems. During the following years several examples of such ciphers have been developed, some of them were found insecure soon after, the security of others is still open.

There are some interesting applications. If a person B wants to receive messages he places his enciphering method, E_B, in a public directory (like a telephone directory), enabling anyone else to communicate with him in guaranteed privacy, since only B has the deciphering method, D_B.

B can also "sign" messages he sends to another person A, so that A is assured of their authenticity. To sign a message M, B first computes the "digital signature"

$$S = D_B(M);$$

then, if he is communicating with A, he looks up E_A in the directory and sends $E_A(S)$. He also sends the uncoded message of which he, B, is the author. A will be able to check the truth of that, as we shall see. Since A has the deciphering procedure D_A, she can first compute

$$D_A(E_A(S)) = S,$$

obtaining the signature S, and then use E_B from the directory to obtain the message M:

$$E_B(S) = E_B(D_B(M)) = M.$$

Since only B can have created an S which deciphers to M by E_B, A knows that M can only have come from B.

(Here we are assuming that only a tiny fraction of symbol sequences actually are meaningful messages M, as is the case for sequences of letters in the English alphabet. Then any forgery S' is likely to be detected, because of the miniscule probability that $E_B(S')$ will be a meaningful message M'. This protection against forgery is analogous to the way error correcting codes give protection against random errors, namely, by having only a small fraction of possible binary blocks actually in the code cf. STILLWELL.)

As a concrete example suppose a test ban treaty for monitoring nuclear tests between country A and country B proposes that each nation place seismic instruments in each other's territory, to record any disturbances and hence detect underground tests. It is possible to protect the instruments (in the sense that they can be made to self-destruct if anyone tampers with them) but not the channel which sends their information (the host nation could cut the wires and send false information). Furthermore, if information is sent in ciphered form, the host nation may suspect that unauthorized information is also being sent, in addition to the agreed-on seismic data.

A digital signature system is the ideal solution to this problem. Nation B's seismic station contains a computer which converts the message M to

$$S = D_B(M).$$

Nation A cannot substitute any S' for S, because of the overwhelming probability that $E_B(S') = M'$ will not be meaningful, and hence nation B will detect the forgery. However, nation B *can* supply A with the procedure E_B, which A can then use to recover

$$M = E_B(S),$$

and thus be reassured that only an authorized message M is being sent.

The trapdoor one-way function is crucial for the construction of public-key cryptosystems: A function $f(x)$ is called a *one-way function* if $f(x)$ can be computed for given x, but for almost all values of y in the range of f it is computationally infeasible or impracticable to find x such that $y = f(x)$, that is, the computation of the inverse of $f(x)$ can be said to be impracticable. f is a *trapdoor* one-way function if it is easy to compute its inverse given additional information, namely the secret deciphering key. One-way functions must lack all obvious structure in the table of $f(x)$ against x. The finite exponential used in the RSA cryptosystem, to be described below, is an example of such a function.

The first public-key cryptosystem we consider is based on the difficult problem of selecting from a collection of objects of various different sizes a set which will exactly fit a defined space. Expressed in one-dimensional terms this is equivalent to selecting rods of different lengths which, put end-to-end, will exactly equal a specified length K. This problem is called the *knapsack problem*. First we express the knapsack problem in symbolic form. Let $\mathbf{y} = (y_1, \ldots, y_n)$ be the components of the knapsack written as a vector. In order to fill a specified knapsack a selection of these numbers is added together to form the total. This selection is a binary array $\mathbf{a} = (a_1, \ldots, a_n)$, in which $a_i = 0$ means that the ith value of $\mathbf{y}$ is not chosen, $a_i = 1$ means that it is chosen. If this selection of components fits the size of the knapsack then

$$K = \mathbf{y} \cdot \mathbf{a}$$

where $\cdot$ denotes the scalar product of vectors. Usually it is not easy to find

a when only K and **y** are given. In the application to cryptography **a** is the plaintext and K is the ciphertext. Cryptanalysis means solving the knapsack problem, that is given K and **y**, find **a**. Deciphering means knowing a secret way to find **a** much more easily than by solving the general problem. It is known that determining a given knapsack problem with sum K and vector $\mathbf{y} = (y_1, \ldots, y_n)$, such that K is the sum of some y_i's, belongs to a class of problems for which no polynomial time algorithm is known (one says it is an NP-*complete problem*). However, the complexity of a knapsack depends largely on the choice of **y**. A simple knapsack is one where $y_i > y_1 + \ldots + y_{i-1}$, for $i = 2, \ldots, n$, and in this case it can be decided in linear time whether there is a solution and a simple algorithm will find it, if it exists. The basic idea of using a knapsack for a cryptosystem is to convert a simple knapsack into a complex one, a trapdoor knapsack. This is done by choosing two large integers k and q, $(k, q) = 1$, and then forming a new trapdoor knapsack $\mathbf{z} = (z_1, \ldots, z_n)$ from a given knapsack vector **y** by letting $z_i \equiv ky_i \pmod{q}$. Knowing k and q makes it easy to compute $K \equiv k^{-1}L \pmod{q}$, where L is assumed to be the sum of some of the z_i, forming the given knapsack problem. Multiplying L by $1/k$ transforms L into K and reduces the given difficult knapsack with sum L and vector $\mathbf{z}$ into the easy knapsack with sum K and vector **y**. This method can be used for a cryptosystem, where $\mathbf{z} = (z_1, \ldots, z_n)$ is the public key, the private keys are k and q in the calculation $z_i \equiv ky_i \pmod{q}$ for a given trivially solvable knapsack vector **y**, with $y_i > y_1 + \ldots + y_{i-1}$ for $i = 2, \ldots, n$. A message $\mathbf{a} = (a_1, \ldots, a_n)$ in binary form is enciphered by computing $L = a_1z_1 + \ldots + a_nz_n$, which is transmitted.

It is thought that the security of some cryptosystems can be increased by repeating the enciphering procedure using new parameters. *Iteration* (or multiple encryption) does not always enhance security but in the case of the trapdoor knapsack it was initially thought to work by constructing doubly (or multiply) iterated knapsacks. But today attacks on simple and doubly iterated knapsacks are known; for multiple knapsacks used for constructing cryptosystems security is also in serious doubt. To obtain a doubly iterated knapsack, we choose large coprime integers k_1 and q_1 and construct a new knapsack vector $\mathbf{u} = (u_1, \ldots, u_n)$ where $u_i \equiv k_1z_i \pmod{q_1}$. This process can be repeated as often as desired. It is not clear to an unauthorized receiver whether one or more iterations are used, since all keys $k, q, k_1, q_1, \ldots$ are secret. In practical applications it has been suggested that n be at least 200, and that other conditions on q, y_i and k are imposed. However, in 1982 Adi Shamir made a successful attack on a basic variant of the trapdoor knapsack system, whose security has been suspect already since its introduction in 1976 by Merkle and Hellman. Nevertheless, we describe the knapsack cryptosystem more specifically. First the letters of the alphabet A are replaced by the elements of $(\mathbb{Z}_2)^t$, for suitable values of t. Then we subdivide the resulting 0–1-sequences into blocks $(a_1, a_2, \ldots, a_n)$. Let $q \in \mathbb{N}$ and let $\mathbf{x} = (x_1, \ldots, x_n)$ be a fixed vector in $(\mathbb{Z}_q)^n$. A block

$\mathbf{a} = (a_1, \ldots, a_n)$ is enciphered as an element $b \in \mathbb{Z}_q$ by means of:

$$\mathbf{a} \to b := \mathbf{a} \cdot \mathbf{x} = a_1 x_1 + \ldots + a_n x_n.$$

Here we have a mapping from $A = (\mathbb{Z}_q)_n^n$ into $B = \mathbb{Z}_q$; this is not a key in the sense of 2.1. In order to make this enciphering method into a key (to enable deciphering), the authorized receiver has to possess additional information. There are several ways to do this. We give an example.

2.13 Example. Let $t = 5$, i.e. we replace the letters of the alphabet by 5-tuples of elements in $\mathbb{Z}_2$.

A	B	C	D	E	F	G	H	I	J	K
00000	00001	00010	00011	00100	00101	00110	00111	01000	01001	01010
L	M	N	O	P	Q	R	S	T	U	V
01011	01100	01101	01110	01111	10000	10001	10010	10011	10100	10101
W	X	Y	Z	,	.	!	?	"	+	
10110	10111	11000	11001	11010	11011	11100	11101	11110	11111	

Here $+$ represents a space. Now we choose $n = 10$, $q = 19999$ and a vector $\mathbf{y} = (4, 10, 64, 101, 200, 400, 800, 1980, 4000, 9000) \in (\mathbb{Z}_{19999})^{10}$. In $\mathbf{y}$ each coordinate is larger than the sum of the preceding coordinates if we regard the elements as integers. We form a new vector

$$\mathbf{x} = 200\mathbf{y} = (800, 2000, 12800, 201, 2, 4, 8, 16019, 40, 90) \in (\mathbb{Z}_{19999})^{10}.$$

If we wish to encipher the word CRYPTOLOGY we subdivide the message into groups of two: CR YP TO LO GY and encode:

$$CR = 0001010001$$

$$\to (0, 0, 0, 1, 0, 1, 0, 0, 0, 1)$$

$$\cdot (800, 2000, 12800, 201, 2, 4, 8, 16019, 40, 90)$$

$$= 295.$$

Also:

$$YP \to 18957, \quad TO \to 17070, \quad LO \to 18270, \quad GY \to 13013.$$

Thus the enciphered plaintext is

$$(295, 18957, 17070, 18270, 13013) \in (\mathbb{Z}_{19999})^5.$$

If the receiver obtains the message

$$\mathbf{w} = (2130, 18067, 17034) \in (\mathbb{Z}_{19999})^3$$

then we decipher it as follows. We evaluate $200^{-1}\mathbf{w} = 100\mathbf{w} \in (\mathbb{Z}_{19999})^3$. For the first component, say, we obtain $13010 = \mathbf{a} \cdot \mathbf{y} = 4a_1 + 10a_2 + \ldots +$

$9000a_{10}$, where $\mathbf{a} = (a_1, \ldots, a_{10})$ with $a_i \in \{0, 1\}$ represents the original first part of the message in numerical form.

13010 can only be represented in the form

$$1 \cdot 9000 + 1 \cdot 4000 + 1 \cdot 10 = 10a_2 + 4000a_9 + 9000a_{10},$$

if we subtract from 13010 the numbers 9000, 4000, etc. Thus one obtains $\mathbf{a} = (0, 1, 0, 0, 0, 0, 0, 0, 1, 1) \leftrightarrow 01000\ 00011 \leftrightarrow ID$. The surprised receiver continues deciphering $6790 = a^* \cdot y$ and $3485 = a^{**} \cdot y$ and indignantly reads the message (a, a^*, a^{**}):

$$a^* = 6790 \leftrightarrow IO, \qquad a^{**} = 3485 \leftrightarrow T! \qquad\qquad \square$$

We summarize these steps in the following remark.

2.14 A Trapdoor-Knapsack Method. For sending a plaintext message, choose a correspondence between the letters of the alphabet and $(\mathbb{Z}_2)'$, i.e. replace letters by t-tuples of zeros and ones. Then choose numbers q and n such that $t \mid n$ and a vector $\mathbf{y} \in (\mathbb{Z}_q)^n$ such that $y_i > y_1 + \ldots + y_{i-1}$ for $2 \le i \le n$. Also choose a number k with $(k, q) = 1$. Then we encipher and decipher as follows:

Enciphering. Form $\mathbf{x} := k\mathbf{y}$, divide the original plaintext message into blocks of n/t letters, and replace these by blocks of zeros and ones of length n. Multiply these blocks by the vector $\mathbf{x}$.

Deciphering. Multiply the received sequence by $k^{-1} \in \mathbb{Z}_q$ (which exists, since $(k, q) = 1$). Each term in the sequence corresponds to one 0–1-sequence of length n. Interpret this as a group of n/t blocks of t letters each. $\qquad\qquad \square$

Here n, q and $\mathbf{x}$ are openly published in a directory. The number k is kept secret between the participants. This method has one great advantage. If we want to enable many participants to exchange information (e.g. between say 1000 branches of a banking group) we would then need $\binom{1000}{2}$, i.e. some half million arrangements between 1000 participants. This method makes it possible for each participant to openly announce a vector $\mathbf{x}$ (similar to telephone numbers). The association letters$\leftrightarrow$numbers and the numbers n and q can be openly agreed upon by the participants (e.g. the banking group). If participant A wants to transmit a message to participant B, she looks up B's vector $\mathbf{x}$ (e.g. given in a register) and multiplies her message "piecewise" by $\mathbf{x}$. Then the enciphered message can be transmitted openly. Only B is capable of deciphering the received message, because only he knows k and thus only he can restore the original message.

A similar cryptosystem is based on a *multiplicative knapsack*. Here we try to find a selection from a collection of numbers, the product of which equals a specified number. We impose the condition that the components of the knapsack vector $\mathbf{y}$ should be relatively prime. To represent a binary

288 Chapter 5. Further Applications of Fields and Groups

number $\mathbf{a} = (a_1, \ldots, a_n)$ we could form $(a_{i_1} y_{i_1}) \ldots (a_{i_r} y_{i_r}) = K$, for all those a_i, $1 \le i \le n$, which are nonzero, say a_{i_j}, $1 \le j \le r$. A reversal of the operation would recover $\mathbf{a}$ with ease, by successive division of the elements of $\mathbf{y}$ into K. Let p be a prime (of the form $2^n + 1$) larger than the product of all the vector elements. Then $\mathbf{y}$ is transformed into an additive knapsack vector $\mathbf{y}'$ by choosing a base b and computing

$$b^{y'_i} \equiv y_i \pmod{p}.$$

Let $K' = \mathbf{y}' \cdot \mathbf{a}$ be the additive knapsack and form

$$K \equiv b^{K'} \pmod{p}.$$

Having obtained K with the secret information, p and b, we find that solving for $\mathbf{a}$ is trivial using the secret values of y.

Next we describe a public-key cryptosystem whose security seems to be very high. It is based on the power function as a one-way function. This system is due to RIVEST, SHAMIR and ADLEMAN (therefore called the *RSA cryptosystem*). For the description of this cryptosystem we need an elementary result from number theory. The reduced set of residues (mod n) is the subset of $\{0, 1, \ldots, n - 1\}$ consisting of integers relatively prime to n.

Euler's ϕ-function $\phi(n)$ gives the number of elements in the reduced set of residues (mod n). For a prime p, obviously $\phi(p) = p - 1$. It can be easily verified that $\phi(pq) = (p - 1)(q - 1)$ for distinct primes p and q. If arbitrary integers a and n are relatively prime then *Euler's theorem* says that $a^{\phi(n)} \equiv 1$ (mod n). This is a generalization of *Fermat's (Little) Theorem*: For p prime and for every a relatively *prime* to p we have $a^{p-1} \equiv 1 \pmod{p}$. Euler's theorem implies that if

$$st \equiv 1 \pmod{\phi(n)},$$

then $m \equiv c^t \pmod{n}$ is the inverse of $c \equiv m^s \pmod{n}$. If m is a plaintext number and c the corresponding ciphertext after using the enciphering key s and n, then the deciphering key t restores the original m. t can be calculated as the multiplicative inverse of s modulo $\phi(n)$ by using the extended Euclidean algorithm. Since $st \equiv 1 \pmod{\phi(n)}$ is equivalent to $st = 1 + k\phi(n)$ for a suitable integer k, and since $\gcd(s, \phi(n)) = 1$ we can represent this gcd as a linear combination of s and $\phi(n)$. That is by retracing the steps in Euclid's algorithm for finding gcd we can find integers t and k such that $st = 1 + k\phi(n)$.

First we give a simple version of a cryptosystem for enciphering and deciphering a message given in base 100 digits, which are subdivided into r-digit words by collecting the digits into r-digit base 100 numbers. For instance, if we associate A, B, C, ... with 01, 02, 03, respectively, then ABEL becomes 01 02 05 12. If we want $r = 2$ digit words in base 100 then 0102 0512 are the corresponding "words". For n a prime the system 2.15 is due to Pohlig and Hellman.

2.15 Remark. Choose some numbers $n, s \in \mathbb{N}$ such that $(s, \phi(n)) = 1$ and such that no prime dividing n is smaller than the largest possible message word. Then there exists a $t \in \mathbb{N}$ such that $st \equiv 1 \pmod{\phi(n)}$. Therefore, by elementary number theory, if $(m, n) = 1$, $(m^s)^t \equiv m \pmod{n}$.

Enciphering. A message word m is enciphered as $m^s = v \pmod{n}$.

Deciphering. A received cipherword z is deciphered as $z^t \equiv m \pmod{n}$.

□

2.16 Example. Let $n = 101$ and $s = 3$. Then $\phi(n) = 100$, and we find $t = 67$ such that $st \equiv 1 \pmod{100}$. If the alphabet A, B, C, ..., Z is replaced by the numbers $1, 2, 3, \ldots, 26$ then the word A L G E B R A is replaced by 1 12 7 5 2 18 1. This is enciphered using $1^3 \equiv 1 \pmod{101}$, $12^3 \equiv 11 \pmod{101}$, $7^3 \equiv 40 \pmod{101}$, etc. The corresponding cipher word is 1 11 40 24 8 75 1. Deciphering is performed by raising each number in the received word to the power $t = 67$.   □

The important RSA cryptosystem is based on n being the product of two large prime numbers of approximately equal size. The message gets split up into blocks and each block, regarded as an integer, must lie between 0 and $n - 1$, since we try to perform an encryption $\pmod{n}$. Let $n = pq$ and let s be relatively prime to $\phi(n) = (p - 1)(q - 1)$. Then n and s (but not the prime factorization of n) can be made publicly known, usually by the receiver, and form the encryption key. The method is based on the fact that for large n it is virtually impossible to find t such that $st \equiv 1 \pmod{\phi(n)}$, without knowing the factorization of n. This is a generalization of Fermat's theorem. Only the authorized receiver knows this factorization of n into p and q. Enciphering consists of transforming the message into a number sequence. Each number or block of numbers is raised to the power s and is reduced modulo n. For deciphering we find the element t which is the inverse of s, being in the prime residue class group modulo n of order $\phi(n) = (p - 1)(q - 1)$. We use the extended form of the Euclidean algorithm to find t. This is easy to do if we know p and q, but it is extremely difficult or computationally infeasible if we only know n, for large n. The unauthorized decipherer thus has the problem of factoring a number into prime factors. If n has 200 digits even the fastest computer will (in 1984) need years in order to factorize n. It is believed that the RSA cryptosystem will be secure for the next 10 to 20 years if one uses a well-chosen 512 bit number n, or even a 1024 bit number when the need arises.

We give two numerical examples of the RSA public-key cryptosystem in order to demonstrate this method, but we have to mention that the values of the crucial parameters chosen here are too small to provide any security, i.e. too small to ensure that parts of the calculation are not feasible for an unauthorized interceptor.

The first example is easily done by trivial calculations; the second one involves slightly larger numbers.

2.17 Example. Choose $n = pq = 3 \cdot 11 = 33$ and $s = 3$. Then
A L G E B R A $= 1, 12, 7, 5, 2, 18, 1$ is enciphered as $1, 12, 13, 26, 8, 24, 1$
by computing $1^3 \equiv 1 \pmod{33}$, $12^3 \equiv 12 \pmod{33}$, etc. We find $t = 7$ such
that, by the Euclidean algorithm, $st = 1 + k20$ or equivalently $st \equiv$
$1 \pmod{20}$, 20 being $\phi(33)$. To decipher $17, 26, 12, 12, 9$ we calculate
$17^7 \equiv 8 \pmod{33}$, etc. and find that the message was H E L L O.   □

2.18 Example. Let $n = 11023 = 73 \cdot 151 = p \cdot q$, $s = 11$. Now we assume
that the letters A, B, C, $\ldots$, Z correspond to $00, 01, 02, \ldots, 25$, respectively. The message I L I K E A L G E B R A, then becomes
$08110810040011060401 1700$. We partition this string of numbers into blocks
of equal length, say $0811, 0810, 0400$, etc. This is enciphered by calculating
$811^{11} \equiv 867 \pmod{11023}$, etc.

The authorized receiver knows p and q, and that the exponents in
congruences $\pmod{n}$ are calculated mod $\phi(n)$. He finds $t \equiv s^{-1} \pmod{\phi(n)}$
to be $t = 5891$, evaluates $867^{5891} \equiv 811 \pmod{11023}$, etc. and finds the plain
text by using the agreed upon correspondence between letters and numbers.
□

We note that the choice of p and q (and of the encrypting exponent s)
does have to be conditioned in certain ways (see below) in order to deny
the unauthorized interceptor a means of deciphering, which does not depend
on the factorization of the modulus n. One condition is that $p - 1$ should
have a large prime factor, say u, and that $u - 1$ should have a large prime
factor. This ensures that the plaintext cannot be found by iterating the
enciphering more quickly than by random search. A similar condition must
hold for q.

More generally, instead of the modulus $n = pq$ one can consider an
arbitrary integer n, which is larger than any number representing a message
block. Then Euler's theorem says

$$a^{\phi(n)} \equiv 1 \pmod{n} \quad \text{for } a \in \mathbb{Z}, (a, n) = 1.$$

It can be shown that

$$a^{\lambda(n)} \equiv 1 \pmod{n},$$

where $\lambda(n)$ is the maximum period of the multiplicative group $\mathbb{Z}_n$. Here
$\lambda(n)$ is called the *Carmichael function* and is defined as follows: Let
$n = p_1^{e_1} \ldots p_r^{e_r}$ be the prime factorization of $n \in \mathbb{Z}$, and let $\phi(p^e)$ be Euler's
function, then

$$\lambda(n) = \text{lcm}(\lambda(p_1^{e_1}), \ldots, \lambda(p_r^{e_r})),$$

$$\lambda(p^e) = \begin{cases} \frac{1}{2}\phi(2^e) & \text{for } p = 2 \text{ and } e > 2, \\ \phi(p^e) & \text{for } p \text{ odd prime or } p = 2 \text{ and } e = 1 \text{ or } 2. \end{cases}$$

It can be shown that for all $a \in \mathbb{Z}$:

$$a^{k+u} \equiv a^k \pmod{n}$$

if and only if n is free of $(k+1)$th powers and $\lambda(n) \mid u$. From $a^{st} \equiv a \pmod{n}$ it follows that $st \equiv 1 \pmod{\lambda(n)}$.

Earlier we used $st \equiv 1 \pmod{(p-1)(q-1)}$, which gives correct results, but could result in t being unnecessarily large, given s. In the special case $n = pq$ Carmichael's function is $\lambda(n) = \mathrm{lcm}(p-1, q-1)$ and the RSA cryptosystem is based on the fact that it is impossible to find t from $st \equiv 1 \pmod{\lambda(n)}$, if s is known but the primes p and q in n are unknown.

As of this time it is not known whether the RSA public-key cryptosystem (with additional conditions on the primes p and q) satisfies all of the requirements listed above, making the factorization of $n = pq$ infeasible and thus the cipher unbreakable.

McEliece has given an example of connecting special cyclic codes and cryptography to obtain a safe method for transmission of information. He uses a so-called Goppa code of length $n = 2^m$, which is a t-error-correcting code over $\mathbb{F}_{2^m}$, defined by the parity-check matrix

$$\mathbf{H} = \begin{pmatrix} g(\alpha_1)^{-1} & \cdots & g(\alpha_n)^{-1} \\ \alpha_1 g(\alpha_1)^{-1} & \cdots & \alpha_n g(\alpha_n)^{-1} \\ \cdots\cdots\cdots\cdots\cdots\cdots\cdots\cdots\cdots \\ \alpha_1^{t-1} g(\alpha_1)^{-1} & \cdots & \alpha_n^{t-1} g(\alpha_n)^{-1} \end{pmatrix}.$$

Here $\alpha_1, \ldots, \alpha_n$ are the elements of $\mathbb{F}_{2^m}$ and g is an irreducible polynomial of degree t over $\mathbb{F}_{2^m}$. From this we can find a generator matrix $\mathbf{G}$ (see Chapter 4), which is a $k \times n$ matrix where $k = n - mt$. The message $\mathbf{a}$ is encoded as $\mathbf{c} = \mathbf{a}\mathbf{G}$. $\mathbf{G}$ can be scrambled by using a random invertible $k \times k$ matrix $\mathbf{S}$ and a random $n \times n$ permutation matrix $\mathbf{P}$ and forming $\mathbf{G}' = \mathbf{SGP}$. This matrix $\mathbf{G}'$ is "published" but the matrices $\mathbf{S}, \mathbf{G}, \mathbf{P}$ are kept secret between sender and receiver. A message k-tuple a is enciphered as $\mathbf{u} = \mathbf{a}\mathbf{G}' + \mathbf{z}$, where $\mathbf{z}$ is a random vector containing t ones, chosen by the sender. To decipher the received word $\mathbf{u}$, one has to calculate $\mathbf{u}\mathbf{P}^{-1} = (\mathbf{aS})\mathbf{G} + \mathbf{z}\mathbf{P}^{-1}$. This is decoded by the t-error-correcting Goppa code to obtain $\mathbf{aS}$ and therefore $\mathbf{a}$. An unauthorized interceptor has the task of decoding a code defined by $\mathbf{G}'$, which is a large, random-looking code. SLOANE gives the numerical example that for $n = 2^{10}$, $t = 50$ there are about 10^{149} possible choices for $g(x)$ and an even larger number of ways of choosing $\mathbf{S}$ and $\mathbf{P}$. The dimension of such a code is at least $k = 524$ and the eavesdropper is faced with the problem of decoding an apparently random 50-error-correcting code of length 1024, a virtually impossible task.

In the literature on public-key cryptosystems several aspects are considered in detail, such as the authentication problem and signatures, extra precautions for security, how to calculate powers, and how to find suitable prime numbers.

In conclusion it should be stated that at present the most widely used cryptosystem is the data encryption standard (DES). It is anticipated that it will be safe for a number of years to come (see DIFFIE). To break it is one of the great challenges in cryptography.

DAVIES, PRICE and PARKIN mention that the principal advantage of the public-key cryptosystem for privacy in communication is that it avoids the need to distribute secret keys. The keys which are distributed are public keys and they can be associated with the intended recipient of the message so that, assuming the corresponding secret key used in decipherment is secure, the message can be deciphered only by the intended recipient. Since anyone, with the aid of the public-key, can produce an enciphered message, the cryptosystem in its simple form gives no information about the origin of the message and therefore message authentication is an important extra requirement.

The use of the public-key cryptosystem for signatures is effective only for those versions which introduce no redundancy in the ciphertext and therefore allow their inverses to be used as message transformations. For this subclass of public-key systems a form of authentication can be employed which gives the message an unforgeable transformation that associates it with the sender, since only he could give the transformation which can be inverted to produce the message by means of his public key. The two transformations can be combined, one producing secrecy and the other producing authentication. For signatures, the RSA system is a very good technique. So with the present state of knowledge one could advise use of the RSA system for exchanging or distributing keys with signed messages in combination with the DES or a similar algorithm for encipherment.

PROBLEMS

1. Find the multiplicative inverse of 791 modulo 3120.

2. Use an algorithm given in the Appendix to test 1987 for primality.

*3. Show that primitive roots modulo a prime p can be found by using Exercise 20, assuming the factors of $p - 1$ are known. Furthermore, in the case of $p = 2q + 1$, p and q primes, a primitive root modulo p can be found by calculating $p - b^2$, for a b between 1 and $p - 1$.

4. Let $q = 9291$, $y = (15, 92, 108, 279, 563, 1172, 2243, 4468)$, $k = 2393$. Use a trapdoor–knapsack algorithm with these parameters to encipher a binary message 1 0 1 1 0 0 0 1.

5. Write a computer program for solving a simple knapsack.

6. Let $y = (14, 28, 56, 82, 90, 132, 197, 284, 341, 455)$ be a knapsack vector. How many binary vectors can you find such that $a \cdot y = 515$? Repeat the question for 516 instead of 515.

7. Let $(1, 2, 2^2, \ldots, 2^{n-1})$ be a knapsack vector. For which values of K can we solve the corresponding knapsack problem with sum K?

*8. Let b be a primitive root modulo a prime b then $b^i \pmod{p}$ is easy to compute. Conversely, given $b^k \pmod{p}$ to find k is conjectured to be difficult. (This is called the *logarithm problem*.) Design a public-key cryptosystem that uses the logarithm problem for defining a one-way function.

*9. Given the knapsack vector $(a_1, \ldots, a_n)$, where $a_i > a_1 + \ldots + a_{i-1}$ for $i = 2, \ldots, n$, construct a linear-time algorithm for determining whether the knapsack problem $K = a_{i_1} + \ldots + a_{i_s}$ has a solution for given K. Here $a_{i_1}, \ldots, a_{i_s}$ form a suitable subset of the knapsack vector $a_1, \ldots, a_n$. If there is a solution, the algorithm should output it. Furthermore, show that this knapsack has at most one solution for any integer K.

10. Use $n = pq = 2867$, $s = 167$, in an RSA cryptosystem to encipher the message ALGORITHM.

*11. Describe an RSA cryptosystem which uses an integer $n = \prod_{i=1}^{r} p_i^{e_i}$, p_i primes, instead of $n = pq$. Which conditions do you have to impose on n?

12. How can the RSA cryptosystem be used for digital signatures and authentication?

13. Suppose $n = 2773$ in an RSA cryptosystem. Determine the rest of the public key under the assumption that it should be the smallest three-digit number possible.

*14. Rabin's public key cryptosystem uses an enciphering function which is not bijective. It is defined by enciphering a message M as

$$C \equiv M(M + b) \pmod{n},$$

where n is the product of two approximately equally large primes, M is the message to be enciphered, and b is an integer with $0 \le b < n$. The integers n and b are the public keys, the prime factors p and q of n are private. Show that for deciphering we have four solutions for M, given $C \pmod{n}$. Also show that it suffices to solve the two congruences

$$C \equiv M(M + b) \pmod{p}, \qquad C \equiv M(M + b) \pmod{q}$$

to determine a solution to the original congruence.

EXERCISES (Solutions in Chapter 8, p. 486)

1. Suppose a Caesar cipher has been used and a received word is XTWTEL?C. Determine the original message ("?" represents an unreadable letter).

2. Determine the number of different 2×2 matrix encipherings of $\mathbb{Z}_{26}$.

3. Find the complete enciphering of the word CRYPTOLOGY in 2.8.

*4. Decipher:
> Z UZMZGRZ RH Z KVIHLM DSL RH SRTSOB VMGSFHRZHGRX
> ZYLFG HLNVGSRMT RM DSRXS BLF ZIV MLG VEVM IVNLGVOB
> RMGVIVHGVW.

5. Decipher the following cipher message assuming that a cipher alphabet derived from a normal alphabet by a linear transformation was used:

R FX SNO FGYFRQ NG ONXNYNNZGNY R WFEL TLLS PLTOLYQFP FSQ R CNEL ONQFP.

Note: The frequency of the letters of the alphabet in a sample of 1000 letters in an English article is as follows:

E 130, T 93, N 78, R 77, I 74, O 74, A 73, S 63, D 44, H 35, L 35, C 30, F 28, P 27, U 27, M 25, Y 19, G 16, W 16, V 13, B 9, X 5, K 3, Q 3, J 2, Z 1.

6. Determine the values a which will make the matrix

$$\begin{pmatrix} 3 & 4 \\ a & 23 \end{pmatrix}$$

involutory, over $\mathbb{Z}_{26}$.

*7. The *Playfair cipher* utilizes a so-called *Polybius square*. The alphabet is written into the square (with I–J always combined), usually led off by a key word [PLAYF(A)IR in this example]:

$$\begin{array}{ccccc} P & L & A & Y & F \\ I & R & B & C & D \\ E & G & H & K & M \\ N & O & Q & S & T \\ U & V & W & X & Z \end{array}$$

To prepare the plaintext for encipherment, it is divided into digraphs. Where double letters could occur in the same digraph, a prearranged letter is inserted between the pair. Encipherment is performed for each digraph by using the Polybius square with the following rules:

(1) if the two letters in the plaintext digraph appear in the same line, substitute for each one the letter immediately to its right;

(2) if the two letters appear in the same column, substitute for each the letter immediately below it; Here the first column (row) is considered to be to the right (below) of the last column (row);

(3) if the two letters are not in the same row or column, substitute for them the letters at the remaining two corners of the rectangle or square which they form, using first the letter on the same *row* as the first plaintext letter.

Encipher: "Better late than never".

Deciphering simply uses the reverse of the enciphering steps. Decipher the following ciphertext (where two plaintext words—written above the cipher-text—are known):

```
N R P R U     O B Z D P     U W S E W     G G W H R
E WN R N     S M I N S     A W P L E     C P R C H
A WK D L     I B A B Q     A L B I N     S P I L O
                                 w o m     a n l  e a
E C MS C     R G NN Y     I D O L G     D W B L B
n i n g
S N I K L     I D V P O     U F Q U C     D S B V C
Z A B A C     I L A V D     S B T L B     Q E O A R
A T Z A E     B P WP E     A S H N T     B U T Z P
A R U H U     R O HC N     B Q B Q .
```

8. Determine all involutory matrices (mod 26) of the form

$$H = \begin{pmatrix} a & b \\ c & a \end{pmatrix} \quad \text{with } a^2 + bc \equiv 1 \ (\text{mod } 26).$$

9. Encode and decode ALGEBRA IS FUN by using the 2×2 cipher matrix

$$\begin{pmatrix} 2 & 23 \\ 3 & 1 \end{pmatrix}.$$

10. Find the determinant of

$$\begin{pmatrix} 1 & 0 & 2 \\ 2 & 3 & 2 \\ 0 & 1 & 0 \end{pmatrix},$$

where entries are in $\mathbb{Z}_q$. Also find the inverse of this matrix for odd q.

11. (i) Prove: A 2×2 matrix M with determinant congruent to -1 (mod 26) is involutory if and only if the sum of the elements in the main diagonal is congruent to 0 (mod 26).
    (ii) Find all involutory matrices (with determinant congruent to -1 (mod 26)) with entry $a_{11} = 2$.

12. (i) Prove that x^k is a permutation polynomial of $\mathbb{F}_q$ if and only if $(k, q - 1) = 1$.
    (ii) Given that $x^5 + x^3 + \frac{1}{3}x$ is a permutation polynomial of $\mathbb{F}_{27}$, use this polynomial to encipher the word GALOIS according to the procedure outlined before and in Example 2.12.

*13. Prove that the method of Remark 2.15 is correct.

14. Use $n = 2803$, $r = 2$ and $s = 113$ to encipher the word NO = 1415 according to 2.15.

15. Estimate the number of digits in the following prime numbers:

$$2^{127} - 1, \quad 180(2^{127} - 1)^2 + 1, \quad 2^{607} - 1, \quad 2^{3217} - 1, \quad 2^{19937} - 1, \quad 2^{44497} - 1.$$

16. Encipher ALGEBRA IS GREAT by using the method described in 2.13. Decipher the message (14800, 17152, 17162, 2044, 11913, 9668).

17. Use the knapsack scheme with $\mathbf{y} = (23, 57, 91, 179, 353)$, $n = p = 719$ and $k = 299$ to encipher the binary number 10110, and then decipher that number.

18. Use the multiplicative knapsack scheme with $\mathbf{y} = (2, 3, 5, 7)$, $n = p = 223$ and $b = 11$, to encipher the binary number 0101.

19. Find a numerical example which shows that the statement "$a^{n-1} \equiv 1 \ (\text{mod } n)$ and $a \not\equiv 0 \ (\text{mod } n)$ implies that n is prime" is not true.

20. Give numerical examples for the following test: If for each prime p such that p divides $n - 1$, there exists an a such that $a^{(n-1)/p} \not\equiv 1 \ (\text{mod } n)$ and $a^{n-1} \equiv 1 \ (\text{mod } n)$, then p is prime.

*21. An integer N satisfying $b^{N-1} \equiv 1 \ (\text{mod } N)$ is called a *base b pseudoprime*. Show that there are infinitely many composite base b pseudoprimes. Verify that 341 is a base 2 pseudoprime, also that 91 is a base 3 pseudoprime.

Moreover, N is called a *strong base b pseudoprime* if $N = 2^s t + 1$, for odd t, implies that either $b^t \equiv 1 \pmod{N}$ or $b^{t2^r} \equiv -1 \pmod{N}$, for some $0 \le r < s$. Give an example of such a strong base b pseudoprime. Note that base b pseudoprimes are used in Rabin's primality test, see Appendix B.

22. The *Jacobi symbol* $J(a, n)$ is defined as follows

$$
J(a, n) = \begin{cases}
1 & \text{if } a = 1, \\
J(a/2, n)(-1)^{(n^2-1)/8} & \text{if } a \text{ is even}, \\
J(n \pmod{a}, a)(-1)^{(a-1)(n-1)/4} & \text{if } a > 1 \text{ is odd}.
\end{cases}
$$

Evaluate $J(6, 13)$.

23. Use the following test due to Solovay and Strassen in a numerical example to show that a given number n is not prime. [This is an example of a probabilistic algorithm for checking primality. If the test has been passed t times then the remaining probability that n is nonprime is less than 2^{-t}.] Choose an integer a at random from $1, \ldots, n - 1$, and then test to see if $(a, n) = 1$ and $a^{(n-1)/2} \equiv J(a, n) \pmod{n}$. If n is prime then it always passes the test. If n is not prime it will pass the test for less than half of the values of a which are coprime to n. (See Appendix B for a computer program.)

24. Use the following algorithm to calculate $a^2 \pmod{n}$ for $a = 116, s = 27, n = 39$.
    (a) Let $b_k b_{k-1} \ldots b_1 b_0$ be the binary representation of s.
    (b) Set $c = 1$.
    (c) Set $i = k$.
    (d) Set c equal to the remainder when c^2 is divided by n.
    (e) If $b_i = 1$ then set c equal to the remainder when ca is divided by n.
    (f) Decrease i by 1.
    (g) If $i < 0$ stop, otherwise go to step (d).

25. Let $n = pq = 61 \cdot 53 = 3233$ and $s = 37$. Find t such that $st \equiv 1 \pmod{(p-1)(q-1)}$. Use the enciphering key $(t, 3233)$ to encipher the first two letters of the message A L G E B R A.

§3. Linear Recurring Sequences

Linear recurring sequences and properties of finite fields are closely related. Examples of applications of linear recurring sequences occur in radar and communications systems. Radar is an abbreviation for "radio detection and ranging". It operates by the emission of electromagnetic energy with radio frequency, which is reflected by an object "hit" by radar waves. One possible way of measuring the distance of an object (aeroplane, satellite, etc.) from a radar station consists in emitting a long, periodic sequence of signals of electromagnetic energy which is reflected back by the object and received by the station. The time t, which elapses between transmission of the signals from the radar station and their return after reflection from the object, can be used to calculate the distance of the object from the station. For a discrete system with "period" r we must have $rc > t$, where c is the time unit used. If the transmitter decreases the time interval between sending signals below this elapsed-time between transmission and reception, a condition called "range ambiguity" results, making it impossible to accurately determine the actual range of the target without further information being available. It is also desirable that the transmitted pulses be of short duration. One therefore tries to use encoding techniques for transmitting signals such that these signals may be decoded without range ambiguity and with an increase in the total on-time for the transmitter. So-called maximal-length linear recurring sequences (or M-sequences) present possibly a method for encoding the transmitter for a tracking radar. These sequences have very long "periods". They may be generated by shift registers.

Let $f \in \mathbb{F}_q[x]$ be an irreducible polynomial with $(q^n - 1)$th roots of unity as its roots. In calculating the formal power series

$$\frac{1}{f} = \sum_{i=0}^{\infty} c_i x^i$$

the c_i are in $\mathbb{F}_q$, and they form a pseudo-random-sequence of elements of $\mathbb{F}_q$. $c = (c_0, c_1, \ldots)$ is periodic with period $q^n - 1$. Within a period every subsequence of length n with elements in $\mathbb{F}_q$ occurs exactly once, except the null-sequence.

Consider the primitive polynomial $f = x^3 + x + 1$ over $\mathbb{F}_2$. We find $1 + x^7 = (x^3 + x + 1)(x^3 + x^2 + 1)(x + 1)$ and thus

$$\frac{1}{f} = (1 + x + x^2 + x^4)(1 + x^7 + x^{14} + x^{24} + \ldots),$$

i.e.

$$\frac{1}{f} = (1 + 1x + 1x^2 + 0x^3 + 1x^4 + 0x^5 + 0x^6)\frac{1}{1 + x^7}.$$

The sequence of coefficients c_i in this case is periodic with period 7, the first coefficients are 1110100.

We construct such sequences which are of interest in various branches of information theory and electrical engineering. As before, $\mathbb{F}_q$ denotes the finite field with $q = p^t$ elements.

3.1 Definition. Let $k \in \mathbb{N}$. A sequence $s_0, s_1, \ldots$ of elements in $\mathbb{F}_q$ is called a *linear recurring sequence* of order k in $\mathbb{F}_q$, if there are elements $a, a_0, a_1, \ldots, a_{k-1} \in \mathbb{F}_q$,

$$s_{n+k} = a_{k-1}s_{n+k-1} + a_{k-2}s_{n+k-2} + \ldots + a_0 s_n + a. \tag{1}$$

The terms $s_0, s_1, \ldots, s_{k-1}$ determining the sequence are called *initial values*. The equation (1) is called a *linear recurrence relation* of order k.

Sometimes also the expressions "difference equations" or "linear recursive relation" are used for (1). In the case $a = 0$ the linear recurrence relation (1) is called *homogeneous,* otherwise it is called *inhomogeneous.* The vector $s_n = (s_n, s_{n+1}, \ldots, s_{n+k-1})$ is called an nth *state vector,* and $s_0 = (s_0, s_1, \ldots, s_{k-1})$ is called an *initial state vector.* It is a characteristic of linear recurring sequences in $\mathbb{F}_q$ that they are periodic.

As was indicated in Chapter 4, §2, the generation of linear recurring sequences can be implemented on a feedback shift register. The building blocks of a feedback shift register are delay elements, adders (see Chapter 4, §2 for a description) and constant multipliers and adders, which multiply and add constant elements of $\mathbb{F}_q$ to the input. In the case of binary homogeneous linear recurring sequences one only needs delay elements, adders and wire connections for their implementation. A feedback shift register that generates a linear recurring sequence (1) is given by

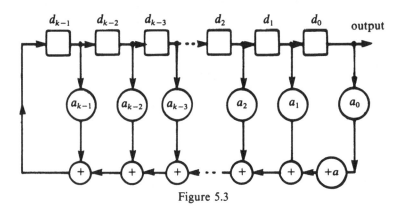

Figure 5.3

At the start each delay element $d_j, j = 0, 1, \ldots, k - 1$, contains the initial values s_j. After one time unit, each d_j will contain s_{j+1}. The output of the feedback shift register is the string of elements $s_0, s_1, s_2, \ldots$ received in intervals of one time unit.

3.2 Example. In order to generate a linear recurring sequence in $\mathbb{F}_5$, which satisfies the homogeneous linear recurrence relation

$$s_{n+5} = s_{n+4} + 2s_{n+3} + s_{n+1} + 3s_n \quad \text{for } n = 0, 1, \ldots,$$

we may use the feedback shift register (see Figure 5.4).   □

3.3 Definition. Let S be an arbitrary set, and let $s_0, s_1, \ldots$ be a sequence in S. The sequence is called *ultimately periodic* if there are numbers $r \in \mathbb{N}$ and $n_0 \in \mathbb{N}_0$, such that $s_{n+r} = s_n$ for all $n \geq n_0$. The number r is called the *period* of the sequence, and the smallest of all possible periods of an ultimately periodic sequence is called the *least period* of the sequence. An ultimately periodic sequence with least period r is called *periodic* if $s_{n+r} = s_n$ holds for all $n = 0, 1, \ldots$. The number n_0 is called *preperiod.*

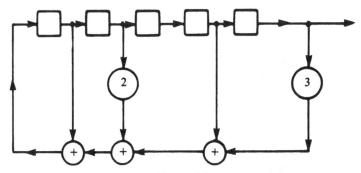

Figure 5.4

It can be verified immediately that a sequence $s_0, s_1, \ldots$ is periodic if and only if there exists an $r > 0$ such that $s_{n+r} = s_n$ for $n = 0, 1, \ldots$.

3.4 Theorem. *Every linear recurring sequence of order k in $\mathbb{F}_q$ is ultimately periodic with least period $r \leq q^k$; and if the sequence is homogeneous, then $r \leq q^k - 1$.*

3.5 Examples. (i) $s_0, s_1, \ldots \in \mathbb{F}_{p^k}$, where $s_{n+1} = s_n + 1$ for $n = 0, 1, \ldots$ has least period $p^k - 1 =: q - 1$.

(ii) It can be shown that for homogeneous linear recurring sequences of first order in $\mathbb{F}_q$ the least period is a divisor of $q - 1$. For $k \geq 2$ the least period of a kth order homogeneous linear recurring sequence need not divide $q^k - 1$. Let $s_0, s_1, \ldots \in \mathbb{F}_5$ with $s_0 = 0$, $s_1 = 1$ and $s_{n+2} = s_{n+1} + s_n$ for $n = 0, 1, \ldots$. The least period for this sequence is 20. $\square$

3.6 Theorem. *If $s_0, s_1, \ldots$ is a linear recurring sequence in $\mathbb{F}_q$ defined by (1) with $a_0 \neq 0$, then the sequence is periodic.*

PROOF. Exercise.

Let $\mathbf{s} = (s_0, s_1, \ldots)$ be a homogeneous recurring sequence of order k in $\mathbb{F}_q$ satisfying the relation

$$s_{n+k} = a_{k-1}s_{n+k-1} + a_{k-2}s_{n+k-2} + \ldots + a_0 s_n \tag{2}$$

for $n = 0, 1, \ldots, a_j \in \mathbb{F}_q, 0 \leq j \leq k - 1$. With this sequence we associate a $k \times k$ matrix

$$\mathbf{A} = \begin{pmatrix} 0 & 0 & 0 & \ldots & 0 & a_0 \\ 1 & 0 & 0 & \ldots & 0 & a_1 \\ 0 & 1 & 0 & \ldots & 0 & a_2 \\ \multicolumn{6}{c}{\dotfill} \\ 0 & 0 & 0 & \ldots & 1 & a_{k-1} \end{pmatrix}. \tag{3}$$

For $k = 1$ let $\mathbf{A} = (a_0)$.

The proof of 3.7 is straightforward.

3.7 Theorem. *Let* $s_0, s_1, \ldots$ *be a homogeneous linear recurring sequence in* $\mathbb{F}_q$ *satisfying* (2), *and let* **A** *be the associated matrix of the form* (3). *Then the state vector of the sequence satisfies*

$$\mathbf{s}_n = \mathbf{s}_0 \mathbf{A}^n \quad \text{for } n = 0, 1, \ldots .$$ □

3.8 Theorem. *Let* $s_0, s_1, \ldots$ *be a homogeneous linear recurring sequence in* $\mathbb{F}_q$ *satisfying* (2) *with* $a_0 \neq 0$. *Then the least period of the sequence is a divisor of the order of the associated matrix* **A** *in* $\mathrm{GL}(k, \mathbb{F}_q)$.

PROOF. We have $\det \mathbf{A} = (-1)^{k-1} a_0 \neq 0$, thus **A** is nonsingular. Let m be the order of **A**, then Theorem 3.7 implies $\mathbf{s}_{n+m} = \mathbf{s}_0 \mathbf{A}^{n+m} = \mathbf{s}_0 \mathbf{A}^n = \mathbf{s}_n$ for all $n \geq 0$. Therefore m is a period of the given sequence. The proof that m is divisible by the least period is left as an exercise. □

3.9 Definition. Let $s_0, s_1, \ldots$ be a homogeneous linear recurring sequence of order k in $\mathbb{F}_q$ satisfying (2). Then the polynomial

$$f = x^k - a_{k-1}x^{k-1} - a_{k-2}x^{k-2} - \ldots - a_0 \in \mathbb{F}_q[x]$$

is called the *characteristic polynomial* of the sequence. The polynomial

$$f^* = 1 - a_{k-1}x - a_{k-2}x^2 - \ldots - a_0 x^k \in \mathbb{F}_q[x]$$

is called the *reciprocal characteristic polynomial.* We have $f^* = x^k (f \circ x^{-1})$.

It can be shown that the characteristic polynomial f is also the characteristic polynomial of the associated matrix, so that, in the sense of linear algebra,

$$f = \det(x\mathbf{I} - \mathbf{A}).$$

We can evaluate the terms of a linear recurring sequence by using the characteristic polynomial.

3.10 Theorem. *Let* $s_0, s_1, \ldots$ *be a homogeneous linear recurring sequence of order* k *over* $\mathbb{F}_q$ *with characteristic polynomial* f. *If all zeros* $\alpha_1, \ldots, \alpha_k$ *of* f *are distinct then*

$$s_n = \sum_{j=1}^{k} \beta_j \alpha_j^n \quad \text{for } n = 0, 1, \ldots$$

where the elements β_j *are uniquely determined by the initial values of the sequence. Moreover, they are elements of the splitting field of* f *over* $\mathbb{F}_q$.

PROOF. The elements $\beta_1, \ldots, \beta_k$ can be determined from the system of linear equations

$$\sum_{j=1}^{k} \alpha_j^n \beta_j = s_n. \tag{4}$$

Cramer's rule shows that $\beta_1, \ldots, \beta_k$ are elements in the splitting field $\mathbb{F}_q(\alpha_1, \ldots, \alpha_k)$ of f over $\mathbb{F}_q$. Since the determinant of (4) is the VanderMonde determinant, which is nonzero, the elements β_j are uniquely determined. Direct substitution shows that the elements s_n in (4) satisfy the recurrence relation (2). $\qquad \square$

3.11 Example. Consider the linear recurring sequence $s_0, s_1, \ldots$ in $\mathbb{F}_q$, defined by $s_0 = s_1 = 1$ and $s_{n+2} = s_{n+1} + s_n$ for $n = 0, 1, \ldots$. The characteristic polynomial is $f = x^2 - x - 1 = x^2 + x + 1 \in \mathbb{F}_2[x]$. Let $\mathbb{F}_{2^2} = \mathbb{F}_2(\alpha)$, then $\alpha_1 = \alpha$ and $\alpha_2 = 1 + \alpha$ are zeros of f. Then we obtain $\beta_1 + \beta_2 = 1$ and $\beta_1 \alpha + \beta_2(1 + \alpha) = 1$, thus $\beta_1 = \alpha$ and $\beta_2 = 1 + \alpha$. Theorem 3.10 implies $s_n = \alpha^{n+1} + (1 + \alpha)^{n+1}$ for all $n \geq 0$. We have $\beta^3 = 1$ for all $0 \neq \beta \in \mathbb{F}_{2^2}$, and therefore $s_{n+3} = s_n$ for all $n \geq 0$, which agrees with the fact that the least period of the sequence is 3. $\qquad \square$

Some rather exhausting comparisons of coefficients prove

3.12 Theorem. *Let $s_0, s_1, \ldots$ be a homogeneous linear recurring sequence of order k and period n over $\mathbb{F}_q$ satisfying (2). Let f be its characteristic polynomial. Then*

$$fs = (1 - x^r)h, \qquad (5)$$

with $s = s_0 x^{r-1} + s_1 x^{r-2} + \ldots + s_{r-1} \in \mathbb{F}_q[x]$ and

$$h = \sum_{j=0}^{k-1} \sum_{i=0}^{k-1-j} a_{i+j+1} s_i x^j \in \mathbb{F}_q[x], \qquad a_k = -1. \qquad \square$$

We recall from 3.3.9 that the order of $f \in \mathbb{F}_q[x]$ is the least natural number e such that $f | x^e - 1$, if f is not divisible by x. The case $x | f$ can be dealt with separately.

3.13 Theorem. *Let $s_0, s_1, \ldots$ be a homogeneous linear recurring sequence in $\mathbb{F}_q$ with characteristic polynomial $f \in \mathbb{F}_q[x]$. Then the least period of the sequence is a divisor of the order of f.*

PROOF. It is easily verified that the order of f equals the order of the associated matrix (3). The result then follows from 3.8. $\qquad \square$

3.14 Theorem. *Let $s_0, s_1, \ldots$ be a homogeneous linear recurring sequence in $\mathbb{F}_q$ with nonzero initial state vector and irreducible characteristic polynomial f over $\mathbb{F}_q$ not divisible by x. Then the sequence is periodic with least period equal to the order of f.*

PROOF. Theorem 3.13 shows that the sequence is periodic with least period r being a divisor of the order of f. On the other hand (5) implies $f | (x^r - 1)h$. Since $s, h \neq 0$ and $\deg h < \deg f$, the irreducibility of f ensures that $f | x^r - 1$ and thus $r \geq$ order of f. $\qquad \square$

In Chapter 3, §3 we saw that the order of an irreducible polynomial f of degree k over $\mathbb{F}_q$ is a divisor of $q^k - 1$. Hence 3.4 implies that the least period of a homogeneous linear recurring sequence in $\mathbb{F}_q$ can be at most $q^k - 1$. In order to construct sequences with least period equal to $q^k - 1$, we use primitive polynomials (see 3.3.12). Such sequences with "long" periods are of particular importance in applications.

3.15 Definition. A homogeneous linear recurring sequence in $\mathbb{F}_q$ is called a *sequence with maximal period*, if its characteristic polynomial is primitive over $\mathbb{F}_q$ and its initial state vector is nonzero.

3.16 Theorem. *A sequence of order k with maximal period in $\mathbb{F}_q$ is periodic with least period $q^k - 1$.*

PROOF. We need to know (see 3.3.13) that a monic polynomial f of degree k over $\mathbb{F}_q$ is primitive if and only if $f(0) = 0$ and the order of f is $q^k - 1$. The result then follows from 3.14 and 3.4.   □

3.17 Example. The sequence $s_{n+7} = s_{n+4} + s_{n+3} + s_{n+2} + s_n, n = 0, 1, \ldots$ in $\mathbb{F}_2$ has the characteristic polynomial $f = x^7 - x^4 - x^3 - x^2 - 1 \in \mathbb{F}_2[x]$. Since f is primitive over $\mathbb{F}_2$, any sequence defined by the linear recurrence equation and having nonzero initial state vector is a sequence of maximal period. The least period of such sequences is $2^7 - 1 = 127$. All possible nonzero vectors of $\mathbb{F}_2^7$ are state vectors of this sequence. Any other sequence of maximal period which can be obtained from the given linear recurrence is a "shifted" copy of the sequence $s_0, s_1, \ldots$.   □

Next we formalize the approach indicated at the beginning of this section by using the concept of formal power series (see Chapter 3, §1).

3.18 Definition. Let $s_0, s_1, \ldots$ be an arbitrary sequence of elements in $\mathbb{F}_q$, then the formal power series

$$G = s_0 + s_1 x + s_2 x^2 + \ldots + s_n x^n + \ldots = \sum_{n=0}^{\infty} s_n x^n$$

is called the *generating function associated with* $s_0, s_1, \ldots$.

We know that power series form an integral domain with respect to the operations addition and multiplication. It can be shown that the power series $\sum_{n=0}^{\infty} b_n x^n \in \mathbb{F}_q[[x]]$ has a multiplicative inverse if and only if $b_0 \neq 0$.

3.19 Theorem. *Let $s_0, s_1, \ldots$ be a homogeneous linear recurring sequence of order k in $\mathbb{F}_q$ satisfying (2), let f^* be its reciprocal characteristic polynomial in $\mathbb{F}_q[x]$ and let $G \in \mathbb{F}_q[[x]]$ be its generating function. Then*

$$G = \frac{g}{f^*}, \tag{6}$$

where

$$g = -\sum_{j=0}^{k-1} \sum_{i=0}^{j} a_{i+k-j} s_i x^j \in \mathbb{F}_q[x], \quad a_k := -1.$$

Conversely, if g is an arbitrary polynomial over $\mathbb{F}_q$ *of degree less than k and if* $f^* \in \mathbb{F}_q[x]$ *is as in Definition 3.9, then the formal power series G defined by* (6) *is the generating function of a homogeneous linear recurring sequence of order k in* $\mathbb{F}_q$, *which satisfies* (2).

PROOF. We have

$$f^*G = g - \sum_{j=k}^{\infty} \left(\sum_{i=0}^{k} a_i s_{j-k+i} \right) x^j. \tag{7}$$

If the sequence $s_0, s_1, \ldots$ satisfies (2) then $f^*G = g$ holds, because $\sum_{i=0}^{k} a_i s_{n+i} = 0$. Since f^* has a multiplicative inverse, equation (6) follows.

Conversely, (7) implies that f^*G can only be a polynomial of degree $< k$ if

$$\sum_{i=0}^{k} a_i s_{j-k+i} = 0 \quad \text{for all } j \geq k.$$

This means that the sequence of coefficients $s_0, s_1, \ldots$ of G satisfies (2). $\square$

Theorem 3.19 implies that there is a one-to-one correspondence between the homogeneous linear recurring sequence of order k with reciprocal characteristic polynomial f^* and the rational functions g/f^* with degree $g < k$.

3.20 Example. Consider $s_{n+4} = s_{n+3} + s_{n+1} + s_n, n = 0, 1, \ldots$, in $\mathbb{F}_2$. The reciprocal characteristic polynomial is $f^* = 1 + x + x^3 + x^4 \in \mathbb{F}_2[x]$. For the initial state vector $(1, 1, 0, 1)$ we obtain $g = 1 + x^2$ from Theorem 3.19. Thus the generating function G is of the form

$$G = \frac{1 + x^2}{1 + x + x^3 + x^4} = 1 + x + x^3 + x^4 + x^6 + \ldots,$$

which corresponds to the binary sequence $1101101 \ldots$ with least period 3. $\square$

Theorem 3.16 has practical applications in systems of information transmission, in locating satellites. For instance, a primitive polynomial of degree 20 over $\mathbb{F}_2$, such as $x^{20} + x^3 + 1$, generates a periodic sequence of period $2^{20} - 1$, which is approximately of length 10^6, just suitable for radar observations of the moon. For satellite communication systems, primitive polynomials of degree 50 or higher must be used.

There is an interesting connection between linear recurring sequences and cryptography, given by stream ciphers. A *stream cipher* consists of a key which is fed to an algorithm for the generation of an infinite sequence. A further connection with another concept of algebra-applications is established, since this algorithm is an example of a finite state machine. Stream ciphers provide one of the most important methods of modern encipherment. There are two major requirements: the input sequence must have a guaranteed minimum length for its period; the ciphertext must appear to be random. Then we only encipher messages which are shorter than the period of the sequence.

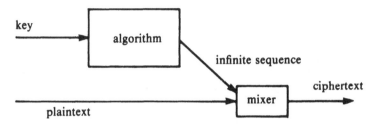

Figure 5.5. Diagram of a stream cipher

Details of stream ciphers are given in BEKER and PIPER.

A linear recurring sequence is defined by a specific linear recurrence relation, but satisfies many other linear recurrence relations as well.

The next theorem establishes the connection between the various linear recurrence relations valid for a given homogeneous linear recurring sequence.

3.21 Theorem. *Let $s_0, s_1, \ldots$ be a homogeneous linear recurring sequence in $\mathbb{F}_q$. Then there exists a uniquely determined monic polynomial $m \in \mathbb{F}_q[x]$ with the following property: a monic polynomial $f \in \mathbb{F}_q[x]$ of positive degree is a characteristic polynomial of $s_0, s_1, \ldots$ if and only if m divides f.*

PROOF. Let $f_0 \in \mathbb{F}_q[x]$ be the characteristic polynomial of a homogeneous linear recurrence relation satisfied by the sequence, and let $h_0 \in \mathbb{F}_q[x]$ be the polynomial h in Theorem 3.12 determined by f_0 and the sequence. If d is the (monic) greatest common divisor of f_0 and h_0, then we can write $f_0 = md$ and $h_0 = bd$ with $m, b \in \mathbb{F}_q[x]$. We shall prove that m is the desired polynomial. Clearly, m is monic. Now let $f \in \mathbb{F}_q[x]$ be an arbitrary characteristic polynomial of the given sequence, and let $h \in \mathbb{F}_q[x]$ be the polynomial h in Theorem 3.12 determined by f and the sequence. By applying Theorem 3.19, we find that the generating function G of the sequence satisfies

$$G = \frac{g_0}{f_0^*} = \frac{g}{f^*},$$

with g_0 and g determined by the expression following (6).

If $s_0, s_1, \ldots$ is periodic with period r then the generating function G can be written as

$$G = \frac{s^*(x)}{1 - x^r},$$

where $s^*(x) = s_0 + s_1 x + \ldots + s_{r-1} x^{r-1}$. From Theorem 3.19 we know that $G = g/f^*$. Equating these expressions for G we obtain

$$f^* s^* = (1 - x^r) g.$$

If f and s are as in (5) (Theorem 3.12) then

$$f(x) s(s) = x^k f^*\left(\frac{1}{x}\right) x^{r-1} s^*\left(\frac{1}{x}\right) = (x^r - 1) x^{k-1} g\left(\frac{1}{x}\right).$$

Comparison of h in Theorem 3.12 and g in Theorem 3.19 shows that

$$x^{k-1} g\left(\frac{1}{x}\right) = -h(x). \tag{8}$$

Therefore $g f_0^* = g_0 f^*$. We have

$$h(x) f_0(x) = -x^{\deg(f(x))-1} g\left(\frac{1}{x}\right) x^{\deg(f_0(x))} f_0^*\left(\frac{1}{x}\right)$$

$$= -x^{\deg(f_0(x))-1} g_0\left(\frac{1}{x}\right) x^{\deg(f(x))} f^*\left(\frac{1}{x}\right)$$

$$= h_0(x) f(x).$$

After division by d we have $hm = bf$, and since m and b are relatively prime, it follows that m divides f.

Now suppose that $f \in \mathbb{F}_q[x]$ is a monic polynomial of positive degree which is divisible by m, say $f = mc$ with $c \in \mathbb{F}_q[x]$. Passing to reciprocal polynomials, we get $f^* = m^* c^*$ in an obvious notation. We also have $h_0 m = b f_0$, so that, using the relation (8), we obtain

$$g_0(x) m^*(x) = -x^{\deg(f_0(x))-1} h_0\left(\frac{1}{x}\right) x^{\deg(m(x))} m\left(\frac{1}{x}\right)$$

$$= -x^{\deg(m(x))-1} b\left(\frac{1}{x}\right) x^{\deg(f_0(x))} f_0\left(\frac{1}{x}\right).$$

Since $\deg(b) < \deg(m)$, the product of the first two factors on the right-hand side (negative sign included) is a polynomial $a \in \mathbb{F}_q[x]$.

Therefore, we have $g_0 m^* = a f_0^*$. It follows then from Theorem 3.19 that the generating function G of the sequence satisfies

$$G = \frac{g_0}{f_0^*} = \frac{a}{m^*} = \frac{a c^*}{m^* c^*} = \frac{a c^*}{f^*}.$$

Since $\deg(ac^*) = \deg(a) + \deg(c^*) < \deg(m) + \deg(c) = \deg(f)$,   the second part of Theorem 3.19 shows that f is a characteristic polynomial of the sequence. It is clear that there can only be one polynomial m with the indicated properties. $\square$

3.22 Definition. The uniquely determined polynomial $m(x)$ over $\mathbb{F}_q$ associated with the sequence $s_0, s_1, \ldots$ as described in Theorem 3.21 is called the *minimal polynomial* of the sequence.

If $s_n = 0$ for all $n \geq 0$, the minimal polynomial is equal to the constant polynomial 1. For all other homogeneous linear recurring sequences, m is that monic polynomial with $\deg(m) > 0$ which is the characteristic polynomial of the linear recurrence relation of least possible order satisfied by the sequence.

3.23 Example. Let $s_0, s_1, \ldots$ be the linear recurring sequence in $\mathbb{F}_2$ with $s_{n+4} = s_{n+3} + s_{n+1} + s_n$, $n = 0, 1, \ldots$, and initial state vector $(1, 1, 0, 1)$. To find the minimal polynomial, we proceed as in the proof of Theorem 3.21. We may take $f_0 = x^4 - x^3 - x - 1 = x^4 + x^3 + x + 1 \in \mathbb{F}_2[x]$. Then by Theorem 3.12 the polynomial h_0 is given by $h_0 = x^3 + x$. The greatest common divisor of f_0 and h_0 is $d = x^2 + 1$, and so the minimal polynomial of the sequence is $m = f_0/d = x^2 + x + 1$. One checks easily that the sequence satisfies the linear recurrence relation $s_{n+2} = s_{n+1} + s_n$, $n = 0, 1, \ldots$, as it should according to the general theory. We note that $\text{ord}(m) = 3$, which is identical with the least period of the sequence. We shall see in Theorem 3.24 below that this is true in general. $\square$

The minimal polynomial plays an important role in the determination of the least period of a linear recurring sequence. This is shown in the following result.

3.24 Theorem. *Let $s_0, s_1, \ldots$ be a homogeneous linear recurring sequence in $\mathbb{F}_q$ with minimal polynomial $m \in \mathbb{F}_q[x]$. Then the least period of the sequence is equal to $\text{ord}(m)$.*

PROOF. If r is the least period of the sequence and n_0 is its preperiod, then we have $s_{n+r} = s_n$ for all $n \geq n_0$. Therefore, the sequence satisfies the homogeneous linear recurrence relation $s_{n+n_0+r} = s_{n+n_0}$ for $n = 0, 1, \ldots$. Then, according to Theorem 3.21, m divides $x^{n_0+r} - x^{n_0} = x^{n_0}(x^r - 1)$, so that m is of the form $m = x^h g$ with $h \leq n_0$ and $g \in \mathbb{F}_q[x]$, where $g(0) \neq 0$ and g divides $x^r - 1$. It follows from the definition of the order of a polynomial that $\text{ord}(m) = \text{ord}(g) \leq r$. On the other hand, r divides $\text{ord}(m)$ by Theorem 3.13, and so $r = \text{ord}(m)$. $\square$

3.25 Example. Let $s_0, s_1, \ldots$ be the linear recurring sequence in $\mathbb{F}_2$ with $s_{n+5} = s_{n+1} + s_n$, $n = 0, 1, \ldots$, and initial state vector $(1, 1, 1, 0, 1)$. Following the method in the proof of Theorem 3.21, we take $f_0 = x^5 - x - 1 = x^5 + x + 1 \in \mathbb{F}_2[x]$ and get $h_0 = x^4 + x^3 + x^2$ from Theorem 3.12. Then $d = x^2 + x + 1$, and so the minimal polynomial m of the sequence is given by $m = f_0/d = x^3 + x^2 + 1$. We have $\operatorname{ord}(m) = 7$, and so Theorem 3.24 implies that the least period of the sequence is 7. □

The details in the example above show how to find the least period of a linear recurring sequence without evaluating its terms. The method is particularly effective if a table of the order of polynomials is available.

3.26 Theorem. *Let $s_0, s_1, \ldots$ be a homogeneous linear recurring sequence in $\mathbb{F}_q$ and let b be a positive integer. Then the minimal polynomial m_1 of the shifted sequence $s_b, s_{b+1}, \ldots$ divides the minimal polynomial m of the original sequence. If $s_0, s_1, \ldots$ is periodic, then $m_1 = m$.*

PROOF. To prove the first part, it is sufficient to show because of Theorem 3.21 that every homogeneous linear recurrence relation satisfied by the original sequence is also satisfied by the shifted sequence. But this is obvious. For the second part, let $s_{n+b+k} = a_{k-1}s_{n+b+k-1} + \ldots + a_0 s_{n+b}$, $n = 0, 1, \ldots$, be a homogeneous linear recurrence relation satisfied by the shifted sequence. Let r be a period of $s_0, s_1, \ldots$, so that $s_{n+r} = s_n$ for all $n \geq 0$, and choose an integer c with $cr \geq b$. Then, by using the linear recurrence relation with n replaced by $n + cr - b$ and invoking the periodicity property, we find that $s_{n+k} = a_{k-1}s_{n+k-1} + \ldots + a_0 s_n$ for all $n \geq 0$, i.e., that the sequence $s_0, s_1, \ldots$ satisfies the same linear recurrence relation as the shifted sequence. By again applying Theorem 3.21, we conclude that $m_1 = m$. □

3.27 Theorem. *Let $f \in \mathbb{F}_q[x]$ be monic and irreducible over $\mathbb{F}_q$, and let $s_0, s_1, \ldots$ be a homogeneous linear recurring sequence in $\mathbb{F}_q$ not all of whose terms are 0. If the sequence has f as a characteristic polynomial, then the minimal polynomial of the sequence is equal to f.*

PROOF. Since, according to 3.21, the minimal polynomial m of the sequence divides f, the irreducibility of f implies that either $m = 1$ or $m = f$. But $m = 1$ holds only for the sequence all of whose terms are 0, and so the result follows. □

There is a general criterion for deciding whether the characteristic polynomial of the linear recurrence relation defining a given linear recurring sequence is already the minimal polynomial of the sequence. For a proof see LIDL and NIEDERREITER.

3.28 Theorem. *Let $s_0, s_1, \ldots$ be a sequence in $\mathbb{F}_q$ satisfying a kth order homogeneous linear recurrence relation with characteristic polynomial $f \in \mathbb{F}_q[x]$. Then f is the minimal polynomial of the sequence if and only if the state vectors $\mathbf{s}_0, \mathbf{s}_1, \ldots, \mathbf{s}_{k-1}$ are linearly independent over $\mathbb{F}_q$.* □

3.29 Corollary. *If $s_0, s_1, \ldots$ is an impulse response sequence for some homogeneous linear recurrence relation in $\mathbb{F}_q$, then its minimal polynomial is equal to the characteristic polynomial of that linear recurrence relation.*   □

(For the definition of an impulse response sequence, cf. Exercise 2.)

PROBLEMS

1. Determine a binary pseudo-random-number of period $2^{15} - 1$ by using a linear recurring sequence.

2. Determine the linear recurring sequence which generates 110001001101011 and design a suitable linear shift register.

3. An n-stage shift register produces a sequence of state vectors $(s_0(t), s_1(t), \ldots, s_{n-1}(t))$, $t = 0, 1, 2, \ldots$, such that $s_i(t + 1) = s_{i+1}(t)$ for $i = 0, 1, \ldots, n - 1$ and $s_{n-1}(t + 1) = f(s_0(t), s_1(t), \ldots, s_{n-1}(t))$. If $s_t := s_0(t)$, then we obtain the shift register sequence $s_0 s_1 s_2 \ldots$ and the function f is called the *feedback function* of the register. The shift register is nonlinear if f is nonlinear. f can be regarded as a Boolean polynomial on $s_0, s_1, \ldots, s_{n-1}$, its values can be given in form of a table (truth table).
   (i) How many feedback functions are there for an n-stage shift register?
   (ii) Let the inputs and outputs be as follows

	Input		Output
s_0	s_1	s_2	s_3
0	0	0	1
0	0	1	0
0	1	0	0
0	1	1	0
1	0	0	0
1	0	1	0
1	1	0	0
1	1	1	1

Find the corresponding f as a polynomial in s_0, s_1, s_2.

4. Design feedback shift registers which (i) multiply polynomials in $\mathbb{F}_q[x]$, (ii) divide polynomials in $\mathbb{F}_q[x]$. Design a feedback shift register which multiplies and divides polynomials in $\mathbb{F}_q[x]$.

*5. Design shift registers that compute discrete logarithms (and their inverses, antilogarithms) for $\mathbb{F}_{3^4}$. (See Chapter 3, §2 for the definition of discrete logarithms.)

6. Design a shift register that divides $x^5 + x^4 + 3x^2 + x$ by $x^3 + x^2 + 1$ in $\mathbb{F}_3[x]$.

*7. Let $\alpha_1, \ldots, \alpha_m$ be the distinct roots of the characteristic polynomial f of a kth order homogeneous linear recurring sequence in F_q and suppose that each α_i has multiplicity $e_i \leq \text{char } F_q$, $i = 1, 2, \ldots, m$. Let P_i be a polynomial of degree less than e_i whose coefficients are uniquely determined by the initial values of the sequence and belong to the splitting field of f over F_q. Prove that $s_n = \sum_{i=1}^{m} P_i(n)\alpha_i^n$ for $n = 0, 1, \ldots$ satisfy the homogeneous linear recurring relation with characteristic polynomial f. (The integer n is identified with an element of F_q.)

8. Find a linear recurring sequence of least order in F_2 whose least period is 24.

9. Given the linear recurring sequence in F_3 with $s_0 = s_1 = s_2 = 1$, $s_3 = s_4 = -1$, and $s_{n+5} = s_{n+4} + s_{n+2} - s_{n+1} + s_n$ for $n = 0, 1, \ldots$, represent the generating function of this sequence in the form (6) of Theorem 3.19.

10. Let $s_0, s_1, \ldots$ be a kth order inhomogeneous linear recurring sequence in F_q satisfying (1). Show how this sequence can be interpreted as a $(k+1)$th order homogeneous linear recurring sequence in F_q.

11. If f is any polynomial over F_2 and f^* is its reciprocal show that (i) $(f^*)^* = f$ iff $f(0) = 1$, (ii) $f^*(0) = 1$ if $\deg f \geq 1$.

12. A nonlinear shift register can be obtained from two linear ones by multiplication that is, if one linear shift register generates a sequence s_i and the other generates a sequence t_i, then the output of their product is $s_i t_i$. Prove: If $(m, n) = 1$ and the sequences generated by two linear shift registers have periods $2^m - 1$ and $2^n - 1$, respectively, then the output sequence of their product has period $(2^m - 1)(2^n - 1)$. Verify this in case $m = 4$, $n = 3$, and characteristic polynomials $f(x) = x^4 + x^3 + 1$, $g(x) = x^3 + x^2 + 1$. Why is multiplication of two linear shift registers not good for cryptanalytic purposes? (Another operation of combining linear shift registers is multiplexing, see BEKER and PIPER.)

*13. A k-stage circulating shift register over F_q is a feedback shift register over F_q with characteristic polynomial $x^k - 1$. Design its circuit, find its period and its associated matrix.

14. Calculate the first five terms of $(1 - 2x + x^3)^{-1}$ in $F_7[[x]]$.

15. Use Theorem 3.21 to determine the minimal polynomial of the linear recurring sequence over F_2 with $s_0 = s_2 = s_4 = s_5 = s_6 = 0$, $s_1 = s_3 = s_7 = 1$, and $s_{n+8} = s_{n+7} + s_{n+6} + s_{n+5} + s_n$ for $n = 0, 1, \ldots$.

*16. Prove Corollary 3.29 by using the methods of the proof of 3.21.

17. The nonzero elements of a finite field F_{p^n} can be generated by using linear recurring sequences. Let $(s_r, \ldots, s_{r+n-1})$ be the nth state vector of a linear recurring sequence in F_p of maximal period $p^n - 1$. Then $a^r = (s_r, \ldots, s_{r+n-1})$, $1 \leq r \leq p^n - 1$, are the nonzero elements of F_{p^n}. Multiplication of two such elements a^r and a^t is defined by the relation

$$a^r \cdot a^t = a^{r+t(\text{mod } p^n - 1)}.$$

Use this approach to generate the elements of the finite field with nine elements. Use the linear recurring sequence defined by the primitive polynomial $x^2 + x + 2$ over F_3 and the initial values $s_0 = 1$, $s_1 = 0$.

EXERCISES (Solutions in Chapter 8, p. 492)

1. Give an example of a linear recurring sequence in F_q, which is ultimately periodic but not periodic.

2. A homogeneous linear recurring sequence $d_0, d_1, \ldots$ of order k in F_q is called the *impulse response sequence* corresponding to the sequence $s_0, s_1, \ldots$ with recursion (2), if

$$d_0 = d_1 = \ldots = d_{k-2} = 0, \; d_{k-1} = 1 \qquad (d_0 = 1 \text{ for } k = 1),$$

and

$$d_{n+k} = a_{k-1}d_{n+k-1} + a_{k-2}d_{n+k-2} + \ldots + a_0 d_n$$

for $n = 0, 1, \ldots$.
   (i) Find the impulse response sequence $d_0, d_1, \ldots$ for the linear recurring sequence $s_{n+5} = s_{n+1} + s_n \; (n = 0, 1, \ldots)$ in F_2.
   (ii) Determine the least period.
   *(iii) Let $d_0, d_1, \ldots$ be an impulse response sequence and let A be the matrix defined in (3) after 3.6. Show that two state vectors $\mathbf{d}_m$ and $\mathbf{d}_n$ are identical if and only if $A^m = A^n$. Determine A for the sequence in (i) and discuss the sequences resulting from all possible state vectors.

3. Design a feedback shift register which yields the elements of F_{2^4}, if ζ is a primitive element, where $\zeta^4 = \zeta + 1$.

4. Determine the impulse response sequence $d_0, d_1, \ldots$ corresponding to the linear recurring sequence of Example 3.20. Find the generating polynomials of this sequence $d_0, d_1, \ldots$. (See Exercise 2 for the definition.)

5. Consider the impulse response sequence (see Exercise 2) $d_0, d_1, \ldots$ corresponding to the linear recurring sequence $s_{n+5} = s_{n+1} + s_n, n = 0, 1, \ldots$ in F_2. Design a feedback shift register generating the sequence.

6. Find the characteristic polynomial f of the recurring sequence $s_{n+6} = s_{n+4} + s_{n+2} + s_{n+1} + s_n, n = 0, 1, \ldots$, in F_2. Determine the order of f and find the impulse response sequence corresponding to the given linear recurrence relation. Take 000011 as the initial state vector and find the corresponding output sequence.

*7. Find the order of $f = x^4 + x^3 + 1$ over F_2, determine the homogeneous linear recurrence and the corresponding sequence which characteristic polynomial f and initial values $s_0 = 1, s_1 = s_2 = s_3 = 0$.

8. Determine the multiplicative inverse of $3 + x + x^2$ in $F_5[[x]]$.

9. Compute a/b in $F_2[x]$, where $a = \sum_{n=0}^{\infty} x^n, b = 1 + x + x^3$.

10. Let $s_0, s_1, \ldots$ be an inhomogeneous linear recurring sequence in F_2 with

$$s_{n+4} = s_{n+3} + s_{n+1} + s_n + 1 \quad \text{for } n = 0, 1, \ldots$$

and initial state vector $(1, 1, 0, 1)$. Find its characteristic polynomial, minimal polynomial and least period.

11. Is the linear recurring sequence $s_0, s_1, \ldots$ in F_2, defined by $s_{n+4} = s_{n+2} + s_{n+1}$ for $n = 0, 1, \ldots$ and initial state vector $(1, 0, 1, 0)$, a periodic sequence?

*12. Use the shifted sequence $s_1, s_2, \ldots$ of the sequence of Exercise 11 to verify the second assertion of Theorem 3.26.

§4. Fast Adding

Group and ring theory turn out to have interesting applications in designs of computer software. They result in computing techniques which speed up calculations considerably. If one considers two numbers such as $a = 37$ and $b = 56$, it makes no difference whether one adds them as natural numbers or as members of some $\mathbb{Z}_n$ with n sufficiently large, say $n = 140$ in our case. The only requirement is that $a + b < n$.

We now decompose n canonically as

$$n = p_1^{t_1} p_2^{t_2} \ldots p_k^{t_k}.$$

The theorem on finite abelian groups (see before 3.2.2) shows that

$$\mathbb{Z}_n \cong \mathbb{Z}_{p_1^{t_1}} \oplus \ldots \oplus \mathbb{Z}_{p_k^{t_k}}.$$

An isomorphism is given by $h: [x]_n \mapsto ([x]_{p_1^{t_1}}, \ldots, [x]_{p_k^{t_k}})$, where $[x]_m$ denotes the residue class of x modulo m. Surjectivity of h means that for all numbers $y_1, \ldots, y_k \in \mathbb{Z}$ some $x \in \mathbb{Z}$ can be found with $x \equiv y_1$ (mod $p_1^{t_1}$), $\ldots$, $x \equiv y_k$ (mod $p_k^{t_k}$), a result which was already basically known in ancient China. It is therefore known as the "Chinese Remainder Theorem" (cf. 3.4.1).

It is not hard to find this solution x explicitly. Form $q_i := n \cdot p_i^{-t_i}$. Because $\gcd(p_i^{t_i}, q_i) = 1$, q_i has an inverse r_i in $\mathbb{Z}_{p_i^{t_i}}$. x is then given by

$$x := y_1 q_1 r_1 + \ldots + y_k q_k r_k.$$

It is unique modulo n.

The importance of this Theorem lies in the fact that one can replace the addition of big natural numbers by a series of k "small" simultaneous additions. We illustrate this by the example mentioned above.

$$n = 140 = 2^2 \cdot 5 \cdot 7$$

$$a = 37 \rightarrow [37]_{140} \rightarrow ([37]_4, [37]_5, [37]_7) = ([1]_4, [2]_5, [2]_7)$$

$$b = 56 \rightarrow [56]_{140} \rightarrow ([56]_4, [56]_5, [56]_7) = ([0]_4, [1]_5, [0]_7)$$

$$\overline{a + b \qquad\qquad\qquad\qquad\qquad\qquad ([1]_4, [3]_5, [2]_7)}$$

Now we have to solve

$$x \equiv 1 \ (\text{mod } 4),$$

$$x \equiv 3 \ (\text{mod } 5),$$

$$x \equiv 2 \ (\text{mod } 7),$$

by the method mentioned above. We get $x = 93$, hence $37 + 56 = 93$.

Of course, using this method does not make sense if one just has to add two numbers. If, however, some numbers are given to a computer which has to work with them a great number of times (as in approximation problems, for instance), it definitely does make sense to transform the numbers to residue classes and to calculate simultaneously in small $\mathbb{Z}_{p_i}$'s.

Before we really adopt this system, we estimate the time we save by using it. The following gives a rough estimate. Adding devices in computers consist of a great number of "gates". Each gate has a small number r of inlets ($r \leq 4$ is typical), and one outlet. Each gate needs a certain time (10^{-7} seconds, say) to produce the output.

4.1 Notation. For $x \in \mathbb{R}$, let $\lceil x \rceil$ be the smallest natural number $\geq x$.

4.2 Theorem. *The time required to produce an output from m inputs by means of r gates is at least $\lceil \log_r m \rceil$.*

PROOF. In one time unit r inputs can be processed, in two time units r^2 inputs can be processed and so on. Hence in t time units, r^t inputs can be processed. Since we have to handle m inputs, we must have $r^t > m = r^{\log_r m}$, whence $t \geq \log_r m$, so $t \geq \lceil \log_r m \rceil$. $\square$

4.3 Theorem. *Usual addition of two m-digit binary numbers needs at least $\lceil \log_r 2m \rceil$ time units.*

PROOF. This follows from 4.2 by noting that the sum of two m-digit numbers $a_1 a_2 \ldots a_m$ and $b_1 b_2 \ldots b_m$ depends on the $2m$ inputs $a_1, \ldots, a_m, b_1, \ldots, b_m$. $\square$

4.4 Theorem. *The addition modulo n (i.e. in $\mathbb{Z}_n$) in binary form consumes at least $\lceil \log_r(2\lceil \log_2 n \rceil) \rceil$ time units.*

PROOF. In order to write $0, 1, \ldots, n - 1$ in binary form one needs m-ary binary numbers with $2^m \geq n$, whence $m \geq \log_2 n$, so $m \geq \lceil \log_2 n \rceil$. Now apply 4.3. $\square$

Similarly, we get the result for the proposed adding method.

4.5 Theorem. *Addition of two (binary) numbers in $\mathbb{Z}_n$ by the methods described above 4.1 needs at least $\lceil \log_r(2\lceil \log_2 n' \rceil) \rceil$ time units. Here, n' is the greatest prime power in the decomposition of n.* $\square$

Hence we will chose n in such a way that n' is as small as possible. It is wise to start from the other end, i.e. to fix n' and look for a large n.

4.6 Example. We want $n' \leq 50$ and can choose n in an optimal way as

$$n := 2^5 \cdot 3^3 \cdot 5^2 \cdot 7^2 \cdot 11 \cdot 13 \cdot 17 \cdot 19 \cdot 23 \cdot 29 \cdot 31 \cdot 37 \cdot 41 \cdot 43 \cdot 47.$$

Now $n > 3 \cdot 10^{21}$ and $n' = 49$. So we can even add numbers with 20 digits. If we choose $r = 3$, the method in 4.4 needs at least

$$\lceil \log_3(2) \lceil \log_2 3 \cdot 10^{21} \rceil) \rceil = \lceil 4.52 \rceil = 5 \text{ time units.}$$

With 4.5 we get (again with $r = 3$)

$$\lceil \log_3(2\lceil \log_2 49\rceil)\rceil = \lceil 2.16\rceil = 3 \text{ time units.}$$

Thus we can add nearly twice as fast.   □

4.7 Remark. Since the Chinese Remainder Theorem holds for the rings $\mathbb{Z}_n$ as well as for their additive groups, we get a method for fast multiplication as well. In general, this method is applicable to all unique factorization domains such as $F[x]$ (F a field).

PROBLEMS

1. Compute "quickly" $8 \cdot (6.43 + 7.17 - 15^2)$ modulo 2520.

2. What is the minimal choice for $\mathbf{n}$ if we want $n > 2 \cdot 10^5$? Estimate the computing times in both cases as in 4.6 (with $r = 4$).

3. Try to compute $(3x + 4)(x^2 - 1)^2 + x^3 + x + 1$ modulo $x^2(x + 1)(x + 2)(x + 3)$ $(x + 4)$ instead of in $\mathbb{R}[x]$.

4. Add 516 and 1687 in two different ways, using the Chinese Remainder Theorem or performing usual addition in $\mathbb{Z}$, and compare the times required to obtain the sum.

5. Explain a method of "fast multiplication" in detail.

6. Determine the number of time units required to multiply two m-digit integers. Similarly for "fast multiplication".

*§5. Pólya's Theory of Enumeration

Groups of permutations (i.e. subgroups of the symmetric group S_M) were the first groups which were studied. N. H. Abel and E. Galois, for instance, studied groups of permutations of zeros of polynomials. Moreover, it is a big understatement to say that one is "only" studying permutation groups. The following classical theorem tells us that all groups are—up to isomorphism—permutation groups.

5.1 Theorem ("Cayley's Theorem"). *If G is a group then*

$$G \leqq S_G.$$

PROOF. The map $h: G \to S_G, g \mapsto \phi_g$ with $\phi_g: G \to G, x \mapsto gx$ does the embedding job: $\phi_{gg'} = \phi_g \circ \phi_{g'}$ ensures that h is a homomorphism, and $\text{Ker } h = \{g \in G | \phi_g = \text{id}\} = \{1\}$, so h is a monomorphism.   □

It follows that every group of order $n \in \mathbb{N}$ can be embedded in S_n. In writing products $\pi\sigma$ of permutations $\pi, \sigma \in S_n$ we consider π, σ to be functions. Hence in $\pi\sigma$ we first perform σ, then π.

5.2 Definition. $\pi \in S_n$ is called a *cycle of length r* if there is a subset $\{i_1, \ldots, i_r\}$ of $\{1, \ldots, n\}$ with $\pi(i_1) = i_2, \pi(i_2) = i_3, \ldots, \pi(i_r) = i_1$ and $\pi(j) = j$ for all $j \notin \{i_1, \ldots, i_r\}$. We will then write $\pi = (i_1, i_2, \ldots, i_r)$. Cycles of length 2 are called *transpositions*.

5.3 Examples. (i) $\begin{pmatrix} 1 & 2 & 3 & 4 & 5 \\ 1 & 4 & 3 & 2 & 5 \end{pmatrix} = (2, 4)$ is a transposition in S_5.

(ii) $\begin{pmatrix} 1 & 2 & 3 & 4 & 5 \\ 1 & 5 & 4 & 2 & 3 \end{pmatrix} = (2, 5, 3, 4)$ is a 4-cycle in S_5.

(iii) $S_3 = \{\text{id}, (1, 2), (1, 3), (2, 3), (1, 2, 3), (1, 3, 2)\}$. □

So every $\pi \in S_3$ is a cycle. This is not true any more from S_4 upwards. But we will show in 5.6 that every permutation in S_n is a product of cycles.

5.4 Definition. For $\pi \in S_n$ let $W_\pi := \{i \in \{1, \ldots, n\} | \pi(i) \neq i\}$ be the *domain of action* of π.

The following result shows that permutations with disjoint domains of action commute. The easy proof is left to the reader.

5.5 Theorem. *If $\pi, \sigma \in S_n$ with $W_\pi \cap W_\sigma = \varnothing$ then $\pi\sigma = \sigma\pi$.* □

5.6 Theorem. *Every $\pi \in S_n \setminus \{\text{id}\}$ can—up to order—be written uniquely as a product of cycles with disjoint domains of action.* □

The decomposition of 5.6 is called "*canonical*". We will give an example; the proof of 5.6 is just a formalization of this example and is omitted.

5.7 Example. Let

$$\pi = \begin{pmatrix} 1 & 2 & 3 & 4 & 5 & 6 & 7 & 8 \\ 2 & 1 & 5 & 6 & 7 & 4 & 3 & 8 \end{pmatrix} \in S_8.$$

π moves 1 into 2 and 2 into 1; this gives the first cycle $(1, 2)$. 3 is transferred into 5, 5 into 7, 7 into 3; second cycle: $(3, 5, 7)$. 4 and 6 are transposed, which yields $(4, 6)$ as the third cycle. Finally, 8 is left fixed. Hence $\pi = (1, 2)(3, 5, 7)(4, 6)$. (By 5.5, there is no reason to worry about the order of these cycles.) □

Without loss of generality, we might assume that the cycles in a canonical decomposition are ordered in such a way that their length is not decreasing. This gives, for instance, $\pi = (1, 2)(4, 6)(3, 5, 7)$ in 5.7. Two canonical decompositions are called *similar* if the sequences of the lengths of the cycles involved are equal. Hence $\pi = (1,2)(4,6)(3,5,7)$ and $\sigma = (1,3)(2,4)(5,8,7)$ are different elements of S_8 having similar decompositions. These considerations prove very useful in looking at the structure of S_n: Let $n \in \mathbb{N}$. A partition of n is a sequence $(a_1, a_2, \ldots, a_s) \in \mathbb{N}^s$ with $s \in \mathbb{N}$, $a_1 \geq a_2 \geq \ldots \geq a_s$ and $a_1 + a_2 + \ldots + a_s = n$. Let $P(n)$ be the number of all different partitions of n.

5.8 Theorem. (i) *If $\pi, \sigma \in S_n$ then $\sigma\pi\sigma^{-1}$ can be obtained from the canonical decomposition of π by replacing every i in its cycles by $\sigma(i)$.*
(ii) *Two cycles are conjugate if and only if they are of the same length.*
(iii) *$\pi_1, \pi_2 \in S_n$ are conjugate if and only if they have similar canonical decomposition.*
(iv) *$P(n)$ is the class number of S_n.*
(v) *$\forall \, n \geq 3$: $Z(S_n) = \{\mathrm{id}\}$. But $Z(S_n) = S_n$ for $n = 1, 2$. Here $Z(S_n) = \{s \in S_n | \forall \, g \in S_n: sg = gs\}$ is the center of S_n.*

PROOF. (i) If $\pi = \xi_1\xi_2 \ldots \xi_m$ is the canonical decomposition of π into cycles ξ_i then $\sigma\pi\sigma^{-1} = (\sigma\xi_1\sigma^{-1})(\sigma\xi_2\sigma^{-1}) \ldots (\sigma\xi_m\sigma^{-1})$, so that it suffices to look at a cycle ξ. Let $\xi = (i_1, \ldots, i_r)$. If $1 \leq k \leq r - 1$ then $\xi(i_k) = i_{k+1}$, hence $(\sigma\xi\sigma^{-1})(\sigma(i_k)) = \sigma(i_{k+1})$. If $k = r$ then $\xi(i_k) = i_1$, whence $(\sigma\xi\sigma^{-1})(\sigma(i_k)) = \sigma(i_1)$. If $i \notin W$ then $\xi(i) = i$, and so $(\sigma\xi\sigma^{-1})(\sigma(i)) = \sigma(i)$. Thus $\sigma\xi\sigma^{-1} = (\sigma(i_1), \ldots, \sigma(i_r))$.
(ii), (iii) and (iv) now follow from (i).
(v) Let $\pi \in S_n$, $\pi \neq \mathrm{id}$. Then there is some i with $\pi(i) \neq i$. If $n \geq 3$ then there is some $k \in \{1, \ldots, n\}$ with $i \neq k \neq \pi(i)$. By (i) we get $\pi(i,k)\pi^{-1} = (\pi(i), \pi(k)) \neq (i, k)$, whence $\pi \notin Z(S_n)$. Hence $Z(S_n) = \{\mathrm{id}\}$ for $n \geq 3$. Since S_1 and S_2 are abelian, $Z(S_n) = S_n$ in these cases. □

Since $(i_1, i_2, \ldots, i_r) = (i_1, i_r)(i_1, i_{r-1}) \ldots (i_1, i_2)$ we get from 5.6:

5.9 Theorem. *Every permutation is a product of transpositions.* □

In contrast to 5.6, this decomposition is not unique.

5.10 Definition. For $\pi \in S_n$,

$$\mathrm{sign}(\pi) := \prod_{i>j} \frac{\pi(i) - \pi(j)}{i - j}$$

is called the *signature* of π.
It is easy to compute $\mathrm{sign}(\pi)$ by means of 5.6 or 5.9:

5.11 Theorem. *Let $n > 1$.*
(i) *If ξ is a cycle of length k then $\mathrm{sign}(\xi) = (-1)^{k-1}$.*
(ii) *$\mathrm{sign}: S_n \to (\{1, -1\}, \cdot)$ is an epimorphism.*

(iii) *If $\pi = \xi_1\xi_2\ldots\xi_r$ is a canonical decomposition of π into cycles of length $k_1,\ldots,k_r$, respectively, then* $\text{sign}(\pi) = (-1)^{k_1+k_2+\ldots+k_r-r}$.

(iv) *If $\pi = \tau_1\tau_2\ldots\tau_s$ with τ_i a transposition then* $\text{sign}(\pi) = (-1)^s$.

(v) $A_n := \text{Ker sign} = \{\pi \in S_n | \text{sign}(\pi) = 1\}$ *is a normal subgroup of S_n.*

(vi) $[S_n : A_n] = 2$, *so* $|A_n| = n!/2$.

PROOF. (i) Let $\xi = (i_1,\ldots,i_k)$. In

$$\text{sign}(\xi) = \prod_{j=1}^{k} \frac{\xi(i_j) - \xi(i_{j-1})}{i_j - i_{j-1}},$$

all factors with $j < k$ can be canceled and there remains

$$\text{sign}(\xi) = \frac{(i_1 - i_2)(i_1 - i_3)\ldots(i_1 - i_k)}{(i_2 - i_1)(i_3 - i_1)\ldots(i_k - i_1)} = (-1)^{k-1}.$$

(ii) For $\pi, \sigma \in S_n$ we get

$$\text{sign}(\pi\sigma) = \prod_{i>j} \frac{\pi(\sigma(i)) - \pi(\sigma(j))}{i - j}$$

$$= \prod_{i>j} \frac{\pi(\sigma(i)) - \pi(\sigma(j))}{\sigma(i) - \sigma(j)} \cdot \frac{\sigma(i) - \sigma(j)}{i - j}$$

$$= \left(\prod_{\sigma(i)>\sigma(j)} \frac{\pi(\sigma(i)) - \pi(\sigma(j))}{\sigma(i) - \sigma(j)} \right)\left(\prod_{i>j} \frac{\sigma(i) - \sigma(j)}{i - j} \right)$$

$$= \text{sign}(\pi)\,\text{sign}(\sigma).$$

By (i) and 5.6, Im sign = $\{1, -1\}$.

(iii) If $\pi = \xi_1 \ldots \xi_r$ then

$$\text{sign}(\pi) = (\text{sign } \xi_1)\ldots(\text{sign } \xi_r) = (-1)^{k_1-1}\ldots(-1)^{k_r-1} = (-1)^{k_1+\ldots+k_r-r}.$$

(iv) follows from (ii) or (iii) since $\text{sign}(\tau) = -1$ for every transposition τ. We leave (v) and (vi) as exercises. $\square$

5.12 Definition. $A_n = \text{Ker sign}$ is called the *alternating group*. Let G be a group. Then $aba^{-1}b^{-1}$ with $a, b \in G$ is called the *commutator* of a and b. $K(G)$ denotes the subgroup of G generated by all commutators.

Note that A_3 is abelian because $|A_3| = 6!/2 = 3$, whence $A_3 \simeq \mathbb{Z}_3$. But A_n is non-abelian if $n \geq 4$.

We list several important properties of these alternating groups (without proofs).

5.13 Theorem. (i) A_n *is the subgroup of S_n generated by the 3-cycles.*

(ii) A_4 *is not simple, since it has a normal subgroup isomorphic to $\mathbb{Z}_2 \times \mathbb{Z}_2$. But A_4, a group of order 12, has no subgroup of order 6.*

(iii) *For $n \geq 5$, A_n is simple and nonabelian. This is the deep reason for the fact that there cannot exist "solution formulas" for equations of degree ≥ 5. Also, A_5 has order 60 and is the smallest nonabelian simple group.*

(iv) $K(S_n) = A_n$ *for $n \geq 3$; $K(S_n) = \{\text{id}\}$ for $n \in \{1, 2\}$.*

(v) $K(A_n) = A_n$ *for* $n \geq 5$; $K(A_n) = \{id\}$ *for* $n \in \{1, 2, 3\}$; $K(A_4) \cong$ $\mathbb{Z}_2 \times \mathbb{Z}_2$.

(vi) $Z(A_n) = \{id\}$ *for* $n \geq 4$; $Z(A_n) = A_n$ *for* $n \in \{1, 2, 3\}$.

(vii) $\bigcup_{n \in \mathbb{N}} A_n =: A_\infty$ *is an infinite (nonabelian) simple group.*   □

Now suppose that X is a set and $G \leq S_X$. Then every $\pi \in G$ can be thought of as being an operator acting on X by sending $x \in X$ into $\pi(x)$. We are interested in what happens to a fixed $x \in X$ under all $\pi \in G$. The easy proofs of the following assertions are left to the reader.

5.14 Theorem and Definition. *Let $G \leq S_X$ and $x, y \in X$. Then*

(i) *x and y are called G-equivalent (denoted by $x \sim_G y$) if $\exists \pi \in G$: $\pi(x) = y$.*

(ii) *$\sim_G$ is an equivalence relation on X.*

(iii) *The equivalence classes $\mathrm{Orb}(x) := \{y \in X \mid x \sim_G y\}$ are called orbits (of G on X).*

(iv) *For every $x \in X$, $\mathrm{Stab}(x) := \{\pi \in G \mid \pi(x) = x\}$ is a subgroup of G, called the stabilizer of x. Let $\pi \sim_{\mathrm{Stab}(x)} \tau$ if $\pi\tau^{-1} \in \mathrm{Stab}(x)$.*

5.15 Examples. (i) Let G be a group, $S \leq G$ and $X = G$. By 5.1, S can be considered as a subgroup of $S_G = S_X$. If $g \in G$ then $\mathrm{Orb}(g) = Sg$ (the right coset of g with respect to S) and $\mathrm{Stab}(g) = \{s \in S \mid sg = g\} = \{1\}$.

(ii) Let G be a group, $X = G$ and $\mathrm{Inn}\, G \leq S_G$, where $\mathrm{Inn}\, G = \{\phi_x \mid x \in G\}$ for all inner automorphisms $\phi_x: G \to G$, $g \mapsto xgx^{-1}$. Then for each g in G we get $\mathrm{Orb}(g) = \{\phi_x(g) \mid \phi_x \in \mathrm{Inn}\, G\} = $ conjugacy class of g, and $\mathrm{Stab}(g) = \{\phi_x \in \mathrm{Inn}\, G \mid xgx^{-1} = \phi_x(g) = g\} = \{\phi_x \in \mathrm{Inn}\, G \mid xg = gx\}$.   □

These concepts turn out to be very useful in several applications. As an example, we mention an application to chemistry, due to G. Pólya.

5.16 Example. From the carbon ring (a) one can obtain a number of molecules by attaching hydrogen (H-) atoms or CH_3-groups in the places ①–⑥ in (b). For instance, one can obtain xylole (c) and benzene (d).

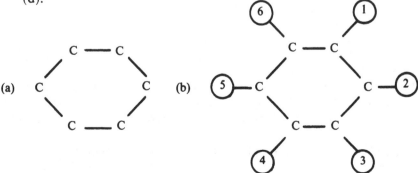

Figure 5.6

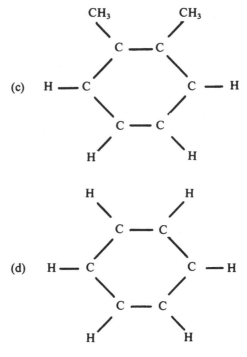

Figure 5.6 (*continued*)

Obviously, (c') gives xylole as well:

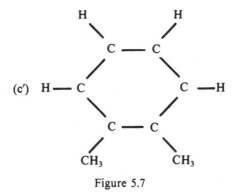

Figure 5.7

The problem which arises is: how many chemically different molecules can be obtained in this way? Altogether, there are 2^6 possibilities to attach either H or CH_3 for ①–⑥. But how many attachments coincide chemically? □

In order to solve this problem we can employ the following result of Burnside.

5.17 Theorem. *Let X be finite and $G \leq S_X$. For every $x \in X$, $|\text{Orb}(x)|$ divides $|G|$ and the number n of different orbits of X under G is given by*

$$n = \frac{1}{|G|} \sum_{g \in G} |\text{Fix}(g)| = \frac{1}{|G|} \sum_{x \in X} |\text{Stab}(x)|,$$

where $\text{Fix}(g) := \{x \in X | g(x) = x\}$. Also $[G : \text{Stab}(x)] = |\text{Orb}(x)|$, and hence $|\text{Stab}(x)| \, |\text{Orb}(x)| = |G|$.

PROOF. First we compute $|\text{Orb}(x)|$. Let $x \in X$ and $f : \text{Orb}(x) \to G/\!\sim_{\text{Stab}(x)}$, $g(x) \mapsto g\,\text{Stab}(x)$. Since

$$g_1(x) = g_2(x) \Leftrightarrow g_2^{-1}g_1(x) = x \Leftrightarrow g_2^{-1}g_1 \in \text{Stab}(x) \Leftrightarrow g_1 \sim_{\text{Stab}(x)} g_2$$

$$\Leftrightarrow g_1 \, \text{Stab}(x) = g_2 \, \text{Stab}(x),$$

f is well defined. Obviously f is bijective and hence $|\text{Orb}(x)| = |G/\!\sim_{\text{Stab}(x)}| = [G : \text{Stab}(x)]$, so $|\text{Orb}(x)|$ divides $|G|$.

Now we compute the cardinality of $\{(g, x) | g \in G, x \in X, g(x) = x\}$ in two different ways:

$$\sum_{x \in X} |\text{Stab}(x)| = |\{(g, x) | g \in G, x \in X, g(x) = x\}| = \sum_{g \in G} |\text{Fix}(g)|.$$

Let us choose representations $x_1, \ldots, x_n$ from the n orbits. Since $\text{Orb}(x) = \text{Orb}(y)$ implies

$$|\text{Stab}(x)| = \frac{|G|}{[G : \text{Stab}(x)]} = \frac{|G|}{|\text{Orb}(x)|} = \frac{|G|}{|\text{Orb}(y)|} = \frac{|G|}{[G : \text{Stab}(y)]}$$

$$= |\text{Stab}(y)|,$$

we get

$$\sum_{x \in X} |\text{Stab}(x)| = \sum_{i=1}^{n} |\text{Orb}(x_i)| \cdot |\text{Stab}(x_i)| = \sum_{i=1}^{n} \frac{|G|}{|\text{Stab}(x_i)|} \cdot |\text{Stab}(x_i)| = n \cdot |G|.$$

$\square$

Now we use 5.17 for our problem 5.16.

5.16 Example (continued). Let us denote the said $2^6 = 64$ attachments by $\{x_1, \ldots, x_{64}\} =: X$. Attaching x_i and x_j will yield the same molecule if and only if x_j can be obtained from x_i by means of a symmetry operation of the hexagon ①–⑥, i.e. by means of an element of D_6, the dihedral group of order 12. Hence the number n of different possible molecules we are looking for is just the number of different orbits of X under D_6. From 5.17 we get

$$n = \frac{1}{|D_6|} \sum_{g \in D_6} |\text{Fix}(g)| = \tfrac{1}{12} \sum_{g \in D_6} |\text{Fix}(g)|.$$

Now id fixes all elements, whence $|\text{Fix}(\text{id})| = 64$. A reflection r on the axis ①–④ in (b) fixes the four attachments possible in ① and ④ and also the four other possible attachments in ② and ③ if they are the same as those in ⑥ and ⑤, respectively. Hence $|\text{Fix}(r)| = 4 \cdot 4 = 16$, and so on. Altogether we get $n = \frac{1}{12} \cdot 156 = 13$ different molecules. $\qquad\qquad\square$

One sees that this enumeration can be applied to situations where one looks for the number of possible "attachments". The result in 5.17 shows that n is the "arithmetic mean" of the $|\text{Fix}(g)|$'s (and also of the $|\text{Stab}(x)|$'s in G).

One can improve the formula in 5.17 slightly by the remark that if g_1 and g_2 are conjugate then $|\text{Fix}(g_1)| = |\text{Fix}(g_2)|$. Of course, this only helps in the nonabelian case. So we get

5.18 Theorem. *Let X be finite and $G \le S_X$. Let $g_1, \ldots, g_r$ be a complete set of representatives for the conjugacy classes in $G/\sim$ and let k_i be the number of elements conjugate to g_i. Then the number n of orbits of X under G is given by*

$$n = \frac{1}{|G|}(k_1|\text{Fix}(g_1)| + \ldots + k_r|\text{Fix}(g_r)|).$$

We give a simple example in which we can use our knowledge about the conjugacy classes of S_3.

5.19 Example. Find the number of essentially different possibilities, n, that there are of placing three different elements from the set $\{A, B, C, D, E\}$ in the three corners $1, 2, 3$ of an equilateral triangle, such that at least two letters are distinct.

SOLUTION. $G = S_3$ acts on $\{1, 2, 3\}$ as the group of symmetries. Take, in the language of 5.18, $g_1 = \text{id}$, $g_2 = (1, 2)$ and $g_3 = (1, 2, 3)$. Then $k_1 = 1$, $k_2 = 3$, $k_3 = 2$. Now X contains all triples (a, b, c) with pairwise different $a, b, c \in \{A, B, C, D, E\}$. Hence

$|X| = 5(4 + 4.5) = 120,$
$|\text{Fix}(g_1)| = 120,$
$|\text{Fix}(g_2)| = 20,$ since g_2 fixes precisely all (a, a, b),
$|\text{Fix}(g_3)| = 0,$ since g_3 fixes exactly all (a, a, a); these combinations are not
$\qquad\qquad$ allowed.

Hence 5.18 gives us
$$n = \tfrac{1}{6}(120 + 60 + 0) = 30. \qquad\qquad\square$$

The results 5.17/5.18 are indeed useful for 5.19. A direct treatment of 5.19 would require an examination of all $\dbinom{120}{2} = 7260$ pairs of attachments with respect to being essentially different. And that's quite a job. There might, however, remain quite a bit to do in 5.17/5.18, especially if G is big and if there are many conjugacy classes. So we might still be dissatisfied with what we have accomplished so far. Also, we still have no tool for finding a representative in each class of essentially equal attachments.

5.20 Definition. Suppose that $\pi \in S_n$ decomposes into j_1 cycles of length 1, j_2 cycles of length $2, \ldots, j_n$ cycles of length n according to 5.6 (we then have $1j_2 + 2j_2 + \ldots + nj_n = n$). $(j_1, \ldots, j_n)$ is then called the *cycle index* of π. If $G \le S_n$ then

$$Z(G) := \frac{1}{|G|} \sum_{\pi \in G} x_1^{j_1} x_2^{j_2} \ldots x_n^{j_n} \in \mathbb{Q}[x_1, \ldots, x_n]$$

(where $(j_1, \ldots, j_n)$ is the cycle index of π) is called the *cycle index polynomial* of G.

5.21 Example. In S_3, we have one permutation (namely id) with cycle index $(3, 0, 0)$, three permutations $((1, 2), (1, 3), (2, 3))$ with cycle index $x_1^2 x_2$ (since $(1, 2) = (1, 2)(3)$, and so on) and two permutations with cycle index x_3. Hence

$$Z(S_3) = \tfrac{1}{6}(x_1 + 3x_1^2 x_2 + 2x_3). \qquad \square$$

It is not hard, but it is lengthy, to determine the cycle index polynomial of the following groups which mostly appear as groups of symmetries. We give the following list without proof.

5.22 List of Some Cycle Index Polynomials.

(i)
$$Z(S_n) = \sum \frac{1}{(1^{j_1} j_i!)(2^{j_2} j_2!) \ldots (n^{j_n} j_n!)} x_1^{j_1} x_2^{j_2} \ldots x_n^{j_n},$$

where the summation goes over all $(j_1, \ldots, j_n) \in \mathbb{N}_0^n$ with $1j_1 + 2j_2 + \ldots + nj_n = n$. In particular, we get

$$Z(S_1) = x_1,$$
$$Z(S_2) = \tfrac{1}{2}(x_1^2 + x_2),$$
$$Z(S_3) = \tfrac{1}{6}(x_1^3 + 3x_1^2 x_2 + 2x_3),$$
$$Z(S_4) = \tfrac{1}{24}(x_1^4 + 6x_1^2 + 3x_2^2 + 8x_1 x_3 + 6x_4).$$

(ii)
$$Z(A_n) = \sum \frac{1 + (-1)^{j_2 + j_4 + \ldots}}{2(1^{j_1} j_1!)(2^{j_2} j_2!) \ldots (n^{j_n} j_n!)} x_1^{j_1} x_2^{j_2} \ldots x_n^{j_n},$$

where the summation is as in (i). In particular,

$$Z(A_1) = x_1,$$
$$Z(A_2) = \tfrac{1}{2} x_1^2,$$
$$Z(A_3) = \tfrac{1}{2}(x_1^3 + 2x_3).$$
$$Z(A_4) = \tfrac{1}{12}(x_1^4 + 8x_1 x_3 + 3x_2^2).$$

(iii) $$Z(\mathbb{Z}_n) = \frac{1}{n} \sum_{d|n} \varphi(d) x_d^{n/d} \quad (\varphi = \text{Euler's } \varphi\text{-function}).$$

For instance,

$$Z(\mathbb{Z}_7) = \tfrac{1}{7}(x_1^7 + 6x_7),$$

$$Z(\mathbb{Z}_8) = \tfrac{1}{8}(x_1^8 + x_2^4 + 2x_4^2 + 4x_8),$$

$$Z(\mathbb{Z}_9) = \tfrac{1}{9}(x_1^9 + 2x_3^3 + 6x_9).$$

(iv) $$Z(D_n) = \tfrac{1}{2}Z(\mathbb{Z}_n) + \tfrac{1}{2}x_1 x_2^{(n-1)/2} \quad \text{if } n \text{ is odd.}$$

$$Z(D_n) = \tfrac{1}{2}Z(\mathbb{Z}_n) + \tfrac{1}{4}(x_2^{n/2} + x_1^2 x_2^{(n-1)/2}) \quad \text{if } n \text{ is even.}$$

Here D_n is the *dihedral group*.

For instance,

$$Z(D_4) = \tfrac{1}{8}(x_1^4 + 2x_1^2 x_2 + 3x_2^2 + 2x_4),$$

$$Z(D_5) = \tfrac{1}{10}(x_1^5 + 5x_1 x_2^2 + 4x_5),$$

$$Z(D_6) = \tfrac{1}{12}(x_1^6 + 3x_1^2 x_2^2 + 4x_2^3 + 2x_3^2 + 2x_6).$$

(v) Let G be the group of all rotations mapping a given cube (or a given octahedron) into itself. Then $|G| = 24$. Consider G as a group C_c of permutations on the eight corners of the cube. Then

$$Z(C_c) = \tfrac{1}{24}(x_1^8 + 9x_2^4 + 6x_4^2 + 8x_1^2 x_3^2).$$

(vi) Let G be as in (v). Now let G act as a group C_e on the twelve edges of the cube. Then

$$Z(C_e) = \tfrac{1}{24}(x_1^{12} + 3x_2^6 + 8x_3^4 + 6x_1^2 x_2^5 + 6x_4^3).$$

(vii) Again, let G be as above. Now let G act as a permutation group C_f on the six faces of the cube. Then

$$Z(C_f) = \tfrac{1}{24}(x_1^6 + 3x_1^2 x_2^2 + 6x_1^2 x_4 + 6x_2^3 + 8x_3^2). \qquad \square$$

The reader might be wondering, what these cycle index polynomials might be good for. Recall our examples, in which we wanted to assign certain "figures" $f_1, \ldots, f_r$ (H- or CH-modules in 5.16 or letters $A, \ldots, E$ in 5.19) to a number n of "places" $1, 2, \ldots, n$ (free places in the carbon ring, corners of a triangle, ..). We build up a mathematical model for these situations.

5.23 Definition. Let F be a set of r figures f_i. Let $m \in \mathbb{N}$. If $G \le S_m$ then G can be thought of as a permutation group on F^m via $g(f_1, \ldots, f_m) := (f_{g(1)}, \ldots, f_{g(m)})$.

5.24 Theorem ("Pólya's Theorem"). *In the situation of 5.23, the number n of different orbits on $X = F^m$ under G is given by*

$$n = Z(G)(r, r, \ldots, r).$$

(This equals the value of the induced polynomial function of $Z(G)$ at $x_1 = r, \ldots, x_m = r$.)

PROOF. If $g \in G$ then $(f_1, \ldots, f_m) \in \text{Fix}(g)$ if and only if all f_i, where i runs through the elements of a cycle of g, are equal. Hence $|\text{Fix}(g)| = r^{j_1 + j_2 + \cdots + j_m}$, where $(j_1, \ldots, j_m)$ is the cycle index of g. Now Burnside's Theorem 5.17 gives us the desired result

$$n = \frac{1}{|G|} \sum_{g \in G} |\text{Fix}(g)| = \frac{1}{|G|} \sum_{g \in G} r^{j_1} r^{j_2} \ldots r^{j_m}. \qquad \square$$

5.16 Example (Revisited). We have $F = \{H\text{-}, CH_3\text{-}\}$, $n = 2$, $m = 6$ and $G = D_6$. From 5.22(iv) we get

$$Z(D_6) = \tfrac{1}{12}(x_1^6 + 3x_1^2 x_2^2 + 4x_2^3 + 2x_3^2 + 2x_6),$$

hence

$$n = Z(D_6)(2, 2, 2, 2, 2, 2) = \tfrac{1}{12}(64 + 48 + 32 + 8 + 4) = \tfrac{1}{12} \cdot 156 = 13. \qquad \square$$

5.25 Example. We want to color the faces of a cube with two colors. How many essentially different colorings can we get?

SOLUTION. By 5.22(vii) we get C_f, $m = 6$, $r = 2$, hence $n = \tfrac{1}{24}(64 + 48 + 48 + 48 + 32) = 10$ different colorings. $\qquad \square$

Hence the solution becomes as short as the question. If we actually want to find a representative in each orbit, we can simply try to find one. If one dislikes this brute force method then some more theory is needed. For a proof and a detailed account, see, e.g. STONE.

5.26 Theorem ("Redfield–Pólya Theorem"). *Recall the situation in 5.23. Let us "invent" formal products of the figures $f_1, \ldots, f_n$ and write f^2 for $f \cdot f$, etc. If one now substitutes $f_1 + \ldots + f_r$ for x_1, $f_1^2 + f_2^2 + \ldots + f_r^2$ for x_2, and so on, in $Z(G)$, then one gets, by expanding the products, sums of the form $n_i f_1^{i_1} f_2^{i_2} \ldots f_r^{i_r}$ with $i_1 + \ldots + i_r = m$. This means that there are precisely n_i ways to put i_1 figures f_1 into the place number 1, i_2 of the figures f_2 into number 2, and so on.* $\qquad \square$

5.16 Example (Revisited). Let f_1 be the H- and f_2 the CH_3-group. If we expand as in 5.26, we get $\tfrac{1}{12}(f_1 + f_2)^6 + 3(f_1 + f_2)^2(f_1^2 + f_2^2)^2 + 4(f_1^2 + f_2^2)^3 + 2(f_1^3 + f_2^3)^2 + 2(f_1^6 + f_2^6)) = f_1^6 + f_1^5 f_2 + 3f_1^4 f_2^2 + 3f_1^3 f_2^3 + 3f_1^2 f_2^4 + f_1 f_2^5 + f_2^6$.

Hence there are:

one possibility to give only H-atoms,
one possibility to give five H- and one CH_3-groups,
three possibilities to take four H- and two CH_3-groups, and so on.

In order to find a complete set of representatives, one has to collect these (altogether 13) possibilities. This is very easy.   □

Of course, if one replaces all f_i by 1, one gets Polya's Theorem 5.23 as a corollary of 5.26. Even after the discovery of these two powerful results it was not always realized that these methods are available. Let's give a final example.

5.27 Example. Let us call two truth functions (or switching functions) $f_1, f_2: \{0, 1\}^n \rightarrow \{0, 1\}$ *essentially similar*, if, after a suitable relabeling $(i_1, \ldots, i_n)$ of $(1, \ldots, n)$ we have

$$f_1(b_1, \ldots, b_n) = f_2(b_{i_1}, \ldots, b_{i_n})$$

for all $(b_1, \ldots, b_n) \in \{0, 1\}^n$. For switching theory, this means that the corresponding switching circuits of f_1 and f_2 "work identically" after a suitable permutation of the input wires.

Problem. How many essentially different such functions exist?

History. This problem was explicitly carried out and solved by means of a gigantic computer program in 1951 for $n = 4$. The total number of these functions is already $2^{2^4} = 65,536$.

SOLUTION. Our solution is rather immediate. The group G is basically S_n. However, care must be taken, since G acts as described above on $\{0, 1\}^n$ and not on $\{1, \ldots, n\}$. If we take $n = 4$, for instance, the effect of $g = (1, 2)(3, 4)$ on the quadruple $(a, b, c, d) \in \{0, 1\}^4$ is given by (b, a, d, c). Obviously, Fix(g) consists of precisely those functions which are constant on each cycle of g. In our case for $g = (1, 2)(3, 4)$ we get Fix(g) = $2 \cdot 2 = 4$. Now S_n decomposes into the following congruence classes (see 5.8):

(i) id;
(ii) six 2-cycles;
(iii) three products of two 2-cycles;
(iv) eight 3-cycles;
(v) six 4-cycles;

Now (i) fixes all 16 (a, b, c, d), yielding x_1^{16} in the cycle index polynomial. Also, (ii) contributes $6x_1^8 x_2^4$, since for instance $(1, 2)$ yields the four 2-cycles $((0, 1, c, d), (1, 0, c, d))$ and fixes all $(0, 0, c, d)$ and $(1, 1, c, d)$, thus producing eight 1-cycles, and so on. The cycle index polynomial for G acting on $\{0, 1\}^4$ is then given by

$$Z(G) = \tfrac{1}{24}(x_1^{16} + 6x_1^8 x_2^4 + 3x_1^4 x_2^6 + 8x_1^4 x_3^4 + 6x_1^2 x_2 x_4^3).$$

Hence $x_1 = x_2 = x_3 = x_4 = 2$ gives 3948 equivalence classes of functions from $\{0, 1\}^4$ to $\{0, 1\}$.   □

5.28 Remark. If one adds complementation to G as a symmetry operation (i.e. $f_1 \sim f_2 \Leftrightarrow f_1(b_1, \ldots, b_n) = f_2(b_1', \ldots, b_n')$) then this number reduces further from 3948 to just 222.

PROBLEMS

1. Show that each of the following three subsets of S_n ($n \geq 2$) generate S_n:
   (i) $\{(1, n), (1, n - 1), \ldots, (1, 2)\}$ (cf. 5.9);
   (ii) $\{(1, 2), (2, 3), \ldots, (n - 1, n), (n, 1)\}$;
   (iii) $\{(1, 2), (1, 2, \ldots, n)\}$.
   Remark. You can consider this as a "game", in which you put the numbers $1, 2, \ldots, n$ to the vertices of a regular n gon on the unit circle. In (i), you can interchange 1 with any other number, in (ii) you can transpose every number with the one following it, and in (iii) you can interchange 1 and 2 and turn the whole circle by $360/n$ degrees. The question is always: Can you get every arbitrary order of $1, \ldots, n$ on the circle by means of these "allowed" manipulations?

2. Determine $K(S_n)$ for all $n \in \mathbb{N}$.

3. If $|G| = n$, we know from 5.1 that $G \leqslant S_n$. By (iii) of Problem 1, S_n can be generated by two elements. Hence every finite group can be generated by two elements. Is this correct (where is the error)?

4. Let G be the group of all orthogonal transformations in the $\mathbb{R}^3$ which move the edges of a cube (with center in the origin) into edges. Determine $|G|$ via 5.17.

*5. Consider the class of organic molecules of the form

$$
\begin{array}{c}
X \\
| \\
X-C-X \\
| \\
X
\end{array}
$$

where C is a carbon atom, and each X can be any one of the following: CH_3, C_2H_5, H or Cl. Model each molecule of this class as a regular tetrahedron with the carbon atom at the centre and the X-components at its vertices. Find the number of different molecules of this class. (Hint: Find the number of equivalence classes of maps from the four vertices of the tetrahedron as the domain to the range $\{CH_3, C_2H_5. H, Cl\}$. A suitable permutation group is the group of the permutations which correspond to all possible rotations of the tetrahedron.)

6. Two Boolean functions over $\mathbb{B}$ in n variables are said to be equivalent if one can be obtained from the other by permuting and/or complementing the variables. The set of all possible permutations and/or complementations of the n variables can be represented by a permutation group G acting on the domain $\mathbb{B}^n$ of the Boolean function.

    (i) Determine the number of equivalence classes of Boolean functions for $n = 2$.

  *(ii) Also for $n = 3$, where it is known that the cycle index for G is

$$\tfrac{1}{48}(x_1^8 + 13x_2^4 + 8x_1^2x_3^2 + 8x_2x_6 + 6x_1^4x_2^2 + 12x_4^2).$$

7. Let k be a given integer and let $n = 2k + 1$. Each of the 10^n numbers of n digits is printed on a separate slip of paper. Leading zeros are always filled in. We regard two slips to be the same if one of them represents the other one if the slip is held upside down. So we assume that the digits $0, 6, 9, 8, 1$ cannot be distinguished from the upside down counterparts $0, 9, 6, 8, 1$. Determine the number of different slips of paper.

EXERCISES (Solutions in Chapter 8, p. 494)

1. Find the number of distinct bracelets of five beads made up of red, blue and white beads.

2. Three big and six small pearls are connected onto a circular chain. The chain can be rotated and turned. How many different chains can be obtained assuming that pearls of the same size are indistinguishable?

3. Find the number of ways of painting the four faces a, b, c and d of a tetrahedron with two colors of paints.

4. If n colors are available, in how many different ways is it possible to color the vertices of a cube?

5. Determine the number of switching functions using three nonequivalent switches.

NOTES

§1

There is a very large body of mathematical knowledge which is usually referred to as combinatorics; most of such combinatorial topics are part of discrete mathematics with applications to computing. Some of the standard texts on combinatorics are BERMAN and FRYER, HALL, VILENKIN, RYSER, STREET and WALLIS; one of the advanced books is AIGNER. There is a considerable number of conference proceedings, outlining topics in the theory and applications of combinatorics proceedings, outlining topics in the theory and applications of combinatorics; we mention WELSH. The proceedings of the Australian Conferences on Combinatorial Mathematics, an annual and ongoing event, are published in the series "Lecture Notes in Mathematics" by Springer-Verlag. The *British Combinatorial Bulletin* gives an excellent survey of recently submitted or published combinatorial books and papers. MACWILLIAMS and SLOANE and SLOANE give several applications of Hadamard matrices, the theory of these matrices is developed in detail in WALLIS, STREET and WALLIS. The standard references for Latin squares is DENES and KEEDWELL; applications to the design of statistical experiments are discussed in MANN, VAJDA, RAGHAVARAO. Finite projective planes are considered in most books on projective geometry, e.g. HUGHES and PIPER, HORADAM. The most comprehensive account of projective geometries over finite fields is given in HIRSCHFELD. In the notes to Chapter 4 we referred to the applications of finite geometries to coding theory. Balanced incomplete block designs, tactical configurations, Steiner systems and difference equations are all standard topics of combinatorics, and the following references contain applications of some of these topics: BOGART, VAJDA, WELSH.

§2

Cryptography is concerned with the protection of an established communication channel against (passive) eavesdropping and (active) forging, thus secrecy and authenticity being the two main concerns. Today cryptography draws heavily from mathematics and computer science. The importance of mathematics in cryptography was already recognized by the famous algebraist A. Adrian Albert, who said almost half a century ago: "It would not be an exaggeration to state that abstract cryptography is identical with abstract mathematics."

Cryptography is a subject which has evolved rapidly recently, particularly in its applications in data security. The number of publications in cryptography has also grown rapidly. There are now several excellent books on crytography available: BEKER and PIPER includes interesting applications of finite fields, DENNING and KONHEIM present a wealth of stimulating mathematical aspects of the subject. DENNING includes an excellent introduction to encryption algorithms, cryptographic techniques, and chapters on access control, information flow control and inference control together with extensive bibliographical references. MYER and MATYAS contains many references to applications. Whereas GALLAND and SINKOV cover the older literature and the mathematically more elementary aspects of cryptography. Other books, some of them of a survey nature, are: BETH, SIMMONS, LEISS, MERKLE and BETH, HESS and WIRL. SIMMONS, SLOANE and GARDNER are survey articles on various aspects of cryptography. The NPL published a series of annotated bibliographies on data security and cryptology (see PRICE). There is also a journal, *Cryptologia*, devoted entirely to this subject area. The event of an annual conference and workshop on cryptography give an indication of the vigor and fast development of this discipline. Since this is a rather new and in our view fascinating subject, we give more extensive notes and mention a few aspects in the history of cryptography with some further references to the bibliography. KAHN gives an exciting and mostly nonmathematical account of the historical development of cryptography. The first system of military cryptography was established by the Spartans in ancient Greece in the fifth century B.C. They used a transposition cipher which was implemented on an apparatus called "skytale"—a staff of wood around which a strip of papyrus or leather was wrapped. The secret message is written on the papyrus down the length of the staff, the papyrus is then unwound and sent to the recipient. The Greek writer Polybius devised a cryptosystem which has as one of its characteristics the conversion of letters to numbers. Cryptology proper, which comprises both cryptography and cryptanalysis, was first discovered and written down by the Arabs in 1412. Kahn traces the history of cryptography in detail over the following centuries, with emphasis on the important developments earlier in the twentieth century, particularly resulting from two World Wars.

It was recognized by Alberti (1466) and Viète (1589) many centuries ago that monoalphabetic substitution offers no security. F. W. Kasiski (1863) laid the foundation for breaking polyalphabetic substitutions with periodic or nonperiodic keys, if the same key is used several times. This lead to the development of running keys for one-time use. J. C. Mauborgne, of the U.S. Army Signal Corps, introduced such random keys in 1918 based on Vernam ciphers, which had been invented for teletype machines by G. S. Vernam a year before. Claude Shannon studied the probabilistic nature of the one-time key and defined the concept of "perfect secrecy". The theoretical basis of this and other systems was established by information theory. One time pads require extremely long keys and are therefore prohibitively expensive in most applications.

The period after World War I saw the development of special purpose machines for enciphering. Later the development of digital computers has greatly facilitated the search for better cryptographic methods. World War II brought mathematicians in close contact with cryptology; for example, Alan Turing in England and A. A. Albert in the USA, who organized mathematicians for the work in cryptology. In 1940 the British were able to break the German ciphers produced by the ENIGMA machines, which are rotor machines for polyalphabetic enciphering. It is said that some 200,000 of these machines were in use. Another example of a cryptographic machine is the American Hagelin C-36 machine or M-209 converter. It was used by the U.S. Army until the early 1950s.

United States cryptologists under the leadership of W. F. Friedman succeeded in cracking the Japanese PURPLE ciphers in 1941. German cryptanalysts under H. Rohrbach were able to break one of the early American M-138-A ciphers, which were then changed into new cipher machines, such as the M-134-C.

Only from the mid-1970's on was there also a wide-spread commercial interest in cryptography. Incentives for this development came from the need for protecting data, computer-supported message systems such as electronic mail or industrial espionage, to mention a few.

In the early 1970's IBM developed and implemented the system LUZIFER, which uses a transformation that alternatively applies substitution and transposition. This system was then modified into DES, the National Bureau of Standard's Data Encryption Standard. An interesting and rather new area of cryptography is the problem of secure speech transmission. There are, essentially, two techniques for encrypting speech: digital and analogues. Both techniques of voice scrambling are described in BEKER and PIPER.

SIMMONS gives a survey of the state of cryptography, comparing classical methods (symmetric encryption) with newer developments of public-key cryptography (asymmetric encryption). Symmetric cryptosystems involve substitution and transposition as the primary operation; Caesar ciphers, Vigenères, Hill ciphers and the DES fall into this category. One important advantage of the DES is that it is very fast and thus can be used for high-volume, high speed enciphering.

The concept of a public-key cryptosystem is due to DIFFIE and HELLMAN: as an important implementation the RSA cryptosystem was developed by RIVEST, SHAMIR and ADLEMAN. The trapdoor knapsack is another implementation of a public key cryptosystem given by MERKLE and HELL-MANN. Public-key cryptosystems have the advantage that there is no need for an initial exchange of keys via a secure channel between two users who wish to communicate with each other. Secondly, only two keys are required per user (the public and the private key) independent of the number of participants in such a system.

It is still open if the RSA system can be broken. So far there is no proof that deciphering in the RSA system is computationally as hard as factorization of an integer. It is conceivable that some day a method will be developed that factors in polynomial time. Today it appears that a length of 400 digits for the integer n, and p and q primes of length approximately 200 each are sufficient.

The security of most cryptographic systems depends on the computational difficulty to discover the plaintext without knowledge of the key. The two modern disciplines which study the difficulty of solving computational problems are complexity theory and analysis of algorithms. The ultimate test of a system is the attack by a skilled cryptanalyst under very favorable conditions such as a *chosen plaintext attack*, where the cryptanalyst can submit an unlimited number of plaintext messages of his choice and examine the resulting cryptograms.

Today the need for governmental (diplomatic and military) cryptography and for private and commercial cryptography is recognized. The times are now over when the Secretary of State under President Hoover of the USA in 1929 dismissed on ethical grounds the Black Chamber of the State Department, with the explanation "Gentlemen do not read each other's mail."

§3

The history of linear recurring sequences from 1202 to 1918 has been traced by DICKSON. The celebrated *Fibonacci sequence* $F_0, F_1, F_2, \ldots$ defined by

$F_0 = 0$, $F_1 = 1$, $F_{n+2} = F_{n+1} + F_n$ for $n = 0, 1, \ldots$ attracted a great deal of attention initially. Then linear recurring sequences of real and complex numbers were considered, and more recently linear sequences in $\mathbb{Z}$ have been considered modulo a prime p. Such sequences became important in electrical engineering because of their connection with switching circuits and coding theory. SELMER gives a brief survey of the history of the subject with particular emphasis on the developments after 1918. The basic paper for the modern theory of linear recurring sequences in finite fields is ZIERLER, and expository accounts of this theory can be found in: BIRKHOFF and BARTEE, DORNHOFF and HOHN, GOLOMB, GILL, LIDL and NIEDER-REITER, LÜNEBURG, PETERSON and WELDON. Applications of linear recurring sequences in coding theory were mentioned in the notes for Chapter 4. Some of the pioneering work on the interplay between linear recurring sequences, feedback shift registers and coding theory is presented in ASH, LIN, PETERSON and WELDON, ABRAMSON. Linear recurring sequences in cryptography are discussed in BEKER and PIPER; other applications of such sequences are presented in GOLOMB. Linear recurring sequences are also used in the construction of Hadamard matrices, in the theory of difference sets, in pseudo-random number generators and in simplifying computations in $\mathbb{F}_q$ and $\mathbb{F}_q[x]$.

§4

The area of algebraic computing is a young and growing discipline. Many aspects of it are dealt with in computer science courses rather than algebra courses. The book by LIPSON tries to build a bridge between these two disciplines as an alternative viewpoint to modern algebra. More useful algorithms and programs for algebraic calculations can be found in AHO, HOPCROFT and ULLMAN and also BORODIN and MUNRO. Our example is contained in SZABO and TANAKA.

§5

The central position in the theory of counting is taken by Pólya's theorem, a very elegant result contained in a paper published in 1937 in *Acta Mathematica 68*. A survey of Pólya's theory of counting is given in DE BRUIJN, with many further references. Some of the texts on applied algebra also contain a section on applications of the central theory of this paragraph, e.g. GILBERT, DORNHOFF and HOHN.

CHAPTER 6
Automata

One meets automata or machines in various forms such as calculating machines, computers, money changing devices, telephone switch boards and elevator or lift switchings. All of the above have one aspect in common, namely a "box" which can assume various "states". These states can be transformed into other states by outside influences (called "inputs"), for instance by electrical or mechanical impulses. Often the automaton "reacts" and produces "outputs" like results of computations or change.

We shall indicate what is common to all automata and describe an abstract model which will be amenable to mathematical treatment. We shall see that there is a close relationship between automata and semigroups. Essential terms are a set of states and a set of inputs which operate on the states and transform them into other states (and sometimes yield an output). If outputs occur then we shall speak of automata, otherwise we speak of semiautomata.

§1. Semiautomata and Automata

1.1 Definition. A *semiautomaton* is a triple $\mathscr{S} = (Z, A, \delta)$, consisting of two nonempty sets Z and A and a function $\delta: Z \times A \to Z$. Z is called the *set of states*, A the *input alphabet* and δ the "*next-state function*" of $\mathscr{S}$.

The above definition is very much an abstraction of automata in the usual sense. Historically, the theory of automata developed from concrete automata in communication techniques, nowadays it is a fundamental science. If we want "outputs", then we have to study automata rather than semiautomata.

1.2 Definition. An *automaton* is a quintupel $\mathscr{A} = (Z, A, B, \delta, \lambda)$ where (Z, A, δ) is a semiautomaton, B is a nonempty set called the *output alphabet* and $\lambda: Z \times A \to B$ is the "*output function*".

If $z \in Z$ and $a \in A$, then we interpret $\delta(z, a) \in Z$ as the next state into which z is transformed by the input a. $\lambda(z, a) \in B$ is the output of z resulting from the input a. Thus if the automaton is in state z and receives input a, then it changes to state $\delta(z, a)$ with output $\lambda(z, a)$.

1.3 Definition. A (semi-) automaton is *finite*, if all sets Z, A and B are finite; finite automata are also called *Mealy automata*. If a special state $z_0 \in Z$ is fixed, then the (semi-) automaton is called *initial* and z_0 is the *initial state*. We write (Z, A, δ, z_0) in this case. An automaton with λ depending only on z is called a *Moore automaton*.

In practical examples it often happens that states are realized by collections of switching elements each of which has only two states (e.g. current–no current), denoted by 0 and 1. Thus Z will be the cartesian product of several copies of $\mathbb{Z}_2$. Similarly for A and B.

PROBLEMS

1–4. Give interpretations of the situations described as (semi-) automata by specifying Z, A and δ. When does it make sense to consider these semiautomata as automata, Mealy automata, Moore automata, initial automata?

1. A computer.

2. A money changing device.

3. An elevator switching circuit.

4. A mouse trap.

EXERCISES (Solutions in Chapter 8, p. 494)

1. Describe parts of your brain as a semiautomaton, such that A is the set of theorems in this book and δ describes "studying".

2. Could you interpret the semiautomaton in Exercise 1 as a computer (brain = storage, etc.)?

3. What would be a possible interpretation if the semiautomaton in Exercise 1 should be considered as an automaton?

4. Same as Exercise 3, but with the interpretation as an initial automaton.

§2. Description of Automata; Examples

Now we investigate how to describe a finite (semi-) automaton explicitly. There are several ways of doing this, two of which seem to be particularly appropriate. Let $A = \{a_1, \ldots, a_n\}$, $B = \{b_1, \ldots, b_m\}$ and $Z = \{z_1, \ldots, z_k\}$.

I. *Description by Tables*
In order to describe δ we can simply give a table (the "*transition table*")

δ	a_1	$\ldots$	a_n	
z_1	$\delta(z_1, a_1)$	$\ldots$	$\delta(z_1, a_n)$	
$\vdots$	$\vdots$		$\vdots$	(all $\delta(z_i, a_j) \in Z$).
z_k	$\delta(z_k, a_1)$	$\ldots$	$\delta(z_k, a_n)$	

In case of automata we also need an "output table":

λ	a_1	$\ldots$	a_n	
z_1	$\lambda(z_1, a_1)$	$\ldots$	$\lambda(z_1, a_n)$	
$\vdots$	$\vdots$		$\vdots$	(all $\lambda(z_i, a_j) \in B$).
z_k	$\lambda(z_k, a_1)$	$\ldots$	$\lambda(z_k, a_n)$	

II. *Description by Graphs*
We depict $z_1, \ldots, z_k$ as "discs" in the plane and draw an arrow labeled a_i from z_r to z_s if $\delta(z_r, a_i) = z_s$. In case of an automaton we denote the arrow also by $\lambda(z_r, a_i)$. This graph is called the *state graph*:

Figure 6.1

For further descriptions by means of incidence matrices see ARBIB. We are now in a position to consider the first example.

2.1 Example (Marriage Semiautomaton). We consider the following situation of a somewhat idealized marriage: The husband is angry or bored or happy; the wife is quiet or shouts or cooks his favorite dish. Silence on her part does not change the husband's mood, shouting "lowers" it by one "degree" (if he is already angry, then no change), cooking of his favorite dish creates general happiness for him. We try to describe this rather limited view of a marriage in terms of a semiautomaton $\mathscr{S} = (Z, A, \delta)$. We define $Z = \{z_1, z_2, z_3\}$ and $A = \{a_1, a_2, a_3\}$, with

$z_1 \ldots$ "husband is angry", $a_1 \ldots$ "wife is quiet",

$z_2 \ldots$ "husband is bored", $a_2 \ldots$ "wife shouts",

$z_3 \ldots$ "husband is happy", $a_3 \ldots$ "wife cooks".

Using description I, we obtain

δ	a_1	a_2	a_3
z_1	z_1	z_1	z_3
z_2	z_2	z_1	z_3
z_3	z_3	z_2	z_3

Description II yields

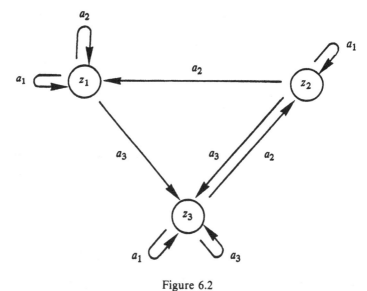

Figure 6.2

2.2 Example (Marriage Automaton). Let the situation be as in 2.1 and add the outputs $B = \{b_1, b_2\}$ with the interpretation

$b_1 \ldots$ "husband shouts", $b_2 \ldots$ "husband is quiet".

We assume that the husband only shouts if he is angry and his wife shouts. Otherwise he is quiet, even in state z_3. In description I we have to add the output table

λ	a_1	a_2	a_3
z_1	b_2	b_1	b_2
z_2	b_2	b_2	b_2
z_3	b_2	b_2	b_2

In II we add

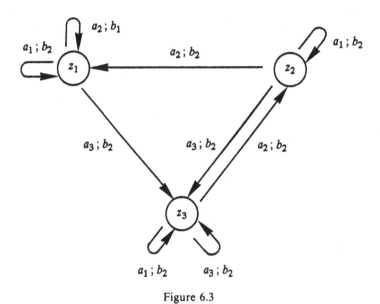

Figure 6.3   □

2.3 Example ("Parity-Check Automaton"). Let $Z = \{z_0, z_1\}$, $A = B = \{0, 1\}$ and

δ	0	1
z_0	z_0	z_1
z_1	z_1	z_0

λ	0	1
z_0	0	1
z_1	0	1

With II we have

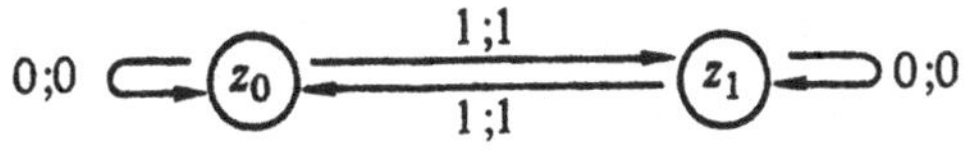

$$0;0 \;\; z_0 \quad \xrightarrow[\;1;1\;]{\;1;1\;} \quad z_1 \;\; 0;0$$

Figure 6.4

A possible interpretation would be: z_0 and z_1 are the two states of a number storage with

$$z_0 \ldots \text{"machine} \quad \text{stores 0"},$$

$$z_1 \ldots \text{"machine} \quad \text{stores 1"}.$$

For $i, j \in \{0, 1\}$ the input i operates on the state z_j by creating the state z_{j+1} (addition modulo 2). The output is identical to the input.   □

2.4 Example ("Flip-Flop"). The following three (mini-) automata are particularly important, since they are often used as building blocks for bigger automata. Let $Z = B = \{0, 1\}$.

(i) Shift register: $A = \{0, 1\}$

δ	0	1		λ	0	1
0	0	1		0	0	0
1	0	1		1	1	1

or

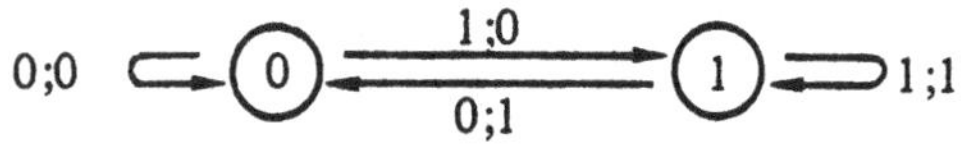

Figure 6.5

(ii) Trigger flip-flop: $A = \{0, 1\}$

δ	0	1		λ	0	1
0	0	1		0	0	0
1	1	0		1	1	1

or

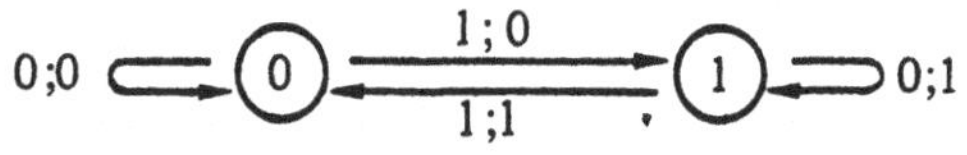

Figure 6.6

(iii) IR flip-flop: $A = \{e, 0, 1\}$

δ	e	0	1		λ	e	0	1
0	0	0	1		0	0	0	0
1	1	0	1		1	1	1	1

or

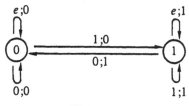

Figure 6.7

The input e does not change anything, the inputs $i \in \{0, 1\}$ change all states into i, the output is equal to the corresponding input.   □

All three miniautomata are Moore automata. We shall clarify how they work after the introduction of semigroups into the theory of automata.

PROBLEMS

1–4. Describe simple cases of the (semi-) automata in Problems 1–4 of §1 by means of tables and graphs.

5. Do the same for Exercise 1 of §1.

6. Let $Z = A = \mathbf{Z}_5$ and $\delta(z, a) = z + a$ (addition modulo 5). Describe (Z, A, δ) by a table and by a graph.

7. As in Problem 6, but now with $\delta(z, a) = za$ (multiplication modulo 5).

8. As in Problem 6, now with $\delta(z, a) = z^a$ (powers in $\mathbf{Z}_5$).

EXERCISES (Solutions see Chapter 8, p. 494)

1. A stamp automaton $\mathscr{S}$ has a capacity of ten stamps. Define it as a semiautomaton according to 1.1, such that the state z means "there are z stamps in $\mathscr{S}$", the inputs are "no coin is inserted", "a correct coin is inserted", "a wrong coin is inserted". Describe this semiautomaton by a table and a state graph.

2. Generalize the semiautomaton of Exercise 1 to an automaton by adding the outputs "no output", "a stamp is output", "a coin is output". Complete the table and state graph.

3. Given $A = B = Z = \mathbf{Z}_4$ and $\delta(z, a) = z + a$, $\lambda(z, a) = z \cdot a$, draw the graph of this automaton.

4. Let $A = \{0, 1, 2, 3, 4\} \le \mathbf{Z}_{10}$, $B = \mathbf{Z}_{10}$, $Z = $ all multiples of 4 in $\mathbf{Z}_{10}$. Let $\delta(z, a) = za$ and $\lambda(z, a) = z$. Determine Z, give the tables for δ and λ, and draw the graph of this automaton.

5. More generally, given a semiautomaton (A, Z, δ) with $A = Z$, will we always get $(A, \circ)$ as a semigroup via $a \circ b := \delta(a, b)$? Conversely, given a semigroup H, can it always be considered as a semiautomaton (A, Z, δ) with $A = Z$?

6. In a metabolic cycle, let $Z = \{z_1, z_2, z_3\}$ consist of three organic acids. Two enzymes a_1, a_2 act on Z. a_1 changes z_1 into z_2, z_2 into z_3, and z_3 into z_1. a_2 does the exact inverse. Interpret this situation by a semiautomaton and draw its graph.

§3. Semigroups

Algebraic automata theory and the theory of formal languages (cf. Chapter 7) make extensive use of semigroups. In this paragraph we summarize those results of semigroup theory, which will be needed for the description of automata. Some of the results given early in this section are elementary and are probably known to many readers, however these introductory results are included for the benefit of those who are unfamiliar with semigroups.

A. Fundamental Concepts

3.1 Definition. A *semigroup* is a set S together with an associative binary operation $\circ$ defined on it. We shall write $(S, \circ)$ or simply S for a semigroup.

Here are some examples of semigroups.

3.2 Examples. (i) $(\mathbb{N}, +)$, $(\mathbb{N}_0, +)$, $(\mathbb{N}, \cdot)$, $(\mathbb{Z}, +)$, $(\mathbb{Z}, \cdot)$, $(\mathbb{R}, +)$ and $(\mathbb{R}, \cdot)$ are prominent examples of semigroups. These examples will show up frequently.

(ii) $(\mathbb{R}, \max)$ with $x(\max)y := \max(x, y)$ is another example of a semi-group on the set $\mathbb{R}$.

(iii) Let $\mathbb{R}_n^n$ denote the set of all $n \times n$-matrices with entries in $\mathbb{R}$. Then $(\mathbb{R}_n^n, +)$ and $(\mathbb{R}_n^n, \cdot)$ are semigroups.

(iv) Let $\mathbb{R}[x]$ be the set of all real polynomials. If $+, \cdot, \circ$ denote addition, multiplication and substitution, respectively, then $(\mathbb{R}[x], +)$, $(\mathbb{R}[x], \cdot)$ and $(\mathbb{R}[x], \circ)$ are examples of semigroups. Again one can see that many different operations can be defined on a set, turning this set into (different) semigroups.

(v) Let S be any nonempty set. Let S^S be the set of all mappings from S to S and let $\circ$ denote the composition of mappings. Then $(S^S, \circ)$ is a semigroup. These types of semigroups will turn out to be "universal examples" in the sense that every semigroup "sits inside" such a $(S^S, \circ)$ for a properly chosen S.

(vi) Let V be a vector space and $\mathrm{Hom}(V, V)$ the set of all linear maps from V to V. Then $(\mathrm{Hom}(V, V), +)$ and $(\mathrm{Hom}(V, V), \circ)$ (with $\circ$ as in (v)) are semigroups.

(vii) Let S again be a set (now S might be empty). Then $(\mathscr{P}(M), \cap)$, $(\mathscr{P}(M), \cup)$ and $(\mathscr{P}(M), \Delta)$ are semigroups whose elements are sets.

(viii) We now show that every nonempty set S can be turned into a semigroup. Define in S, $s_1 * s_2 =: s_1$. Then $(S, *)$ is a semigroup. $\square$

More examples will follow. Sets with operations which together do not form a semigroup are $(\{1, 2, 3, 4, 5\}, +)$ since $+$ is not a closed operation on the given set, and $(\mathbb{Z}, -)$ since $-$ is not associative. If the underlying set S is finite we speak of a *finite semigroup* $(S, \circ)$, say $(\{x_1, \ldots, x_n\}, \circ)$. Such a semigroup can be characterized by a semigroup table,

3.3

$\circ$	x_1	$\cdots$	x_j	$\cdots$	x_n
x_1	$x_1 \circ x_1$	$\cdots$	$x_1 \circ x_j$	$\cdots$	$x_1 \circ x_n$
$\vdots$	$\vdots$		$\vdots$		$\vdots$
x_i	$x_i \circ x_1$	$\cdots$	$x_i \circ x_j$	$\cdots$	$x_i \circ x_n$
$\vdots$	$\vdots$		$\vdots$		$\vdots$
x_n	$x_n \circ x_1$	$\cdots$	$x_n \circ x_j$	$\cdots$	$x_n \circ x_n$

where all "products" $x_i \circ x_j$ are elements of S.

However, if a table such as 3.3 is given, where all $x_i \circ x_j$ are in $\{x_1, \ldots, x_n\}$, then $(\{x_1, \ldots, x_n\}, \circ)$ is not necessarily a semigroup, since the operation $\circ$ need not be associative. To show that $a \circ (b \circ c) = (a \circ b) \circ c$ for all $a, b, c \in \{x_1, \ldots, x_n\}$, one has to check this for all n^3 possible choices for a, b, c. In 7.2.2 we describe a considerably quicker method for testing associativity.

We now introduce another important class of semigroups. If we regard the subsets of $M \times M$ as relations we write $\mathcal{R}(M)$ instead of $\mathcal{P}(M \times M)$ and define for $R, S \in \mathcal{R}(M)$ a relation $R \,\square\, S$ by

$$x(R \,\square\, S)y :\Leftrightarrow \exists\, z \in M : x R z \wedge z S y.$$

The following result is straightforward.

3.4 Theorem. $(\mathcal{R}(M), \square)$ *is a semigroup for any set M.* $\square$

3.5 Definition. $(\mathcal{R}(M), \square)$ is called the *relation semigroup* on M. The operation $\square$ is called the *relation product*.

3.6 Remark. $R \in \mathcal{R}(M)$ is transitive if and only if $R \,\square\, R \subseteq R$.

3.7 Definition. Let $(S, \circ)$ be a semigroup and $s \in S$.
 (i) s is called a *zero element* of $(S, \circ)$ if $x \circ s = s \circ x = s$ holds for all x in S;
 (ii) s is called an *identity element* of $(S, \circ)$ provided that $x \circ s = s \circ x = x$ holds for all $x \in S$;
 (iii) s is called an *idempotent* of $(S, \circ)$ if $s \circ s = s$.

If we have only $x \circ s = s$ for all $x \in S$ in (i), then s is called a *right zero element*. *Right identity elements, left zero elements*, etc. are defined analogously.

In general we write s^2 instead of $s \circ s$, s^3 instead of $s \circ s \circ s$, etc. s is an idempotent if and only if $s = s^2 = s^3 = \dots$. Not every semigroup has a zero or an identity. However, no semigroup can have more than one of each of these. More precisely we have

3.8 Theorem. *Let* $(S, \circ)$ *be a semigroup.*

  (i) *If l is a left zero element and r is a right zero element then they are equal and therefore the uniquely determined zero element.*

 (ii) *If l is a left identity element and r is a right identity element then they are equal and therefore the uniquely determined identity element.*

(iii) *Any left or right zero element is an idempotent.*

(iv) *Any left or right identity element is an idempotent.*

PROOF. (i) and (ii) follow from $l = l$ or $= r$; (iii) and (iv) are trivial consequences.   □

3.9 Definition. If the semigroup $(S, \circ)$ has an identity element (which is unique by 3.8(ii)) then $(S, \circ)$ is called a *monoid*. If $(S, \circ)$ is a monoid and if $x \in S$ then x is called *invertible* if there is some $y \in S$ such that $x \circ y$ and $y \circ x$ are the identity element.

Left and right invertible elements in a monoid are defined similarly.

3.10 Theorem and Definition. *The element y in 3.9 is uniquely determined by x and is called the inverse of x, denoted by $y = x^{-1}$ or by $y = -x$.*

3.11 Examples. (i) $(\mathbb{N}, +)$ is a semigroup which has neither a zero nor an identity element.

 (ii) $(\mathbb{N}_0, +)$ is a monoid with identity 0; only 0 is invertible.

(iii) $(\mathbb{N}, \cdot)$ is a monoid with identity 1; only 1 is invertible.

(iv) $(\mathbb{Z}, \cdot)$ is a monoid with identity 1, in which precisely 1 and -1 are invertible. The only idempotents are given by 0 and 1.

 (v) $(\mathbb{Q}, \cdot)$ is also a monoid with 1 as the identity, but this time every element except 0 is invertible. 0 and 1 are still the only idempotents.

(vi) The $n \times n$-matrices over the reals form a monoid $(\mathbb{R}_n^n, \cdot)$ with the unit matrix as an identity. The invertible elements are precisely the regular matrices. $A = \begin{pmatrix} 1 & 0 \\ 0 & 0 \end{pmatrix}$ is an example of a "nontrivial" idempotent.

(vii) $(S^S, \circ)$, as in 3.2(v), is a monoid with the identity function as the identity element. Invertible are exactly the invertible ($=$ bijective)

functions. The projection $f: \mathbb{R} \times \mathbb{R} \to \mathbb{R}$, $(x, y) \to (x, 0)$ is an example
of an idempotent element (with $S = \mathbb{R} \times \mathbb{R}$).

(viii) $(\mathscr{P}(M), \cap)$ is a monoid for every set M. This set M, considered as
an element of $\mathscr{P}(M)$, serves as the identity; it is the only invertible
element. All elements are idempotent.

(ix) The relation semigroup $\mathscr{R}(M)$ on the nonempty set M is a monoid
with the equality relation $\{(x, x) | x \in M\}$ (sometimes referred to as
the diagonal of M) as identity. The graphs of the bijective functions
from M to M are precisely the invertible elements. The empty set $\varnothing$
is the zero element and one of many idempotent elements.

(x) Let $(G, \circ)$ be a group. Then $(G, \circ)$ is a monoid in which every element
is invertible. No element beside the identity can be idempotent.

$\square$

PROBLEMS

1–8. Prove all assertions in 3.2(i)–(viii).

9. Show that $(\mathscr{R}(M), \square)$ is a semigroup for every set M.

10–19. Check all details in 3.11(i)–(x).

20. Find five idempotent elements in $(\mathscr{R}(\mathbb{N}), \square)$.

B. Subsemigroups, Homomorphisms

Within a semigroup we can have smaller semigroups, the substructures or
subsemigroups, e.g. $(\mathbb{N}, \cdot)$ in $(\mathbb{Z}, \cdot)$.

3.12 Definition. Let $(S, \circ)$ be a semigroup and $A, B \subseteq S$. Then

$$A \circ B := \{a \circ b \,|\, a \in A, b \in B\}.$$

Instead of $A \circ A$, $A \circ A \circ A$ etc. we write $A^2, A^3, \ldots$, and for $\{a\} \circ B$ we
use the notation $a \circ B$.

3.13 Definition. Let $(S, \circ)$ be a semigroup and $U \subseteq S$, $U \neq \varnothing$. U is a
subsemigroup if $U^2 \subseteq U$.

In this case $(U, \circ)$, or rather $(U, \circ|_{U \times U})$, is a semigroup, called a subsemi-
group of $(S, \circ)$. We write $(U, \circ) \leq (S, \circ)$ or $U \leq S$ if the operations are
obvious.

3.14 Examples. $(\mathbb{N}, \cdot) \leq (\mathbb{Z}, \cdot)$, $(\mathbb{N}, +) \leq (\mathbb{Z}, +)$, but $(\mathbb{N}, +) \nleq (\mathbb{Z}, \cdot)$. $\square$

A subsemigroup U of a monoid S in general is not a monoid, e.g.
$(\mathbb{N}, +) \leq (\mathbb{N}_0, +)$. If U is a submonoid then the identities of U and S may
differ, e.g. $(\{0\}, \cdot) \leq (\mathbb{Z}, \cdot)$. Alternatively, in a semigroup or a monoid we
can have a group. In order to show this we "collect" the invertible elements.

3.15 Definition. Let $(S, \circ)$ be a monoid. $G_S := \{x \in S \mid x \text{ is invertible}\}$ is called the *group kernel* (or unit group) of $(S, \circ)$.

3.16 Theorem. G_S *is a group within* $(S, \circ)$.

PROOF. (i) If $x, y \in G_S$ then $x \circ y$ has $y^{-1} \circ x^{-1}$ as an inverse (why?). Hence $x \circ y \in G_S$.

(ii) Since the identity element e is invertible $(e = e^{-1})$, e belongs to G_S.

(iii) If $x \in G_S$ then $x^{-1} \circ (x^{-1})^{-1} = x^{-1} \circ x = e$ and also $(x^{-1})^{-1} \circ x = e$. Hence $x^{-1} \in G_S$ and so every element in G_S is invertible (in G_S).   □

3.17 Examples. The following list gives the group kernels of the monoids in 3.11.

Monoid	Group kernel	Monoid	Group kernel
$(\mathbb{N}_0, +)$	$\{0\}$	$(S^S, \circ)$	$\{f \in S^S \mid f \text{ bijective}\}$
$(\mathbb{N}, \cdot)$	$\{1\}$	$(\mathcal{P}(M), \cap)$	$\{M\}$
$(\mathbb{Z}, \cdot)$	$\{1, -1\}$	$(\mathcal{R}(M), \square)$	$\{\{(m, f(m) \mid m$
$(\mathbb{Q}, \cdot)$	$\mathbb{Q}^* = \mathbb{Q} \setminus \{0\}$		$\in M\} \mid f \text{ bijective}\}$
$(\mathbb{R}_n^n, \cdot)$	$\mathrm{Gl}(n, \mathbb{R})$	$(G, \circ)$, a group	G

□

In a certain sense, the larger the group kernel, the closer a monoid resembles a group.

3.18 Theorem. *Any nonempty intersections of subsemigroups S_i of a semigroup $(S, \circ)$ is again a subsemigroup. In general, however, the union of subsemigroups of $(S, \circ)$ is not a subsemigroup any more.*

PROOF. If x and y are in $\bigcap S_i$ then $x \in S_i$ and $y \in S_i$, hence $x \circ y \in S_i$ hold for all i. This shows that $x \circ y \in \bigcap S_i$. The second part of the statement will be shown in the exercises.   □

If $(S, \circ)$ is a semigroup and $\varnothing \neq T \subseteq S$, then the intersection of all subsemigroups containing T forms a subsemigroup, according to 3.18. This is the smallest subsemigroup which contains T and is denoted by $\langle T \rangle$. Since the subsemigroups of S, along with $\varnothing$, form an (inductive) Moore system, it is possible to speak about generated subsemigroups.

3.19 Definition. $\langle T \rangle$ is called the *subsemigroup* of $(S, \circ)$ *generated* by T.

We shall write $\langle t \rangle$ instead of $\langle \{t\} \rangle$. It is obvious that T forms a subsemigroup if and only if $T = \langle T \rangle$. We always have $T \subseteq \langle T \rangle$. Next we find $\langle T \rangle$ for $T \neq \varnothing$.

3.20 Theorem. $\forall\; T \subseteq S,\; T \neq \varnothing\colon \langle T \rangle = \bigcup_{n \in \mathbb{N}} T^n$.

PROOF. $\bigcup_{n \in \mathbb{N}} T^n$ is the set of all finite products of elements of T. This set forms a subsemigroup of $(S, \circ)$ and any subsemigroup containing T must contain all finite products of elements of T. Thus $\bigcup T^n$ is the smallest subsemigroup of $(S, \circ)$, containing T, and this is $\langle T \rangle$. $\qquad\square$

Every semigroup $(S, \circ)$ is generated by "something", e.g. by itself, since $S = \langle S \rangle$. If $T \subseteq V$ and T generates S, then V generates S. Therefore we want to find generating systems which are as small as possible.

3.21 Definition. $(S, \circ)$ is called *finitely generated* if there exists $T \subseteq S$ with T finite and $\langle T \rangle = S$. $(S, \circ)$ is *cyclic* if there is a $t \in S\colon \langle t \rangle = S$.

3.22 Examples. (i) $(\mathbb{N}, +) = \langle 1 \rangle$; $(\mathbb{N}, +)$ is cyclic.
(ii) $(\mathbb{N}, \cdot) = \langle P \cup \{1\} \rangle$.
(iii) $(\mathbb{N}_0, +) = \langle \{0, 1\} \rangle$. $\qquad\square$

We mention without proof the following result due to Vorobev, which shows that the bijective functions on a finite set N, together with one single nonbijective function, generate already all of S^S.

3.23 Theorem. *Let* $|N| = n \in \{2, 3, \ldots\}$ *and* $f \in N^N$ *with* $|f(N)| = n - 1$. *Then* $\langle S_N \cup \{f\} \rangle = (N^N, \circ)$. $\qquad\square$

3.24 Theorem and Definition. *Let* $R \in \mathscr{R}(M)$, *then the union* R' *of the sets in* $\langle R \rangle$ *is the smallest transitive relation of* $\mathscr{R}(M)$ *containing* R. R' *is called the transitive hull of* R. $xR'y \Leftrightarrow \exists\, n \in \mathbb{N},\, \exists\, x_1, \ldots, x_n \in M\colon xRx_1 \wedge x_1 R x_2 \wedge \ldots \wedge x_n Ry$.

PROOF. We have $R' \square R' \subseteq R'$, because $\langle R \rangle \leq \mathscr{R}(M)$, and hence $\langle R \rangle$ is transitive. The characterization of $xR'y$ follows from 3.18. Let $U \in \mathscr{R}(M)$ be transitive with $R \subseteq U$. Then $xR'y \Rightarrow \exists\, n \in \mathbb{N},\, \exists\, x_1, \ldots, x_n \in M\colon xRx_1 \wedge x_1 R x_2 \wedge \ldots \wedge x_n Ry \Rightarrow \exists\, n \in \mathbb{N},\, \exists\, x_1, \ldots, x_n \in M\colon xUx_1 \wedge x_1 U x_2 \wedge \ldots \wedge x_n Uy \Rightarrow xUy$. Therefore $R' \subseteq U$. $\qquad\square$

As an application of this result we obtain

3.25 Corollary. *If* $R \in \mathscr{R}(M)$ *then the smallest equivalence relation* $\bar{R}$ *containing* R *is given by the union* U *of the sets in* $\langle R \cup R^{-1} \cup \{(x, x) \mid x \in M\} \rangle$. *It is called the equivalence relation generated by* R.

PROOF. $\bar{R}$ must contain R, R^{-1} and the diagonal $D = \{(x, x) \mid x \in M\}$, since $\bar{R}$ is symmetric and reflexive. $\bar{R}$ is also transitive, thus $\bar{R} = R'$ contains U because of 3.24. This is equal to $\bar{R}$, since U is an equivalence relation. $\square$

By this time, the reader will certainly have discovered some of the similarities and also some of the differences between semigroup theory and group theory. Many more topics from group theory can be carried over, for instance the concepts of homomorphisms and factor (semi-) groups which we shall discuss now.

3.26 Definition. Let $(S, \circ)$ and $(S', \circ')$ be semigroups and $f: S \to S'$ a mapping. f is called a *homomorphism* if

$$\forall \, x, y \in S: f(x \circ y) = f(x) \circ' f(y).$$

Injective homomorphisms are called *monomorphisms* (or *embeddings*), surjective homomorphisms are called *epimorphisms*, bijective homomorphisms are called *isomorphisms*. If f is an isomorphism then $(S, \circ)$ and $(S', \circ')$ are called *isomorphic*, in symbols $(S, \circ) \cong (S', \circ')$. If there exists a monomorphism from $(S, \circ)$ into $(S', \circ')$, then $(S, \circ)$ is *embeddable* in $(S', \circ')$, in symbols $(S, \circ) \hookrightarrow (S', \circ')$. We denote by $\mathrm{Hom}(S, S')$ the set $\{f \mid f$ is a homomorphism from $(S, \circ)$ into $(S', \circ')\}$.

3.27 Examples. (i) $f: (\mathbb{Z}, \cdot) \to (\mathbb{N}_0, \cdot)$, $x \mapsto |x|$ is a homomorphism, because $f(x \circ y) = |x \cdot y| = |x| \cdot |y| = f(x) \cdot f(y)$. f is also an epimorphism, but not a monomorphism, since $f(3) = f(-3)$.

(ii) $g: (\mathbb{Z}, \cdot) \to (\mathbb{N}_0, \cdot)$, $x \mapsto 0$ is a homomorphism.

(iii) $(\mathscr{P}(\{1, 2, 3\}, \cap)$ and $(\mathscr{P}(\{a, b, c\}, \cap)$ are isomorphic semigroups.   □

Isomorphic semigroups can be regarded as being "essentially equal". In fact, one can be obtained from the other by relabeling the elements and the operation.

3.28 Definition. Let $(S, \circ)$ be a semigroup and $\sim$ an equivalence relation on S. $\sim$ is called *congruence relation* on $(S, \circ)$ if $\sim$ is compatible with $\circ$, i.e. $a \sim a'$ and $b \sim b' \Rightarrow a \circ a' \sim b \circ b'$.

For congruence relations on $(S, \circ)$ we define on the factor set $S/\!\!\sim: [a] \circledcirc [b] := [a \circ b]$. This is well defined: If $[a] = [a']$, $[b] = [b']$ then $a \sim a'$, $b \sim b'$, hence $a \circ b \sim a' \circ b'$, whence $[a \circ b] = [a' \circ b']$.

3.29 Theorem and Definition. *Let $\sim$ be a congruence relation on the semigroup $(S, \circ)$. Then $(S/\!\!\sim, \circledcirc)$ is a semigroup, called the factor semigroup (or quotient semigroup) of $(S, \circ)$ over $\sim$. The mapping $\pi: S \to S/\!\!\sim, s \mapsto [s]$ is an epimorphism, called the canonical epimorphism.*   □

For $(S/\!\!\sim, \circledcirc)$ we write $(S/\!\!\sim, \circ)$ or simply $S/\!\!\sim$. Conversely, the following homomorphism theorem holds (the proof is left to the reader).

3.30 Theorem. *Let $f: (S, \circ) \to (S', \circ')$ be an epimorphism. Then $\sim$ defined by $x \sim y :\Leftrightarrow f(x) = f(y)$ is a congruence relation on $(S, \circ)$ and $(S/\sim, \circ) \cong (S', \circ')$.* $\square$

This means that all homomorphic images $(S', \circ')$ of $(S, \circ)$ can be found "within" $(S, \circ)$, just by finding all suitable congruence relations. A most important special case is the following. Let $(S, \circ)$ be a semigroup and R a relation on S. Modifying 3.25 we see that there is (with respect to $\subseteq$) a smallest congruence relation $\tilde{R}$ in $(S, \circ)$ which contains R. Thus we define

3.31 Definition. $\tilde{R}$ is the *congruence relation generated* by R.

Let $\sim_1$ and $\sim_2$ be two congruences on a semigroup $(S, \circ)$, such that $\sim_1 \subseteq \sim_2$ (hence for all $x, y \in S: x \sim_1 y \Rightarrow x \sim_2 y$). We say that $\sim_1$ is "finer" than $\sim_2$. The equivalence classes with respect to $\sim_1$ and $\sim_2$ are denoted by $[x]_1$ and $[x]_2$.

3.32 Theorem. *Let $\sim_1$ and $\sim_2$ be two congruence relations on the semigroup $(S, \circ)$. Then*

$$\sim_1 \subseteq \sim_2 \text{ if and only if}$$
$$[x]_1 \mapsto [x]_2 \text{ is an epimorphism of } (S/\sim_1, \circ) \text{ onto } (S/\sim_2, \circ).$$

PROOF. "$\Rightarrow$" If $\sim_1 \subseteq \sim_2$ then $f: [x]_1 \to [x]_2$ is well defined, since $[x]_1 = [y]_1 \Rightarrow x \sim_1 y$, thus $x \sim_2 y$ and therefore $[x]_2 = [y]_2$. f is clearly surjective and is an epimorphism, since $f([x]_1 \odot [y]_1) = f([x \circ y]_1) = [x \circ y]_2 = f([x]_1) \odot f([y]_1)$ for all $[x]_1, [y]_1 \in S/\sim_1$. i.e. if $[x]_1 = [y]_1$ then $[x]_2 = [y]_2$. It follows that $x \sim_1 y$ implies $x \sim_2 y$, as required.

"$\Leftarrow$" If $[x]_1 \to [x]_2$ is an epimorphism, then f is a well-defined mapping, since $x \sim_1 y$, i.e. $[x]_1 = [y]_1$ implies $[x]_2 = [y]_2$, thus $x \sim_2 y$. $\square$

In particular, $\sim_1 \subseteq \sim_2$ implies that $(S/\sim_2, \circ)$ is a homomorphic image of $(S/\sim_1, \circ)$ ("*Dyck's Theorem*").

3.33 Example. Let $(S, \circ) = (\mathbb{Z}, \cdot)$ and $\sim_1, \sim_2$ be the congruence relations $\equiv_4, \equiv_2$, respectively. Here we have $\equiv_4 \subseteq \equiv_2$ and also $(S/\sim_1, \circ) = (\mathbb{Z}/\equiv_4, \cdot) \cong (\mathbb{Z}_4, \cdot)$ and $(S/\sim_2, \circ) = (\mathbb{Z}/\equiv_2, \cdot) \cong (\mathbb{Z}_2, \cdot)$. In fact $h: (\mathbb{Z}_4, \odot) \to (\mathbb{Z}_2, \cdot), [x]_1 \to [x]_2$ is an epimorphism. $\square$

3.34 Definition. Let $\sim_1 \subseteq \sim_2$ be congruence relations on $(S, \circ)$. Then the mapping $f: [x]_1 \to [x]_2$ of 3.32 is called the *standard epimorphism* of $(S/\sim_1, \circ)$ onto $(S/\sim_2, \circ)$.

We describe a further way of obtaining a new semigroup from a given one, which generalizes the construction of direct products of groups.

3.35 Theorem and Definition. *Let* $(S_1, \circ_1)$ *and* $(S_2, \circ_2)$ *be semigroups. The semigroup* $(S_1, \circ_1) \times (S_2, \circ_2) := (S_1 \times S_2, \circ)$ *with* $(s_1, s_2) \circ (s_1', s_2') :=$ $(s_1 \circ_1 s_1', s_2 \circ_2 s_2')$ *is called the direct product of* $(S_1, \circ_1)$ *and* $(S_2, \circ_2)$. *The product of finitely many semigroups is defined inductively.* $\square$

3.36 Example. $(\mathbb{Z}_2, +) \times (\mathbb{Z}_3, +)$ consists of six elements $([0], [0])$, $([0], [1])$, $([0], [2])$, $([1], [0])$, $([1], [1])$, $([1], [2])$. Note that we should write $([0]_2, [0]_3)$, etc. For instance, we have $([1], [1]) + ([1], [1]) = ([0], [2])$. $\square$

3.37 Theorem and Definition. *Let* $(S, \circ) = (S_1, \circ_1) \times (S_2, \circ_2)$. *Then* $\pi_1 \colon (S, \circ) \to$ $(S_1, \circ_1)$, $(s_1, s_2) \mapsto s_1$ *is an epimorphism called the first projection. The second projection* $\pi_2 \colon S \to S_2$ *is defined analogously.* $\square$

We note that f in 3.27(i) takes the zero element (identity element) of $(\mathbb{Z}, \cdot)$ to the zero (identity) of $(\mathbb{N}_0, \cdot)$. This is not always true. The reason why it works in this case is that there f is surjective.

3.38 Theorem. *Let* $f \colon (S, \circ) \to (S', \circ')$ *be an epimorphism. Let* n *and* e *be the zero and identity elements, respectively, in* $(S, \circ)$. *Then* $f(n)$ *and* $f(e)$ *are the zero and identity elements, respectively, in* $(S', \circ')$.

PROOF. Let $f(x)$ be an arbitrary element of S'. Then $f(n) \circ' f(x) = f(n \circ x) =$ $f(n) = f(x \circ n) = f(x) \circ' f(n)$. Thus $f(n)$ is a zero element of $(S', \circ')$. The proof for the identity element is analogous. $\square$

The following constructions show how to adjoin a zero and an identity element to a given semigroup.

3.39 Theorem. *Let* $(S, \circ)$ *be a semigroup. Let* 0 *and* 1 *be elements which are not contained in* S *(otherwise we relabel* 0, 1 *with* $0'$, $1'$). *We define in* $S \cup \{0\}$: $0 \circ x = x \circ 0 := 0$ *and* $0 \circ 0 := 0$ *for all* $x \in S$. *In* $S \cup \{1\}$ *we define:* $1 \circ x =$ $x \circ 1 := x$ *and* $1 \circ 1 := 1$ *for all* $x \in S$. *Then* $S^{(0)} := (S \cup \{0\}, \circ)$ *is a semigroup with zero element and* $S^{(1)} := (S \cup \{1\}, \circ)$ *is a semigroup with identity element.*

The proof of this theorem is left as an exercise. With the above construction any semigroup can be made into a monoid. If a zero or identity already exist in $(S, \circ)$ then it must relinquish this property to the additional elements 0 or 1, respectively.

The next result (which generalizes Cayley's Theorem) enables us to stress the importance of semigroups of the type $(N^N, \circ)$ as "universal examples". In short: All semigroups can be found among the subsemigroups of $(N^N, \circ)$ for suitably chosen sets N.

3.40 Theorem. *For any semigroup* $(S, \circ)$ *there is a set* N *such that* $(S, \circ) \hookrightarrow (N^N, \circ)$.

PROOF. If $(S, \circ)$ is a monoid, choose $N := S$. If $(S, \circ)$ is not a monoid then construct $S^{(1)}$ and set $N := S \cup \{1\}$. For $s \in S$ define $f_s: N \to N$, $x \mapsto s \circ x$. Let ϕ be the mapping $\phi: S \to N^N$, $s \mapsto f_s$. ϕ is a monomorphism: $\forall s, s' \in S$, $\forall x \in N: \phi(s \circ s')(x) = f_{s \circ s'}(x) = (s \circ s') \circ x = s \circ (s' \circ x) = f_s(s' \circ x)$
$= f_s(f_{s'}(x)) = (f_s \circ f_{s'})(x) = (\phi(s) \circ \phi(s'))(x)$, thus $\forall s, s' \in S: \phi(s \circ s') = \phi(s) \circ \phi(s')$. If $\phi(s) = \phi(s')$ then $f_s = f_{s'}$, therefore $\forall x \in N: s \circ x = s' \circ x$. If x is the identity element of N then $s = s'$ follows.   □

Therefore any semigroup is a semigroup of transformations, but this property is usually neither particularly useful nor natural: only a few people will regard $(\mathbb{N}, +)$ as a subsemigroup of $(\mathbb{N}_0^{\mathbb{N}_0}, \circ)$. From Theorem 3.40, for each semigroup we have a monomorphism of the semigroup into a "transformation semigroup". Shortly we shall see that any semigroup is the epimorphic image of a so-called "free" semigroup.

3.41 Theorem. *The mapping* $\psi: (N^N, \circ) \to (\mathcal{R}(N), \square)$, *which assigns to each* $f \in N^N$ *its graph* G_f *is a monomorphism. Here* $G_f := \{(x, f(x)) \mid x, f(x) \in N\}$.

PROOF. Since any graph G_f uniquely determines its function f and since $\psi(f \circ g) = G_{f \circ g} = G_f \square G_g = \psi(f) \square \psi(g)$ we have

$$G_f = G_g \Leftrightarrow f = g.$$ □

3.42 Corollary. *For any semigroup* $(S, \circ)$ *there exist sets* M, N *such that* $(S, \circ) \leqq (N^N, \circ)$ *and* $(S, \circ) \leqq (\mathcal{R}(M), \square)$.   □

Thus any $(N^N, \circ)$ can be embedded in $(\mathcal{R}(N), \square)$ and any $(\mathcal{R}(N), \square)$ can be embedded in $(M^M, \circ)$ with $M = \mathcal{R}(N)$.

PROBLEMS

1–9. Check the correctness of the results obtained in 3.17.

10. Let $S = \left\{ \begin{pmatrix} a & 0 \\ 0 & 0 \end{pmatrix} \middle| a \in \mathbb{Z}_5 - \{0\} \right\} \cup \left\{ \begin{pmatrix} 1 & 0 \\ 0 & 1 \end{pmatrix} \right\}$.

    (i) Show that $(S, \cdot)$ is a semigroup and that $H := \left\{ \begin{pmatrix} a & 0 \\ 0 & 0 \end{pmatrix} \middle| a \in \mathbb{Z}_5^* \right\}$ is a
       group. Is S a monoid?

    (ii) Compute G_S and compare it with G_H. (See 3.15.)

    (iii) Is G_S the largest subgroup in S? Is G_S a maximal subgroup?

11. If S is a submonoid of the monoid M, can one say that G_S is a subgroup of G_M? (Hint: use Problem 10.)

12. Find two subsemigroups S_1, S_2 of a semigroup S such that $S_1 \cup S_2$ is not a subsemigroup of S.

13. Show that every subsemigroup of a finite group G is a subgroup of G.

14. Show that the result in Problem 13 is not necessarily true if G is infinite.

15. Show 3.23 for $N = \{1, 2, 3\}$.

16. Prove 3.22.

17. Check the details in 3.27.

*18. Prove 3.39.

19. Let $S = \{a, b, c\}$ and $x \circ y = x$ for all $x, y \in S$. Give the multiplication tables for $S^{(0)}$ and $S^{(1)}$.

C. Free Semigroups

In linear algebra one very often uses the important result that any mapping of a basis of a vector space V into a vector space W can be uniquely extended to a linear mapping from V into W. This can also be done for semigroups. Semigroups with a "basis" B will be called "free on B".

3.43 Definition. Let $\mathscr{S}$ be the class of all semigroups, $F \in \mathscr{S}$ and $\varnothing \neq B \subseteq F$. F is called a *free semigroup* (on B), if each mapping f from B into a semigroup G can be extended to a unique homomorphism h of F into G. B is called a basis of F.

We should really write $(F, \circ)$ but we shall omit this from now on. Definition 3.43 can be formally restated as

3.43' Definition. $F \in \mathscr{S}$ is called *free on the basis* $B \subseteq F$ if

$$\forall\, G \in \mathscr{S}, \forall f\colon B \to G, \exists\, |\, h \in \mathrm{Hom}(F, G)\colon h|_B = f.$$

A diagram for this situation is of the form

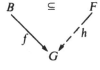

Figure 6.8

Many questions arise immediately, such as: Is there a basis for any semigroup; if yes, how many; are all bases of the same cardinality? We can show that not every semigroup has a basis. However, if it has bases, then all its bases have the same cardinality.

3.44 Theorem. *Let F be free with basis B. Then $\langle B \rangle = F$.*

PROOF. Consider the diagram (and note that $B \subseteq F$)

Figure 6.9

with $\iota: B \to \langle B \rangle$, $b \mapsto b$. h exists and extends ι. Now look at

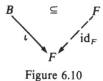

Figure 6.10

Now id_F also extends ι, hence $h = \mathrm{id}_F$ and $\langle B \rangle = F$. $\qquad \square$

It is useful to start with B and try to find a semigroup F such that B is a basis of F.

3.45 Theorem (Existence Theorem). *For any set $B \neq \varnothing$ there exists a semigroup F which is free on B.*

PROOF. Let F be the set of all finite sequences $(b_1, b_2, \ldots, b_n)$ of elements in B, also written as $b_1 b_2 \ldots b_n$. Then we have

$$b_1 b_2 \ldots b_n = b_1' b_2' \ldots b_m' \Leftrightarrow n = m \wedge b_1 = b_1' \wedge \ldots \wedge b_n = b_n'.$$

We define $b_1 b_2 \ldots b_n * c_1' c_2' \ldots c_m' := b_1 b_2 \ldots b_n c_1' c_2' \ldots c_m'$. The operation $*$ on sequences simply consists of writing the two sequences next to each other (called "*concatenation*" or "*juxtaposition*"). $(F, *)$ is a semigroup, again denoted by F. Now we have to show that F satisfies the conditions in 3.43. Let

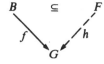

Figure 6.11

where h is defined by $h: F \to G$ by $h(b_1 b_2 \ldots b_n) := f(b_1) \circ f(b_2) \circ \ldots \circ f(b_n)$. Then $h|_B = f$; h is a homomorphism, if $h' \in \mathrm{Hom}(F, G)$ is an extension of f, then $\forall b_1 b_2 \ldots b_n \in F: h'(b_1 b_2 \ldots b_n) = h'(b_1) \circ h'(b_2) \circ \ldots \circ h'(b_n) = f(b_1) \circ f(b_2) \circ \ldots \circ f(b_n) = h(b_1 b_2 \ldots b_n)$, therefore $h = h'$. $\qquad \square$

3.46 Theorem. *Let F and F′ be free on B and B′, respectively. Then $F \cong F′ \Leftrightarrow |B| = |B′|$.*

PROOF. Let h be an isomorphism from F onto $F′$. For $b \in B$ let $h(b) = b_1′b_2′ \ldots b_n′ \in F′$. Then $b = h^{-1}(b_1′) \circ h^{-1}(b_2′) \circ \ldots \circ h^{-1}(b_n′)$, thus $n = 1$, since the representation is unique. Therefore $h|_B$ is a bijection from B onto $B′$ and $|B| = |B′|$. The converse follows by the reverse argument.   □

3.47 Corollary. *Let F and F′ be free on B. Then $F \cong F′$.*

Therefore we can speak of "the" free semigroup on B. Moreover

3.48 Corollary. *If F has two bases B and B′ then $|B| = |B′|$.*

Thus all bases for a free semigroup are of the same cardinality.

3.49 Notation. If F is free on B, F in 3.45 is also called the "*word semigroup over B*", the elements of F are *words* in the *alphabet B*. We denote by F_B the free semigroup on B and sometimes write F_B as F_β, where $\beta = |B|$, using 3.47.

Thus for any cardinal β, the free semigroup F_β with a basis of cardinality β exists.

3.50 Remarks. (i) If $B = \{b\}$ then $F_B = \{b, b \circ b = b^2, b^3, \ldots\}$ is isomorphic to $(\mathbb{N}, +)$ and is therefore commutative.

(ii) If $|B| > 1$ then F_B is not commutative, since $b_1, b_2 \in B$, $b_1 \neq b_2 \Rightarrow b_1 * b_2 = b_1 b_2 \neq b_2 b_1 = b_2 * b_1$.

(iii) F_B is infinite for every $B \neq \varnothing$.

No finite semigroup can be free, therefore no finite semigroup can have a basis. Thus free semigroups are rather "rare", but there are still plenty of them, since every semigroup is the homomorphic image of a free semigroup:

3.51 Theorem. *Let S be a semigroup with generating system E. Then S is a homomorphic image of F_E.*

PROOF. Consider

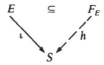

Figure 6.12

Then there exists an extension $h \in \text{Hom}(F_E, S)$ of ι. Therefore

$$E = \iota(E) = h(E) \subseteq h(F_E) \le S$$

and thus $S = \langle E \rangle \subseteq h(F_E) \subseteq S$. Hence h is an epimorphism. $\qquad\square$

This explains the name "free semigroup": F_B is "free" of all restrictive relations, such as commutativity (see 3.50(ii)) etc. By "adding" such relations we can obtain any semigroup. F_β is the most general semigroup which can be obtained from a generating set of cardinality β. Theorem 3.51 enables us to describe a semigroup in terms of its generating set and relations.

3.52 Definition. Let S be a semigroup generated by X, let R be a relation on F_X. Then $(X, R) := F_X/\tilde{R}$. If $S = (X, R)$, the pair (X, R) is called a *presentation* of S.

Elements of R are often called "defining relations", despite the fact that they are not relations at all. We note that R is by no means uniquely determined. The homomorphism theorem guarantees a presentation for every semigroup.

3.53 Example. Let $(S, \circ) = (\mathbb{Z}_4, \cdot)$. We choose $X = \{2, 3\}$ and let $x_1 = 2$, $x_2 = 3$. Let $R = \{(x_1x_2, x_2x_1), (x_2x_2x_1, x_1), (x_2x_2x_2, x_2), (x_1x_1x_1, x_1x_1), (x_1x_2, x_1)\}$. We consider the equivalence classes $a_2 := [x_1]$, $a_3 := [x_2]$, $a_0 := [x_1x_1]$, $a_1 := [x_2x_2]$ in $F_X/\tilde{R}$. Operation on these classes is as follows: $a_0a_0 = [x_1x_1][x_1x_1] = [x_1x_1x_1x_1] = [x_1x_1x_1] = [x_1x_1] = a_0$ (where the third and fourth equations hold because $(x_1x_1x_1, x_1x_1) \in \tilde{R}$); $a_0a_1 = [x_1x_1][x_2x_2] = [x_1x_1x_2x_2] = [x_1x_1x_2] = [x_1x_1] = a_0$ (where the third and fourth equations hold because $(x_1x_2, x_1) \in \tilde{R}$), etc. We obtain:

	a_0	a_1	a_2	a_3
a_0	a_0	a_0	a_0	a_0
a_1	a_0	a_1	a_2	a_3
a_2	a_0	a_2	a_0	a_2
a_3	a_0	a_3	a_2	a_1

Since the products of a_0, a_1, a_2, a_3 do not yield any new elements, we have $F_X/\tilde{R} \cong \{a_0, a_1, a_2, a_3\}$. Comparison with the operation table of $(\mathbb{Z}_4, \cdot)$ shows that $\{a_0, a_1, a_2, a_3\} \cong \mathbb{Z}_4$, therefore $F_X/\tilde{R} \cong \mathbb{Z}_4$ and $[\{x_1, x_2\}, R]$ is a presentation of $(\mathbb{Z}_4, \cdot)$. Theoretically we could have $a_0 = a_1$, etc., but the existence of $(\mathbb{Z}_4, \cdot)$ shows that (x_1x_1, x_2x_2) cannot be in $\tilde{R}$. $\qquad\square$

3.54 Definition. If we replace $\mathscr{S}$ in 3.43 by the class of all monoids then we obtain *free monoids*.

3.55 Theorem. *To any set $B \neq \varnothing$ there is (up to isomorphism) exactly one free monoid on B and any monoid is a homomorphic image of a free monoid.*

PROOF. It can be easily verified that $F_B^{(1)}$ is a (hence: *the*) free monoid on B. The adjoined identity element is often called the *empty word* since it can be interpreted as the "word which consists of none of the $b \in B$".   □

3.56 Example. While the free semigroup over $B = \{b\}$ is given by $(\mathbb{N}, +)$, the free monoid over $\{b\}$ is (isomorphic to) $(\mathbb{N}_0, +)$.   □

Let S_1 and S_2 be two semigroups presented by $[X, R_1]$ and $[X, R_2]$, respectively, then we have for $S = F_X$ that $[X, R_2]$ is a standard epimorphic image of $[X, R_1] \Leftrightarrow \tilde{R}_1 \subseteq \tilde{R}_2$. For this case we write $[X, R_2] \leqslant [X, R_1]$. For $\tilde{R}_1$ and $\tilde{R}_2$ there is a least congruence relation on F_X containing $\tilde{R}_1$ and $\tilde{R}_2$, namely the one generated by $\tilde{R}_1 \cup \tilde{R}_2$. Also there is exactly a greatest congruence relation on F_X which is contained in $\tilde{R}_1$ and $\tilde{R}_2$, namely $\tilde{R}_1 \cap \tilde{R}_2$. Since also $[X, R_2] \leq [X, R_1] \Rightarrow \tilde{R}_1 \subseteq \tilde{R}_2$. We deduce the so-called "*theorem of the standard epimorphic reduction and production*".

3.57 Theorem. *In the notation above, there exists for all semigroups S_1 and S_2 a least semigroup $\bar{S} = [X, \bar{R}]$ (under $\leqslant$) such that $[X, R_1] \leqslant [X, \bar{R}]$ and $[X, R_2] \leqslant [X, \bar{R}]$. Also, there exists a greatest semigroup $\underline{S} = [X, \underline{R}]$ with $[X, \underline{R}] \leqslant [X, R_1]$ and $[X, \underline{R}] \leqslant [X, R_2]$.*   □

$\bar{S}$ could be described as the "*least common multiple*" and $\underline{S}$ as the "*greatest common divisor*" of S_1 and S_2. A diagram illustrates the situation, the arrows indicate standard epimorphisms:

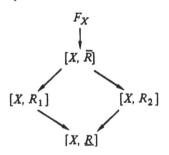

Figure 6.13

Theorem 3.57 makes it possible to measure a "distance" between finite semigroups with the same generating system. This can be done in several ways. We shall use the method, described by BOORMAN and WHITE, for sociological applications (see Chapter 7, §3). First we define a distance between homomorphic images. Let S_2 be the homomorphic image of a finite semigroup S_1. Then S_2 is finite and, by the homomorphism theorem, there is a congruence $\sim$ in S_1 with $S_1/\sim \cong S_2$. We assume $|S_1| > 1$.

Let $S_1 = K_1 \cup \ldots \cup K_n$ be the partition of S_1 into disjoint $\sim$-equivalence classes. Then we define

3.58 Definition.

$$\delta(S_1, S_2) := \binom{|S_1|}{2}^{-1} \sum_{i=1}^{n} \binom{|K_i|}{2}.$$

It is easily seen that $0 \le \delta(S_1, S_2) \le 1$. Moreover, $\delta(S_1, S_2) = 0 \Leftrightarrow$ all $|K_i| = 1 \Leftrightarrow S_1 \cong S_2$ and $\delta(S_1, S_2) = 1 \Leftrightarrow n = 1 \Leftrightarrow \sim$ is the universal relation $\Leftrightarrow |S_2| = 1$.

3.59 Example. Let $S_1 = (\mathbb{Z}_4, \cdot)$, $S_2 = (\mathbb{Z}_2, \cdot)$ and $H: \mathbb{Z}_4 \to \mathbb{Z}_2, [x]_4 \to [x]_2$. $S_1 = \{0, 2\} \cap \{1, 3\}$ is the corresponding partition of S_1 into classes K_1, K_2. We have $n = 2$, $|S_1| = 4$, $|K_1| = |K_2| = 2$ and thus

$$\delta(S_1, S_2) = \binom{4}{2}^{-1}\left(\binom{2}{2} + \binom{2}{2}\right) = \tfrac{1}{3}. \qquad \square$$

3.60 Definition. Let S_1, S_2 be finite semigroups with more than one element and the same generating set X, let S be their "greatest common divisor" (see 3.57). Then

$$d(S_1, S_2) := \tfrac{1}{2}(\delta(S_1, S) + \delta(S_2, S))$$

is called the *distance* between S_1 and S_2.

As above we have $0 \le d(S_1, S_2) \le 1$, $d(S_1, S_2) = 0 \Leftrightarrow S_1 \cong S_2$ and $d(S_1, S_2) = d(S_2, S_1)$. In general, the triangle inequality $d(S_1, S_3) \le d(S_1, S_2) + d(S_2, S_3)$ does not hold. Further, one can show that d is independent of the choice of generators. By relabeling the elements in S_1 or S_2 we can extend Definition 3.60 to finite semigroups with generating sets of the same cardinality.

PROBLEMS

1. Check the remarks in 3.50.

2. Give the multiplication table for the semigroup which is presented by $X = \{x\}$, $R = \{(x^5, x^3)\}$.

3. As in Problem 2, but now for $X = \{x, y\}$, $R = \{(xy, yx), (x^3, x), (y^2, y)\}$.

4. Find a presentation for $(\mathbb{Z}_4, +)$ with a one-element set X.

5. As in Problem 4, but now with a two-element set X.

6. As in Problem 4, now with a four-element set X.

7. Call a semigroup $(S, \circ)$ commutative if $\circ$ is commutative (hence $x \circ y = y \circ x$ holds for all $x, y \in S$). Define the concept of a free commutative semigroup on a set B.

8. Find the free commutative semigroup on $\{b\}$ and on $\{a, b\}$.

9. Show that $(\{2, 3, 4, \ldots\}, \cdot)$ is the free commutative semigroup over $B = \mathbb{P}$.

10. Compute the distance between $(\mathbb{Z}_3, +)$ and $(\mathbb{Z}_6, +)$.

11. Compute the distance between $(\mathbb{Z}_6, +)$ and $(\mathbb{Z}_8, +)$.

12. Compute the distance between $(\mathbb{Z}_3, +)$ and $(\mathbb{Z}_8, +)$ and check by looking at Problems 10 and 11 if the triangle inequality holds in this special case.

EXERCISES (Solutions in Chapter 8, p. 496)

1. (i) Find an isomorphism from $(\mathscr{P}(M), \cap)$ onto $(\mathscr{P}(N), \cap)$, if $|M| = |N|$.
    (ii) For $M \subseteq N$ find an embedding of $(\mathscr{P}(M), \cap)$ into $(\mathscr{P}(N), \cap)$.

2. Determine if $h: (\mathbb{Z}, +) \to (\mathbb{Z}_n, +), x \mapsto [x]$ is a homomorphism, an epimorphism, or a monomorphism. What does the homomorphism theorem 3.30 mean in this case?

*3. Let $(S, \circ)$ be a semigroup and $\varnothing \neq I \subseteq S$ with $IS \subseteq I$ and $SI \subseteq I$. In this case I is called an *ideal* of S. Show that $\sim$ defined by

$$x \sim y :\Leftrightarrow (x = y) \vee (x \in I \wedge y \in I)$$

is a congruence relation (called the *Rees congruence* with respect to I). Describe $(S/\sim, \odot)$.

4. Adjoin a zero element and an identity element to $(\mathscr{P}(\{0, 1\}), \cap)$ and find the resulting operation tables.

5. Give a proof of Theorem 3.30.

6. How many words of length $\leq n$ are in the free semigroup (the free monoid) over a set B with seven elements?

7. Let $S = (\mathbb{Z}_n, \cdot)$. Find a free semigroup F and an epimorphism from F onto S.

8. Are $(\mathbb{N}, +)$, $(\mathbb{N}_0, +)$, $(\mathbb{N}, \cdot)$, $(\mathbb{Q}, \cdot)$, $(\mathbb{R}, +)$ free semigroups?

9. Let H be presented by $[X, R]$ with $X = \{x_1, x_2\}$. What does it mean for H if (x_1x_2, x_2x_1) is in R?

*10. Find the semigroup H presented by X if $X = \{x_1, x_2, x_3\}$,

$$R = \{(x_1x_1, x_1), (x_2x_2, x_1), (x_3x_3, x_1), (x_1x_2, x_1), (x_1x_3, x_1), (x_2x_1, x_2),$$

$$(x_3x_1, x_1), (x_3x_2, x_2x_3)\}.$$

§4. Input Sequences

In a computer it would be rather artificial to consider only single input signals. We know that (calculating) programs consist of a sequence of elements of an input alphabet. Thus it is reasonable to consider the set of all finite sequences of elements of the set A, including the empty sequence Λ. In other words, in our study of automata we extend the input set A to the free monoid $\bar{A} := F_A^{(1)}$ (see 3.54) with Λ as identity.

We also extend δ and λ from $Z \times A$ to $Z \times \bar{A}$ by defining for $z \in Z$ and $a_1, a_2, \ldots, a_r \in A$:

$$\bar{\delta}(z, \Lambda) := z,$$

$$\bar{\delta}(z, a_1) := \delta(z, a_1),$$

$$\bar{\delta}(z, a_1 a_2) := \delta(\bar{\delta}(z, a_1), a_2),$$

$$\vdots$$

$$\bar{\delta}(z, a_1 a_2 \ldots a_r) := \delta(\bar{\delta}(z, a_1 a_2 \ldots a_{r-1}), a_r),$$

and

$$\bar{\lambda}(z, \Lambda) := \Lambda,$$

$$\bar{\lambda}(z, a_1) := \lambda(z, a_1),$$

$$\bar{\lambda}(z, a_1 a_2) := \lambda(z, a_1)\bar{\lambda}(\delta(z, a_1), a_2),$$

$$\vdots$$

$$\bar{\lambda}(z, a_1 a_2 \ldots a_r) := \lambda(z, a_1)\bar{\lambda}(\delta(z, a_1), a_2 a_3 \ldots a_r).$$

In this way we obtain functions $\bar{\delta}: Z \times \bar{A} \to Z$ and $\bar{\lambda}: Z \times \bar{A} \to \bar{B}$. The semiautomaton $\mathscr{S} = (Z, A, \delta)$ (the automaton $\mathscr{A} = (Z, A, B, \delta, \lambda)$) is thus generalized to the new semiautomaton $\bar{\mathscr{S}} := (Z, \bar{A}, \bar{\delta})$ (automaton $\bar{\mathscr{A}} = (Z, \bar{A}, \bar{B}, \bar{\delta}, \bar{\lambda})$, respectively). We can easily describe the operation of $\mathscr{S}$ and $\mathscr{A}$ if we let $z \in Z$ and $a_1, a_2 \ldots \in A$:

$$z_1 := z,$$

$$z_2 := \delta(z_1, a_1),$$

$$z_3 := \bar{\delta}(z_1, a_1 a_2) = \bar{\delta}(\delta(z_1, a_1), a_2) = \delta(z_2, a_2),$$

$$z_4 := \delta(z_3, a_3), \ldots.$$

If the (semi-) automaton is in state z and an input sequence $a_1 a_2 \ldots a_r \in \bar{A}$ operates, then the states are changed from $z = z_1$ to $z_2, z_3 \ldots$ until the final state z_{r+1} is obtained. As a result the output sequence is $\lambda(z_1, a_1)\lambda(z_2, a_2) \ldots \lambda(z_r, a_r)$.

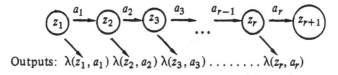

Outputs: $\lambda(z_1, a_1)\,\lambda(z_2, a_2)\,\lambda(z_3, a_3) \ldots \ldots \ldots \lambda(z_r, a_r)$

Figure 6.14

4.1 Example. Suppose the husband in Example 2.2 is bored; at first his wife is quiet (Input a_1), then she cooks (a_3) and then she shouts at her husband (a_2). Thus the input sequence $a_1a_3a_2$ necessitates a state sequence z_2, z_2, z_3, z_2 ($=$ final state); the output sequence is $b_2b_2b_2$ (very quiet indeed). Diagrammatically:

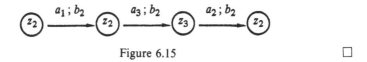

Figure 6.15   □

4.2 Example. Let $\mathscr{A}$ be the parity-check automaton 2.3 and let the input sequence 01101 operate on z_0. Then we have

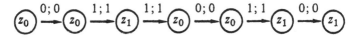

Figure 6.16

The final state is z_1, output 011010 (identical to the input).

In general we see: If this automaton starts in z_0, then its final state after an input sequence is z_0 if and only if the number of 1's in the input sequence is even (otherwise it terminates in z_1). This is the reason for the name "parity-check".   □

4.3 Examples.
   (i) *Shift register.* If the input is 011010 then, starting at $z = 0$, we have:

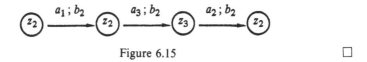

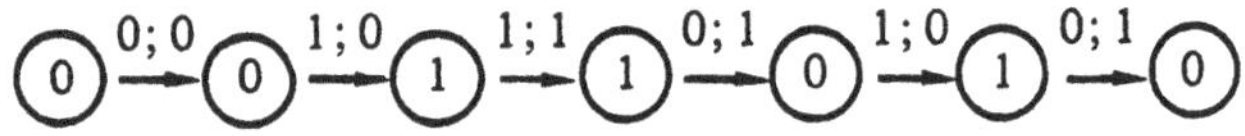

Figure 6.17

The final state is 0, the output is 001101. If we start at $z = 1$ we obtain:

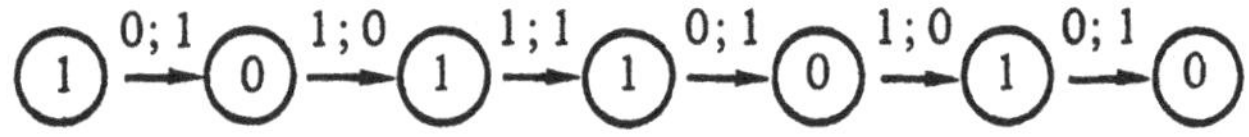

Figure 6.18

The final state is 0, output is 101101.

More generally: An input sequence $a_1a_2 \ldots a_r$ yields the output sequence $za_1a_2 \ldots a_{r-1}$, if we start in z. The $(i+1)$th state is $z_{i+1} = a_i$, the final state. This justifies the name "shift register", since this automaton "registers" z and "shifts" the input sequence by one place.

   (ii) *Trigger flip-flop.* Similar to (i) we see that if we start in z_1 an input sequence $a_1a_2 \ldots a_r$ induces new states $z_{i+1} = z_i + a_i$ (addition modulo 2 on the right-hand side), if we start in z_1. Thus each state is obtained from the previous one by adding the inputs. The output sequence is given as $z_1 z_2 \ldots z_r$. It is clear that such miniautomata can be used in building adding machines.

(iii) *IR flip-flop.* If we start in z then an input sequence $a_1 a_2 \ldots a_r$ gives us the final state a_k, where a_k is the last of the inputs a_i with $a_i \neq e$. If all $a_i = e$, then the automaton ends in z.   □

We see that the transition from $\mathcal{S}$ to $\bar{\mathcal{S}}$ or from $\mathcal{A}$ to $\bar{\mathcal{A}}$ greatly increases the operating power of the semiautomaton or automaton. Now we shall introduce some more algebraic concepts for automata.

PROBLEMS

1–4. Give interpretations of input/output sequences in the automata of Problems 1–4 in §1.

5. In Exercise 2 to §2, consider the input sequences $a_1 a_1 a_1$, $a_1 a_2 a_3$, $a_1 a_2 a_3 a_3 a_2 a_1$. Determine the corresponding output sequences (when started in z_1), and the corresponding final states.

6. Which input sequences in the preceding Problem 5 transfer the state z_1 into z_1?

7. Which input sequences in Problem 5 transfer each state into itself?

EXERCISES (Solutions in Chapter 8, p. 497)

1. An automaton $\mathcal{A}$ is given by the state graph

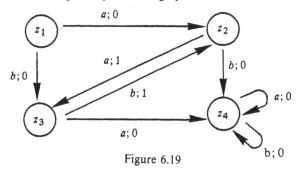

Figure 6.19

(i) Describe this automaton by finding Z, A, B, δ, λ (in form of a table).
(ii) Let $\mathcal{A}$ be in state z_2, when the input sequence *aaabbaa* begins. What is the final state and what is the output sequence?
(iii) Find all input sequences of shortest length, such that z_1 is transformed into z_4 and the corresponding output sequences have exactly two 1's.
(iv) Why can we describe z_4 as the "dead end" of this automaton?
(v) Find all input sequences which transform z_2 into z_3.
(vi) Let 01000 be an ouput sequence. Find which state has been transformed into which state.

*2. Using the notation introduced at the beginning of §4, show that

$$\bar{\lambda}(z, a_1 a_2 \dots a_r) = \bar{\lambda}(z, a_1 a_2 \dots a_{r-1}) \lambda(\bar{\delta}(z, a_1 a_2 \dots a_{r-1}), a_r).$$

Interpret both formulas for $\bar{\lambda}(z, a_1 a_2 \dots a_r)$.

3. In $\mathscr{A} = (Z, A, B, \delta, \lambda)$ let $\bar{\lambda}$ be given by $\bar{\lambda}: Z \times (\bar{A} \setminus \{\Lambda\}) \to B$, $(z, a_1 a_2 \dots a_n) \to \lambda(\delta(z, a_1 a_2 \dots a_{n-1}), a_n)$. Give a good interpretation of $\bar{\lambda}$ and show that $\bar{\lambda}(z, a_1 a_2 \dots a_n a_{n+1}) = \bar{\lambda}(\delta(z, a_1, \dots, a_{n-1}), a_n a_{n+1})$ holds for all $z \in Z$, $n \in \mathbb{N}$ and $a_i \in A$.

*4. For a set A and for $\alpha = a_1 a_2 \dots a_n \in \bar{A}$ let $|\alpha| = n$ denote the *length* of α. A function $f: \bar{A} \to \bar{B}$ is called *sequential* if $|f(\alpha)| = |\alpha|$ for all $\alpha \in \bar{A}$ and if for $\alpha_1, \alpha_2 \in \bar{A}$, $f(\alpha_1 \alpha_2)$ is of the form $f(\alpha_1) a_1' \dots a_m'$ with $m \in \mathbb{N}_0$ and $a_1', \dots, a_m' \in A$. Further, $f: \bar{A} \to \bar{B}$ is said to be *generated* by $(Z, A, B, \delta, \lambda)$ with respect to $z \in Z$, if $f(\alpha) = \bar{\lambda}(z, \alpha)$ holds for all $\alpha \in A$. Show that $f: \bar{A} \to \bar{B}$ can be generated by an automaton with respect to one of its states iff f is sequential.

5. For which δ will $(\{z_0, z_1, z_2, z_3\}, \{a, b, c\}, \delta)$ be a semiautomaton, such that exactly those input sequences $a_1 a_2 \dots a_n$ change z_1 into z_0, in which $a_i = a$ for every odd i?

§5. The Monoid of a (Semi-) Automaton and the (Semi-) Automaton of a Monoid

We now show that any (semi-) automaton determines a certain monoid and conversely that any monoid gives rise to a certain (semi-) automaton. Let $\mathscr{S} = (Z, A, \delta)$ be a semiautomaton. We consider $\bar{\mathscr{S}} = (Z, \bar{A}, \bar{\delta})$ as introduced in §4.

5.1 Notation. For $\bar{a} \in \bar{A}$ let $f_{\bar{a}}: Z \to Z$, $z \to \bar{\delta}(z, \bar{a})$.

5.2 Theorem and Definition. $(\{f_{\bar{a}} | \bar{a} \in \bar{A}\}, \circ) =: M_{\mathscr{S}}$ *is a monoid* (*submonoid of* $(Z^Z, \circ)$), *called the monoid of* $\mathscr{S}$. *We have*

$$\forall \, \bar{a}, \bar{a}' \in \bar{A}: f_{\bar{a}} \circ f_{\bar{a}'} = f_{\bar{a}'\bar{a}}.$$

The monoid of the automaton $(Z, A, B, \delta, \lambda)$ *is the monoid of* (Z, A, δ).

PROOF. Composition of mappings is associative. For $\bar{a}, \bar{a}' \in \bar{A}$ we have

$$(f_{\bar{a}} \circ f_{\bar{a}'})(z) = f_{\bar{a}}(f_{\bar{a}'}(z)) = f_{\bar{a}}(\bar{\delta}(z, \bar{a}')) = \delta(\bar{\delta}(z, \bar{a}'), \bar{a}) = f_{\bar{a}'\bar{a}}(z),$$

thus $f_{\bar{a}} \circ f_{\bar{a}'} = f_{\bar{a}'\bar{a}}$. Here $\bar{a}'\bar{a}$ denotes the "product" in $\bar{A}$, i.e. concatenation of sequences. $\{f_{\bar{a}} | \bar{a} \in \bar{A}\}$ is closed with respect to $\circ$, therefore is a semigroup and because of $\mathrm{id}_Z = f_\Lambda$ it is a monoid. $\qquad \square$

5.3 Definition. Let $(S, \circ)$ and $(S', \circ)$ be semigroups and $f: S \to S'$. f is called an *antihomomorphism*, if $\forall \, x, y \in S: f(x \circ y) = f(y) \circ f(x)$. An *antiisomorphism* is a bijective antihomomorphism.

If $\mathscr{S}$ is finite (it is sufficient to require that Z is finite) then so is $M_{\mathscr{S}}$. It is an antihomomorphic image of $\bar{A}$ as the following result shows.

5.4 Theorem. *In A let $\bar{a}_1 \equiv \bar{a}_2 :\Leftrightarrow f_{\bar{a}_1} = f_{\bar{a}_2}$. Then $\equiv$ is a congruence relation in $\bar{A}$ and $\bar{A}/\equiv$ is antiisomorphic to $M_{\mathscr{S}}$.*

PROOF. The mapping $f: \bar{A} \to M_{\mathscr{S}}, \bar{a} \mapsto f_{\bar{a}}$ is an antihomomorphism, by 5.2, therefore a homomorphism from A into $M'_{\mathscr{S}}$, where $M'_{\mathscr{S}} := (\{f_{\bar{a}} | \bar{a} \in A\}, \circ')$ with $f_{\bar{a}_1} \circ' f_{\bar{a}_2} := f_{\bar{a}_2} \circ f_{\bar{a}_1}$. The homomorphism theorem 3.30 implies that $\bar{A}/\equiv$ is isomorphic to $M'_{\mathscr{S}}$, therefore $M_{\mathscr{S}}$ is antiisomorphic to $\bar{A}/\equiv$.   □

What is the meaning of $\equiv$? We have

$$\bar{a}_1 \equiv \bar{a}_2 \Leftrightarrow f_{\bar{a}_1} = f_{\bar{a}_2} \Leftrightarrow \forall\, z \in Z: f_{\bar{a}_1}(z) = f_{\bar{a}_2}(z)$$
$$\Leftrightarrow \forall\, z \in Z: \delta(z, \bar{a}_1) = \delta(z, \bar{a}_2).$$

Thus $\bar{a}_1$ and $\bar{a}_2$ are equivalent with respect to $\equiv$ if and only if they operate in the same way on each state. If Z is finite then so is $M_{\mathscr{S}}$ and also $\bar{A}/\equiv$. How can we calculate $M_{\mathscr{S}}$ explicitly? We start with an example.

5.5 Example. Let $\mathscr{S}$ be the automaton of Example 2.1. First we construct the table $f_\Lambda, f_{a_1}, f_{a_2}, f_{a_3}$:

	f_Λ	f_{a_1}	f_{a_2}	f_{a_3}
$z_1 \mapsto$	z_1	z_1	z_1	z_3
$z_2 \mapsto$	z_2	z_2	z_1	z_3
$z_3 \mapsto$	z_3	z_3	z_2	z_3

Since $f_\Lambda = f_{a_1}$ we delete, say, f_{a_1} and test to see if $\{f_\Lambda, f_{a_2}, f_{a_3}\}$ forms a monoid. The operations are given as follows:

$\circ$	f_Λ	f_{a_2}	f_{a_3}
f_Λ	f_Λ	f_{a_2}	f_{a_3}
f_{a_2}	f_{a_2}	$f_{a_2 a_2}$	$f_{a_3 a_2}$
f_{a_3}	f_{a_3}	f_{a_3}	f_{a_3}

$f_{a_2 a_2}$ and $f_{a_3 a_2}$ are not elements of $\{f_\Lambda, f_{a_2}, f_{a_3}\}$ because

	$f_{a_2 a_2}$	$f_{a_3 a_2}$
$z_1 \mapsto$	z_1	z_2
$z_2 \mapsto$	z_1	z_2
$z_3 \mapsto$	z_1	z_2

Thus $\{f_\Lambda, f_{a_2}, f_{a_3}\}$ is not a monoid. We extend it to $\{f_\Lambda, f_{a_2}, f_{a_3}, f_{a_2 a_2}, f_{a_3 a_2}\}$ and verify that this set is a monoid. This must be the monoid we are looking for, since all $f_{\bar{a}}$ ($\bar{a} \in \bar{A}$) can be obtained by composition of the given functions (this can be shown by induction). $\quad\square$

From this we deduce an often cumbersome method: We construct the table of all f_a, $a \in A$ and $f_\Lambda = \mathrm{id}_Z$, take multiple copies only once and form the operation table. If we obtain new functions in the table, then we have to extend the table until we obtain the operation table of a semigroup. This then is the monoid $M_{\mathcal{S}}$. Since $|Z|$ can be large, say 10^{10}, the corresponding monoid can have up to $10^{10^{11}}$ elements, which obviously makes an explicit table nearly impossible.

5.6 Example. The parity-check automaton and the trigger flip-flop have (isomorphic) monoids, with two elements. The shift-register and the IR flip-flop have the monoid $\{f_\Lambda, f_0, f_1\}$. $\quad\square$

Conversely, to any monoid there corresponds an automaton:

5.7 Theorem. *For any monoid* $(S, \circ)$ *there exists a* (*semi-*) *automaton whose monoid is isomorphic to* $(S, \circ)$.

PROOF. Let $(S, \circ)$ be a monoid with identity e. We form the semiautomaton $\mathcal{S}_S := (S, S, \delta)$ with $\delta \colon S \times S \to S$, $(s_1, s_2) \mapsto s_2 \circ s_1$. Now we find the monoid of $\mathcal{S}_S$ according to 5.5. For f_s, $s \in S$, we have: $\forall\, s' \in S \colon f_s(s') = \delta(s', s) = s \circ s'$. For $s_1 \neq s_2$ we have $f_{s_1}(e) = s_1 \circ e = s_1 \neq s_2 = s_2 \circ e = f_{s_2}(e)$, thus $f_{s_1} \neq f_{s_2}$. Moreover we have $f_{s_1} \circ f_{s_2} = f_{s_1 \circ s_2}$ for all $s_1, s_2 \in S$. Therefore $(\{f_s | s \in S\}, \circ)$ is a monoid, namely the monoid of $\mathcal{S}_S$. The mapping $g \colon S \to M_{\mathcal{S}_S}$, $s \mapsto f_s$ is injective and surjective by definition, and because $g(s_1 \circ s_2) = f_{s_1 \circ s_2} = g(s_1) \circ g(s_2)$ it is an isomorphism. In the case of automata we consider $\mathcal{A}_S := (S, S, S, \delta, \lambda)$ with δ as above and $\lambda \colon S \times S \to S$, $(s_1, s_2) \mapsto s_1$; this is a Moore automaton. $\quad\square$

We see that $\mathcal{S}_S$ is finite if and only if S is finite. Note that $\{\lambda(z, \bar{a}) | z \in Z \wedge \bar{a} \in \bar{A}\}$, the set of obtainable output sequences, does not have to be a monoid (submonoid of $\bar{B}$) for an automaton $\mathcal{A}$.

PROBLEMS

1–4. Find the monoids of the automata in Exercises 1, 3, 4, 6 to §2.
5. Let $(S, \circ)$ and $(S', \circ')$ be semigroups and let $x \circ_{\mathrm{op}} y := y \circ x$.
   (i) Show that $(S, \circ_{\mathrm{op}})$ is a semigroup (called the opposite of $(S, \circ)$).
   (ii) If $h \colon (S, \circ) \to (S', \circ')$ is a homomorphism, show that $h_{\mathrm{op}} \colon (S, \circ_{\mathrm{op}}) \to (S', \circ')$ is an antihomomorphism.

6. What is the maximal number of elements in the monoid of a semiautomaton $\mathcal{S}$ if the state set in $\mathcal{S}$ has 20 elements? Does the size of the input set have any influence on the answer?

EXERCISES (Solutions in Chapter 8, p. 498)

1. Let $\mathcal{A}$ be as in Exercise 1 in §4. Find its monoid.

2. Prove the statements made in Example 5.6.

3. Find an automaton for each of the monoids $(\mathbb{Z}_3, +)$ and $(\mathbb{Z}_3, \cdot)$.

4. Start with an automaton $\mathcal{A}$, form $M_\mathcal{A}$ and then $\mathcal{A}_{M_\mathcal{A}}$. Is this the same as $\mathcal{A}$?

5. Find a semiautomaton whose monoid is a group.

6. Can there exist infinite automata whose monoid is finite?

§6. Composition and Decomposition

In sets, semigroups, groups, rings, vector spaces etc. we have three important constructions, the formation of substructures (subsets, subsemigroups, etc.), factor structures, and direct (cartesian) products. We shall introduce similar concepts for (semi-) automata and consider the respective results in the corresponding monoids. Moreover we shall find other constructions. Unless otherwise stated the material will be given for automata. The corresponding definitions and theorems for semiautomata are obtained by omitting outputs and output functions.

A. Elementary Constructions of Automata

6.1 Definition. $\mathcal{A}_1 = (Z_1, A, B, \delta_1, \lambda_1)$ is called a *subautomaton* of $\mathcal{A}_2 = (Z_2, A, B, \delta_2, \lambda_2)$ (in symbols: $\mathcal{A}_1 \leq \mathcal{A}_2$), if $Z_1 \subseteq Z_2$ and δ_1 and λ_1 are the restrictions of δ_2 and λ_2, respectively, on $Z_1 \times A$.

Subautomata of $\mathcal{A}$ thus have the same input and output alphabets as $\mathcal{A}$. The state set of $\mathcal{A}_1 (\leq \mathcal{A}_2)$ must be a subset of the state set of $\mathcal{A}_2$, such that in the diagram representation no arrow goes outside the subset.

6.2 Example.

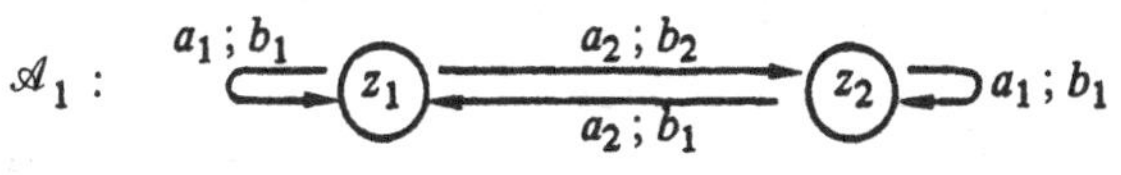

Figure 6.20

is a subautomaton of

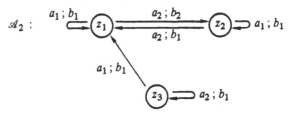

Figure 6.21

$$\mathcal{A}_1 = (\{z_1, z_2\}, \{a_1, a_2\}, \{b_1, b_2\}, \delta_1, \lambda_1)$$
$$\leq \mathcal{A}_2 = (\{z_1, z_2, z_3\}, \{a_1, a_2\}, \{b_1, b_2\}, \delta_2, \lambda_2). \qquad \square$$

What is the meaning of $\mathcal{A}_1 \leq \mathcal{A}_2$ for $M_{\mathcal{A}_1}$ and $M_{\mathcal{A}_2}$? Initially we might guess that $M_{\mathcal{A}_1} \leq M_{\mathcal{A}_2}$. However, this is impossible if $Z_1 \neq Z_2$ for $M_{\mathcal{A}_1} \leq Z_1^{Z_1}$, but $M_{\mathcal{A}_2} \leq Z_2^{Z_2}$, therefore $M_{\mathcal{A}_1}$ cannot even be a subset of $M_{\mathcal{A}_2}$.

The correct answer is:

6.3 Theorem. $\mathcal{A}_1 \leq \mathcal{A}_2 \Rightarrow M_{\mathcal{A}_1}$ is a homomorphic image of $M_{\mathcal{A}_2}$.

PROOF. Let $M_{\mathcal{A}_1} := \{f_{\bar{a}}^{(1)} | \bar{a} \in \bar{A}\}$ and $M_{\mathcal{A}_2} := \{f_{\bar{a}}^{(2)} | \bar{a} \in \bar{A}\}$. The mapping $\varphi: M_{\mathcal{A}_2} \to M_{\mathcal{A}_1}; f_{\bar{a}}^{(2)} \mapsto f_{\bar{a}}^{(1)}$ is well defined, since

$$f_{\bar{a}}^{(2)} = f_{\bar{a}'}^{(2)} \Rightarrow \forall\, z \in Z_2: \bar{\delta}_2(z, \bar{a}) = \bar{\delta}_2(z, \bar{a}') \Rightarrow \forall\, z \in Z_1: \bar{\delta}_1(z, \bar{a})$$
$$= \bar{\delta}_2(z, \bar{a}) = \bar{\delta}_2(z, \bar{a}') = \bar{\delta}_1(z, \bar{a}') \Rightarrow f_{\bar{a}}^{(1)} = f_{\bar{a}'}^{(1)}.$$

φ is surjective and φ is a homomorphism of monoids, since

$$\forall\, f_{\bar{a}_1}^{(2)}, f_{\bar{a}_2}^{(2)} \in M_{\mathcal{A}_2}: \varphi(f_{\bar{a}_1}^{(2)} \circ f_{\bar{a}_2}^{(2)}) = \varphi(f_{\bar{a}_2 \circ \bar{a}_1}^{(2)})$$
$$= f_{\bar{a}_2 \circ \bar{a}_1}^{(1)} = f_{\bar{a}_1}^{(1)} \circ f_{\bar{a}_2}^{(1)} = \varphi(f_{\bar{a}_1}^{(2)}) \circ \varphi(f_{\bar{a}_2}^{(2)}).$$

Finally, $\varphi(\mathrm{id}_{Z_2}) = \varphi(f_{\Lambda}^{(2)}) = f_{\Lambda}^{(1)} = \mathrm{id}_{Z_1}. \qquad \square$

A comparison of the parity-check and the trigger flip-flop (5.6) shows that the converse of 6.3 is incorrect. Even isomorphism of $M_{\mathcal{A}_1}$ and $M_{\mathcal{A}_2}$ gives very little information about the relationship between $\mathcal{A}_1$ and $\mathcal{A}_2$.

To construct a factor automaton, we first need the definition of an (automaton) homomorphism.

6.4 Definition. Let $\mathcal{A}_1 = (Z_1, A_1, B_1, \delta_1, \lambda_1)$ and $\mathcal{A}_2 = (Z_2, A_2, B_2, \delta_2, \lambda_2)$ be automata. An (*automata-*) *homomorphism* $\Phi: \mathcal{A}_1 \to \mathcal{A}_2$ is a triple $\Phi = (\zeta, \alpha, \beta) \in Z_2^{Z_1} \times A_2^{A_1} \times B_2^{B_1}$ with the property

$$\zeta(\delta_1(z, a)) = \delta_2(\zeta(z), \alpha(a)),$$
$$\beta(\delta_1(z, a)) = \lambda_2(\zeta(z), \alpha(a)), \qquad (z \in Z, a \in A_1).$$

Φ is called a *monomorphism* (*epimorphism, isomorphism*) if all functions ζ, α and β are injective (surjective, bijective). If there is an isomorphism from $\mathcal{A}_1$ onto $\mathcal{A}_2$, then $\mathcal{A}_1$ and $\mathcal{A}_2$ are called isomorphic (in symbols $\mathcal{A}_1 \cong \mathcal{A}_2$).

6.5 Theorem. *Isomorphism of automata is an equivalence relation on any set of automata.*

PROOF. (1) The mapping $(\mathrm{id}, \mathrm{id}, \mathrm{id}): \mathcal{A} \to \mathcal{A}$ is always an isomorphism, therefore $\cong$ is reflexive.

(2) If $\Phi = (\zeta, \alpha, \beta)$ is an isomorphism from $\mathcal{A}_1$ onto $\mathcal{A}_2$ then $\Phi^{-1} := (\zeta^{-1}, \alpha^{-1}, \beta^{-1})$ exists and is an isomorphism from $\mathcal{A}_2$ onto $\mathcal{A}_1$. Thus $\cong$ is symmetric.

(3) It is easy to see that the composition of isomorphisms is an isomorphism, thus $\cong$ is transitive. $\square$

Without undue complications, the following is immediately applicable only to semiautomata. Therefore we restrict our attention to semiautomata. Let $\mathcal{S} = (Z, A, \delta)$ and $\mathcal{S}' = (Z', A', \delta')$ be semiautomata and $\Phi = (\zeta, \alpha)$ an epimorphism from $\mathcal{S}$ onto $\mathcal{S}'$. We define a relation $\sim_\zeta$ on Z as follows:

$$z_1 \sim_\zeta z_2 :\Leftrightarrow \zeta(z_1) = \zeta(z_2).$$

$\sim_\zeta$ is an equivalence relation. Let $[z]$ be the equivalence class of z with respect to $\sim_\zeta$. Define

$$\delta_\zeta: Z/\sim_\zeta \times A \to Z/\sim_\zeta, \qquad ([z], a) \mapsto [\delta(z, a)],$$

$$\pi: Z \to Z/\sim_\zeta, \qquad z \mapsto [z],$$

$$\bar{\zeta}: Z/\sim_\zeta \to Z', \qquad [z] \mapsto \zeta(z).$$

6.6 Theorem and Definition ("Homomorphism Theorem"). *Using the notation above we have $\mathcal{S}/\sim_\zeta := (Z/\sim_\zeta, A, \delta_\zeta)$ is a semiautomaton, called the factor semiautomaton of $\mathcal{S}$ over $\sim_\zeta$. (π, id_A) is an epimorphism from $\mathcal{S}$ onto $\mathcal{S}/\sim_\zeta$ and $(\bar{\zeta}, \alpha)$ is an epimorphism from $\mathcal{S}/\sim_\zeta$ onto $\mathcal{S}'$.*

PROOF. First we show that $\bar{\zeta}$ is well defined. $[z_1] = [z_2] \Rightarrow z_1 \sim_\zeta z_2$, therefore $\zeta(z_1) = \zeta(z_2)$. Now we get for all $a \in A$:

$$\zeta(\delta(z_1, a)) = \delta'(\zeta(z_1), \alpha(a)) = \delta'(\zeta(z_2), \alpha(a))$$

$$= \zeta(\delta(z_2, a)), \quad \text{thus} \quad [\delta(z_1, a)] = [\delta(z_2, a)],$$

we see that δ_ζ is well defined. That $\pi, \mathrm{id}_A, \bar{\zeta}$ and α are surjective is easily verified as also are the homomorphism properties of (π, id_A) and $(\bar{\zeta}, \alpha)$. $\square$

6.7 Example. Let

$$\mathcal{S}_2 := (\{z_1, z_2, z_3\}, \{a_1, a_2\}, \delta_2) \quad \text{and} \quad \mathcal{S}_1 := (\{z_1, z_2\}, \{a_1, a_2\}, \delta_1)$$

be the semiautomata corresponding to the automata $\mathcal{A}_2$ and $\mathcal{A}_1$ in 6.2. Let $\zeta: \{z_1, z_2, z_3\} \mapsto \{z_1, z_2\}$ be defined as $\zeta(z_1) = z_1$, $\zeta(z_2) = z_2$, $\zeta(z_3) = z_1$. Then (ζ, id) is an epimorphism from $\mathcal{S}_2$ onto $\mathcal{S}_1$. Therefore $\mathcal{S}_1$ is a subsemiautomaton and a homomorphic image of $\mathcal{S}_2$. $\square$

6.8 Theorem. *If the automaton $\mathcal{A}'$ is a homomorphic image of $\mathcal{A}$, then $M_{\mathcal{A}'}$ is a homomorphic image of $M_{\mathcal{A}}$.*

PROOF. Let (ζ, α, β) be an epimorphism from $\mathcal{A} = (Z, A, B, \delta, \lambda)$ onto $\mathcal{A}' = (Z', A', B', \delta', \lambda')$. We regard $\alpha: A \to A'$ as a mapping from A into $\bar{A}'$. Then there exists a uniquely determined extension to a homomorphism $\bar{\alpha}: \bar{A} \to \bar{A}'$, since $\bar{A}$ is free (see 3.43'):

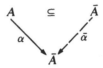

Figure 6.22

We show that $\varphi: M_{\mathcal{A}} \to M_{\mathcal{A}'}, f_{\bar{a}} \mapsto f_{\bar{\alpha}(\bar{a})}$ is a monoid epimorphism.

(1) φ is well defined, since $f_{\bar{a}_1} = f_{\bar{a}_2}$ implies $\forall z \in Z: \bar{\delta}(z, \bar{a}_1) = \bar{\delta}(z, \bar{a}_2)$, therefore $\zeta(\bar{\delta}(z, \bar{a}_1)) = \zeta(\bar{\delta}(z, \bar{a}_2))$. By induction on the "length" of $\bar{a} \in \bar{A}$ we obtain $\zeta(\bar{\delta}(z, \bar{a})) = \bar{\delta}'(\zeta(z), \bar{\alpha}(\bar{a}))$, therefore $\bar{\delta}'(\zeta(z), \bar{\alpha}(\bar{a}_1)) = \bar{\delta}'(\zeta(z), \bar{\alpha}(\bar{a}_2))$. Since ζ is surjective, we have $\forall z' \in Z': \bar{\delta}'(z', \bar{\alpha}(\bar{a}_1)) = \bar{\delta}'(z', \bar{\alpha}(\bar{a}_2))$ and thus $f_{\bar{\alpha}(\bar{a}_1)} = f_{\bar{\alpha}(\bar{a}_2)}$.

(2) $\forall \bar{a}_1, \bar{a}_2 \in \bar{A}$:

$$\varphi(f_{\bar{a}_1} \circ f_{\bar{a}_2}) = \varphi(f_{\bar{a}_2\bar{a}_1}) = f_{\bar{\alpha}(\bar{a}_2\bar{a}_1)} = f_{\bar{\alpha}(\bar{a}_2)\bar{\alpha}(\bar{a}_1)} = f_{\bar{\alpha}(\bar{a}_1)} \circ f_{\bar{\alpha}(\bar{a}_2)} = \varphi(f_{\bar{a}_1}) \circ \varphi(f_{\bar{a}_2}).$$

(3) $\varphi(f_\Lambda) = f_{\bar{\alpha}(\Lambda)} = f_\Lambda$; and so because of (2) and (3) φ is a homomorphism.

(4) Because of 3.51, $\bar{\alpha}$ is surjective. Also $\bar{\alpha}(\bar{A}) = \bar{A}'$, thus φ is surjective. $\square$

Next we consider the third type of constructions of new structures from given ones, namely direct products.

6.9 Definition. Let $\mathcal{A}_i := (Z_i, A_i, B_i, \delta_i, \lambda_i)(i \in \{1, 2\})$ be automata. The *direct product* $\mathcal{A}_1 \times \mathcal{A}_2$ of $\mathcal{A}_1$ and $\mathcal{A}_2$ is defined by $\mathcal{A}_1 \times \mathcal{A}_2 := (Z_1 \times Z_2, A_1 \times A_2, B_1 \times B_2, \delta, \lambda)$ with

$$\delta((z_1, z_2), (a_1, a_2)) := (\delta_1(z_1, a_1), \delta_2(z_2, a_2)),$$

$$\lambda((z_1, z_2), (a_1, a_2)) := (\lambda_1(z_1, a_1), \lambda_2(z_2, a_2)),$$

for all $(z_1, z_2) \in Z_1 \times Z_2$ and $(a_1, a_2) \in A_1 \times A_2$.

6.10 Remark. $\mathcal{A}_1 \times \mathcal{A}_2$ is often called the *parallel composition* of $\mathcal{A}_1$ and $\mathcal{A}_2$, since Z_1 and Z_2 can be interpreted as two parallel blocks. A_i operates

on Z_i with output B_i ($i \in \{1, 2\}$), $A_1 \times A_2$ operates on $Z_1 \times Z_2$, the outputs
are in $B_1 \times B_2$ (see Figure 6.23).

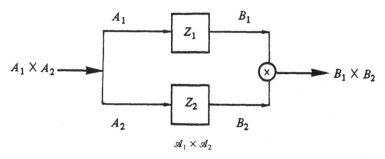

Figure 6.23

6.11 Remark. Sometimes the direct product of semiautomata is defined in
the following manner: For $\mathscr{S}_1 = (Z_1, A, \delta_1)$ and $\mathscr{S}_2 = (Z_2, A, \delta_2)$ (the input
alphabets must be equal) let $\mathscr{S}_1 \otimes \mathscr{S}_2 := (Z_1 \times Z_2, A, \delta)$ with $\delta((z_1, z_2), a) :=$
$(\delta_1(z_1, a), \delta_2(z_2, a))$ for $(z_1, z_2) \in Z_1 \times Z_2$ and $a \in A$. However, this definition is a special case of the above. In fact if we replace $\mathscr{S}_1 = (Z_1, A_1, \delta_1)$
and $\mathscr{S}_2 = (Z_2, A_2, \delta_2)$, by $\mathscr{S}_1' := (Z_1, A_1 \times A_2, \delta_1')$ and $\mathscr{S}_2' := (Z_2, A_1 \times
A_2, \delta_2')$, with $\delta_1'(z_1, (a_1, a_2)) := \delta_1(z_1, a_1)$ and $\delta_2'(z_2, (a_1, a_2)) := \delta_2(z_2, a_2)$,
then $\mathscr{S}_1 \times \mathscr{S}_2 = \mathscr{S}_1' \otimes \mathscr{S}_2'$.

The parallel composition prompts the question; "Is there a series composition?" Such a composition is easy to obtain if there is a "connection"
between $\mathscr{A}_1$ and $\mathscr{A}_2$.

6.12 Definition. Let $\mathscr{A}_1$, $\mathscr{A}_2$ be as in 6.9, with the additional assumption
$A_2 = B_1$. The *series composition* $\mathscr{A}_1 \not\parallel \mathscr{A}_2$ of $\mathscr{A}_1$ and $\mathscr{A}_2$ is defined as the
automaton $(Z_1 \times Z_2, A_1, B_2, \delta, \lambda)$ with

$$\delta(z_1, z_2), a_1) := (\delta_1(z_1, a_1), \delta_2(z_2, \lambda_1(a_1, a_1))),$$

$$\lambda((z_1, z_2), a_1) := \lambda_2(z_2, \lambda_1(z_1, a_1)), \qquad ((z_1, z_2) \in Z_1 \times Z_2, a_1 \in A).$$

This automaton operates as follows: An input $a_1 \in A_1$ operates on z_1
and gives a state transition into $z_1' := \delta_1(z_1, a_1)$ and an output $b_1 :=$
$\lambda_1(z_1, a_2) \in B_1 = A_2$. This output b_1 operates on Z_2, transforms a $z_2 \in Z_2$
into $z_2' := \delta_2(z_2, b_1)$ and produces the output $\lambda_2(z_2, b_1)$. Then $\mathscr{A}_1 \not\parallel \mathscr{A}_2$ is
in the next state (z_1', z_2') (see Figure 6.24).

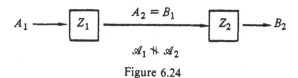

Figure 6.24

6.13 Remark. In the definition of series composition of semiautomata we have obvious difficulties, since the prerequisite $B_1 = A_2$ does not make sense. Here we assume $B_1 := Z_1 \times A_1 = A_2$ and define $\lambda_1 := \mathrm{id}_{B_1}$.

We would expect that we could construct many complicated automata by using parallel and series compositions. However, we can do more. We can obtain all finite automata in this way, starting from a few basic types of automata. We return to this in 6.26.

Now we ask for $M_{\mathscr{A}_1 \times \mathscr{A}_2}$ and $M_{\mathscr{A}_1 \# \mathscr{A}_2}$. It is not obvious how to proceed in the case of $M_{\mathscr{A}_1 \# \mathscr{A}_2}$. We might expect something like $M_{\mathscr{A}_1 \times \mathscr{A}_2} = M_{\mathscr{A}_1} \times M_{\mathscr{A}_2}$ in case of the product. However, ARBIB[1] gives an example for $M_{\mathscr{A}_1} = M_{\mathscr{A}_2} = M_{\mathscr{A}_1 \times \mathscr{A}_2} = \mathbb{Z}_2$. Thus the situation is more complicated than one might think, and we need more tools to answer this question.

PROBLEMS

1. In the notation of 6.1–6.3, can one embed $Z_1^{Z_1}$ in $Z_2^{Z_2}$? Does this show that $M_{\mathscr{A}_1}$ can be "embedded" in $M_{\mathscr{A}_2}$?

2. In Example 6.2, construct an epimorphism from $M_{\mathscr{A}_1}$ to $M_{\mathscr{A}_2}$.

3. Construct the corresponding factor automaton to $\mathscr{A}_1$ and $\mathscr{A}_2$ in 6.2.

4. Let $\mathscr{A}_1$ be the shift register and $\mathscr{A}_2$ the trigger flip-flop. Does there exist a homomorphism from $\mathscr{A}_1$ to $\mathscr{A}_2$? An epimorphism? An isomorphism?

5. Let $\mathscr{A}_1$, $\mathscr{A}_2$ be as in Problem 4. Answer the same questions as in Problem 4 for the monoids of these two automata.

6. Again, let $\mathscr{A}_1$ and $\mathscr{A}_2$ be as in Problem 4. Construct $\mathscr{A}_1 \times \mathscr{A}_1$, $\mathscr{A}_1 \times \mathscr{A}_2$, $\mathscr{A}_2 \times \mathscr{A}_1$ and $\mathscr{A}_2 \times \mathscr{A}_2$. Compare the results.

7. Do the same as in Problem 6 for the four corresponding series compositions of $\mathscr{A}_1$ and $\mathscr{A}_2$.

*B. Cascades

We introduce a further method of constructing automata and show that this provides a generalization of series and parallel decomposition.

6.14 Definition. Let $\mathscr{A}_1$ and $\mathscr{A}_2$ be as in 6.9. Let $X \neq \varnothing$ be a set and $\varphi: X \times B_1 \to A_2$ and $\eta: X \to A_1$ mappings. Then the automaton

$$\mathscr{A}_1 \times_{\varphi}^{\eta} \mathscr{A}_2 := (Z_1 \times Z_2, X, B_1 \times B_2, \delta_{\varphi}^{\eta}, \lambda_{\varphi}^{\eta}),$$

with

$$\delta_{\varphi}^{\eta}((z_1, z_2), x) := (\delta_1(z_1, \eta(x)), \delta_2(z_2, \varphi(x, \lambda_1(z_1, \eta(x))))),$$

$$\lambda_{\varphi}^{\eta}((z_1, z_2), x) := (\lambda_1(z_1, \eta(x)), \lambda_2(z_2, \varphi(x, \lambda_1(z_1, \eta(x))))),$$

is called the *cascade* of $\mathscr{A}_1$ and $\mathscr{A}_2$ (relative to X, φ and η).

The following description explains how $\mathscr{A}_1 \times_\varphi^\eta \mathscr{A}_2$ works: First, the inputs in X are "encoded" by η and operate on Z_1. The resulting output represents the first component of the output of $\mathscr{A}_1 \times_\varphi^\eta \mathscr{A}_2$; at the same time it operates, together with the input from X and encoded by φ, on Z_2 and thus produces the second component of the output; see Figure 6.25.

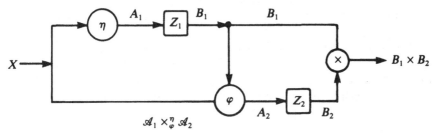

Figure 6.25

6.15 Remark. For semiautomata cascades are defined by making them into (Moore) automata with $B := Z$ and $\lambda(z, a) := z$.

6.16 Examples. We specialize X, φ and η in two ways. Let $\mathscr{A}_1$ and $\mathscr{A}_2$ be as above. Two of the ways we can choose X, φ and η are the following.
(a) Let

$$X := A_1 \times A_2, \qquad \varphi: A_1 \times A_2 \times B_1 \to A_2, \qquad (a_1, a_2, b_1) \mapsto a_2$$

and $\eta: A_1 \times A_2 \to A_1;$ $(a_1, a_2) \mapsto a_1$. Then the equations of 6.14 are

$$\delta_\varphi^\eta((z_1, z_2), (a_1, a_2)) = (\delta_1(z_1, a_1), \delta_2(z_2, a_2)),$$

and

$$\lambda_\varphi^\eta((z_1, z_2), (a_1, a_2)) = (\lambda_1(z_1, a_1), \lambda_2(z_2, a_2)).$$

$$((z_1, z_2) \in Z_1 \times Z_2, (a_1, a_2) \in A_1 \times A_2 = X).$$

Thus in this case the cascade is the parallel decomposition.
(b) Let $X := A_1$, $\varphi: A_1 \times B_1 \to B_1 = A_2$, $(a_1, b_1) \mapsto b_1$ and $\eta := \mathrm{id}_A$. For all $(z_1, z_2) \in Z_1 \times Z_2$ and $a_1 \in A_1$ we obtain:

$$\delta_\varphi^\eta((z_1, z_2), a_1) = (\delta_1(z_1, a_1), \delta_2(z_2, \lambda_1(z_1, a_1))),$$

$$\lambda_\varphi^\eta((z_1, z_2), a_1) = (\lambda_1(z_1, a_1), \lambda_2(z_2, \lambda_1(z_1, a_1))).$$

This is almost the series decomposition, except that the first output is carried forward. If we wanted the series decomposition exactly, then we would have to introduce a "decoder" $\psi: B_1 \times B_2 \to Y$ in 6.14 after $A_1 \times B_2$ such that the cascade is dependent on X, Y, φ, ψ and η.   □

Thus we see that cascades are a generalization of parallel and series decompositions. From the engineer's point of view they are much more satisfactory than series/parallel decompositions.
In order to find the monoid of $\mathscr{A}_1 \times_\varphi^\eta \mathscr{A}_2$ we still need more concepts.

6.17 Definition. (i) A semigroup S_1 *divides* a semigroup S_2, if S_1 is a homomorphic image of a subsemigroup of S_2. In symbols: $S_1|S_2$.

(ii) An automaton $\mathscr{A}_1 = (Z_1, A, B, \delta_1, \lambda_1)$ *divides* an automaton $\mathscr{A}_2 = (Z_2, A, B, \delta_2, \lambda_2)$ (equal input and output alphabets) if $\mathscr{A}_1$ is a homomorphic image of a subautomaton of $\mathscr{A}_2$. In symbols: $\mathscr{A}_1|\mathscr{A}_2$.

(iii) Two semigroups or automata are called *equivalent*, if they divide each other. In symbols: $\mathscr{A}_1 \sim \mathscr{A}_2$.

6.18 Remark. If S_1 divides S_2 then we also say that "S_2 *covers* S_1" or "S_2 *simulates* S_1". In this case the multiplication in S_1 is determined by part of the multiplication in S_2 via an epimorphism. Similarly for automata: "$\mathscr{A}_2$ covers $\mathscr{A}_1$" means that $\mathscr{A}_2$ can "do at least as much as" $\mathscr{A}_1$.

We use the notation introduced in 6.17.

6.19 Theorem. (i) *On any set of automata the relation $|$ is reflexive and transitive and $\sim$ is an equivalence relation.*

(ii) *Isomorphic automata are equivalent (but not conversely).*

(iii) *$\mathscr{A}$ and the automaton of $M_{\mathscr{A}}$ (see 5.7) are always equivalent.*

(iv) *$\mathscr{A}_1|\mathscr{A}_2 \Leftrightarrow M_{\mathscr{A}_1}|M_{\mathscr{A}_2}$.*

(v) *$\mathscr{A}_1 \sim \mathscr{A}_2 \Leftrightarrow M_{\mathscr{A}_1} \sim M_{\mathscr{A}_2}$.*

PROOF SKETCH. (i) and (ii) are obvious. For (iii) see KALMAN, FALB and ARBIB. "$\Rightarrow$" in (iv) follows from 6.3 and 6.8, the converse follows from (iii). (v) is obtained by applying (iv) twice.   □

It turns out that the monoid of $\mathscr{A}_1 \times_\varphi^\eta \mathscr{A}_2$ divides a certain "composition" of $M_{\mathscr{A}_1}$ and $M_{\mathscr{A}_2}$. This composition is the wreath product.

6.20 Definition. Let S, T be two semigroups. The *wreath product* $S \wr T$ is given as $(S^T \times T, \diamondsuit)$, where $\diamondsuit$ is defined by $(f_1, t_1) \diamondsuit (f_2, t_2) := (f_{f_1, f_2, t_1}, t_1 t_2)$ with $f_{f_1, f_2, t_1}: T \to S$, $x \mapsto f_1(x) f_2(xt_1)$.

6.21 Remark. Since $S \cong \{\bar{s}: S \to S | \bar{s}$ constant map with value $s\}$, we have $S \times T \cong \{(\bar{s}, t) | \bar{s}$ constant with values equal to $s, t \in T\} \leq S \wr T$. (Moreover we can verify $S \times T \subsetneq S \wr T$.) This implies $S \times T | S \wr T$ and furthermore $S|S \wr T$ and $T|S \wr T$. If S contains n elements and T contains m elements $(m, n \in \mathbb{N})$, then $S \wr T$, has $n^m m$ elements. We note that $\wr$ is neither commutative nor associative. More information about wreath products can be found in WELLS, for instance.

6.22 Theorem. *Let $\mathscr{A} = \mathscr{A}_1 \times_\varphi^\eta \mathscr{A}_2$. Then $M_{\mathscr{A}}|M_{\mathscr{A}_2} \wr M_{\mathscr{A}_1}$.*

PROOF SKETCH (See ARBIB[1]). Let $f_{\bar{x}} \in M_{\mathscr{A}}$. The result of the input sequence $\bar{x} \in \bar{X}$ on $z_1 \in Z_1$ can be expressed by $f_{\eta(\bar{x})}^{(1)}$. The result on $z_2 \in Z_2$ depends on $\bar{x}$ and on the output of $\mathscr{A}_1$ under the influence of $\bar{x}$ on z_1. Thus it can be expressed as a function of z_1 in $M_{\mathscr{A}_2}$. It is possible to describe the influence of $\bar{x}$ on $\mathscr{A}_2$ by a function $F: M_{\mathscr{A}_1} \to M_{\mathscr{A}_2}$, altogether $f_{\bar{x}}$ can be described by the pair $(F, f_{\bar{\eta}(\bar{x})}^{(1)}) \in M_{\mathscr{A}_2} \curvearrowright M_{\mathscr{A}_1}$. This yields an epimorphism from the subsemigroup of $M_{\mathscr{A}_1} \curvearrowright M_{\mathscr{A}_2}$ consisting of all pairs $(F, f_{\bar{\eta}(\bar{x})}^{(1)})$ onto $M_{\mathscr{A}}$. $\qquad\square$

Next we consider when an automaton $\mathscr{A}$ cannot be decomposed into cascades of "smaller" automata, i.e. that in any cascade $\mathscr{A} = \mathscr{A}_1 \times_{\varphi}^{\eta} \mathscr{A}_2$ either $\mathscr{A}_1$ or $\mathscr{A}_2$ already achieve the same as $\mathscr{A}$.

6.23 Definition. (i) A monoid S is called *irreducible*, if for all monoids S_1, S_2

$$S|S_1 \curvearrowright S_2 \Rightarrow S|S_1 \vee S|S_2.$$

(ii) An automaton $\mathscr{A}$ is called irreducible, if for all automata $\mathscr{A}_1, \mathscr{A}_2$

$$\mathscr{A}|\mathscr{A}_1 \times_{\varphi}^{\eta} \mathscr{A}_2 \Rightarrow \mathscr{A}|\mathscr{A}_1 \vee \mathscr{A}|\mathscr{A}_2.$$

Being irreducible, however, does not exclude divisors (see ARBIB[1]).

6.24 Theorem. *An automaton $\mathscr{A}$ is irreducible if and only if $M_{\mathscr{A}}$ is irreducible.*

PROOF. (a) Let $\mathscr{A}$ be irreducible and S_1, S_2 be monoids with $M_{\mathscr{A}}|S_1 \curvearrowright S_2$. Let $\mathscr{A}_1, \mathscr{A}_2$ be the automata of S_1, S_2, respectively, according to 5.7. One can show that $\mathscr{A}|\mathscr{A}_1 \times_{\varphi}^{\eta} \mathscr{A}_2$ for suitable φ and η. Irreducibility of $\mathscr{A}$ implies $\mathscr{A}|\mathscr{A}_1 \vee \mathscr{A}|\mathscr{A}_2$. Suppose $\mathscr{A}|\mathscr{A}_1$, then $M_{\mathscr{A}}|M_{\mathscr{A}_1}$, because of 6.19(iv). $M_{\mathscr{A}_1} \sim S_1$ (see 6.19(iii)) implies $M_{\mathscr{A}}|S_1$. The case $\mathscr{A}|\mathscr{A}_2$ is treated similarly.

(b) Conversely, let $M_{\mathscr{A}}$ be irreducible, $\mathscr{A}_1, \mathscr{A}_2$ be automata and $\mathscr{A}|\mathscr{A}_1 \times_{\varphi}^{\eta} \mathscr{A}_2$ for suitable φ, η. Theorem 6.19(iv) implies $M_{\mathscr{A}}|M_{\mathscr{A}_1 \times_{\varphi}^{\eta} \mathscr{A}_2}$ and 6.22 implies $M_{\mathscr{A}_1 \times_{\varphi}^{\eta} \mathscr{A}_2}|M_{\mathscr{A}_1} \curvearrowright M_{\mathscr{A}_2}$, therefore $M_{\mathscr{A}}|M_{\mathscr{A}_1} \curvearrowright M_{\mathscr{A}_2}$ (by 6.19(i)). $M_{\mathscr{A}}$ is irreducible, thus $M_{\mathscr{A}}|M_{\mathscr{A}_1}$ or $M_{\mathscr{A}}|M_{\mathscr{A}_2}$. Theorem 6.19(iv) implies that $\mathscr{A}$ divides either $\mathscr{A}_1$ or $\mathscr{A}_2$. $\qquad\square$

Now we have all the necessary tools to formulate and sketch the proof of a fundamental theorem for automata. For details see ARBIB (several versions of the theorem), EILENBERG (together with an algorithm for the decomposition) or WELLS.

6.25 Theorem ("Decomposition Theorem of Krohn–Rhodes"). *A finite (semi-) automaton can be simulated by cascades of the following basic types ("atoms") of irreducible (semi-) automata:*

(1) *IR flip-flops.*
(2) *(Semi-) automata, whose monoids are finite, simple groups dividing $M_{\mathscr{A}}$.*

PROOF SKETCH. We list the major steps in this outline.

(a) Any finite monoid is isomorphic to a wreath product of monoids for which all elements outside the group kernel (see 3.15) are left identity elements. This means (see ZEIGER) that any finite (semi-) automaton can be simulated by cascades of PF (semi-) automata (permutation-fixing (semi-) automata). These are automata with $f_{\bar{a}}$ either bijective (i.e. permutations) or constants (i.e. fixing the elements).

(b) The monoids of (a) are wreath products of groups and monoids with trivial group kernel and otherwise consisting only of left identity elements. In other words: Any PF (semi-) automaton can be decomposed by cascades into group (semi-) automata (where the monoid of the automaton is a group) and into IF (semi-) automata (where all $f_{\bar{a}}$ are identities or constant, identity-fixing (semi-) automaton).

(c) Any finite group can be covered by wreath products of simple groups, such that any group (semi-) automaton can be simulated by a cascade of simple group (semi-) automata. To do this one needs the Jordan–Hölder theory.

(d) Next, it is to show that any IF (semi-) automaton can be decomposed into IR flip-flops by cascades.

(e) Finally, one shows that IR flip-flops and finite simple groups (thus also finite simple group (semi-) automata) are irreducible.   □

The number of basic automata used in 6.25 is called the complexity of $\mathscr{A}$; since cascades can be simulated by series/parallel composition (although with some "delays", see ARBIB$_1$, and also our remark after 6.16), we obtain the original result by Krohn–Rhodes:

6.26 Corollary. *A finite (semi-) automaton can be simulated by series/parallel decomposition of the basic types of irreducible automata given in 6.25.*   □

ARBIB$_1$ gives applications of these results to combinatorial semigroups (i.e. semigroups with all proper subsemigroups being singletons; subsemigroups, homomorphic images, finite direct products and wreath products of combinatorial semigroups are also combinatorial).

PROBLEMS

1. Let $\mathscr{A}_1, \mathscr{A}_2$ be as in Problem 4 of part A. Construct the cascade product $\mathscr{A}_1 \times_\varphi^\eta \mathscr{A}_2$ for three different choices of X, φ and η.

2. In Problem 1, specify the maps φ and η which yield the parallel and the series connection of $\mathscr{A}_1$ and $\mathscr{A}_2$.

3. Suppose $S_1 | S_2$ and that S_2 is a monoid. Is S_1 a monoid, too?

4. As in Problem 3 with "commutative semigroup" instead of "monoid".

5. If S, T are semigroups, show that their wreath product $S \wr T$ is again a semigroup.

6. If S, T are groups, show that $S \wr T$ is again a group.

7. If S, T are commutative semigroups, is $S \curlyvee T$ again commutative?

*8. Prove all assertions in 6.21.

9. Give three examples of irreducible monoids.

10. Using Problem 9, find three irreducible automata.

EXERCISES (Solutions in Chapter 8, p. 499)

1. Let $\mathscr{A}$ be the automaton of Exercise 1 in §4. Determine if $\{z_1\}$ or $\{z_2, z_3, z_4\}$ or $\{z_4\}$ are subautomata. For those which are, find their epimorphisms from $M_{\mathscr{A}}$ onto these monoids (see 6.3).

2. As in Exercise 1 for $M_{\mathscr{A}_1}$ and $M_{\mathscr{A}_2}$ in Example 6.2.

3. In the notation of Example 6.2, is $\mathscr{A}_1$ a homomorphic image of $\mathscr{A}_2$?

4. Give the tables of parallel and series compositions of two IR flip-flops.

5. Let $X = \{0, 1\}$, $\varphi: \{0, 1\}^2 \to \{0, 1\}, (x, b) \mapsto x + b$ (addition mod 2) and $\eta: \{0, 1\} \to \{0, 1\}, x \mapsto x + 1$. Give the table and state graph of the cascade of the shift-register with the trigger flip-flop.

6. Find the operation table of the wreath product of $(\mathbb{Z}_2, \oplus)$ by itself. Is this wreath product a group? If it is, is it abelian?

7. Show that $(\mathbb{Z}_n, +) | (\mathbb{Z}_m, +) \Rightarrow n | m$.

8. If $\mathscr{A}_1 \leq \mathscr{A}_2$ and $M_{\mathscr{A}_2}$ is a group, does this hold for $M_{\mathscr{A}_1}$, too?

9. If $\mathscr{A}$ is the homomorphic image of $\mathscr{A}'$ and if $M_{\mathscr{A}'}$ is a simple group, what can one say about $M_{\mathscr{A}}$? Give examples for this case.

*10. If the semigroup H divides a group G, is H a group, too?

11. Find all semigroups which divide $(\{a, b, c\}, *)$ with $x * y := x$.

12. Find all semigroups which divide $(\mathbb{Z}_5, +)$.

§7. Minimal Automata

In many automata one can find different states which show the same "input–output behavior", which means that they yield the same outputs for the same inputs. For most applications it means some waste of time and space to have such "equivalent" states in the automaton. We shall call an automaton "minimal", if no two different states are equivalent, and we shall present an algorithm to transfer each finite automaton into an equivalent minimal one.

7.1 Definition. Let $\mathscr{A} = (Z, A, B, \delta, \lambda)$ be an automaton and $z, z' \in Z$. Then z and z' are called *equivalent* (in symbols $z \sim z'$) if $\forall \bar{a} \in \bar{A}: \bar{\lambda}(z, \bar{a}) = \bar{\lambda}(z', \bar{a})$.

Henceforth we write λ instead of $\bar{\lambda}$. It is obvious that $\sim$ is an equivalence relation on Z. Unfortunately, since $\bar{A}$ has infinitely many elements it is difficult to decide in practice if two states are equivalent or not. We shall overcome this difficulty with the help of the following idea.

7.2 Definition. Let $k \in \mathbb{N}$ and $\mathcal{A}, z, z'$ be as above. z and z' are called *k-equivalent* (in symbols $z \sim_k z'$), if

$$\forall a_1, \ldots, a_k \in A: \lambda(z, a_1 a_2 \ldots a_k) = \lambda(z', a_1 a_2 \ldots a_k).$$

$\sim_k$ is also an equivalence relation on Z, for all $k \in \mathbb{N}$. We have $\forall z, z' \in Z: z \sim z' \Rightarrow z \sim_k z'$. But this result is of minor interest to us. We want to find a $k \in \mathbb{N}$ such that $\sim = \sim_k$, in this case we can verify the equivalence of states in finitely many steps (if $\mathcal{A}$ is finite). First we try to jump from $\sim_k$ to $\sim_{k+1}$:

7.3 Remark. In the notation above, we have:

$$\lambda(z, a_1 a_2 \ldots a_k) = \lambda(z, a_1)\lambda(\delta(z, a_1), a_2 a_3 \ldots a_k) = \ldots$$

$$= \lambda(z_1, a_2) \ldots \lambda(z_k, a_k) \in \bar{B},$$

with $z_1, \ldots, z_k$ as before (see 4.1). Because of the definition of equality in $\bar{B}$ we have:

$$\lambda(z, a_1 \ldots a_k) = \lambda(z', a_1 a_2 \ldots a_k) \Leftrightarrow \lambda(z_1, a_1)\lambda(z_2, a_2) \ldots \lambda(z_k, a_k)$$

$$= \lambda(z_1', a_1)\lambda(z_2', a_2) \ldots \lambda(z_k', a_k) \Leftrightarrow \lambda(z_1, a_1) = \lambda(z_1', a_1)$$

$$\wedge \ldots \wedge \lambda(z_k, a_k) = \lambda(z_k', a_k).$$

7.4 Lemma. $\forall k \in \mathbb{N} \, \forall z, z' \in Z: z \sim_k z' \Leftrightarrow \forall n \leq k: z \sim_n z'$.

PROOF. Let $z \sim_k z'$ and $a_1, \ldots, a_n \in A$ with $n \leq k$. We extend $a_1, \ldots, a_n$ with a_n's to obtain a sequence of k terms. Then we have $\lambda(z, a_1 a_2 \ldots a_n a_n \ldots a_n) = \lambda(z', a_1 a_2 \ldots a_n a_n \ldots a_n)$ and, because of 7.3, $\lambda(z, a_1 \ldots a_n) = \lambda(z', a_1 \ldots a_n)$, thus $z \sim_n z'$. The converse implication of the theorem is obvious. $\square$

7.5 Lemma. $\forall k \in \mathbb{N}, \forall z, z' \in Z:$

$$z \sim_{k+1} z' \Leftrightarrow (z \sim_k z' \wedge \forall a \in A: \delta(z, a) \sim_k \delta(z', a)).$$

PROOF. "$\Rightarrow$": $z \sim_{k+1} z'$ implies $z \sim_k z'$, because of 7.4. Let $a \in A$ and $a_1, \ldots, a_k \in A$, then $\lambda(z, a)\lambda(\delta(z, a), a_1 \ldots a_k) = \lambda(z', a a_1 \ldots a_k) = \lambda(z', a)\lambda(\delta(z', a), a_1 \ldots a_k)$. Hence $\lambda(\delta(z, a), a_1 \ldots a_k) = \lambda(\delta(z', a), a_1 \ldots a_k)$, because of equality in $\bar{B}$. Therefore $\delta(z, a) \sim_k \delta(z', a)$.

"$\Leftarrow$": Let $z \sim_k z'$ and $\forall a \in A: \delta(z, a) \sim_k \delta(z', a)$. Given $a_1, \ldots, a_{k+1} \in A$ we have from 7.4 that $z \sim_k z'$ implies $z \sim_1 z'$, thus $\lambda(z, a_1) = \lambda(z', a_1)$. Since $\delta(z, a_1) \sim_k \delta(a', a_1)$, $\lambda(\delta(z, a_1), a_2 \ldots a_{k+1}) = \lambda(\delta(z', a_1), a_2 \ldots a_{k+1})$.

Therefore (by 7.3) $\lambda(z, a_1 a_2 \ldots a_{k+1}) = \lambda(z', a_1 a_2 \ldots a_{k+1})$, which yields $z \sim_{k+1} z'$. $\quad\square$

7.6 Lemma. $\sim_k = \sim_{k+1} \Rightarrow \sim_k = \sim_{k+1} = \sim_{k+2} = \ldots = \sim$.

PROOF. Let $\sim_k = \sim_{k+1}$, $z, z' \in Z$ with $z \sim_{k+1} z'$. Then $\forall\, a \in A$ $\delta(z, a) \sim_k \delta(z', a)$ and also $\delta(z, a) \sim_{k+1} \delta(z', a)$. By 7.5 this yields $z \sim_{k+2} z'$. Lemma 7.4 establishes $\sim_k = \sim_{k+1} = \sim_{k+2}$ and by induction we conclude $\sim_{k+2} = \sim_{k+3} = \ldots$. If $z \sim z'$ then $z \sim_k z'$. Conversely, let $a_1, \ldots, a_m \in A$, $m \in \mathbb{N}$ and let $z, z' \in Z$ with $z \sim_k z'$. Then $\lambda(z, a_1 a_2 \ldots a_m) = \lambda(z', a_1 a_2 \ldots a_m)$ for any m. $\quad\square$

Now we can prove the main result:

7.7 Theorem. *Let* $|Z| = m \in \mathbb{N}$. *Then* $\sim_{m-1} = \sim$.

PROOF. Since $z \sim_1 z' \Rightarrow z \sim_2 z' \Rightarrow \ldots$ we have $m \ge |Z/\sim_1| \ge |Z/\sim_2| \ge \ldots \ge 1$. At the worst this chain will become stationary at $\sim_{m-1}$; we can apply 7.6 to obtain the result. $\quad\square$

This gives us a method for determining $\sim$.

7.8 Reduction Method. Let $\mathscr{A} = (Z, A, B, \delta, \lambda)$ be a finite automaton with $|Z| = m$.

(1) Determine the equivalence classes of Z with respect to $\sim_1$.
(2) Decompose the elements of these classes into classes with respect to $\sim_2$, and so on, until

$\quad\vdots$

(k) $\sim_k = \sim_{k+1}$. Then $\sim_k = \sim$. (This happens no later than $k = m - 1$.)

7.9 Example. Let $\mathscr{A} = (\{z_1, z_2, \ldots, z_7\}, \{a_1, a_2\}, \{0, 1\}, \delta, \lambda)$ with

δ	a_1	a_2		λ	a_1	a_2
z_1	z_2	z_7		z_1	0	0
z_2	z_2	z_7		z_2	0	0
z_3	z_5	z_1		z_3	0	0
z_4	z_6	z_2		z_4	0	0
z_5	z_5	z_3		z_5	1	0
z_6	z_6	z_4		z_6	1	0
z_7	z_1	z_6		z_7	0	1

We obtain $Z/\sim_1 = \{\{z_1, z_2, z_3, z_4\}, \{z_5, z_6\}, \{z_7\}\}$ and try $\sim_2$. For $\sim_2$:

$$Z/\sim_2 = \{\{z_1, z_2\}, \{z_3, z_4\}, \{z_5\}, \{z_6\}, \{z_7\}\}.$$

For $\sim_3$:

$$Z/\sim_3 = \{\{z_1, z_2\}, \{z_3, z_4\}, \{z_5\}, \{z_6\}, \{z_7\}\}.$$

Therefore $\sim_2 = \sim_3$, thus $\sim_2 = \sim$ and $Z/\sim = Z/\sim_2$. $\qquad \square$

Here $k = 2$ suffices instead of $k = m - 1 = 6$. As the next example shows, there are instances in which one has to calculate up to $m - 1$:

7.10 Example ("Modulo m Counter").

$$\mathcal{A} = (\{z_1, z_2, \ldots, z_m\}, \{a\}, \{0, 1\}, \delta, \lambda)$$

with

δ	a
z_1	z_2
z_2	z_3
$\vdots$	$\vdots$
z_{m-1}	z_m
z_m	z_1

λ	a
z_1	0
z_2	0
$\vdots$	$\vdots$
z_{m-1}	0
z_m	1

For all $k \le m - 2$ we have

$$\lambda(z_1, \underbrace{aa\ldots a}_{k\text{-times}}) = \underbrace{000\ldots 0}_{k\text{-times}} = \lambda(z_2, \underbrace{aa\ldots a}_{k\text{-times}}),$$

but

$$\lambda(z_1, \underbrace{aa\ldots a}_{m-1\text{-times}}) = \underbrace{00\ldots 0}_{m-1\text{-times}} = \underbrace{00\ldots 01}_{m-2\text{-times}} = \lambda(z_2, \underbrace{aa\ldots a}_{m-1\text{-times}}),$$

therefore $\sim_{m-2} \ne \sim_{m-1} = \sim$. $\qquad \square$

Now we want to show how to "reduce" a given automaton. We shall replace this given automaton by an equivalent one having a state set as small as possible. As mentioned earlier, such considerations are important from an economical point of view (although the advances in computer technology have somewhat modified this problem).

7.11 Definition. An automaton $\mathcal{A}$ is called a *minimal automaton* if $\sim$ is the identity.

7.12 Theorem. *For any automaton there is (up to isomorphisms) a uniquely determined equivalent minimal automaton.*

PROOF SKETCH. Let $\mathcal{A} = (Z, A, B, \delta, \lambda)$. We form $\mathcal{A}^* := (Z/\sim, A, B, \delta^*, \lambda^*)$, where $\delta^*: Z/\sim \times A \to Z/\sim$, $([z], a) \mapsto [\delta(z, a)]$, $\lambda^*: Z/\sim \times A \to B$, $([z], a) \mapsto \lambda(z, a)$. δ^* and λ^* are well defined (see 7.5 and 7.2).

One can show that $\mathscr{A}^*$ is the desired equivalent minimal automaton of $\mathscr{A}$ and any other automaton with this property is isomorphic to $\mathscr{A}^*$.   □

7.13 Example. Let $\mathscr{A}$ be as in 7.9. Then the minimal automaton equivalent to $\mathscr{A}$ is

$$\mathscr{A}^*/\sim = (\{[z_1], [z_3], [z_5], [z_7]\}, \{a_1, a_2\}, \{0, 1\}, \delta^*, \lambda^*)$$

with

$$[z_1] = \{z_1, z_2\},$$

$$[z_3] = \{z_3, z_4\},$$

$$[z_5] = \{z_5, z_6\},$$

$$[z_7] = \{z_7\},$$

δ^*	a_1	a_2
$[z_1]$	$[z_1]$	$[z_7]$
$[z_3]$	$[z_5]$	$[z_1]$
$[z_5]$	$[z_5]$	$[z_3]$
$[z_7]$	$[z_1]$	$[z_5]$

λ^*	a_1	a_2
$[z_1]$	0	0
$[z_3]$	0	0
$[z_5]$	1	0
$[z_7]$	0	1

□

7.14 Remarks. 1. Since any set M can be regarded as an automaton (e.g. $(M, M, M, \delta, \lambda)$ with arbitrary δ, λ), we would only expect good concrete results in the case of special automata (thus the generality of Theorem 6.25 is impressive). An important special class of automata are the *"linear"* automata. These are automata $(Z, A, B, \delta, \lambda)$, where Z, A, B are vector spaces and δ, λ are linear mappings, see, e.g. DORNHOFF and HOHN.

2. For the concrete design of automata see ARBIB or BIRKHOFF and BARTEE.

3. Let $\mathscr{A} = (Z, A, B, \delta, \lambda, z_0)$ be an initial Moore automaton. $f_{\mathscr{A}}: \bar{A} \mapsto B, \bar{a} \mapsto \lambda(\delta(z_0, \bar{a})$, for any $a \in A$, is called the function of $\mathscr{A}$. An arbitrary mapping f of a free monoid F_X into a set Y is called computable if there is an initial Moore automaton $\mathscr{A}$ with $f = f_{\mathscr{A}}$. Let $M_f := F_X/\equiv_f$ be the monoid of f, where $w_1 \equiv_f w_2 :\Leftrightarrow \forall v, w \in F_X: f(vw_1w) = f(vw_2w)$ is the so-called *Myhill equivalence*. Then, as an example of one result in this context, f is computable if and only if M_f is finite (see RABIN and SCOTT).

PROBLEMS

1–3. Recall Problems 7–9 to §2 and let $B = A$, $\lambda = \delta$. Find the corresponding minimal automata.

4. Let $\mathscr{A}$ and $\mathscr{A}'$ be automata such that to every state in $\mathscr{A}$ there exists an equivalent state in $\mathscr{A}'$. Are $\mathscr{A}$ and $\mathscr{A}'$ then isomorphic? Are they isomorphic if one of these automata is minimal?

*5. Recall Exercise 4 of § 4. For $\alpha_1, \alpha_2 \in \bar{A}$ let for $f: \bar{A} \mapsto \bar{B}$

$$f(\alpha_1\alpha_2) =: f(\alpha_1)f_{\alpha_1}(\alpha_2).$$

Show that f is a well-defined function from $\bar{A}$ to $\bar{B}$. Let $f, g: \bar{A} \mapsto \bar{B}$ be sequential such that there are n different f_α's and m different g_α's ($n, m \in \mathbb{N}$). Show that $f = g$ iff $f(\alpha) = g(\alpha)$ holds for all α of length $n + m - 1$.

6. Construct an automaton with $\sim_3 = \sim$, but $\sim_2 \neq \sim$.

7. Let G be a group and $\mathscr{A}_G$ be the corresponding automaton. Do these assumptions simplify the reduction process 7.8?

EXERCISES (Solutions in Chapter 8, p. 500)

1. Show that $\sim$ and $\sim_k$ ($k \in \mathbb{N}$) in 7.1 and 7.2 are equivalence relations.

2. Find the minimal automaton to the marriage automaton in 2.2.

3. Do the same for the parity-check automaton in 2.3.

4. Do the same for the automaton in Exercise 1 to §4.

5. Does a minimal automaton have to be irreducible? Is an irreducible automaton always minimal?

*6. Prove 7.12 in detail.

7. If $\mathscr{A}^*$ is the reduced automaton associated with $\mathscr{A}$, is $\mathscr{A}^*$ a subautomaton of $\mathscr{A}$? Is $\mathscr{A}^*$ a homomorphic image of $\mathscr{A}$?

8. If $\mathscr{A}_1 | \mathscr{A}_2$ and $\mathscr{A}_2$ is minimal, are $\mathscr{A}_1$ and $\mathscr{A}_2$ isomorphic?

NOTES

The theory of automata has its origins in the papers by A. M. Turing (1936), C. E. Shannon (1938) and W. S. McCulloch and W. Pitts (1943). Turing developed the theoretical concept of what is now called Turing machines, in order to give computability a more concrete and precise meaning. Shannon investigated the analysis and synthesis of electrical contact circuits using switching algebra. The work of McCulloch and Pitts centred on neuron models to explain brain functions and neural networks by using finite automata. Their work was continued by Kleene. Our description of a finite automaton is due to Moore and Melay. The development of technology in the areas of electro-mechanical and electronic machines and particularly computers has had a great influence on automaton theory. From about the mid-1950's on, one speaks of the discipline of automaton theory. Many different parts of pure mathematics are used as tools, such as abstract algebra, universal algebra, lattice theory, category theory, graphs, mathematical logic and the theory of algorithms. In turn, automaton theory can be used, for example, in economics, linguistics and in learning processes.

In Chapter 6 we are concerned mainly with one subdiscipline of automaton theory, namely the algebraic theory of automata, which uses algebraic concepts to formalize and study certain types of finite-state machines. One of the main algebraic tools used to do this is the theory of semigroups.

The concept of a semigroup is relatively young, the first, often fragmentary, studies were carried out early in the twentieth century. Then the necessity of studying general transformations, rather than only invertible transformations (which played a large role in the development of group theory) became clear. During the past few decades connections in the theory of semigroups and the theory of machines became of increasing importance, both theories enriching each other. In association with the study of machines and automata, other areas of applications such as formal languages and software, use the language of modern algebra in terms of Boolean algebras, semigroups and others. But also parts of other areas, such as biology, psychology, biochemistry and sociology make use of semigroups.

The term "semigroup" first appeared in the literature in J.-A. de Séguier's book, *Éléments de la Théorie des Groupes Abstraits*, in 1904. The first paper of fundamental importance for semigroup theory was a publication by A. K. Suschkewitsch in 1928, in which he showed that every finite semigroup contains a simple ideal. D. Rees in 1940 introduced the concept of a matrix over a group with zero and studied infinite simple semigroups, and from then on the work on semigroups increased through contributions by Clifford, Dubreil and many others. In 1948 Hille published a book on the analytic theory of semigroups. The first two monographs covering many of the main aspects of the subject are LJAPIN, *Semigroups* and CLIFFORD and PRESTON, *The Algebraic Theory of Semigroups*.

Other books on semigroups which are of interest to research workers in the field as well as being suitable for classroom use are books by PETRICH, HOWIE, and LALLEMENT.

The contents of Chapters 6 and 7 on automata and formal languages, and actually much more, are part of what is vaguely described as computer science. EILENBERG gives the theory of automata and the theory of formal languages a coherent mathematical presentation, in a work of four projected volumes, the first two of which have appeared. The following books, some of which are amongst the first publications on the subject, include the material of Chapter 6 and much more: ARBIB$_1$, ARBIB$_2$, ARBIB$_3$, GINZBURG, GINSBURG, RABIN and SCOTT, MINSKY. Parts of KALMAN, FALB and ARBIB and also ROSEN are also relevant. The book BOBROW and ARBIB gives a nice introduction for the beginner. A good introduction to algebraic automata theory is the recent book by HOLCOMBE.

Systems theory is a branch of applied mathematics which is in some parts very similar to automata theory. A system is "an automaton in which time plays a role". More precisely, a (dynamical) system is a 7-tuple consisting of a state set Z, input sets A and B, an ordered time set T, a set Ω of possible input functions (one thinks of inputs to come as functions of the time), an output function and finally (but most essentially) a "state-transition function" $f: T \times T \times Z \times \Omega \to Z$. One interprets $f(t_2, t_1, z, \omega)$ as "the state at time t_2, if the system has been in state z at time $t_1 \leq t_2$ and if between t_1 and t_2 the function ω was supplying inputs". Differential equations, difference equations and algebra (especially ring and module theory) are the main tools in systems theory. See, for instance, KALMAN, FALB and ARBIB for a very good introduction to this field.

CHAPTER 7

Further Applications of Semigroups

In this chapter we investigate three more instances in which semigroups occur naturally.

§1. Formal Languages

Besides the "natural" languages such as English, German, French or Chinese, the so-called formal languages are of importance in the formal sciences, especially in computing or information sciences. The underlying principle is always the same: Given an "alphabet" or "vocabulary" (consisting of letters, words, punctuation symbols, number symbols, programming instructions, etc.) one has a method ("grammar") for constructing "meaningful" words or sentences (i.e. the "language") from this alphabet. This immediately reminds us of the term "word semigroup" and, indeed, these free semigroups will play a major role, the language L constructed will be a subset of the free semigroup F_A on the alphabet A.

There are essentially three ways to construct a language L.

(a) Approach via grammar: given a collection of rules ("grammar"), generate L from A.
(b) Approach via automata: consider an initial semiautomaton which processes the elements of L in a suitable way.
(c) Algebraic approach: L is constructed by the algebraic combination of certain subsets of F_A.

In all three approaches we shall use algebraic methods. First we study the different approaches, then investigate connections between them. The whole theory is part of mathematical linguistics, which also uses probabilistic methods. We use the following notation: A_* is the free semigroup over a set A, A^* is the free monoid over A, so $A^* = A_* \cup \{\Lambda\}$, where Λ denotes the empty word.

A. Approach via Grammar

This approach is largely based on the work of CHOMSKY.

1.1 Definition. A *phrase-structure grammar* (in short: *grammar*) is a quadrupel $\mathscr{G} = (A, G, \to, g_0)$, where A and G are nonempty, disjoint, finite sets, $g_0 \in G$ and $\to$ is a relation from A_* into $(A \cup G)^*$.

1.2 Definition. Let $\mathscr{G}$ be a grammar.

- (i) A is called the *alphabet* of $\mathscr{G}$.
- (ii) G is called the set of *grammar symbols* (or metalinguistic symbols).
- (iii) $V := A \cup G$ is the *complete vocabulary*.
- (iv) The elements (x, y) in $\to$ (which is a subset of $A_* \times V^*$) are also written in the form $x \to y$ and are called *productions* or *rewriting rules*.
- (v) g_0 is called the *initial symbol*.

$\mathscr{G}$ determines the grammatical rules. We now show how to obtain a suitable subset of A^*, namely the desired language, by using $\mathscr{G}$. In order to do this we need another relation, this time on V^*.

1.3 Definition. Let $\mathscr{G} = (A, G, \to, g_0)$ be a grammar. For $y, z, \in V^*$ we define

$$y \Mapsto z :\Leftrightarrow \exists\, u \in G_* \,\exists\, u_1, u_2, w \in V^*: y = u_1 u u_2 \wedge z = u_1 w u_2 \wedge u \to w.$$

The reason for introducing $\Mapsto$ is to obtain a new word $v_1 v_2 \ldots v_r' \ldots v_n$ from a given word $v_1 v_2 \ldots v_r \ldots v_n$ and a rule $v_r \to v_r'$. Thus we extend $\to$ to a compatible relation $\Mapsto$ on V^*. We recall from 6.3.24 that the transitive hull $\Mapsto'$ of $\Mapsto$ is given by

$$y \Mapsto' z \Leftrightarrow \exists\, n \in \mathbb{N}\, \exists\, x_1, \ldots, x_n \in V^*: y \Mapsto x_1 \Mapsto x_2 \Mapsto \ldots \Mapsto x_n \Mapsto z.$$

The sequence $x_0, x_1, \ldots, x_n$ is called a *derivation* (or *generation*) *of* z from y; n is the *length* of the derivation.

1.4 Definition. Let $\mathscr{G} = (A, G, \to, g_0)$ be a grammar.

$$L(\mathscr{G}) := \{l \in A^* \,|\, g_0 \Mapsto' l\}(\subseteq A^*),$$

is called the *language generated by* $\mathscr{G}$ (also the *phrase-structure language*).

We give a few examples.

1.5 Example. Let $A = \{a\}$, $G = \{g_0\}$, $g_0 \neq a$, and $\to = \{g_0 \to a\}$. $\mathcal{G} :=$ $(A, G, \to, g_0)$; we find $L(\mathcal{G})$. Let $z \in A^*$ with $g_0 \mapsto' z$. Then there are some $u \in \{g_0\}_*$ and $v_1, v_2, w \in \{a, g_0\}^*$ with $g_0 = v_1 v v_2$, $z = v_1 w v_2$ and $v \to w$. This implies $v = g_0$ and $w = a$. From $g_0 = v_1 g_0 v_2$ we obtain $v_1 = v_2 = \Lambda$ and also $z = w = a$. From a we cannot generate anything new, since $a \mapsto x$, $g_0 \mapsto' x$ means we had to find $v' \in \{g_0\}_*$, $v_1, v_2 \in \{a, g_0\}^*$ such that $a = v_1' v' v_2'$; this is impossible because of $g_0 \neq a$. Hence $L(\mathcal{G}) = \{a\}$, a very simple language. □

The following language can do a little more.

1.6 Example. Let A and G be as in 1.5 and $\to := \{g_0 \to a, g_0 \to a g_0\}$, $\mathcal{G} :=$ $(A, G, \to, g_0)$. Then $g_0 \mapsto z$ implies $z = a$ or $z = a g_0$. Again there is no x with $a \mapsto x$. $a g_0 \mapsto y$ is only possible for $y = aa$ or $y = aa g_0$. Thus $aa g_0 \mapsto aaa$ and $aa g_0 \mapsto aaa g_0$, etc. In a diagram

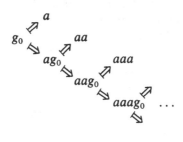

Cancellation of the elements outside A^* gives us the result $L(\mathcal{G}) = \{a, aa, aaa, \ldots\} = A_*$. □

The next example is more interesting. We describe the calculation of $L(\mathcal{G})$ in detail.

1.7 Example. *Let $A = \{0, 1, \ldots, 9\}$, g_0 an arbitrary element not in A, $G =$ $\{g_0\}$, $\to := \{g_0 \to 0, g_0 \to 1, \ldots, g_0 \to 9, \quad g_0 \to g_0 0, \ldots, g_0 \to g_0 9\}$, $\mathcal{G} =$ $(A, G, \to, g_0)$.* First we calculate

$$\bar{L} := \{l \in V^* | g_0 \mapsto' l\}.$$

We begin by determining all $l \in V^*$ with $g_0 \mapsto l$. Thus we have to find $v \in G_*$ and $v_1, v_2, w \in V^*$ such that

$$g_0 = v_1 v v_2, \qquad l = v_1 w v_2, \qquad v \to w.$$

We find $v = g_0$ and $v_1 = v_2 = \Lambda$, $g_0 \to w$ implies $l = w = z$ or l or $l = w = g_0 z$ with $z \in \{0, 1, \ldots, 9\} = A$.

From this stage we conjecture that all $z_1 z_2 \ldots z_k$, $z_i \in A$, i.e. all $n \in \mathbb{N}_0$ and all $g_0 n$ can be obtained in this way. (Note that $z_1 z_2$ is not a product of integers but a sequence of numerals. For $z_1 = 4$ and $z_2 = 6$ we obtain the number 46 and not $4 \cdot 6 = 24$.) We claim $\bar{L} = \{g_0\} \cup \mathbb{N}_0 \cup \{g_0 n | n \in \mathbb{N}_0\} := N$. To show $\bar{L} \subseteq N$, we use induction with respect to the shortest lengths k_l of all derivations of $l \in \bar{L}$ from g_0, for all $l \in \bar{L}$.

From above we already know that the only element with shortest length is g_0, the elements with $k_l = 2$ are the 20 elements z and $g_0 z$, $z \in A$. Thus the elements $l \in \bar{L}$ with $k_l = 1$ and 2 are in N.

Let $l \in N$ with $k_l = n \in \mathbb{N}$, $l \Rightarrow l' \in V^*$. We have to show $l' \in N$. There are elements $v \in G_*$, $v_1, v_2, w \in V^*$ with $l = v_1 v v_2$, $l' = v_1 w v_2$ and $v \to w$. Since $l \in \{g_0\} \cup \mathbb{N}_0 \cup \{g_0 n | n \in \mathbb{N}_0\}$ and $v \in G_* = \{g_0\}_*$, we can only have $l = \Lambda g_0 \Lambda$ or $l = \Lambda g_0 n$. This yields for $l' = v_1 w v_2$ either $l' = z$ or $l' = g_0 z$, or $l' = zn$, or $l' = g_0 zn$; in any case $l' \in N$. As the result of the induction we obtain $\bar{L} \subseteq N$. It is easily verified that $N \subseteq \bar{L}$, which yields $N = \bar{L}$. Finally, $L(\mathcal{G}) = \bar{L} \cap A^* = \bar{L} \cap \mathbb{N}_0 = \mathbb{N}_0$. $\qquad\square$

In "natural" languages (more precisely: in excerpts of natural languages), the set A consists of the words of the language (e.g. all words of a dictionary, together with all declensions, etc.), G consists of the names for terms of the grammar (e.g. "sentence", "noun", "adverb", ...), $\to$ consists of rules such as "sentence $\to$ subject, predicate" and substitution rules, e.g. "subject $\to$ wine bottle". In this way we obtain in (partial) languages, sets of grammatically correct sentences, which however do not have to be meaningful with respect to content. For more details on the formation of languages and grammar and interesting psychological rules see GREEN.

As another illustration we show how to construct approximations to programming languages. Let A consist of the letters, numbers, algebraic symbols and special programming words used, let G be a collection of certain subsets of A^*, such as INTEGER, REAL, and further let it contain the grammatical rules of the programming languages. We describe the situation, in particular, for the programming language ALGOL 60, one of the important languages for scientific purposes. The exact description of ALGOL 60 would be far beyond the scope of this book, but we give an indication of the basic idea.

1.8 Example. Let

$$A = \{A, B, \ldots, Z, a, b, \ldots, z, 0, 1, \ldots, 10, 11, \ldots, +, -, \times, /,$$

$$\div, \uparrow, =, \neq, <, \leq, >, \geq, \wedge, \vee, \neg, \equiv, ;, \cdot, :, ::=, (,), [,],$$

$$\text{BEGIN, TRUE, FALSE, GO TO, FOR, STEP, UNTIL, END}\},$$

G consists of the sets $\text{I, U, R, B, D, L}, \ldots \in \mathcal{G}(A^*)$, which symbolize "IDENTIFIER", "UNSIGNED INTEGER", "REAL", "BOOLEAN", "DIGIT", "LETTER", ... They are implicitly defined by a sequence of equations, e.g.

$$I = L \cup IL \cup ID, \tag{1}$$

$$U = D \cup UD, \tag{2}$$

etc.

Here IL is the product of subsets in the monoid A^*, etc. . . . . . These equations arise from rules in ALGOL:

$$\langle\text{IDENTIFIER}\rangle::=\langle\text{LETTER}\rangle|\langle\text{IDENTIFIER}\rangle$$
$$\langle\text{LETTER}\rangle|\langle\text{IDENTIFIER}\rangle|\langle\text{DIGIT}\rangle \qquad (1)'$$

$$\langle\text{UNSIGNED INTEGER}\rangle::=\langle\text{DIGIT}\rangle|\langle\text{UNSIGNED INTEGER}\rangle$$
$$\langle\text{DIGIT}\rangle \qquad (2)'$$

etc.

g_0 is an arbitrary initial symbol $(\notin A)$ and $\rightarrow$ contains $g_0 \rightarrow I$, $g_0 \rightarrow g_0 D$, $g_0 \rightarrow g_0 7$, etc. Languages defined in this way by an arbitrary finite A and a $G \subseteq (A^*)$, which is defined by equations, are called ALGOL-like. $\qquad\square$

The class of all possible languages is very large and coincides with the class of recursively countable sets, which arise in mathematical logic. Therefore one often studies special classes of languages.

1.9 Definition. Let A be a set and $x \in A^*$, $x = a_1 a_2 \ldots a_n$. n is called the *length* of x, in symbols $n = l(x)$.

If in a grammar $\mathscr{G}$ $x \rightarrow y$, then in general we don't know anything about the connection between $l(x)$ and $l(y)$.

1.10 Definition. Let $\mathscr{G} = (A, G, \rightarrow, g_0)$ be a grammar.

(i) $\mathscr{G}$ is called *context-sensitive:* $\Leftrightarrow \forall\, x \rightarrow y \in \rightarrow: l(x) \leq l(y)$.
(ii) $\mathscr{G}$ is called *context-free:* $\Leftrightarrow \forall\, x \rightarrow y \in \rightarrow: l(x) = 1$.
(iii) $\mathscr{G}$ is called *right linear:* $\Leftrightarrow$ all elements in $\rightarrow$ are of the form $g \rightarrow a$ or $g \rightarrow ag'$ with $g, g' \in G, a \in A^*$.
(iv) A language L is called context-sensitive (context-free, right linear), if there is a context-sensitive (context-free, right linear) grammar with $L = L(\mathscr{G})$.

In a context-sensitive language the length of a word is never shortened by any derivation, in a context-free language all productions are of the form $g \rightarrow v$ with $g \in G$ (and not only $g \in G_*$). All right linear languages are context-free. We show the connections in Figure 7.1.

1.11 Example. Examples 1.5 to 1.8 are context-free; 1.5 and 1.6 are right linear (1.7 would be "left linear"). We shall see after 1.13 that 1.5–1.8 are also context-sensitive. $\qquad\square$

1.12 Remark. A language $L(\mathscr{G})$ is context-sensitive, if any rule $x \rightarrow y$ is of the form $x = v_1 x' v_2 \rightarrow v_1 y' v_2 = y$, for $x' \in G, y' \neq \Lambda$. This follows from $l(x) = l(v_1) + l(x') + l(v_2) = l(v_1) + 1 + l(v_2) \leq l(v_1) + l(y') + l(v_2) = l(y)$.

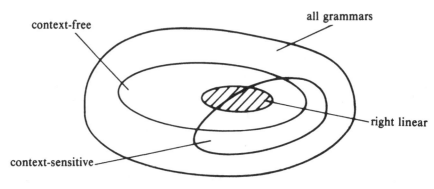

Figure 7.1

Conversely it can be shown that any context-sensitive language can be obtained in this way. The inequality in 1.10(i) also shows that in context-sensitive languages $y \to z$ implies $l(y) \le l(z)$. ☐

In languages with a rule of the form $v_1 x v_2 \to v_1 y v_2$ we can only derive y from x in the "context" of v_1 and v_2. In context-free languages $x \to y$ holds "independent of the context". We also note that one and the same language L can be described as $L = L(\mathcal{G})$ or also as $L = L(\mathcal{G}')$, where $\mathcal{G}$ is context-free but $\mathcal{G}'$ is not. The same applies to context-sensitive and right linear languages.

A context-free language may contain Λ (if $\to$ contains $g_0 \to \Lambda$, for example). The following theorem shows that most context-free languages are also context-sensitive.

1.13 Theorem. *Let* $\mathcal{G} = (A, G, \to, g_0)$ *be a grammar and* $L = L(\mathcal{G})$.

(i) *L is context-sensitive* $\Rightarrow \Lambda \notin L$.
(ii) *L is context-free with no $g \to \Lambda$ in $\to$, for some $g \in G$, implies that L is context-sensitive.*
(iii) *L is context-free* $\Rightarrow L \setminus \{\Lambda\}$ *is context-sensitive.*

PROOF. (i) Let L be context-sensitive. Suppose $\Lambda \in L$, then there exist $z_0, \dots, z_n \in V^*$ with $z_0 = g_0$, $z_n = \Lambda$ and $g_0 \Rightarrow z_0 \Rightarrow \dots \Rightarrow z_n = \Lambda$. Hence $1 = l(z_0) \le l(z_1) \le \dots \le l(z_n) = 0$, a contradiction.

(ii) Let L be context-free with no $g \to \Lambda$ ($g \in G$). Let $x \to y$ in $\to$. Since L is context-free, we have $x \in G$. Hence $y \ne \Lambda$ and $l(x) = 1 \le l(y)$. Thus L is context-sensitive.

(iii) This can be derived from (ii) by deleting all productions of the form $g \to \Lambda$ ($g \in G$) from $\to$. This "shrunken" grammar generates $L \setminus \{\Lambda\}$. ☐

Part (ii) shows the following

1.14 Corollary. *The examples* 1.5 *to* 1.8 *are context-sensitive.* ☐

If we could describe the natural languages in the form $L(\mathscr{G})$, they would not be context-free but context-sensitive. Without proof we state the following result (see GINSBURG):

1.15 Theorem. *A language is context-free if and only if it is ALGOL-like.*

□

Next, by means of an example we give an important class of languages, for which we can use semigroup theoretical methods to describe $L(\mathscr{G})$.

1.16 Example. Let $A = \{a_1, a_1', a_2, a_2', \ldots, a_n, a_n'\}$, $g_0 \notin A$,

$$W := \{a_1a_1', a_2a_2', \ldots, a_na_n'\} \subseteq A_*, \qquad G = \{g_0\},$$

$$\rightarrow := \{g_0 \rightarrow \Lambda\} \cup \{g_0 \rightarrow g_0x_1g_0x_2\ldots g_0x_rg_0 | r \in \mathbb{N} \wedge x_1, x_2, \ldots, x_r \in W\}$$

and

$$\mathscr{G} = (A, G, \rightarrow, g_0).$$

Furthermore, let S be a semigroup with presentation $[A, R]$, $R := \{(w, \Lambda) | w \in W\}$, $f: A^* \rightarrow S$, $x \mapsto [x]$. Here $[x]$ is the equivalence class of x with respect to the congruence relation $\tilde{R}$ of A_* generated by R. Then it can be shown that $L(\mathscr{G}) = \{x \in A^* | f(x) = [\Lambda]\}$. Languages constructed in this form are called Dyck languages. Since $g_0 \rightarrow \Lambda \in \rightarrow$ we see that $\mathscr{G}$ is context-free, but not context-sensitive. For details see GINSBURG, Section 3.7. □

PROBLEMS

1–5. Let $A = \{a, b\}$, $G = \{g_0\}$ $(g_0 \notin A)$. Determine $L(\mathscr{G})$ for the following relations $\rightarrow$.

1. $\rightarrow = \{g_0 \rightarrow a, g_0 \rightarrow b\}$.

2. $\rightarrow = \{g_0 \rightarrow a, g_0 \rightarrow aba, g_0 \rightarrow ag_0a\}$.

3. $\rightarrow = \{g_0 \rightarrow \Lambda, g_0 \rightarrow a, g_0 \rightarrow b\}$.

4. $\rightarrow = \{g_0 \rightarrow a, g_0 \rightarrow ag_0, g_0 \rightarrow g_0b\}$.

5. $\rightarrow = \{g_0 \rightarrow g_0\}$.

6–10. Which of the languages in Problems 1–5 are context-free? context-sensitive? right linear? ALGOL-like?

11. Find a grammar that generates $\{aba, aabaa, aaabaaa, \ldots\}$.

12. Find a grammar that generates A^* and another one that generates A_*.

B. Approach via Automata and Semigroups

We now define special semiautomata and introduce the concept of "accepting", so that we obtain "accepted" subsets of free monoids, which turn out to be languages $L(\mathscr{G})$. The notation is the same as in Chapter 6.

1.17 Definition. A finite, initial semiautomaton $\mathscr{A} = (Z, A, \delta, z_0)$ is called an *acceptor*, if, together with $\mathscr{A}$, a set $E \subseteq Z$ is given. We write $\mathscr{A} = (Z, A, \delta, z_0, E)$ and call E the set of designated final states. $W(\mathscr{A}) := \{w \in A^* | \bar{\delta}(z_0, w) \in E\}$ is called the *set of words accepted by* $\mathscr{A}$.

1.18 Definition. Let A be a set and $L \subseteq A^*$. L is called an *acceptor language* if there is an acceptor $\mathscr{A}$ with $W(\mathscr{A}) = L$.

1.19 Example. Let $\mathscr{A} = \{\{z_0, z_1\}, \{0, 1\}, \delta, z_0, \{z_1\}\}$ and δ as in Example 7.2.3. Then

$$W(\mathscr{A}) = \{w \in \{0, 1\}^* | \bar{\delta}(z_0, w) = z_1\} = \{1, 10, 01, 100, 010, 001, 111, \ldots\}$$

$$= \{a_1 a_2 \ldots a_n \in \{0, 1\}^* | a_1 + a_2 + \ldots + a_n \text{ is odd}\}. \qquad \square$$

1.20 Definition. Let A be a set and $R \subseteq A^*$. R is called *regular* or a regular language if R can be obtained from one-element subsets of A^* by a finite number of "admissible operations". Admissible operations are the formations of unions, products (in A^*) and generated submonoids.

1.21 Example. Let $A = \{a_1, a_2, a_3\}$. The following subsets of A^* can be obtained using the operations mentioned in 1.20.

$$R_1 = \{a_1 a_2 a_1, a_1 a_2 a_3, a_1 a_1 a_2 a_2 a_3 a_3\},$$

since R_1 is finite. $R_2 = \{\Lambda, a_1, a_1 a_1, a_1 a_1 a_1, \ldots\}$, since R_2 is the submonoid generated by $\{a_1\}$. $R_3 = \{a_1^n a_2^m | n, m \in \mathbb{N}_0\}$, since R_3 is the product of submonoids generated by a_1 and a_2, respectively. $\qquad \square$

General characterizations of regular sets and results on regular sets are difficult to obtain at this stage. We shall return to this after 1.25.

PROBLEMS

1. Let $\mathscr{A}$ be the marriage automaton (2.2 in Chapter 6), turned into an acceptor by choosing z_2 as initial state and $E = \{z_3\}$. Compute $W(\mathscr{A})$ and find out, what a married woman "has to do".

2. Let A be a nonempty set. Find (if possible) acceptors $\mathscr{A}_1, \mathscr{A}_2$ such that $W(\mathscr{A}_1) = A^*$ and $W(\mathscr{A}_2) = A_*$ hold.

3. For $A \neq \varnothing$, find (if possible) an acceptor $\mathscr{A}$ such that $W(\mathscr{A})$ is the set of all words of length ≤ 5.

4–6. Are the languages in Problems 1–3 right regular? If so, demonstrate how they can be obtained by using 1.20.

7. Is $\{ab, aabb, aaabbb, \ldots\}$ a regular set? A regular language?

C. Connections Between the Different Approaches

We give some connections between grammar languages, acceptor languages and regular languages. Not every (not even every context-free) grammar language is connected with an acceptor language. But for right linear languages we have

1.22 Theorem. *Let A be a set and $L \subseteq A^*$. L is an acceptor language if and only if L is a right linear language.*

PROOF IDEA (See GINSBURG, 2.2.1). Let L be an acceptor language. Then there is an acceptor $\mathscr{A} = (Z, A, \delta, z_0, E)$ with $W(\mathscr{A}) = L$. Without loss of generality let $Z \cap A = \varnothing$. We choose $\mathscr{G} := (A, Z, \rightarrow, z_0)$ with $\rightarrow :=$ $\{z \rightarrow a\delta(z, a) | z \in Z, a \in A\} \cup \{e \rightarrow \Lambda | e \in E\}$. Obviously $\mathscr{G}$ is right linear. Then $L(\mathscr{G}) = L = W(\mathscr{A})$.

Conversely, if $\mathscr{G}$ is a right linear grammar, then we can find an acceptor $\mathscr{A}$ with $L(\mathscr{G}) = W(\mathscr{A})$. Since this is rather complicated to show, we leave it out and refer the reader to GINSBURG. □

From acceptor languages we can go to the regular languages and back.

1.23 Theorem (Kleene). *Let A be a set and $L \subseteq A^*$. L is an acceptor language if and only if L is a regular language.*

PROOF IDEA (See ARBIB₁). Let L be an acceptor language. $L = W(\mathscr{A})$, $\mathscr{A} = (Z, A, \delta, z_0, E)$. One shows that L can be built up from singletons of A^* by means of admissible operations. Thus L is regular.

Conversely, let L be regular. For $\varnothing$ and for singletons of A^* there are acceptors, which accept precisely these sets. It can be shown that the result of each admissible operation is again a set of words defined by a suitable acceptor. So by induction we arrive at an acceptor $\mathscr{A}$ to L such that $L = W(\mathscr{A})$, L is an acceptor language. □

Using 1.22 and 1.23, we have a connection between right linear languages and regular languages. Figure 7.2 is a sketch of the situation.

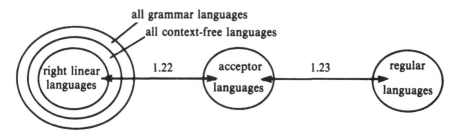

Figure 7.2

Therefore we can "identify" the right linear languages, acceptor languages and regular languages and give them the neutral name "formal languages". Since there are context-free languages, which are not right linear, we see that not every context-free language is regular. Context-free languages L can be described in the form $L = W(\mathscr{A})$, by using a more general class of automata, the so-called Keller automata $\mathscr{A}$.

We mention that not all subsets of A^* are regular. See 1.26 for such an example. We also mention that regular sets and related concepts are applicable in mathematical logic (computability, decidability, etc.), where so-called "Turing machines" are studied. See ARBIB$_1$, or BIRKHOFF and BARTEE.

We close this topic with a characterization of regular languages over singleton alphabets $A = \{a\}$. The proof of the result would be cumbersome without 1.23.

1.24 Definition. A subset $P = \{p_1, p_2, \ldots\}$ of $\mathbb{N}_0$ is called *periodic*, if there are numbers $q, k, n_0 \in \mathbb{N}_0$ such that for all $n \geq n_0$: $p_{n+k} - p_n = q$.

1.25 Theorem. *Let* $A = \{a\}$ *and* $L \subseteq A^*$. L *is regular if and only if there is a periodic subset* P *of* $\mathbb{N}_0$ *with* $L = \{a^p \mid p \in P\}$.

PROOF. (a) Let L be regular. Then there is an acceptor $\mathscr{A} = (Z, \{a\}, \delta, z_0, E)$ with $L = W(\mathscr{A})$. For $i \in \mathbb{N}_0$ let $z_i = \bar{\delta}(z_0, a^i)$. Since Z is finite, there exists $m, n \in \mathbb{N}_0$ with $z_m = z_n$ and, say, $m \leq n$. Then

$$z_{m+1} = \bar{\delta}(z_0, a^{m+1}) = \delta(\bar{\delta}(z_0, a^m), a) = \delta(z_m, a) = \delta(z_n, a) = \ldots = z_{n+1}.$$

In general, $z_{m+j} = z_{n+j}$ for all $j \in \mathbb{N}_0$, i.e. $\forall\, l \geq m$: $z_{l+(m-n)} = z_l$ hence $\forall\, l \geq m, \forall\, r \in \mathbb{N}_0$: $z_{l+r(m-n)} = z_l$. Let

$$M := \{k \in \mathbb{N}_0 \mid k < m \wedge z^k \in E\} \quad \text{and} \quad N := \{l \in \mathbb{N}_0 \mid m \leq l < n \wedge z^l \in E\}.$$

Then

$$W(\mathscr{A}) = \{a^k \mid k \in M\} \cup \{a^{l+r(m-n)} \mid l \in N \wedge r \in \mathbb{N}_0\},$$

and $P = \{p \in \mathbb{N}_0 \mid a^p \in W(\mathscr{A})\} = M \cup \{l + r(m - n) \mid l \in N \wedge r \in \mathbb{N}_0\}$. Therefore P is periodic.

(b) Let $L = \{a^p \mid p \in P\}$, where P is periodic. Then there are $m, n_1 \in \mathbb{N}_0$ with $\forall\, j \geq m$: $j \in P \Rightarrow j + n_1 \in P$. Let $M := P \cap \{0, 1, \ldots, m - 1\}$ and $N := P \cap \{m, m + 1, \ldots, m + n_1\}$. Then $P = M \cup \{l + rn_1 \mid l \in N \wedge r \in \mathbb{N}_0\}$. Hence

$$L = \{a^p \mid p \in P\} = \{a^m \mid m \in M\} \cup \{a^{l+rn_1} \mid l \in N \wedge r \in \mathbb{N}_0\}$$

$$= \{a^m \mid m \in M\} \cup \bigcup_{l \in L} \{a^l\} \cdot \langle a^{n_1} \rangle,$$

where $\langle a^{n_1} \rangle$ denotes the submonoid of A^* generated by a^{n_1}. This is the union of a finite set with a union of finitely many sets, which are products of singletons with submonoids generated by singletons. By Definition 1.20, L is regular.   □

1.26 Corollary. *Let $A = \{a\}$. Then $\{a^{n^2}|n \in \mathbb{N}_0\}$ is not regular.*

PROOF. This follows from 1.25 and the fact that $\{1, 4, 9, \ldots\}$ is not periodic.

□

PROBLEMS

1–3. Find acceptors for the regular languages in 1.21.

4–6. Find right linear grammars (if possible) for the languages in 1.21.

7. Which of the languages in Problems 1–5 of Section A are regular?

8. For the regular languages in Problem 7 find corresponding acceptors and show how these languages can be constructed by means of 1.20.

9. Let p_n be the solutions of the difference equation $p_n = 3p_{n-1} + 1$, $p_1 = 0$. Is $\{p_1, p_2, \ldots\}$ periodic? Is, for $A = \{a\}$, $\{a^{p_n}|n \in \mathbb{N}\}$ regular?

EXERCISES (Solutions in Chapter 8, p. 501)

1. Let $A = \{a\}$, $G = \{g_0\}(g_0 \neq a)$, $\rightarrow = \{g_0 \rightarrow aa, g_0 \rightarrow g_0 a\}$, $\mathcal{G} = (A, G, \rightarrow, g_0)$. Find $L(\mathcal{G})$.

2. Let $A = \{a\}$, $G = \{g_0\}$ $(g_0 \neq a)$, $\rightarrow = \{g_0 \rightarrow a g_0, g_0 \rightarrow g_0 a\}$, $\mathcal{G} = (A, G, \rightarrow, g_0)$. Find $L(\mathcal{G})$.

3. Are the languages in the above Exercises 1 and 2 context-sensitive? context-free? right linear? ALGOL-like?

4. Let $\mathcal{A}$ be the marriage automaton (see 6.2.2), which is an acceptor with respect to $z_0 := z_1$, and $E = \{z_3\}$. What is $W(\mathcal{A})$? Interpret $W(\mathcal{A})$ in "everyday language".

5–7. Let $A = \{a, b\}$, $G = \{g_0\}$ $(g_0 \notin A)$. For the following definitions of $\rightarrow$, find the generated language.

5. $\rightarrow = \{g_0 \rightarrow a g_0, g_0 \rightarrow b\}$.

6. $\rightarrow = \{g_0 \rightarrow g_0 g_0, g_0 \rightarrow aa\}$.

7. $\rightarrow = \{g_0 \rightarrow a g_0 a, g_0 \rightarrow aa\}$.

8. Compare the results of Exercises 6 and 7. What does it tell us?

9. Find some grammar $\mathcal{G}$ such that $L(\mathcal{G}) = \{b, aba, aabaa, \ldots\}$.

10. Find $\mathcal{G}$ such that $L(\mathcal{G}) = \{\Lambda, b, aa, aba, aaaa, aabaa, \ldots\}$.

11. Call two grammars equivalent if they generate the same language. Show that this yields an equivalence relation on every set of grammars.

12. Let $A = \{a, b, c\}$, $G = \{g_0, g_1, g_2, g_3\}$, $\to = \{g_0 \to g_1, g_0 \to g_2, g_1 \to abc$,
    $g_2 \to ab, g_2 \to g_3, g_3 \to c\}$.
    (a) Is this grammar $\mathcal{G}$ context-free? (b) Is $abc \in L(\mathcal{G})$?

13. Is $M = \{\Lambda, ab, aabb, aaabbb, \ldots\}$ a regular set? Does there exist an acceptor $\mathcal{A}$ with $L(\mathcal{A}) = M$?

14. Let $T := \{a_1 a_2 a_3 | a_1, a_2, a_3 \in A\}$. Does there exist a grammar $\mathcal{G}$ such that $T = L(\mathcal{G})$? Does there exist an acceptor $\mathcal{A}$ with $T = W(\mathcal{A})$?

15. Is $\{1, 2, 3, 6, 9, 12, 15, \ldots\}$ periodic? If yes, what is its corresponding regular language?

16. Is $\{a^n | n \equiv 3 \pmod 4\}$ a right linear language?

§2. Semigroups in Biology

Semigroups can be used in biology to describe certain aspects in the crossing of organisms, in genetics and in considerations of metabolisms. Before giving some applications we first describe a simple associativity test for binary operations on a set. A set with a binary operation is called *groupoid*. If $(S, \circ)$ is a finite groupoid with $S = \{x_1, \ldots, x_n\}$, then in order for $\circ$ to be associative we must have $\forall\, x, y, z \in S: (x \circ y) \circ z = x \circ (y \circ z)$. We begin by showing that we can restrict our choice of y to be in a generating set of S.

2.1 Theorem. *Let* $(S, \circ)$ *be a groupoid and* $E \subseteq S$, *such that* $\langle E \rangle = S$. *If* $\forall\, x, z \in S, \forall\, e \in E$

$$(x \circ e) \circ z = x \circ (e \circ z)$$

then $\circ$ *is associative.*

PROOF. It is sufficient to show that $S' := \{s \in S | \forall\, x, z \in S: (x \circ s) \circ z = x \circ (s \circ z)\} = S$. S' is a groupoid: let $s, t \in S'$, then $s \circ t \in S'$, because $(x \circ (s \circ t)) \circ z = ((x \circ s) \circ t) \circ z = (x \circ s) \circ (t \circ z) = x \circ (s \circ (t \circ z)) = x \circ ((s \circ t) \circ z)$. Since $E \subseteq S'$ we have $\langle E \rangle \subseteq S'$, thus $S \subseteq S'$ and therefore $S = S'$.   □

2.2 "Light-Test". If we are given the operation table of a finite groupoid $(S, \circ)$ with $S = \{x_1, \ldots, x_n\}$ then to check whether $\circ$ is associative we choose a generating set E of S and compare for all $e \in E$ the two tables.

e	x_1	x_2	$\ldots$	x_n
$x_1 \circ e$	$(x_1 \circ e) \circ x_1$	$(x_1 \circ e) \circ x_2$	$\ldots$	$(x_1 \circ e) \circ x_n$
$x_2 \circ e$	$(x_2 \circ e) \circ x_1$	$(x_2 \circ e) \circ x_2$	$\ldots$	$(x_2 \circ e) \circ x_n$
$\vdots$	$\vdots$	$\vdots$		$\vdots$
$x_n \circ e$	$(x_n \circ e) \circ x_1$	$(x_n \circ e) \circ x_2$	$\ldots$	$(x_n \circ e) \circ x_n$

and

e	$e \circ x_1$	$e \circ x_2$	$\ldots$	$e \circ x_n$
x_1	$x_1 \circ (e \circ x_1)$	$x_1 \circ (e \circ x_2)$	$\ldots$	$x_1 \circ (e \circ x_n)$
x_2	$x_2 \circ (e \circ x_1)$	$x_2 \circ (e \circ x_2)$	$\ldots$	$x_2 \circ (e \circ x_n)$
$\vdots$	$\vdots$	$\vdots$		$\vdots$
x_n	$x_n \circ (e \circ x_1)$	$x_n \circ (e \circ x_2)$	$\ldots$	$x_n \circ (e \circ x_n)$

$\circ$ is associative if for each $e \in E$ these two tables are equal.

2.3 Notation. Method 2.2 is called the Light-test, named after F. W. Light.

2.4 Remarks. The Light-test does not have to be used if either:
(i) the operation table for $(S, \circ)$ can be transformed into the table of a semigroup by relabeling the elements (then $\circ$ is associative); or
(ii) $(S, \circ)$ does not have an idempotent element, in this case $\circ$ can be shown to be nonassociative, since every finite semigroup has an idempotent (why?).

2.5 Examples.

(i) $S = \{a, b, c\}$;

$\circ$	a	b	c
a	a	b	c
b	b	c	a
c	c	a	b

Renaming a as $[0]$, b as $[1]$ and c as $[2]$, we obtain the operation table of $(\mathbb{Z}_3, +)$, which is a group.

(ii) $S = \{a, b, c\}$

$\circ$	a	b	c
a	b	a	c
b	c	a	b
c	b	c	a

No element is idempotent, thus $\circ$ cannot be associative.

(iii) $S = \{a, b, c, d\}$

∘	a	b	c	d
a	a	a	a	a
b	a	a	a	b
c	a	a	c	c
d	a	b	c	d

We choose $E = \{b, c, d\}$ and start with b

b	a	b	c	d
$a \circ b = a$	a	a	a	a
$b \circ b = a$	a	a	a	a
$c \circ b = a$	a	a	a	a
$d \circ b = b$	a	a	a	b

b	$b \circ a = a$	$b \circ b = a$	$b \circ c = a$	$b \circ d = b$
a	a	a	a	a
b	a	a	a	a
c	a	a	a	a
d	a	a	a	b

Similarly we see that the respective tables for c and d are equal. Therefore $(S, \circ)$ is a semigroup.   □

2.6 Remarks. From a given semigroup operation table we can derive the following information:

(i) ∘ is commutative if and only if the table is symmetric with respect to the main diagonal, as in Example 2.5(i).

(ii) $(S, \circ)$ is a monoid if and only if there is an element e such that its row and column is equal to the first (index) row and column. In this case e is the identity element. In 2.5(i) the element a is the identity, in (iii) it is the element d.

(iii) $(S, \circ)$ has n as zero element if all elements in its row and column are n, e.g. 2.5(iii).

(iv) $(S, \circ)$ is a group if and only if ∘ is associative and each element of S occurs exactly once in each row and column of the operation table, see, e.g. 2.5(i).

2.7. Example. In breeding a strain of cattle, which can be black or brown, monochromatic or spotted, it is known that black is dominant and brown recessive and that monochromatic is dominant against spotted. Thus there are four possible types of cattle in this herd:

a ... black, monochromatic, c ... brown, monochromatic,
b ... black, spotted, d ... brown, spotted.

Due to dominance, in crossing a black, spotted one with a brown, monochromatic one we expect a black, monochromatic one. This can be symbolized by "$b * c = a$". The "operation" $*$ can be studied for all possible pairs to obtain the table

$*$	a	b	c	d
a	a	a	a	a
b	a	a	a	b
c	a	a	c	c
d	a	b	c	d

This is the table (iii) in Example 2.5. The Light-test showed that $S :=$ $(\{a, b, c, d\}, *)$ is a semigroup. Moreover the table is symmetric with respect of the main diagonal, therefore $*$ is commutative, d is the identity element (because of 2.6(ii)) and a is the zero element (because of 2.6(iii)). From 2.6(iv) we conclude that S cannot be a group. S is a commutative monoid with zero element.   □

In general the table for breeding operations is more complicated, see NAHIKIAN. We can ask for connections between hereditary laws and the corresponding semigroups. Of course, we need $S \circ S = S = S^2$ for such semigroups S, since $s \in S \setminus S^2$ would vanish after the first generation and would not even be observed (rare cases excepted).

A different biological problem which leads to algebraic problems, is as follows: All genetic information in an organism is given in the so-called deoxyribonucleic acid (DNA). DNA consists of two strands, which are combined together to form the famous double helix. Each strand is made up (a polymer) of four different basic substances, the nucleotides. We concentrate our attention on one strand only. If the nucleotides are denoted by n_1, n_2, n_3, n_4, then the strand can be regarded as a word over $\{n_1, n_2, n_3, n_3\}$. DNA cannot put the genetic information into effect. By means of a messenger ribonucleic acid the information contained in the DNA is copied ("transcription") and then transferred onto the protein chains ("translation"). These protein chains are polymeres consisting of 20 different basic substances, the amino acids, denoted by $a_1, \ldots, a_{20}$. As with the DNA, each protein chain can be regarded as a word over $\{a_1, \ldots, a_{20}\}$.

In general it is assumed that the sequence of nucleotides in DNA is uniquely determined by the sequence of amino acids in a protein chain. In other words it is assumed that there is a monomorphism from the free semigroup $H_{20} := \{a_1, \ldots, a_{20}\}$ into the free semigroup $H_4 := \{n_1, n_2, n_3, n_4\}$. The DNA protein-coding problem is the question, how many, if any, monomorphisms are there from H_{20} into H_4.

A first glance would indicate that the answer is a clear "none". How could we have a monomorphism from the "huge" semigroup H_{20} into the "much smaller" H_4? There are plenty of monomorphisms from H_4 into H_{20}—would this not mean that $H_4 \cong H_{20}$, in contradiction to 6.3.47? The somewhat surprising answer is given in the following theorem.

2.8 Theorem. *There are infinitely many monomorphisms from H_{20} into H_4. Thus the DNA protein-coding problem has infinitely many solutions.*

PROOF. Consider the following diagram

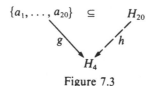

Figure 7.3

with $g(a_1) := n_1 n_1 n_1$, $g(a_2) := n_1 n_1 n_2$, $g(a_3) = n_1 n_1 n_3, \ldots g(a_{20}) := n_2 n_1 n_4$. Then g is injective. From 6.3.47 we know that there is a uniquely determined homomorphism $h; H_{20} \to H_4$ with $h|_{\{a_1,\ldots,a_{20}\}} = g$. We show that h is a monomorphism. $h(a_{i_1} a_{i_2} \ldots a_{i_r}) = h(a_{j_1} a_{j_2} \ldots a_{j_s})$ implies $h(a_{i_1}) h(a_{i_2}) \ldots h(a_{i_r}) = h(a_{j_1}) h(a_{j_2}) \ldots h(a_{j_s})$, therefore $g(a_{i_1}) g(a_{i_2}) \ldots g(a_{i_r}) = g(a_{j_1}) g(a_{j_2}) \ldots g(a_{j_s})$. From the definition of equality in free semigroups and because all $g(a_k)$ have the same length 3, we conclude $r = s$ and $g(a_{i_1}) = g(a_{j_1}), \ldots, g(a_{i_r}) = g(a_{j_r})$. Injectivity of g implies $a_{i_1} = a_{j_1}, \ldots, a_{i_r} = a_{j_r} = a_{j_s}$, therefore $a_{i_1} a_{i_2} \ldots a_{i_r} = a_{j_1} a_{j_2} \ldots a_{j_s}$. Thus h is a monomorphism from H_{20} into H_4. Obviously by different choices of g, e.g. $g(n_1) := n_3 n_1 n_1$, etc. or by using words of length 4 or 5, etc. we can obtain infinitely many such monomorphisms.   □

As a point of information, it appears that the DNA protein-coding in nature is always done by means of words of length 3 (triplet-code), see ROSEN2.

2.9 Remark. We saw that there are monomorphisms from H_{20} into H_4 and also we know there are monomorphisms from H_4 into H_{20}, e.g. via the extension $n_i \to a_i$, $i = 1, 2, 3, 4$. It can be shown that H_4 and H_{20} must be of the same cardinality. By 6.3.47, they are nonisomorphic, and so we see that two nonisomorphic semigroups can be embeddable into each other.

A different aspect of applied algebra appears in problems of metabolisms and processes of cellular growth. Metabolism consists of a sequence of transitions between organic substances (acids), triggered off by enzymes and co-enzymes. Such a process can be described in terms of semiautomata, in which the organic acids form the state set Z, the (co-)enzymes form the inputs A and the chemical reactions "substance + (co-)enzymes → new substance" define the state transition function δ.

This (finite) semiautomaton can be decomposed into IR flip-flop and simple group-semiautomata, using the theorem of Krohn–Rhodes, see 6.6.25. This decomposition illustrates also the structure of the metabolic processes (see KROHN, LANGER and RHODES). Metabolic processes with replacement of destroyed functional parts are studied in ROSEN₂, who also considered the effect of environmental influences. Self-learning aspects of automata theory are studied in SUPPES and KIERAS. As a further illustration we show by means of an example how cellular growth can be described in terms of formal languages:

2.10 Example. Let $A = \{0, 1, 2, 3, 4, 5, (,)\}$, $G = \{g_0\}$ and

$$\rightarrow = \{g_0 \rightarrow 0g_0, g_0 \rightarrow 1g_0, \ldots, g_0 \rightarrow 5g_0, g_0 \rightarrow 3(4)g_0, g_0 \rightarrow (4)(4)g_0, g_0 \rightarrow \Lambda\}.$$

The language $L(\mathscr{G})$ with $\mathscr{G} = (A, G, \rightarrow, g_0)$ is right linear and contains also

2          $(g_0 = \Lambda g_0 \Lambda \Rightarrow \Lambda 2g_0 \Lambda \Rightarrow \Lambda_2 \Lambda\Lambda = 2)$,

233       $(g_0 = \Lambda g_0 \Lambda \Rightarrow \Lambda 2g_0 \Lambda = 2g_0 \Lambda \Rightarrow 23g_0 \Lambda \Rightarrow 233g_0 \Lambda \Rightarrow \ldots \Rightarrow 233)$,

2333(4) $(g_0 = \Lambda g_0 \Lambda \Rightarrow \ldots \Rightarrow 2333(4)g_0 \Lambda \Rightarrow 2333(4)\Lambda\Lambda = 2333(4))$.

We also obtain 2333(4)(4), 2333(4)(4)55, 2333(4)(4)553(4). In Figure 2.1 we interpret $0, 1, \ldots, 5$ as cells number 0 to 5, e.g. 233 is the sequence of cells No. 2, 3, 3 in the direction of growth (for instance, straight upwards). 3(4) is denoted as a branch point, where in cell No. 3 cell No. 4 branches out and forms a "bud" sideways. The last "word" above in this example is of the form

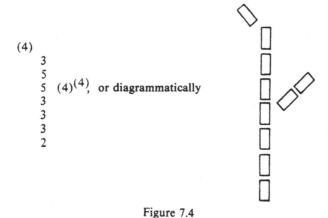

(4)
3
5
5 $(4)^{(4)}$, or diagrammatically
3
3
3
2

Figure 7.4

This looks like the beginning of a plant. In this example only cell No. 3 has the capability to form "buds", the "branches" consist of cells No. 4 only. □

In this way the growth of plants can be described algebraically, see HERMANN and ROSENBERG. Further material on this subject is contained in HOLCOMBE.

PROBLEMS

1.–3. Check for associativity.

1. $\circ_1$	a	b	c	d
a	a	b	c	d
b	a	b	c	d
c	a	b	c	d
d	a	b	c	d

2. $\circ_2$	a	b	c	d
a	a	b	a	a
b	b	a	d	c
c	c	a	a	b
d	d	a	b	a

3. $\circ_3$	a	b	c	d
a	a	b	c	d
b	a	b	c	d
c	c	a	d	b
d	d	c	b	a

4. Which of the structures in Problems 1–3 have a zero element?

5. As in Problem 4, now with unit element.

6. Which of the structures in Problems 1–3 are commutative semigroups, which are groups? abelian groups?

7. Determine all idempotent elements in the structures of Problems 1–3.

8. How many binary operations $\circ$ are there which turn $\{a, b, c, d, e\}$ into a groupoid, such that e is the identity and such that in every row and in every column each element appears precisely once?

9. Which of the groupoids in Problem 8 are groups?

10. Let $a * b := a$ be defined on a finite set N. Why does the Light-test help very little in this case?

11. Find a monomorphism from the H_4 into H_{20} (see 2.9).

12. Is 23(4)3(4)(4) obtainable in 2.10? If so, draw the corresponding "plant".

EXERCISES (Solutions in Chapter 8, p. 502)

1–3. Check for associativity:

1. $\circ_1$	a	b	c	d
a	a	b	c	d
b	b	c	d	a
c	c	d	a	b
d	d	a	b	c

2. $\circ_2$	a	b	c	d
a	a	b	c	d
b	b	a	d	c
c	c	d	a	b
d	d	c	b	a

3. $\circ_3$	a	b	c	d
a	a	b	c	d
b	b	d	a	c
c	c	a	d	b
d	d	c	b	a

4. Which of the structures in Exercises 1–3 have a zero element? which have an identity?

5. Which of the structures in Exercises 1–3 are groups? abelian groups?

6. Find the group kernels in Problems 1–3 (if defined).

7. Determine all idempotent elements in Problems 1–3.

8. Let N be a finite non-empty set, fix a certain element z in N and define $a \# b := z$ for all $a, b \in N$. Is $(N, \#)$ a semigroup? Is it wise to use the Light-test in this case?

9. Let $N = \{e, a, b, c\}$. How many binary operations $\circ$ can be defined on N such that e serves as the identity and such that every element of N appears in every row and every column?

10. Which of the groupoids in Exercise 9 are groups? which of them are abelian groups?

11. How many different solutions to the DNA protein-coding problem by triplet codes are there?

12. Draw for Example 2.4 the "plant" corresponding to the word $123(4)53(4)3(4)3(4)$. Are $(4)(4)(4)(4)$, $(4)(4)(4), \ldots, 12345(4)$ elements of $L(\mathcal{G})$?

§3. Semigroups in Sociology

Are the enemies of my enemies my friends? Is the enemy of my friend a friend of my enemy? We shall see that questions like these can be elegantly formulated in the language of semigroups. There are several approaches to algebraic methods in sociology; three such approaches will be described.

A. Kinship Systems

We continue our investigation of relation semigroups defined in 6.3.5. Kinship relationships such as "being father", "being mother", can be combined to "being the mother of the father", i.e. "grandmother on father's side", etc.

First we define the relation semigroup of all kinship relations. However, we omit the tricky problem of defining the term "kinship relation".

First Attempt. We begin with a few "basic relations", like "being father", "being mother", etc. and draw up rules on how these kinship relationships can be combined, hoping that we can obtain a grammar $\mathcal{G}$ in such a way that $L(\mathcal{G})$ is the set of all "meaningful" kinship relations. If we define a grammar as in 1.1 this approach seems to have failed so far, see BOYD, HAEHL and SAILER.

Second Attempt. Again we start with a selected set X of basic relations such that the set of all their relation products yields all remaining kinship relations. In this way we seem to converge towards the concept of a free semigroup over X. But this approach also fails: a set is a collection of distinctive objects and we have not defined equality of kinship relations. In the free semigroup X_* all formally combined relations would be different so that we obtained infinitely many kinship relations. Irrespective of the fact that there are only finitely many people and names on earth, in X_* equal things are made different, e.g. "being brother" $\square$ "being child" = "being child", since

$$(x \text{ brother of } y) \wedge (y \text{ child of } z) \Rightarrow (x \text{ child of } z)$$

as long as we can regard ourself as brother of ourself. The children of a wife could also be the children of the husband in most cases, but not in all cases, e.g. second marriages. Thus we see that the definition of equality of kinship relationships depends very much on the society in which we study these relations.

Third Attempt. We use the presentation of semigroups (see 6.3.52) and define

3.1 Definition. A *kinship system* is a semigroup $S = [X, R]$, where

(i) X is a set of "kinship relationships".
(ii) R is a relation on X_*, which expresses equality of kinship relationships.

The product in S is always interpreted as relation product.

3.2 Example. Let $X = \{$"is father of", "is mother of"$\}$ and $R = \emptyset$. Then the kinship system S is the semigroup $\{$"is father of", "is mother of", "is grandfather of", ...$\}$. $\square$

Since our interest is not exclusively focused on grandfathers, we consider another example.

3.3 Example. Let $F :=$ "is father of", $M :=$ "is mother of", $S :=$ "is son of", $D :=$ "is daughter of", $B :=$ "is brother of", $Si :=$ "is sister of", $C :=$ "is child of". Then $X := \{F, M, S, D, B, Si, C\}$ and

$$R := \{(CM, CF), (BS, S), (SiD, D), (CBM, CMM), (MC, FC),$$
$$(SB, S), (DSi, D), (MBC, MMC), \ldots\}.$$

See BOYD, HAEHL and SAILER for the complete list of R. The first pair means that in the semigroup we have $CM = CF$, i.e. children of the mother are the same as the children of the father, thus we do not differentiate between brothers and stepbrothers, etc. It can be shown that we obtain a finite semigroup. $\square$

Let $\mathcal{G}$ be a "society", i.e. a nonempty set of people and let $S(\mathcal{G})$ be the finite semigroup of all different kinship relationships of this society.

3.4 Definition. Any set X which generates $S(\mathcal{G})$ is called a *kinship generator*. $S(\mathcal{G})$ is called the kinship system of the society $\mathcal{G}$.

3.5 Remark. In 3.4 we can choose $X = S(\mathcal{G})$, but usually a much smaller generating set suffices.

3.6 Example. It is difficult to imagine that the semigroup in Example 3.2 is the kinship system of an interesting society. But the semigroup S of Example 3.3 is "very close to" the kinship system of the society of the Fox indians in North America, described by S. Tax in 1937. $\square$

What does it mean when we say "it comes very close to the society"?

3.7 Definition. Let $\mathscr{G}_1$ and $\mathscr{G}_2$ be two societies for which there are equal generating systems for $S(\mathscr{G}_1)$ and $S(\mathscr{G}_2)$. Then

$$\delta(\mathscr{G}_1, \mathscr{G}_2) := \delta(S(\mathscr{G}_1), S(\mathscr{G}_2)) \qquad (\text{see } 6.3.60)$$

is called the *distance* between $\mathscr{G}_1$ and $\mathscr{G}_2$.

In particular, societies with isomorphic kinship systems have distance 0. The semigroups $S(\mathscr{G})$ often have special properties, e.g. "is a son of" and "is father of" are nearly inverse relations. The framework for the investigation of special $S(\mathscr{G})$ would be the theory of inverse semigroups, i.e. semigroups S such that $\forall\, s \in S, \exists\, s' \in S\colon ss's = s \wedge s'ss' = s'$. If s denotes "is son of" and s' denotes "is father of", then these two equations hold in most (monogamous) societies.

Next we describe an example where semigroups and sociology meet each other, appropriately or inappropriately, within the confines of a monastery.

PROBLEMS

1. In 3.2, give four more elements which belong to S.

2. Show that one obtains a finite semigroup in 3.3 (recall the presentation of semigroups).

3. What does it mean for a kinship system to be an inverse semigroup?

4. Let $X = \{F, M\}$ and $R = \{(FFF, F), (MMM, M), (FM, MF)\}$. Describe the kinship system S as a semigroup.

5. Let $X = \{F, M\}$ and $R' = \{(FFF, F), (MM, M), (FM, MF)\}$. Give the operation table for the resulting kinship system S'.

6. Let X be as in Problem 5 and $R'' = \{(FF, F), (MM, M), (FM, MF), (MF, F)\}$. Which kinship systems S'' do we get?

7–9. Are the semigroups S, S', S'' in Problems 4–7 monoids? commutative? groups? abelian groups?

10. Compute $\delta(S, S')$.

11. Compute $\delta(S, S'')$.

12. Compute $\delta(S', S'')$.

13. In Problems 10–12, check if the triangle inequality holds.

B. Social Networks

Sociology is the study of human interactive behavior in group situations. In many societies, some underlying structure is what is of interest to the sociologist. Such structures are often revealed by mathematical analysis. This indicates how algebraic techniques may be introduced into studies of this kind.

Imagine for a moment an arbitrary society, such as your family, your circle of friends, your university colleagues, etc. In such societies, provided they are large enough (at least ≥ 3 members), coalitions can be formed consisting of groups of people who like each other, who have similar interests or who behave similarly. How can we recognize such coalitions or formations of "blocks"? One way of finding out is asking, either directly or by questionnaires. Evaluation of the results of such questionnaires is very messy and usually it is not possible to see a pattern of coalitions. For instance, we number the members of a society by $1, 2, \ldots$ and ask each member No. i for his/her opinion about No. j (e.g. praise, fear, love, esteem, etc.). Here is a typical example.

3.8 Example (due to SAMPSON). In an American monastery the degree of integration of newcomers to the monastery has been studied. Numbers were allocated to the monks and novices according to the number of years they have been serving in the monastery. Each of the 18 members was asked to assign 3 or 2 or 1 points according to the esteem in which they hold a person ("no answer" was also permitted as well as the same point value for several people). The result was as follows:

	1	2	3	4	5	6	7	8	9	10	11	12	13	14	15	16	17	18
1					1		3				2							
2	3				1		2											
3	3												2			1	1	
4						1				2	3							
5				3					1		2							
6				3							2							
7		3											1		1	2		
8				3	1	2												
9	1			3				2										
10																		
11					2				3									
12	1	2												3				

	1	2	3	4	5	6	7	8	9	10	11	12	13	14	15	16	17	18
13				3			2				1							
14	3	2										1			2			
15	1	3												2				
16	3						2				1				2			
17	1	2											1					3
18		2	3													1		

The subdivision into blocks makes the table easier to handle. As we might expect, this block design is not very useful as it stands.   □

In the mid-1970's a Harvard University computer program was developed to change the order of $1, 2, \ldots, 18$ in a way which makes it possible to observe a separation into "strong" and "weak" blocks in the table. These algorithms are called BLOCKER and CONCOR, see WHITE, BOORMAN and BREIGER.

3.8 Example (Continued). These algorithms applied to our example of a monastery yield:

	10	5	9	6	4	11	8	12	1	2	14	15	7	16	13	3	17	18
10																		
5			1	3	2													
9				3		2		1										
6				3	2													
4	2			1	3													
11		2	3															
8		1		2	3													
12									3	2	1							
1		1					3	2										
2		1					2	3										
14									1	3	2		2					
15									1	3	2							
7									1	3			1	2				
16									1	3			2	2				
13		3		1										2				
3										3					2		1	1
17										1					1	2		3
18										2						3	1	

This table clearly shows formation of blocks.   □

When a formation of blocks has been obtained by these algorithms, the
"strong" blocks (the ones with "many" elements) are encoded as 1, the
"weak" blocks are encoded as 0. We still have to define "weakness" of
blocks: We can choose between

(i) no assignment at all; except possibly a few ones; or
(ii) the sum of the elements in a block is smaller than half of the "on
average" expected sum within the block.

or possibly other similar conventions.

3.8 Example (Continued). If we use definitions (i) and (ii) for weak blocks
we obtain from the second table above the following matrices, respectively:

$$\begin{pmatrix} 1 & 0 & 0 \\ 1 & 1 & 0 \\ 1 & 1 & 1 \end{pmatrix} \quad \begin{pmatrix} 1 & 0 & 0 \\ 0 & 1 & 0 \\ 0 & 1 & 1 \end{pmatrix}, \quad \text{etc.}$$

$$\text{for (i)} \qquad \text{for (ii)}$$

We can interpet (i) as follows: the first group of monks and novices (No.
10, 5, 9, 6, 4, 11, 8) have high regard for themselves, the second group (No.
12, 1, 2, 14, 15, 7, 16) regards itself and the first group highly, the third group
regards all three groups highly.

If we use the definition (ii) we obtain a slightly different picture. Both
cases matched very well the observed behavior of the members of the
monastery: the first group consisted of the "well-established" monks, the
second comprised the "young radicals" and the third consisted of some
"outsiders".

The question "whom do you like" resulted in the matrices

$$\begin{pmatrix} 1 & 1 & 1 \\ 0 & 1 & 1 \\ 1 & 1 & 1 \end{pmatrix} \quad \begin{pmatrix} 1 & 0 & 0 \\ 0 & 1 & 0 \\ 0 & 1 & 1 \end{pmatrix}.$$

$$\text{for (i)} \qquad \text{for (ii)}$$

The answer to "do not like him" resulted in

$$\begin{pmatrix} 0 & 1 & 1 \\ 1 & 1 & 1 \\ 1 & 1 & 0 \end{pmatrix} \quad \begin{pmatrix} 0 & 1 & 1 \\ 1 & 0 & 1 \\ 1 & 1 & 0 \end{pmatrix}.$$

$$\text{for (i)} \qquad \text{for (ii)} \qquad \square$$

In general, for each relation $R_1, \ldots, R_s$ (each R_i corresponding to questions of attitude towards others) one obtains $k \times k$ matrices $\mathbf{M}_1, \ldots, \mathbf{M}_s$
over $\{0, 1\}$ with relatively small k, e.g. $k = 3$ in 3.8. The relational products
$R_i \square R_j$ correspond to the matrices $\mathbf{M}_i * \mathbf{M}_j$, where $*$ is defined as follows:

3.9 Definition. Let $\mathbf{A} = (a_{ie})$, $\mathbf{B} = (b_{ej})$ be $k \times k$ matrices over $\{0, 1\}$. Then $\mathbf{A} * \mathbf{B}$ is the usual product of matrices if we operate with 0 and 1 in the following way:

+	0	1
0	0	1
1	1	1

·	0	1
0	0	0
1	0	1

3.10 Example. We continue Example 3.8 and choose the variant (ii). R_1 meaning "have esteem for" has the matrix

$$\mathbf{M}_1 = \begin{pmatrix} 1 & 0 & 0 \\ 0 & 1 & 0 \\ 0 & 1 & 1 \end{pmatrix};$$

R_2 meaning "like well" has

$$\mathbf{M}_2 = \mathbf{M}_1;$$

R_3 meaning "do not like" has

$$\mathbf{M}_3 = \begin{pmatrix} 0 & 1 & 1 \\ 1 & 0 & 1 \\ 1 & 1 & 0 \end{pmatrix}$$

as matrix representation.

We omit R_1 and interpret R_2 as "friend" and R_3 as "enemy". "Enemy of an enemy" then corresponds to

$$\mathbf{M}_3 * \mathbf{M}_3 = \begin{pmatrix} 0 & 1 & 1 \\ 1 & 0 & 1 \\ 1 & 1 & 0 \end{pmatrix} * \begin{pmatrix} 0 & 1 & 1 \\ 1 & 0 & 1 \\ 1 & 1 & 0 \end{pmatrix} = \begin{pmatrix} 1 & 1 & 1 \\ 1 & 1 & 1 \\ 1 & 1 & 1 \end{pmatrix} \neq \mathbf{M}_2.$$

Thus, the enemy of an enemy is not necessarily a friend in this society. The matrix corresponding to "friend of an enemy" is

$$\mathbf{M}_2 * \mathbf{M}_3 = \begin{pmatrix} 1 & 0 & 0 \\ 0 & 1 & 0 \\ 0 & 1 & 1 \end{pmatrix} * \begin{pmatrix} 0 & 1 & 1 \\ 1 & 0 & 1 \\ 1 & 1 & 0 \end{pmatrix} = \begin{pmatrix} 0 & 1 & 1 \\ 1 & 0 & 1 \\ 1 & 1 & 1 \end{pmatrix},$$

while the matrix corresponding to "enemy of a friend" is

$$\mathbf{M}_3 * \mathbf{M}_2 = \begin{pmatrix} 0 & 1 & 1 \\ 1 & 0 & 1 \\ 1 & 1 & 0 \end{pmatrix} * \begin{pmatrix} 1 & 0 & 0 \\ 0 & 1 & 0 \\ 0 & 1 & 1 \end{pmatrix} = \begin{pmatrix} 0 & 1 & 1 \\ 1 & 1 & 1 \\ 1 & 1 & 0 \end{pmatrix} \neq \mathbf{M}_2 * \mathbf{M}_3.$$

So in this society "friend of an enemy" is different from "enemy of a friend".

□

The question of whether the enemy of an enemy is a friend is equivalent to $M_3 * M_3 = M_2$; the "equation": friend of enemy = enemy of friend is equivalent to the property that the matrices M_2 and M_3 commute. We can have different answers to these questions according to the different societies we are investigating.

BOORMAN and WHITE introduce the term "role structure", which is an abstraction of concrete situations, like the one in a monastery or in a commune.

3.11 Definition. Let $\mathscr{G}$ be a community partitioned into blocks $B_1, \ldots, B_k$, let $R_1, \ldots, R_s$ be relations on $\mathscr{G}$ with corresponding $k \times k$ matrices $M_1, \ldots, M_s$ over $\{0, 1\}$. The *role structure* corresponding to $\mathscr{G}, B_1, \ldots, B_k$, $R_1, \ldots, R_s, M_1, \ldots, M_s$ is the subsemigroup, generated by $M_1, \ldots, M_s$, of the finite semigroup of all $k \times k$ matrices over $\{0, 1\}$ with respect to $*$.

3.12 Example (3.8 and 3.10 continued). Let

$$\mathscr{G} = \{1, 2, \ldots, 18\} = \{10, 5, 9, 6, 4, 11, 8\} \cup \{12, 1, 2, 14, 15, 7, 16\}$$
$$\cup \{13, 3, 17, 18\} = B_1 \cup B_2 \cup B_3;$$

let R_2 and R_3 be as in 3.10. We use Theorem 6.3.20 to obtain

$$\langle \{M_2, M_3\} \rangle = \{M_2, M_3\} \cup \{M_2, M_3\}^2 \cup \{M_2, M_3\}^3 \cup \ldots.$$

Here

$$\{M_2, M_3\}^2 = \{M_2 * M_2 = M_2, M_2 * M_3, M_3 * M_2, M_3 * M_3 := N\},$$

N is zero element with respect to $*$.

$$\{M_2, M_3\}^3 = \left\{ M_2 * M_2 * M_2 = M_2, M_2 * M_2 * M_3 = M_2 * M_3 * M_2 \right.$$

$$= \begin{pmatrix} 0 & 1 & 1 \\ 1 & 1 & 1 \\ 1 & 1 & 1 \end{pmatrix}, M_2 * M_3 * M_3 = N,$$

$$M_3 * M_2 * M_2 = M_3 * M_2, M_3 * M_2 * M_3 = N,$$

$$\left. M_3 * M_3 * M_2 = N, M_3 * M_3 * M_3 = N \right\}.$$

In $\{M_2, M_3\}^4$ (and in all higher powers) no new matrices occur. The role structure is therefore given as the semigroup

$$\langle \{M_2, M_3\} \rangle = \left(\left\{ \begin{pmatrix} 1 & 0 & 0 \\ 0 & 1 & 0 \\ 0 & 1 & 1 \end{pmatrix}, \begin{pmatrix} 0 & 1 & 1 \\ 1 & 0 & 1 \\ 1 & 1 & 0 \end{pmatrix}, \begin{pmatrix} 0 & 1 & 1 \\ 1 & 0 & 1 \\ 1 & 1 & 1 \end{pmatrix}, \right. \right.$$

$$\left(\begin{pmatrix} 0 & 1 & 1 \\ 1 & 1 & 1 \\ 1 & 1 & 0 \end{pmatrix}, \begin{pmatrix} 0 & 1 & 1 \\ 1 & 1 & 1 \\ 1 & 1 & 1 \end{pmatrix}, \begin{pmatrix} 1 & 1 & 1 \\ 1 & 1 & 1 \\ 1 & 1 & 1 \end{pmatrix} \right\}, * \right)$$

with six elements. ☐

The introduction of finite semigroups for role structures enables one to compare different societies. Fundamental for this is the theorem of the standard epimorphic reduction and production (6.3.57) and the resulting concept of distance between semigroups, which we introduced in Chapter 6, §3.

3.13 Definition. Let $\mathscr{G}_1$, $\mathscr{G}_2$ be blockwise partitioned societies and $R_1, \ldots, R_s$ and $R'_1, \ldots, R'_s$ be relations on $\mathscr{G}_1$ and $\mathscr{G}_2$, respectively. The *distance* between $\mathscr{G}_1$ and $\mathscr{G}_2$ is the distance between the corresponding role structures according to 6.3.60 (if defined).

A detailed analysis of these concepts together with explicit examples is given in BOORMAN and WHITE. For instance, there is a closeness in the role structure of a monastery and the observed situation in a commune (despite all ideological differences), whereas the role structure in a group of biomedical researchers deviates more clearly. The observed role structure of a set of managers in a large company is completely different. It appears that there is only a minor dependence of these results on the definitions of "weak blocks" see 3.8 and the "special relations" R_i.

It is also interesting to observe the situation at various points in time. Here the block partitions and the corresponding matrices may or may not change. In our example of a monastery such observations had been taken over certain time intervals. It is hoped that these methods eventually will enable us to predict future developments. The observed monastery was rapidly polarised. Soon monks No. 2, 3, 17 and 18 were excluded; then No. 1 left of his own accord and within a few days No. 16, 15, 14, 7, 13 and 8 followed. One month later No. 10 left. This is in excellent accordance with the blocks obtained in 3.8.

We conclude and agree with BOORMAN and WHITE that these investigations and methods could also find applications in other areas, possibly in studies of official organizations (public service?) or in legal situations (e.g. in the clarification of complicated conflict situations between individuals or groups).

PROBLEMS

1. Show that the product defined in 3.9 turns the set $S_k(0, 1)$ of all $k \times k$-matrices over $\{0, 1\}$ into a semigroup (for each $k \in \mathbb{N}$).

2. Are the semigroups $S_k(0, 1)$ of Problem 1 commutative? monoids? groups? abelian groups?

3. When does $S_k(0, 1)$ contain idempotent elements? how can they be characterized?

4. If $M_1 = \begin{pmatrix} 0 & 1 \\ 1 & 0 \end{pmatrix}$ and $M_2 = \begin{pmatrix} 1 & 0 \\ 0 & 1 \end{pmatrix}$, compute the subsemigroup S_1 of $S_2(0, 1)$ generated by M_1 and M_2.

5. Do the same as in Problem 4 with $M_1 = \begin{pmatrix} 0 & 0 \\ 0 & 0 \end{pmatrix}$, $M_2 = \begin{pmatrix} 0 & 1 \\ 1 & 0 \end{pmatrix}$, $M_3 = \begin{pmatrix} 0 & 1 \\ 0 & 1 \end{pmatrix}$ in order to get a semigroup S_2.

6–7. Find presentations of the role structures S_1, S_2 of Problems 4 and 5.

8–9. Are S_1, S_2 in Problems 4 and 5 commutative? monoids? groups? abelian groups? Do there exist zero elements?

10. Find the "greatest common divisor" $\underline{S}$ of S_1 and S_2 in the sense of 6.3.57.

11. Compute $\delta(S_1, \underline{S})$.

12. Compute $\delta(S_2, \underline{S})$.

13. Find the distance between S_1 and S_2.

14. Does the triangle inequality hold for δ in the case of S_1, S_2, and $\underline{S}$?

EXERCISES (Solutions in Chapter 8, p. 503)

1. Describe the kinship system S of 3.2 in terms of semigroup theory.

2. Are the sisters of daughters in 3.3 considered to be "the same" as the daughters of sisters?

3. What does it mean for a kinship system to be a group?

4. Let $X = \{F, M\}$ and $R = \{(FFF, F), (MM, M), (FM, MF)\}$. Which kinship system S do we get?

5. Let X be as above and $R' = \{(FF, F), (MM, M), (FM, MF)\}$. Describe the kinship system S' as a semigroup.

6–7. Are the semigroups S, S' in Exercises 4 and 5 commutative? monoids? groups? abelian groups?

8. Compute the distance between the kinship systems S and S' of Exercises 4 and 5.

9. Let the kinship system $S = [X, R]$ be defined by $X = \{P(= \text{"is parent of"}), C(= \text{"is child of"})\}$, $R = \{(PP, P), (CC, C), (PC, CP)\}$. How many elements are in S? Find its operation table. Determine whether S is an (abelian) group.

10. Re-work Example 3.10/3.12 using variant (i) instead of (ii) as in the text. What differences do you observe?

11. The members of a mathematics department belong to two different groups A, B. A hates B and loves itself, while members of B hate everybody (including themselves), but love nobody. What's their semigroup with respect to "love" and "hate"?

12. The members of a department of psychology fall into three coalitions A, B, C. All of them love members of their own group, but nobody else. Members of A hate members of the other groups, and conversely. There is no hate between B and C. Again, work out their role structure.

13–14. Find presentations of the role structures S_1, S_2 in the departments of mathematics, and psychology (Exercises 11 and 12).

15–16. Are the role structures S_1 and S_2 in Problems 13 and 14 commutative? monoids? groups? abelian groups?

17. Find the "greatest common divisor" $\underline{S}$ of S_1 and S_2 in the sense of 6.3.57.

18. In Problem 17, find $\delta(S_1, \underline{S})$ and $\delta(S_2, \underline{S})$.

19. Find the distance between the role structures in the departments of mathematics and psychology (Exercises 11 and 12).

20. Does the triangle inequality hold for S_1, S_2 and $\underline{S}$ in Exercise 17?

21. Does the result in Exercise 19 imply that mathematics and psychology have nothing in common?

NOTES

The beginning of the study of formal languages can be traced to Chomsky, who in 1959 introduced the concept of a context-free language in order to model natural languages. Since the late 1960's there has been considerable activity in the theoretical development of context-free languages both in connection with natural languages and with the programming languages. Chomsky used Semi–Thue systems to define languages, which can be described as certain subsets of finitely generated free monoids. CHOMSKY details a revised approach in the light of experimental evidence and careful consideration of semantic and syntactic structure of sentences.

For a common approach to formal languages and the theory of automata (of Chapter 6) we refer to EILENBERG. Some of the early books on formal languages from a mathematician's or computer scientist's viewpoint are ARBIB and GINSBURG. (See also GINSBURG, SALOMAA₂ and BOBROW and ARBIB for more recent books, the latter is an elementary, first introduction to this subject.) COHN gives an interesting survey from an algebraist's point of view. HOLCOMBE contains some examples of biological applications. For educational aspects and the role of automata and machine theory in mathematics syllabuses we refer to HOLCOMBE. Details on the use of semiautomata in metabolic pathways and the aid of a computer therein, including a theory of scientific experiments, can be found in KROHN, LANGER and RHODES. ROSEN studies ways in which environmental changes can affect the repair capacity of biological systems and considers carcinogenesis and reversibility problems. Language theory is used in cell-development problems, as introduced by LINDENMEYER, see also HERMANN and ROSENBERG. SUPPES and

KIERAS develop a theory of learning in which a subject is instructed to behave like a semiautomaton.

The study of kinship goes back to a study by A. Weil in response to an inquiry by the anthropologist C. Levi-Strauss (see the appendix to Part I of *Elementary Structures of Kinship* by Levi-Strauss, 1949; see also WHITE). The concept of distance in semigroups and other binary structures is investigated in BOGART. For examples similar to the one described in § 3 we refer to WHITE, BOORMAN, BREIGER, and BREIGER, BOORMAN, ARABIE, see also KIM and ROUSH for an elementary account of such examples. BALLONOFF presents several of the fundamental papers on kinship; CARLSON gives elementary examples of applications of groups in anthropology and sociology.

CHAPTER 8

Solutions to the Exercises

This chapter contains solutions or answers to all of the exercises in the text. It should be noted that we did not always choose the shortest or most elegant solution to an exercise question. Some of the solutions have been provided over the years by our students, and will often indicate the most obvious or direct approach for finding the answer to a given question.

Chapter 1, §1

1. The subgroups of G are

$$\{+1\}, \{\pm 1\}, \{\pm 1, \pm i\}, \{\pm 1, \pm j\}, \{\pm 1, \pm k\}, \{\pm 1, \pm i, \pm j, \pm k\}.$$

Hence the Hasse diagram for the lattice of all subgroups of G is

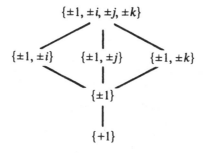

Figure 8.1

2. The proof is an induction proof. Clearly $y \sqcap x_1 \geq y \sqcap x_1$. So we proceed by induction assuming

$$y \sqcap \bigsqcup_{i=1}^{k} x_i \geq \bigsqcup_{i=1}^{k} (y \sqcap x_i).$$

Then

$$y \sqcap \bigsqcup_{i=1}^{k+1} x_i = y \sqcap \left(\bigsqcup_{i=1}^{k} x_i \sqcup x_{k+1} \right)$$

$$\geq \left(y \sqcap \bigsqcup_{i=1}^{k} x_i \right) \sqcup (y \sqcap x_{k+1}) \quad \text{by Theorem 1.14.}$$

$$\geq \bigsqcup_{i=1}^{k+1} (y \sqcap x_i) \quad \text{by induction hypothesis.}$$

The result now follows by the Principle of Mathematical Induction.

3.

⊓	0	a	b	c	d	e	1
0	0	0	0	0	0	0	0
a	0	a	0	a	a	0	a
b	0	0	b	0	b	b	b
c	0	a	0	c	a	0	c
d	0	a	b	a	d	b	d
e	0	0	b	0	b	e	e
1	0	a	b	c	d	e	1

⊔	0	a	b	c	d	e	1
0	0	a	b	c	d	e	1
a	a	a	d	c	d	1	1
b	b	d	b	1	d	e	1
c	c	c	1	c	1	1	1
d	d	d	d	1	d	1	1
e	e	1	e	1	1	e	1
1	1	1	1	1	1	1	1

4. (a) Assume L is metric and $z \leq x$. Then

$$x \sqcap y \sqcap z = y \sqcap z,$$

$$x \sqcup y \sqcup z = x \sqcup z.$$

Then

$$v((x \sqcap y) \sqcup z) = v(x \sqcap y) + v(z) - v(x \sqcap y \sqcap z)$$

$$= v(x) + v(y) - v(x \sqcup y) + v(z) - v(x \sqcap y \sqcap z)$$

$$= v(x) + v(y) + v(z) - v(x \sqcup y \sqcup z) - v(x \sqcap y \sqcap z).$$

Similarly,

$$v(x \sqcap (y \sqcap z)) = v(x) + v(y \sqcup z) - v(x \sqcup y \sqcup z)$$

$$= v(x) + v(y) + v(z) - v(y \sqcap z) - v(x \sqcup y \sqcup z)$$

$$= v(x) + v(y) + v(z) - v(x \sqcap y \sqcap z) - v(x \sqcup y \sqcup z).$$

So $v((x \sqcap y) \sqcup z) = v(x \sqcap (y \sqcup z))$. If L is not modular then x, y, z exist such that $z \leq x$ and $(x \sqcap y) \sqcup z \leq x \sqcap (y \sqcup z)$. Hence $v((x \sqcap y) \sqcup z) \leq v(x \sqcap (y \sqcup z))$, a contradiction. Hence L must be modular.

(b) Since $y \le x \Rightarrow v(y) \le v(x)$ it follows that

$$d(x, y) = 0 \quad \text{iff} \quad v(x \sqcup y) = v(x \sqcap y)$$
$$\text{iff} \quad x \sqcup y = x \sqcap y$$
$$\text{iff} \quad x = y.$$

So d is reflexive. Symmetry follows by the commutativity of $\sqcap, \sqcup$. Finally, since $z \sqcap (x \sqcup y) \le z \le z \sqcup (x \sqcap y)$ implies $v(z \sqcap (x \sqcup y)) \le v(z) \le v(z \sqcup (x \sqcap y))$, we have

$$d(x, y) \le d(x, y) + v(z \sqcup (x \sqcap y)) - v(z \sqcap (x \sqcup y))$$
$$= v(x \sqcup y) - v(x \sqcap y) + v(z \sqcup (x \sqcap y)) - v(z \sqcap (x \sqcup y))$$
$$\le v(x \sqcup y \sqcup z) + v((x \sqcup z) \sqcap (y \sqcup z))$$
$$- (v(x \sqcap y \sqcap z) + v((x \sqcap z) \sqcup (y \sqcap z))$$
$$= v(x \sqcup z) + v(y \sqcup z) - v(x \sqcap z) + v(y \sqcap z))$$
$$= d(x, z) + d(y, z).$$

So d satisfies the triangle inequality and thus (L, d) is a metric space.

5. (i) Let L be modular. If $x \le a \le x \sqcup y$ we have

$$x \sqcap y \le a \sqcap y \le y = (x \sqcup y) \sqcap y.$$

So $f: a \to a \sqcap y$ maps $[x, x \sqcup y]$ into $[x \sqcap y, y]$ since if $a \in [x, x \sqcup y]$ then $a \sqcap y \in [x \sqcap y, y]$. Dually $g: b \to b \sqcup x$ maps $[x \sqcap y, y]$ into $[x, x \sqcup y]$. Now

$$g(f(a)) = x \sqcup (a \sqcap y) = a \sqcap (y \sqcup x) \quad (L \text{ is modular}),$$
$$= a, \qquad\qquad \text{since } a \in [x, x \sqcup y],$$

and if $b \in [x \sqcap y, y]$ then

$$g(f(b)) = (b \sqcup x) \sqcap y = b \sqcup (x \sqcap y) = b.$$

So f and g are inverses and f and g are bijective. If $a_1 \le a_2$ then $a_1 \sqcap y \le a_2 \sqcap y$, i.e. $f(a_1) \le f(a_2)$ and if $b_1 \le b_2$ then $b_1 \sqcup x \le b_2 \sqcup x$, so $g(b_1) \le g(b_2)$. Now if $c = a_1 \sqcup a_2$, $a_1, a_2 \in [x, x \sqcup y]$, then

$$f(a_1), f(a_2) \le f(c),$$

so

$$f(a_1) \sqcup f(a_2) \le f(c).$$

If $f(a_1)$, $f(a_2) \le f(a_1) \sqcup f(a_2) \le f(d) \le f(c)$ (f is bijective) then $g(f(a_1)), g(f(a_2)) \le g(f(d)) \le g(f(c))$, i.e. $a_1, a_2 \le d \le c$. But

$$c = a_1 \sqcup a_2 \quad \text{so} \quad d = c,$$

i.e.

$$f(c) = f(a_1) \sqcup f(a_2).$$

Similarly,

$$f(e) = f(a_1 \sqcap a_2) = f(a_1) \sqcap f(a_2).$$

Therefore f is an isomorphism and

$$[x, x \sqcup y] \cong [x \sqcap y, y].$$

(ii) Conversely, suppose f (and g), defined above are isomorphisms. Let $x \leq z$, then $z \sqcap (y \sqcup x) \in [x, x \sqcup y]$ and

$$z \sqcap (y \sqcup x) = gf(z \sqcap (y \sqcup x))$$
$$= g(z \sqcap (y \sqcup x) \sqcap y)$$
$$= g(z \sqcap y)$$
$$= (z \sqcap y) \sqcup x.$$

So the modular equality holds.

6. For all x, y, z in L we have $[(x \sqcap y) \sqcup (x \sqcap z)] \sqcap [(x \sqcap y) \sqcup (y \sqcap z)] \leq x \sqcap y$, since the first-term square bracket is $\leq x$ and the second-term square bracket is $\leq y$. Also, the left-hand side is $\geq x \sqcap y$, since each expression in square brackets is $\geq x \sqcap y$. This proves the required equality.

7. The Hasse diagram for the lattice of subgroups of A_4 is:

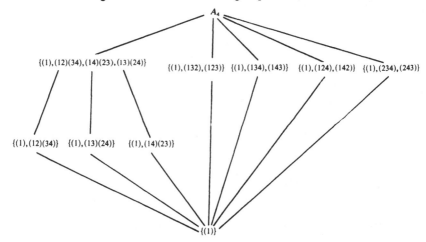

Figure 8.2. This lattice is not modular since there is a sublattice isomorphic to the pentagon lattice.

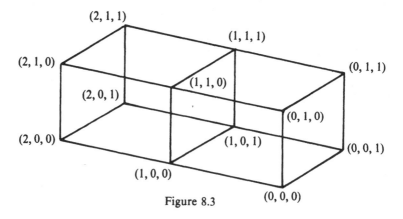

Figure 8.3

9. Order the set of normal subgroups by set-theoretic inclusion, this clearly defines a poset. For A being a normal subgroup of G we write $A \lhd G$. If $A, B \lhd G$ then $A \cap B \lhd G$ is the largest subgroup containing both A and B, so $A \sqcap B = A \cap B$. Also $AB = \{AB \,|\, a \in A, b \in B\}$ is a normal subgroup containing A and B. Suppose $A, B \subseteq C$, then for any $a \in A, b \in B, a, b \in C$. So $ab \in C$, i.e. $AB \subseteq C$. Therefore $AB = A \sqcup B$. So the normal subgroups form a lattice. Let $A \lhd C$; we need to show $(A \sqcup B) \sqcap C \subseteq A \sqcup (B \sqcap C)$. If $x \in (A \sqcup B) \sqcap C$ then $x = ab = c$ for $a \in A$, $b \in B$, $c \in C$. Therefore $b = a^{-1}c \in C$, since $A \lhd C$, i.e. $b \in B \sqcap C$. So $x = ab \in A \sqcup (B \sqcap C)$. Thus $(A \sqcup B) \sqcap C \subseteq A \sqcup (B \sqcap C)$ and the lattice is modular.

10. Let a, b, c be elements of a modular lattice such that $(a \sqcup b) \sqcap c = 0$, so $c \sqcap (b \sqcup a) = 0$ by (L1). Consider

$$b \sqcup (c \sqcap (b \sqcup a)) = (b \sqcup c) \sqcap (b \sqcup a) \quad \text{by Theorem 1.23,}$$

so

$$b \sqcup 0 = (b \sqcup c) \sqcap (b \sqcup a),$$

hence

$$b = (b \sqcup a) \sqcap (b \sqcup c).$$

So

$$a \sqcap b = a \sqcap [(b \sqcup a) \sqcap (b \sqcup c)]$$
$$= [a \sqcap (b \sqcup a)] \sqcap (b \sqcup c) \quad \text{by (L2)}$$
$$= a \sqcap (b \sqcup c) \quad \text{by (L1) and (L3).}$$

11. (i) $\Rightarrow$ (ii) \
 (i) $\Rightarrow$ (iii) $\Big\}$ For any $a, b, c \in L$ we have $a \sqcap b \leq a$ and therefore

$$a \sqcap ((a \sqcap b) \sqcup c) = ((a \sqcap b) \sqcup c) \sqcap a = (a \sqcap b) \sqcup (a \sqcap c).$$

Furthermore, if the assumptions in (iii) hold then

$$a = a \sqcap (a \sqcup c) = a \sqcap (b \sqcup c) = b \sqcup (a \sqcap c) = b \sqcup (b \sqcap c) = b.$$

(ii) $\Rightarrow$ (i) Let $a \leq c$ then

$$(a \sqcup b) \sqcap c = ((a \sqcap c) \sqcup b) \sqcap c = (a \sqcap c) \sqcup (b \sqcap c) = a \sqcup (b \sqcap c),$$

(iii) $\Rightarrow$ (i) If it were nonmodular it would contain such a sublattice in which $a \neq b$ although

$$a \sqcap c = b \sqcap c,$$

$$a \sqcup c = b \sqcup c.$$

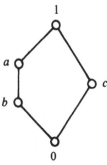

Figure 8.4

12. Let X_n denote a subset with n elements,

$$X_n = \{a_1, \ldots, a_n\}, \qquad a_i \in L.$$

We shall now show by induction that X_n has a least upper bound. For

$$X_1 = \{a_1\}, \qquad a_1 \in L,$$

clearly the least upper bound of X_1 is a_1. Assume that $X_k = \{a_1, \ldots, a_k\}$ has a least upper bound, a_x say, so $a_i \sqcup a_x = a_x$, $\forall\, 1 \le i \le k$. Consider $X_{k+1} = \{a_1, \ldots, a_k, a_{k+1}\}$. Now

$$a_{k+1} \sqcup a_x = \begin{cases} a_x & \text{if } a_x \ge a_{k+1}, \\ a_{k+1} & \text{if } a_{k+1} \ge a_x, \end{cases}$$

as $a_i \in L$. If $a_{k+1} \sqcup a_x = a_x$ then a_x is a least upper bound for X_{k+1}. If $a_{k+1} \sqcup a_x = a_{k+1}$ then $a_{k+1} \sqcup a_i = a_{k+1}$, $\forall\, 1 \le i \le k + 1$ since $a_x \sqcup a_i = a_x$, $\forall\, 1 \le i \le k$. Hence X_{k+1} has a least upper bound (either a_x or a_{k+1}). So by induction hypothesis the proof is complete.

13. Clearly $\le$ is a partial order and $f \sqcup g = \sup(f, g)$ is defined by $(f \sqcup g)(x) = f(x) \sqcup g(x)$, $f \sqcap g = \inf(f, g)$ is defined by

$$(f \sqcap g)(x) = f(x) \sqcap g(x).$$

For all $x \in S$,

$$f(x) \sqcap (g(x) \sqcup h(x)) = (f(x) \sqcap g(x)) \sqcup (f(x) \sqcap h(x))$$
$$= (f \sqcap g)(x) \sqcup (f \sqcap h)(x).$$

So $f \sqcap (g \sqcap h) = (f \sqcap g) \sqcup (f \sqcap h)$.

14. If a has a complement a' then

$$a \sqcup (a' \sqcap b) = (a \sqcup a') \sqcap (a \sqcup b)$$
$$= 1 \sqcap (a \sqcup b)$$
$$= a \sqcup b.$$

15. L is not modular since it contains the sublattice $\{0, a, b, c, d\}$ which is isomorphic to the pentagon lattice V_4^5. L is not distributive since L is not modular. L is not complemented since a does not have a complement.

16. We intersect both sides of the equation with x. Then the right-hand side yields:

$$x \sqcap [(x \sqcup y) \sqcap (y \sqcup z) \sqcap (z \sqcup x)] = [x \sqcap (x \sqcup y)] \sqcap [(y \sqcup z) \sqcap (x \sqcup x)]$$
$$= x \sqcap [(y \sqcup z) \sqcap (z \sqcup x)]$$
$$= [x \sqcap (x \sqcup z)] \sqcap (y \sqcup z)$$
$$= x \sqcap (y \sqcup z).$$

For the left-hand side we have:

$$x \sqcap [(x \sqcap y) \sqcup (y \sqcap z) \sqcup (z \sqcap x)] = x \sqcap [((y \sqcap z) \sqcup (z \sqcup x)) \sqcap (x \sqcup y)]$$
$$= [x \sqcap ((y \sqcap z) \sqcup (z \sqcap x))] \sqcup (x \sqcap y).$$

By the modular equation, since $x \sqcap y \leq x$, this last expression equals

$$[(x \sqcap (y \sqcap z)) \sqcup (z \sqcap x)] \sqcup (x \sqcap y).$$

Using the modular equation, since $z \sqcap x \leq x$, this equals

$$[(x \sqcap z) \sqcup ((x \sqcap z) \sqcap y)] \sqcup (x \sqcup y) = (x \sqcap z) \sqcup (x \sqcap y)$$

$$= (x \sqcap y) \sqcup (x \sqcap z).$$

Given the distributive equation, then

$$(x \sqcap y) \sqcup (y \sqcap z) \sqcup (z \sqcap x) = [((x \sqcap y) \sqcup y) \sqcap ((x \sqcap y) \sqcup z)] \sqcup (z \sqcap x)$$

$$= [y \sqcap ((z \sqcap y) \sqcup z)] \sqcup (z \sqcap x)$$

$$= [y \sqcap ((x \sqcup z) \sqcap (y \sqcup z))] \sqcup (z \sqcap x)$$

$$= [y \sqcap (x \sqcup z)] \sqcup (z \sqcap x)$$

$$= [y \sqcup (z \sqcap x)] \sqcap ((x \sqcup z) \sqcup (z \sqcap x))$$

$$= [(y \sqcup z) \sqcap (y \sqcup x)) \sqcap [x \sqcup (z \sqcup (z \sqcap x))]$$

$$= ((y \sqcup z) \sqcap (y \sqcup x)) \sqcap (z \sqcup x)$$

$$= (x \sqcup y) \sqcap (y \sqcup z) \sqcap (z \sqcup x).$$

17. $\mathbb{N}$ is clearly a lattice, where $x \sqcap y = \gcd(x, y)$ and $x \sqcup y = \text{lcm}(x, y)$. Let

$$x = \prod_{i=1}^{n} p_i^{x_i}, \qquad y = \prod_{i=1}^{n} p_i^{y_i} \quad \text{and} \quad z = \prod_{i=1}^{n} p_i^{z_i}$$

be the decomposition of x, y and z into primes where x_i, y_i and z_i may be zero. Then $y \sqcap z = \prod_{i=1}^{n} p_i^{\min(y_i, z_i)}$ so

$$x \sqcup (y \sqcap z) = \prod_{i=1}^{n} p_i^{\max(x_i, \min(y_i, z_i))}$$

$$= \prod_{i=1}^{n} p_i^{\min(\max(x_i, y_i), \max(x_i, z_i))}.$$

This in turn equals

$$\prod_{i=1}^{n} p_i^{\max(x_i, y_i)} \sqcap \prod_{i=1}^{n} p_i^{\max(x_i, z_i)} = (x \sqcup y) \sqcap (x \sqcup z).$$

18. Let L be a distributive lattice with $a \sqcup b = a \sqcup c$ and $a \sqcap b = a \sqcap c$. Then

$$b = b \sqcap (a \sqcup b) \qquad \text{by (L3)}$$

$$= b \sqcap (a \sqcup c)$$

$$= (b \sqcap a) \sqcup (b \sqcap c) \quad \text{since } L \text{ is distributive}$$

$$= (a \sqcap b) \sqcup (b \sqcap c) \quad \text{by (L1)}$$

$$= (a \sqcap c) \sqcup (b \sqcap c)$$

$$= (a \sqcup b) \sqcap c \qquad \text{since } L \text{ is distributive}$$

$$= (a \sqcup c) \sqcap c$$

$$= c \qquad \text{by (L3)}.$$

19. (i) $\Rightarrow$ (ii)

$$(a \sqcup b) \sqcap (a \sqcup c) = ((a \sqcup b) \sqcap a) \sqcup ((a \sqcup b) \sqcap c) \quad \text{by (i)}$$
$$= a \sqcup ((a \sqcup c) \sqcap b)$$
$$= a \sqcup ((a \sqcap c) \sqcup (b \sqcup c))$$
$$= (a \sqcup (a \sqcap c)) \sqcup (b \sqcap c)$$
$$= a \sqcup (b \sqcap c).$$

Similarly (ii) $\Rightarrow$ (i).

Chapter 1, §2

1. Consider

$$\gcd(6, 2) = 2 \quad \text{and} \quad \text{lcm}(6, 2) = 6,$$
$$\gcd(6, 3) = 3 \quad \text{and} \quad \text{lcm}(6, 3) = 6,$$
$$\gcd(6, 9) = 3 \quad \text{and} \quad \text{lcm}(6, 9) = 18.$$

So there doesn't exist a complement of 6. Hence $(\{1, 2, 3, 6, 9, 18\}, \text{lcm}, \gcd)$ is not a Boolean algebra.

2. Let $B = \{1, 2, 5, 10, 11, 22, 55, 110\}$ be the set of positive divisors of 110. Show that $(B, \text{lcm}, \gcd, ')$ is a Boolean algebra where $x' = 110/x$, the zero is 1, and the unit is 110. The binary operations greatest common divisor (gcd), and least common multiple (lcm) are both associative and commutative. Also $\text{lcm}(1, x) = x$ and $\gcd(110, x) = x$ for all $x \in B$; thus 1 and 110 are the zero and one elements, respectively.

The first distributive law states that

$$\gcd(x, \text{lcm}(y, z)) = \text{lcm}(\gcd(x, y), \gcd(x, z)).$$

To prove this, we factor x, y and z into prime factors to obtain

$$x = 2^{\alpha_2} 5^{\alpha_5} 11^{\alpha_{11}}, \qquad y = 2^{\beta_2} 5^{\beta_5} 11^{\beta_{11}} \quad \text{and} \quad z = 2^{\gamma_2} 5^{\gamma_5} 11^{\gamma_{11}},$$

where the exponents are zero or one. The exponent of the prime p in $\gcd(x, \text{lcm}(y, z))$ is

$$\min(\alpha_p, \max(\beta_p, \gamma_p)) = \max(\min(\alpha_p, \beta_p), \min(\alpha_p, \gamma_p)).$$

This is the same as the exponent of p in $\text{lcm}(\gcd(x, y), \gcd(x, z))$. Hence the first distributive law holds. The other distributive law can be proved in a similar way.

In the factorization of 110, no prime occurs as a square or higher power; thus each of the primes 2, 5 and 11 occur exactly once in one of the numbers x and $110/x$. Hence $\gcd(x, 110/x) = 1$, and $\text{lcm}(x, 110/x) = 110$. This proves that the integer $110/x$ is the complement of x, and $(B, \text{lcm}, \gcd, ')$ is a Boolean algebra.

Alternatively, Exercises 1 and 2 could be answered by constructing the Hasse diagrams and deducing the result from them.

3. Let B be the set of all positive divisors of n. Suppose that $n = p_1 p_2 \ldots p_r$ where the p_i's are distinct primes. Then n and 1 are the upper and lower bounds, respectively. For any $a \in B$ let $b = n/a$. Since a and b are relatively prime we have $\gcd(a, b) = 1$ and $\text{lcm}(a, b) = ab = n$. So b is the complement of a. Suppose that one of the following sublattices exists:

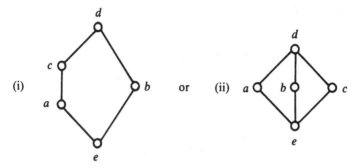

Figure 8.5

In (i) we have

$$\text{lcm}(a, b) = \frac{ab}{e} = d,$$

$$\text{lcm}(c, b) = \frac{cb}{e} = d, \quad \text{so } a = c, \text{ a contradiction.}$$

In (ii) we have

$$\frac{ab}{e} = d = \frac{bc}{e}, \quad \text{so } a = c, \text{ a contradiction.}$$

So neither (i) or (ii) can occur and therefore B is distributive. Thus B is a Boolean algebra.

Conversely, suppose that $n = p_1^{\alpha_1} p_2^{\alpha_2} \ldots p_r^{\alpha_r}$ where the p_i are distinct primes with $\alpha_1 \geq 2$. Now p_1 has no complement, for if b is the complement of p_1 we have $\gcd(p_1, b) = 1$ so $p \nmid b$, thus b is a divisor of $p_2^{\alpha_2} \ldots p_r^{\alpha_r}$. But then

$$\text{lcm}(p_1, b) \leq \text{lcm}(p_1, p_2^{\alpha_2} \ldots p_r^{\alpha_r})$$

$$\leq p_1 p_2^{\alpha_2} \ldots p_r^{\alpha_r}$$

$$\neq n,$$

contradicting the assumption that b is the complement of p_1. Therefore B is not a Boolean algebra in this latter case.

4. Let $x, y \in B$ let $x \leq y$.
   (i) Now

$$y \sqcap y' = 0 \quad \text{by definition}$$

$$\Rightarrow x \sqcap y' = 0 \quad \text{by Lemma 1.13}$$

(ii) If for $x, y \in B$: $x \sqcap y' = 0$, then

$$(x \sqcap y')' = 0'$$
$$\Rightarrow x' \sqcup y'' = 1 \quad \text{by Theorem 2.4}$$
$$\Rightarrow x' \sqcup y = 1.$$

(iii) If for $x, y \in B$: $x' \sqcup y = 1$, then

$$x \sqcap (x' \sqcup y) = x \sqcap 1$$
$$\Rightarrow x \sqcap (x' \sqcup y) = x$$
$$\Rightarrow (x \sqcap x') \sqcup (x \sqcap y) = x \quad \text{as } B \text{ is distributive}$$
$$\Rightarrow 0 \sqcup (x \sqcap y) = x$$
$$\Rightarrow x \sqcap y = x.$$

(iv) If for $x, y \in B$: $x \sqcap y = x$, then

$$y \sqcup (x \sqcap y) = y \sqcup x$$
$$y = y \sqcup x \quad \text{as } B \text{ is a lattice, (L3)}$$
$$y \sqcup x = y.$$

(v) If for $x, y \in B$: $x \sqcup y = y$, then, as B is a lattice,

$$\inf(x, y) = x$$
$$\Rightarrow x \leq y.$$

Hence all statements in the theorem are equivalent.

5. Consider

$$N_d = \left\{ \bigsqcup_{(i)_1, \ldots, i_n) \in \{1, -1\}^n} d_{i_1 \ldots i_n} \sqcap x_1^{i_1} \sqcap x_2^{i_2} \sqcap \ldots \sqcap x_n^{i_n} \text{ where } d_{i_1 \ldots i_n} \in \{0, 1\} \right\}.$$

Since $i_j \in \{1, -1\}$, $\exists\, 2^n$ polynomials of the form

$$x_1^{i_1} \sqcap x_2^{i_2} \sqcap \ldots \sqcap x_n^{i_n}$$

and also there are 2^{2^n} ways of combining these polynomials by $\sqcup$. So $|N_d| = 2^{2^n}$. Hence there are 2^{2^n} different polynomials in B. So, since each polynomial induces a polynomial function, we have that in any Boolean algebra B there are 2^{2^n} different Boolean functions in n variables.

6. (a) $(x \sqcap y) \sqcup (x \sqcap y') \sqcup (x' \sqcup y) = (x \sqcap (y \sqcup y')) \sqcup (x' \sqcup y) \quad \text{by distributivity}$

$$= x \sqcup x' \sqcup y$$
$$= 1.$$

(b) $(x \sqcap y') \sqcup [x \sqcap (y \sqcap z)'] \sqcup z = (x \sqcap y') \sqcup (x \sqcap (y' \sqcup z')) \sqcup z$

$$= (x \sqcap (y' \sqcup z')) \sqcup z \qquad (y' \leq y' \sqcup z')$$
$$= (x \sqcup z) \sqcap (y' \sqcup z' \sqcup z) \quad \text{by distributivity}$$
$$= x \sqcup z.$$

7. (a) $p = (x_1 \sqcap (x_2 \sqcup x_3)') \sqcup ((x_1 \sqcap x_2) \sqcup x_3') \sqcap x_1)$. Since $d_{i_1 i_2 i_3} = \bar{p}_B(1^{i_1}, 1^{i_2}, 1^{i_3})$
and since

$$\bar{p}_B(1, 1, 1) = 1, \qquad \bar{p}_B(0, 1, 1) = 0,$$
$$\bar{p}_B(1, 1, 0) = 1, \qquad \bar{p}_B(0, 1, 0) = 0,$$
$$\bar{p}_B(1, 0, 1) = 0, \qquad \bar{p}_B(0, 0, 1) = 0,$$
$$\bar{p}_B(1, 0, 0) = 1, \qquad \bar{p}_B(0, 0, 0) = 0,$$

the disjunctive normal form of p is

$$(x_1 \sqcap x_2 \sqcap x_3) \sqcup (x_1 \sqcap x_2 \sqcap x_3') \sqcup (x_1 \sqcap x_2' \sqcap x_3').$$

(b) $p = (x_2 \sqcup (x_1 \sqcap x_3)) \sqcap ((x_1 \sqcup x_3) \sqcap x_2)'$. Now

$$\bar{p}_B(1, 1, 1) = 0, \qquad \bar{p}_B(0, 1, 1) = 0,$$
$$\bar{p}_B(1, 1, 0) = 0, \qquad \bar{p}_B(0, 1, 0) = 1,$$
$$\bar{p}_B(1, 0, 0) = 0, \qquad \bar{p}_B(0, 0, 1) = 0,$$
$$\bar{p}_B(1, 0, 1) = 1, \qquad \bar{p}_B(0, 0, 0) = 0.$$

Hence

$$p = (x_1' \sqcap x_2 \sqcap x_3') \sqcup (x_1 \sqcap x_2' \sqcap x_3).$$

8. $p = (x_1 \sqcup x_2 \sqcup x_3) \sqcap ((x_1 \sqcap x_2) \sqcup (x_1' \sqcap x_3))'$. Since $\bar{p}_B(0^{i_1}, 0^{i_2}, 0^{i_3}) = c_{i_1 i_2 i_3}$ and

$$\bar{p}_B(0, 0, 0) = 0, \qquad \bar{p}_B(1, 0, 0) = 1,$$
$$\bar{p}_B(0, 0, 1) = 0, \qquad \bar{p}_B(1, 0, 1) = 1,$$
$$\bar{p}_B(0, 1, 0) = 1, \qquad \bar{p}_B(1, 1, 0) = 0,$$
$$\bar{p}_B(0, 1, 1) = 0, \qquad \bar{p}_B(1, 1, 1) = 0,$$

we have

$$p = (x_1 \sqcup x_2 \sqcup x_3) \sqcap (x_1 \sqcup x_2' \sqcup x_3') \sqcap (x_1' \sqcup x_2' \sqcup x_3')$$
$$\sqcap (x_1 \sqcup x_2' \sqcup x_3) \sqcap (x_1 \sqcup x_2 \sqcup x_3').$$

9. Suppose F is a filter in B. Now $\forall f', g' \in F'$ and $\forall b \in B$ we have $b' \in B$ so that
$f' \sqcup g' = (f \sqcap g)' \in F'$ and $f' \sqcap b = (f \sqcup b')' \in F'$ since F is a filter. $F' = B$.
Conversely, suppose F' is an ideal of B. Now $\forall f, g \in F$ and $\forall b \in B$ we
have $b' \in B$. So $f \sqcap g = (f' \sqcup g')' \in F$ and $f \sqcup b = (f' \sqcap b')' \in F$ since F' is an
ideal of B. So F is a filter.

10. (i) Let $J = \bigcap \{I \trianglelefteq B; S \subseteq I\}$. $\forall i, j \in J$, $\forall b \in B$ we have that $i, j \in I$ for each
$I \trianglelefteq B$ such that $S \subseteq I$.

$$\Rightarrow \text{For each such } I, i \sqcap b \in I \text{ and } i \sqcup j \in I,$$
$$\Rightarrow i \sqcap b \in J \text{ and } i \sqcup j \in J,$$
$$\Rightarrow J \trianglelefteq B.$$

Since each $I \geq S$ we get

$$S \subseteq J.$$

(ii) Let D be the set of elements of the given form. The join of any two elements of D is clearly again of the same form and therefore also in D. In addition, for any $y \in B$, the meet

$$y \sqcap ((b_1 \sqcap s_1) \sqcup \ldots \sqcup (b_k \sqcap s_k))$$

$$= (y \sqcap (b_1 \sqcap s_1)) \sqcup \ldots \sqcup (y \sqcap (b_k \sqcap s_k))$$

$$= ((y \sqcap b_1) \sqcap s_1) \sqcup \ldots \sqcup ((y \sqcap b_k) \sqcap s_k)$$

is again in D. Thus D is an ideal. Since $s = 1 \sqcap s$, every member s of S is in D. Every member $(b_1 \sqcap s_1) \sqcup \ldots \sqcup (b_k \sqcap s_k)$ of D belongs to any ideal I containing S, for, since $s_i \in S_j$ it follows that $b_i \sqcap s_i \in I$ and therefore that $(s_1 \sqcap b_1) \sqcup \ldots \sqcup (s_k \sqcap b_k) \in I$. Hence D is the intersection of all ideals containing S.

(iii) If $y \le s_{i_1} \sqcup \ldots \sqcup s_{i_k}$ and $z \le s_{j_1} \sqcup \ldots \sqcup s_{j_m}$, then

$$y \sqcup z \le s_{i_1} \sqcup \ldots \sqcup s_{i_k} \sqcup s_{j_1} \sqcup \ldots \sqcup s_{j_m}$$

and if $y \le s_1 \sqcup \ldots \sqcup s_k$ and $v \le y$, then $v \le s_1 \sqcup \ldots \sqcup s_k$. Thus A is an ideal. In addition, if s is in S, then $s \le s$ and therefore s is in A. Clearly, every member of A belongs to every ideal containing S. Hence A is generated by S.

(iv) Note first that $1 \le u$ is equivalent to $1 = u$, for any u. Hence by (iii) $1 \in \text{Gen}(S)$ if and only if $1 = s_1 \sqcup \ldots \sqcup s_k$ for some $s_1, \ldots, s_k$ in S. But, $1 \in \text{Gen}(S)$ if and only if $\text{Gen}(S)$ is not a proper ideal.

11. Let $\mathcal{F}$ be the set of all filters in B partially ordered by set-theoretic inclusion. Let $\mathcal{C}$ be any chain in $\mathcal{F}$ and $C = \bigcup \mathcal{C}$. Choose any $x, y \in C$ then $x \in D$, $y \in E$ for some $D, E \in \mathcal{C}$. Since $\mathcal{C}$ is a chain we may assume without loss of generality that $D \subseteq E$ whence $x, y \in E$. Therefore

$$x \sqcap y \in E \subseteq C.$$

If $z \in B$, $x \le z$ then $z \in D \subseteq C$. Therefore C is a filter and is the l.u.b. for $\mathcal{C}$ in $\mathcal{F}$. Zorn's Lemma now tells us that for each filter F in B, $\mathcal{F}$ contains a maximal element, i.e. an ultrafilter, containing F.

12. If $a, b \in H$ then $h(a \sqcap b) = h(a) \sqcap h(b) = 1$. If $c \in B$, $h(a \sqcup c) = h(a) \sqcup h(c) = 1 \sqcup h(c) = 1$. So $a \sqcap b \in H$ and $a \sqcup c \in H$. Therefore H is a filter.

13. We have $x \nleq y$ or $y \nleq x$. Without loss of generality assume $x \nleq y$ then by Theorem 2.6, $x \sqcap y' \ne 0$. So by the ultrafilter theorem there is an ultrafilter F containing the filter generated by $x \sqcap y'$. If $y \in F$ then $(x \sqcap y') \sqcap y = 0 \in F$ which is not possible. Hence $x \in F$, $y \notin F$.

Chapter 1, §3

1. *Steps* 1 *and* 2

				row number
$x'yz'$	0	1	0	(1)
xyz'	1	1	0	(2)
$xy'z$	1	0	1	(3)
xyz	1	1	1	(4)

$$
\begin{array}{lll}
(1)\,(2) & -10 & A \\
(2)\,(4) & 11- & B \\
(3)\,(4) & 1\text{-}1 & C
\end{array}
$$

Thus prime implicants of $x'yz' + xyz' + xy'z + xyz$ are yz', xy, xz.
Prime implicant table

				(1)	(2)	(3)	(4)
				0	1	1	1
				1	1	0	1
				0	0	1	1
–	1	0	A	×	×		
1	1	–	B		×		×
1	–	1	C			×	×

2. Let $p = xy + xy'z + x'y'z$. xy implies p because

$$
(\overline{xy})(1, 1, i_2) = 1, \quad \text{but then } \bar{p}(1, 1, i_2) = 1;
$$

also neither x nor y imply p, hence xy is a prime implicant of p. $y'z$ implies p because

$$
(\overline{y'z})(i_2, 0, 1) = 1, \quad \text{but then } \bar{p}(i_2, 0, 1) = 1;
$$

also neither y' nor z imply p, hence $y'z$ is a prime implicant of p. xz implies p because

$$
\overline{xz}(1, i_2, 1) = 1, \quad \text{but then } \bar{p}(1, i_2, 1) = 1;
$$

also neither x nor z imply p, hence xz is a prime implicant of p.

3. *Steps* 1 *and* 2

$$
\begin{array}{cccc l}
0 & 0 & 0 & 0 & (1) \\
\hline
0 & 1 & 0 & 0 & (2) \\
\hline
1 & 0 & 0 & 0 & (3) \\
\hline
0 & 0 & 1 & 0 & (4) \\
\hline
1 & 0 & 0 & 1 & (5) \\
\hline
0 & 1 & 1 & 0 & (6) \\
\hline
0 & 1 & 1 & 1 & (7) \\
\hline
1 & 0 & 1 & 1 & (8) \\
\hline
\end{array}
$$

Step 3

(1) (2)	0 – 0 0	A
(1) (3)	– 0 0 0	B
(1) (4)	0 0 – 0	

| (2) (6) | 0 1 – 0 | |
| (3) (5) | 1 0 0 – | C |

| (5) (8) | 1 0 – 1 | D |
| (6) (7) | 0 1 1 – | E |

| (7) (9) | – 1 1 1 | F |
| (8) (9) | 1 – 1 1 | G |

| (1) (4) ⎫ (2) (6) ⎭ | 0 – – 0 | H |

Prime implicant table (*Step* 4)

		(1)	(2)	(3)	(4)	(5)	(6)	(7)	(8)	(9)
		0	0	1	0	1	0	0	1	1
		0	1	0	0	0	1	1	0	1
		0	0	0	1	0	1	1	1	1
		0	0	0	0	1	0	1	1	1
0 – 0 0	A	×	×							
– 0 0 0	B	×		×						
1 0 0 –	C			×		×				
1 0 – 1	D					×			×	
0 1 1 –	E						×	×		
– 1 1 1	F							×		×
1 – 1 1	G								×	×
0 – – 0	H	×	×		×		×			

The core is *H*.

The new table is

		(3)	(5)	(7)	(8)	(9)
		1	1	0	1	1
		0	0	1	0	1
		0	0	1	1	1
		0	1	1	1	1
0 – 0 0	A					
– 0 0 0	B	×				
1 0 0 –	C	×	×			
1 0 – 1	D		×		×	
0 1 1 –	E			×		
– 1 1 1	F			×		×
1 – 1 1	G				×	×

This means that the minimal form is

(i) $H + C + E + G$; or

(ii) $H + B + D + F$,

i.e. the minimal form is

(i) $w'z' + wx'y' + w'xy + wyz$; or

(ii) $w'z' + x'y'z' + wx'z + xyz$.

4.

$$
\overset{(1)}{f = w'x'y'z'} + \overset{(2)}{w'x'yx'} + \overset{(3)}{w'xy'z} + \overset{(4)}{w'xyz'} + \overset{(5)}{w'xyz}
$$

$$
+ \overset{(6)}{wx'y'z'} + \overset{(7)}{wx'yz} + \overset{(8)}{wxy'z} + \overset{(9)}{wxyz} + \overset{(10)}{wxyz'}.
$$

First round

from	$w'x'y'z'$	and	$w'x'yz'$	to	$w'x'z'$	$\ldots$	A	
from	$w'x'y'z'$	and	$wx'y'z'$	to	$x'y'z'$	$\ldots$	B	
from	$w'x'y'z'$	and	$w'xyz'$	to	$w'yz'$	$\ldots$	C	
from	$w'xy'z$	and	$w'xyz$	to	$w'xz$	$\checkmark$		
from	$w'xy'z$	and	$wxy'z$	to	$xy'z$	$\checkmark$		
from	$w'xyz'$	and	$w'xyz$	to	$w'xy$	$\checkmark$		
from	$w'xyz'$	and	$wxyz'$	to	xyz'	$\checkmark$		
from	$w'xyz$	and	$wxyz$	to	xyz	$\checkmark$		
from	$wx'yz$	and	$wxyz$	to	wyz	$\ldots$	D	
from	$wxy'z$	and	$wxyz$	to	wxz	$\checkmark$		
from	$wxyz'$	and	$wxyz$	to	wxy	$\checkmark$		

Second round

from	$w'xz$	and	wxz	to	xz	$\ldots$	E
from	$xy'z$	and	xyz	to	xz		
from	$w'xy$	and	wxy	to	xy	$\ldots$	F
from	xyz'	and	xyz	to	$xy.$		

Hence the prime implicants of f are

$$w'x'z', x'y'z', w'yz', wyz, xz, xy.$$

Prime implicant table

			(1)	(2)	(3)	(4)	(5)	(6)	(7)	(8)	(9)	(10)
			0	0	0	0	0	1	1	1	1	1
			0	0	1	1	1	0	0	1	1	1
			0	1	0	1	1	0	1	0	1	1
			0	0	1	0	1	0	1	1	1	0

					(1)	(2)	(3)	(4)	(5)	(6)	(7)	(8)	(9)	(10)
0	0	–	0	A	×	×								
–	0	0	0	B	×					×				
0	–	1	0	C		×		×						
1	–	1	1	D							×		×	
–	1	–	1	E			×		×			×	×	
–	1	1	–	F				×	×				×	×

The core is $E + B + D + F$.
The new table is

					(2)
					0
					0
					1
					0
0 0 – 0	A	×
0 – 1 0	C	×

Therefore the minimal form of f is

(i) $E + B + D + F + A$; or
(ii) $E + B + D + F + C$,

i.e.

(i) $xz + x'y'z' + wyz + xy + w'x'z'$; or
(ii) $xz + x'y'z' + wyz + xy + w'yz'$.

5.
$$f = x'y + x'y'z + xy'z' + xy'z$$
$$= x'y(z' + z) + x'y'z + xy'z' + xy'z$$
$$= x'yz' + x'yz + x'y'z + xy'z' + xy'z.$$

is the disjunctive normal form of f.

Steps 1 *and* 2

0	1	0	(1)
0	0	1	(2)
1	0	0	(3)
0	1	1	(4)
1	0	1	(5)

Step 3

(1) (4)	0	1	–	A
(2) (4)	0	–	1	B
(2) (5)	–	0	1	C
(3) (5)	1	0	–	D

Prime implicant table (*Step* 4)

		(1)	(2)	(3)	(4)	(5)
		0	0	1	0	1
		1	0	0	1	0
		0	1	0	1	1
0 1 –	A	×			×	
0 – 1	B		×		×	
– 0 1	C		×			×
1 0 –	D			×		×

The core is $A + D$.
The new table is

				(2)
				0
				0
				1
0	–	1	B	×
–	0	1	C	×

The minimal form of f is either

(i) $A + D + B$; or
(ii) $A + D + C$,

i.e.

(i) $x'y + xy' + x'z$; or
(ii) $x'y + xy' + y'z$.

Chapter 2, §1

1. The symbolic representation of

$$p = (x_1 + x_2 + x_3)(x_1' + x_2)(x_1x_3 + x_1'x_2)(x_2' + x_3)$$

is given by

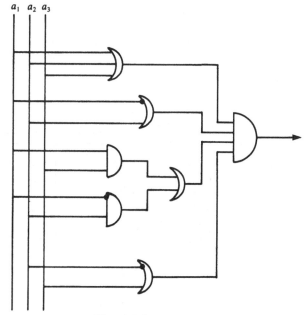

Figure 8.6

2. $p \sim (x_1 + x_3)'((x_2 + x_3)(x_2' + x_3') + x_1')$.

3. f takes the value 1 at $(0,0,0,0)$, $(0,0,0,1)$, $(0,1,0,0)$, $(0,1,0,1)$, $(1,0,1,0)$, $(1,0,1,1)$ and $(1,1,1,0)$. We apply the Quine–McCluskey algorithm to these 4-tuples which correspond to the terms of the disjunctive normal form.

	0	0	0	0	1
	0	0	0	1	2
	0	1	0	0	3
	0	1	0	1	4
	1	0	1	0	5
	1	0	1	1	6
	1	1	1	0	7

1, 2	0	0	0	–	
1, 3	0	–	0	0	
2, 4	0	–	0	1	
3, 4	0	1	0	–	
5, 6	1	0	1	–	B
5, 7	1	–	1	0	C.

1, 2 3, 4	0	–	0	–	A
1, 3 2, 4	0	–	0	–	

$$\begin{array}{ccccccc} 0 & 0 & 0 & 0 & 1 & 1 & 1 \\ 0 & 0 & 1 & 1 & 0 & 0 & 1 \\ 0 & 0 & 0 & 0 & 1 & 1 & 1 \\ 0 & 1 & 0 & 1 & 0 & 1 & 0 \end{array}$$

A	0	–	0	–	×	×	×	×			
B	1	0	1	–					×	×	
C	1	–	1	0					×		×

The core $A + B + C$ covers all expressions. So $f \sim x_1' x_3' + x_1 x_2' x_3 + x_1 x_3 x_4'$

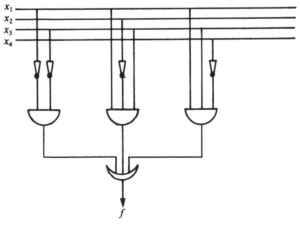

Figure 8.7

4.

$$f \sim x_2'(x_1 + x_1'(x_2 + x_3)) + x_3(x_2(x_1 + x_3') + x_1(x_2' + x_3))$$
$$\sim x_2'x_1 + x_1'x_2x_3 + x_1x_2x_3 + x_1x_2'x_3 + x_1x_3$$
$$\sim x_1x_2'x_3 + x_1x_2'x_3' + x_1'x_2x_3 + x_1x_2x_3.$$

Using the Quine–McCluskey algorithm:

1	0	0	1
0	1	1	2
1	0	1	3
1	1	1	4

1, 3	1	0	–	A
2, 4	–	1	1	B
3, 4	1	–	1	C

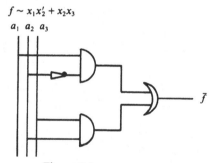

				1 0 1 1	0 1 0 1	0 1 1 1
A	1	0	–	×		×
B	–	1	1		×	×
C	1	–	1		×	×

The core $A + B$ covers all expressions. Hence

$$f \sim x_1x_2' + x_2x_3$$

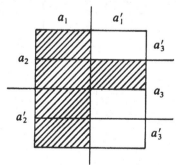

Figure 8.8

5. $p = ((x_1 + x_2)(x_1 + x_3)) + (x_1x_2x_3)$. Its Karnaugh diagram is

Figure 8.9

The diagram gives us that $p \sim x_1 + x_2x_3$.

6. For (i) $p = (x_2' + x_1')x_2'$; for (ii) $p = x_1'x_2 + x_1x_2'$; for (iii) $p = (x_1' + x_2')(x_1 + x_2)$. The Karnaugh diagrams for the above are given by

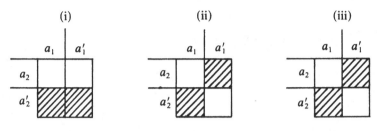

Figure 8.10

So from the diagrams we see that contact diagrams (ii) and (iii) give equivalent circuits.

7. Let

$$x_i = \begin{cases} 1 & \text{if } i \text{ votes Yes,} \\ 0 & \text{if } i \text{ votes No.} \end{cases} \quad i = 1, 2, 3.$$

Then the contact diagram is

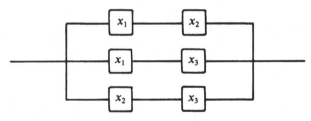

Figure 8.11

So we have $p = x_1x_2 + x_1x_3 + x_2x_3$. The symbolic representation is

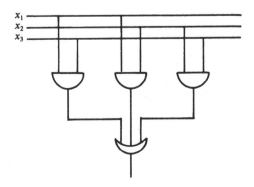

Figure 8.12

8.

b_1	b_2	b_3	$f(b_1, b_2, b_3)$
1	1	1	1
1	1	0	0
1	0	1	0
1	0	0	1
0	1	1	0
0	1	0	1
0	0	1	0
0	0	0	0

$b_i = 1$, S_i open,

$b_i = 0$, S_i closed,

$f(b_1, b_2, b_3) = 1 \ldots$ oil runs,

$f(b_1, b_2, b_3) = 0 \ldots$ no oil runs.

$f = \bar{p}$ for $p = x_1 x_2 x_3 + x_1 x_2' x_3' + x_1' x_2 x_3'$.

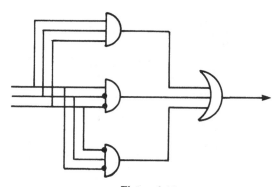

Figure 8.13

9. A hall light is controlled by two switches x_1 and x_2. They take the value 0 when they are up and 1 when they are down. Let l be the function that determines whether the light is on or off. Suppose $l = 0$ when the light is off and that the light is off when both switches are up. The function value table is

x_1	x_2	l
0	0	0
0	1	1
1	0	1
1	1	0

So $l = x_1' x_2 + x_1 x_2'$. A circuit is

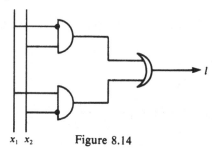

$x_1 \ x_2$ Figure 8.14

10. $p = x_1'x_2 + x_1'x_3 + x_2x_3 + x_1'x_2x_3$

                                      $\sim x_1'x_2 + x_1'x_3 + x_2x_3$

                                      $\sim x_1'(x_2 + x_3) + x_2x_3.$

A series-parallel circuit equivalent to p is

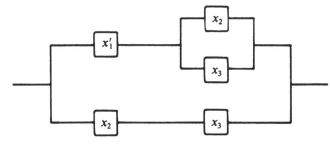

Figure 8.15

11. We have current in the given circuit if and only if either

x_1 and x_4 are closed,   or

x_2 and x_5 are closed,   or

x_1, x_3 and x_5 are closed,   or

x_2, x_3 and x_4 are closed.

A switching circuit with this switching function is given by the polynomial
$p = x_1x_4 + x_2x_5 + x_1x_3x_5 + x_2x_3x_4$. A series-parallel circuit with the same switching function is given by

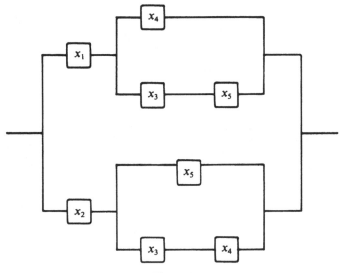

Figure 8.16

12. We have current in the given circuit if and only if either

A and B are closed,   or

A', B and C are closed,   or

B and C are closed.

A switching circuit with this switching function is given by the polynomial $p = AB + A'BC + BC$. The Karnaugh diagram of p is given by

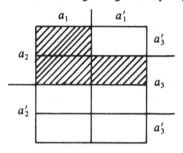

Figure 8.17

So from the diagram we have that $p \sim AB + BC \sim B(A + C)$. A series-parallel circuit with the same switching function is given by

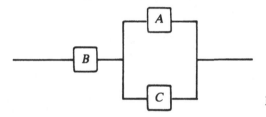

Figure 8.18

13. The circuit is equivalent to

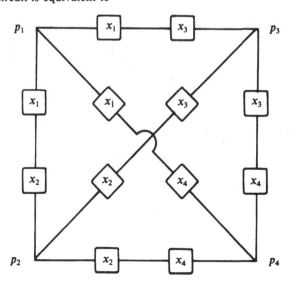

Figure 8.19

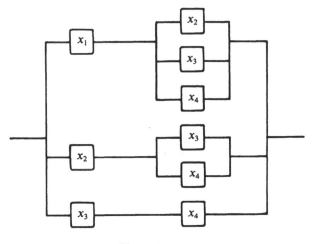

Figure 8.20

14. Using the transformation of the problem above we have

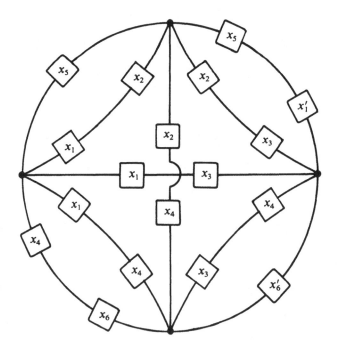

Figure 8.21

which simplifies to

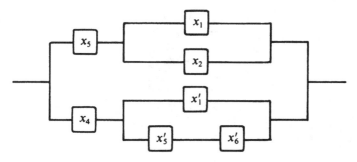

Figure 8.22

15. (i) $p_{12} = x_1 x_2' \ldots$ inhibit function.

row	a_1	a_2	minterm	$\bar{p}(a_1, a_2) = a_1 a_2'$
1	1	1	$x_1 x_2$	0
2	1	0	$x_1 x_2'$	1
3	0	1	$x_1' x_2$	0
4	0	0	$x_1' x_2'$	0

So the Karnaugh diagram for $\bar{p}_{12}$ is given by

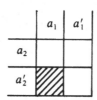

Figure 8.23

(ii) The Karnaugh diagram for $\bar{p}_{14}$ ($p_{14} = x_1' x_2$) is given by

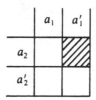

Figure 8.24

(iii) The Karnaugh diagram for the implication function $\bar{p}_5$ ($p_5 = x_1' + x_2$) is given by

Figure 8.25

(iv) For $\bar{p}_3$ it is

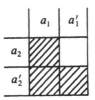

Figure 8.26

16. (i) $p = a_1 a_2 a_3' + a_1' a_2 a_3' + (a_1 + a_2' + a_3')' + (a_1 + a_2 + a_3') + a_3(a_1' + a_2)$. The Karnaugh diagram of p is given by

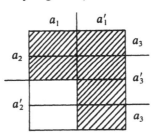

Figure 8.27

Then the diagram gives us $p \sim a_1' + a_2$.

(ii) $p = a_1 a_2 a_3 + a_2 a_3 a_4 + a_1' a_2 a_4' + a_1' a_2 a_3 a_4' + a_1' a_2' a_4'$. The Karnaugh diagram of p is given by

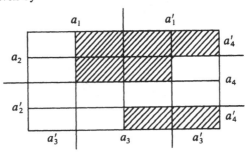

Figure 8.28

Then the diagram gives us $p \sim a_2 a_3 + a_1' a_4'$.

17. (i) We find the following Karnaugh diagram.

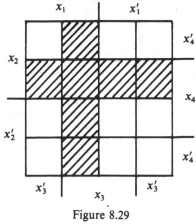

Figure 8.29

So we have the following simplified form $x_3x_4 + x_1x_3 + x_2x_4$.

(ii)

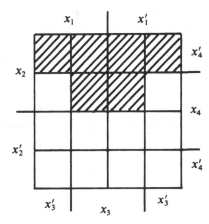

Figure 8.30

We obtain $x_2x_4' + x_2x_3$.

18. (i)

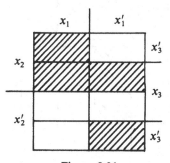

Figure 8.31

(ii)

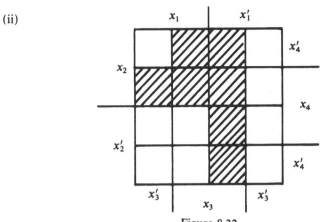

Figure 8.32

(iii)

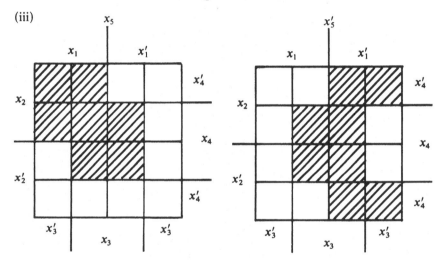

Figure 8.33

(iv)

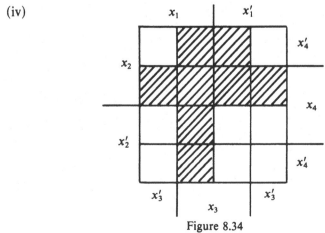

Figure 8.34

19. (a) It is easily shown that the disjunctive normal form is

$$x_1 x_2 x_3' x_4' + x_1 x_2 x_3 x_4' + x_1 x_2 x_3 x_4 + x_1 x_2' x_3' x_4 + x_1 x_2' x_3 x_4 + x_1 x_2' x_3 x_4 + x_1' x_2 x_3 x_4'$$

$$+ x_1' x_2 x_3' x_4 + x_1' x_2 x_3 x_4 + x_1' x_2' x_3 x_4 + x_1' x_2' x_3' x_4$$

which may be written as $\sum_{i=1}^{11} q_i$. Following the Quine–McCluskey procedure we find,

$$
\begin{array}{ccccc}
0 \ 1 \ 0 \ 0 \ \surd & - \ 1 \ 0 \ 0 \ \surd & - \ 1 \ - \ 0 \\
\underline{0 \ 0 \ 0 \ 1} \ \surd & - \ 0 \ 0 \ 1 \ \surd & - \ 1 \ - \ 0 \\
1 \ 1 \ 0 \ 0 \ \surd & 0 \ 1 \ - \ 0 \ \surd & - \ 0 \ - \ 1 \\
1 \ 0 \ 0 \ 1 \ \surd & 0 \ 0 \ - \ 1 \ \surd & - \ 0 \ - \ 1 \\
1 \ 0 \ 1 \ 0 \ \surd & 1 \ 0 \ - \ 1 \ \surd & \\
0 \ 1 \ 1 \ 0 \ \surd & 1 \ 1 \ - \ 0 \ \surd & 1 \ - \ 1 \ - \\
\underline{0 \ 0 \ 1 \ 1} \ \surd & 1 \ - \ 1 \ 0 \ \surd & 1 \ - \ 1 \ - \\
1 \ 1 \ 1 \ 0 \ \surd & 1 \ 0 \ 1 \ - \ \surd & - \ 1 \ 1 \ - \\
1 \ 0 \ 1 \ 1 \ \surd & - \ 1 \ 1 \ 0 \ \surd & - \ 1 \ 1 \ - \\
\underline{0 \ 1 \ 1 \ 0} \ \surd & 0 \ 1 \ 1 \ - \ \surd & - \ - \ 1 \ 1 \\
1 \ 1 \ 1 \ 1 \ \surd & - \ 0 \ 1 \ 1 \ \surd & \\
 & \underline{0 \ - \ 1 \ 1} & \\
 & 1 \ 1 \ 1 \ - \ \surd & \\
 & 1 \ - \ 1 \ 1 \ \surd & \\
 & - \ 1 \ 1 \ 1 \ \surd & \\
\end{array}
$$

So the sum of prime implicants is

$$x_2 x_4' + x_2' x_4 + x_1 x_3 + x_2 x_3 + x_3 x_4.$$

To find the minimal forms we construct the following table

	q_1	q_2	q_3	q_4	q_5	q_6	q_7	q_8	q_9	q_{10}	q_{11}
$x_2 x_4'$	×	×				×	×				
$x_2' x_4$			×	×						×	×
$x_1 x_3$		×	×		×	×					
$x_2 x_3$		×	×				×	×			
$x_3 x_4$			×		×				×	×	
	↑					↑					↑

So the core is $x_2 x_4'$, $x_2' x_4$, $x_1 x_3$. Only q_9 is not covered by the core and we have a choice between $x_2' x_4$ and $x_3 x_4$ to cover it. So the irredundant expressions are

$$x_2 x_4' + x_2' x_4 + x_1 x_3 + x_2 x_3$$

or
$$x_2 x_4' + x_2' x_4 + x_1 x_3 + x_3 x_4.$$

(b) The corresponding Karnaugh diagram is,

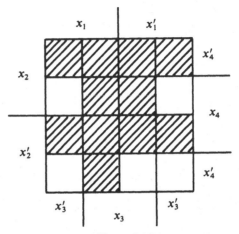

Figure 8.35

By inspection the only possible minimal forms are
$$x_2 x_4' + x_2' x_4 + x_1 x_3 + x_2 x_3$$
and
$$x_2 x_4' + x_2' x_4 + x_1 x_3 + x_3 x_4.$$

20. (a) The disjunctive normal form is
$$x_1 x_2 x_3 x_4' + x_1 x_2 x_3 x_4 + x_1 x_2' x_3 x_4 + x_1 x_2' x_3 x_4' + x_1' x_2' x_3 x_4$$
$$+ x_1' x_2' x_3' x_4 + x_1' x_2' x_3 x_4' + x_1' x_2' x_3' x_4' = \sum_{i=1}^{8} q_i.$$

Following the Quine–McCluskey procedure we have

		0	0	1	–	$\checkmark$				
0 0 0 0	0 0 – 0	$\checkmark$	0 0 – –							
0 0 1 0	0 0 0 –	$\checkmark$	0 0 – –							
0 0 0 1	– 0 1 0	$\checkmark$	– 0 1 –							
0 0 1 1	0 0 – 1	$\checkmark$	– 0 1 –							
1 0 1 0	– 0 1 1	$\checkmark$	1 – 1 –							
1 0 1 1	1 0 1 –	$\checkmark$	1 – 1 –							
1 1 1 0	1 – 1 0	$\checkmark$								
1 1 1 1	1 – 1 1	$\checkmark$								
	1 1 1 –									

Thus the prime implicants are $x_1 x_2 x_3$, $x_1' x_2'$, $x_2' x_3$, $x_1 x_3$. To find the core construct the following table

	q_1	q_2	q_3	q_4	q_5	q_6	q_7	q_8
$x_1'x_2'$					×	×	×	×
$x_2'x_3$			×	×	×		×	
x_1x_3	×	×	×	×				

So the core is $x_1x_3, x_1'x_2'$ and all the q_i are covered. Hence the unique minimal form is

$$x_1x_3 + x_1'x_2'.$$

(b) The corresponding Karnaugh diagram is

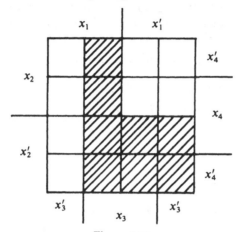

Figure 8.36

By inspection the only possible minimal form is $x_1x_3 + x_1'x_2'$.

21. (i) $x_2x_4' + x_1'x_3 + x_1'x_2'x_3'x_4$;

    (ii) $x_1x_2 + x_1'x_2'x_4 + x_1'x_2'x_3'x_4'$.

Chapter 2, §2

1. $p = ((x_1 \lor x_2) \lor x_3) \lor (x_1 \land x_2)$

x_1	x_2	x_3	p
0	0	0	0
1	0	0	1
0	1	0	1
0	0	1	1
1	1	0	1
1	0	1	1
0	1	1	1
1	1	1	1

2. $p_1 = x_1 \rightarrow (x_2 \wedge (x_1 \rightarrow x_2)) = \neg x_1 \vee (x_2 \wedge (\neg x_1 \vee x_2))$

$$p_2 = x_1 \rightarrow (x_1 \vee x_2)$$

$$= \neg x_1 \vee (x_1 \vee x_2)$$

$$= 1,$$

$$p = (x_1 \rightarrow x_2) \vee (x_2 \rightarrow x_3)$$

$$= (\neg x_1 \vee x_2) \vee (\neg x_2 \vee x_3)$$

$$= 1.$$

So p_2 and p_3 are tautologies. p_1 is not a tautology (e.g. $p_1 = 0$ if $x_2 = 0$, $x_1 = 1$).

3. Denote the statements: "Devaluation will occur"; "exports will increase"; "price controls will be imposed" by p_1, p_2 and p_3, respectively. Then the politician makes three statements:

$$S_1 = p_1 \vee (p_2 \rightarrow p_3) = p_1 \vee \neg p_2 \vee p_3,$$

$$S_2 = \neg p_1 \rightarrow \neg p_2 = p_1 \vee \neg p_2,$$

$$S_3 = p_3 \rightarrow p_2 = \neg p_3 \vee p_2,$$

Consider $S_1 \wedge S_2 \wedge S_3 = (p_1 \vee \neg p_2 \vee p_3) \wedge (p_1 \vee \neg p_2) \wedge (\neg p_3 \vee p_2)$

p_1	p_2	p_3	S_1	S_2	S_3	$S_1 \wedge S_2 \wedge S_3$
0	0	0	1	1	1	1
1	0	0	1	1	1	1
0	1	0	0	0	1	0
0	0	1	1	1	0	0
1	1	0	1	1	1	1
1	0	1	1	1	0	0
0	1	1	1	0	1	0
1	1	1	1	1	1	1

Hence the statements are consistent.

4. Let

$$E = \text{"full employment maintained"};$$
$$T = \text{"taxes must be increased"};$$
$$P = \text{"politicians have to worry about people"}.$$

Then the four statements become

(i) $E \vee T'$;

(ii) $P \rightarrow T \Leftrightarrow P' \vee T$;

(iii) $P \vee E'$;

(iv) $(E \rightarrow T)' \Leftrightarrow (E' \vee T)' \Leftrightarrow E \wedge T'$.

Combining the four statements we have

$$f = (E \vee T') \wedge (P' \vee T) \wedge (P \vee E') \wedge (E \wedge T'),$$

and combining the first three statements we have

$$f^1 = (E \vee T') \wedge (P' \vee T) \wedge (P \vee E').$$

Then the truth table is given by

P	T	E	$\bar{f}$	$\bar{f}^1$
1	1	1	0	1
1	1	0	0	0
1	0	1	0	0
0	1	1	0	0
0	0	1	0	0
0	1	0	0	0
1	0	0	0	0
0	0	0	0	1

So $\bar{f} = f_0$ always so the four statements give a contradiction, whereas $\bar{f}^1 \neq f_0$, so the first three statements are consistent.

5. Let

p_1 be the proposition that A is green.
p_2 be the proposition that A is red.
p_3 be the proposition that B is green.
p_4 be the proposition that B is red.
p_5 be the proposition that C is green.
p_6 be the proposition that C is red.

Then the statements become,

$$S_1: p_3 \wedge (p_5 \vee p_6),$$

$$S_2: p_6 \wedge S_4 = (p_2 \vee p_3 \vee (p_5 \wedge p_6)) \wedge p_6,$$

$$S_3: p_1 \vee p_4,$$

$$S_4: p_2 \vee S_1 = p_2 \vee p_3 \vee (p_5 \wedge p_6),$$

$$S_5: (p_1 \vee p_2) \vee p_3.$$

The answer then is: A is red, B and C are white.

6. We can transform the given system of equations into the single equation of the form

$$(x_1 x_3)'(x_2 + x_1 x_2 x_3') + (x_1 x_3)(x_2 + x_1 x_2 x_3')' + (x_1' + x_3)'(x_1 x_2)$$
$$+ (x_1' + x_3)(x_1 x_2)' = 0.$$

If we express the left-hand side in conjunctive normal form we obtain the equation

$$(x_1' + x_2' + x_3')(x_1' + x_2 + x_3) = 0.$$

So the solutions over $B = \mathbb{B}$ are the zeros of the first and second factor, namely all $(a_1, a_2, a_3) \in B^3$ such that

$$a_1' + a_2' + a_3' = 0 \Rightarrow a_1 = a_2 = a_3 = 1,$$

or

$$a_1' + a_2 + a_3 = 0 \Rightarrow a_1 = 1, a_2 = a_3 = 0.$$

Therefore the system of equations has exactly two solutions in B, namely $(1, 1, 1)$ and $(1, 0, 0)$.

7. If $x = a' \sqcup u$ then $a \sqcup x = a \sqcup a' \sqcup u = 1 \sqcup u = 1$. If $a \sqcup x = 1$, then $a' \sqcap x' = 0$, therefore $x \sqcup (a' \sqcap x') = x \sqcup 0 = x$, thus $(x \sqcup a') \sqcap (x \sqcup x') = x$, i.e. $x = a' \sqcup x$.

8. With the given values for x and y we obtain

$$x \sqcup y = (c \sqcap (u \sqcup v')) \sqcup (c \sqcap (u' \sqcup v))$$
$$= c \sqcap (u \sqcup v' \sqcup u' \sqcup v) = c \sqcap 1 = c.$$

Conversely, if x, y satisfy $x \sqcup y = c$, then

$$x = x \sqcup (y \sqcap y') = (x \sqcup y') \sqcap (x \sqcup y) = c \sqcap (x \sqcup y'),$$
$$y = y \sqcup (x \sqcap x') = (x \sqcup y) \sqcap (x' \sqcup y) = c \sqcap (x' \sqcup y),$$

so that

$$x = c \sqcap (u \sqcup v'),$$
$$y = c \sqcap (u' \sqcup v) \quad \text{with values } x, y \text{ of } u, v.$$

9. If $x = (u \sqcup a') \sqcap b, a \le b$ then $a \sqcup x = a \sqcup b = b$. Conversely, if x is a solution of $a \sqcup x = b$ then $a \le a \sqcup x = b$, $a' \sqcap x' = b'$, thus $a \sqcup x' = a \sqcup (a' \sqcap x') = a \sqcup b'$, so $a' \sqcap x = a' \sqcap b$. Therefore

$$x = x \sqcup (x \sqcap a') = x \sqcup (a' \sqcap b) = (x \sqcup a') \sqcap (x \sqcup b)$$
$$= (x \sqcup a') \sqcap b, \text{ since } x \le b.$$

This shows that $x = (u \sqcup a') \sqcap b$ for any $u = x$.

10. To show that the values of x and y satisfy the equations, consider

$$a \sqcap y = a \sqcap (a' \sqcup b) \sqcap v$$
$$= a \sqcap b \sqcap v.$$

So

$$x \sqcup (a \sqcap y) = b \sqcap [(a \sqcap v) \sqcup u \sqcup (a' \sqcup v')]$$
$$= b \sqcap [u \sqcup (a \sqcap v) \sqcup (a \sqcap v')]$$
$$= b \sqcap 1 = b.$$

Conversely, if $x \sqcup (a \sqcap y) = b$ then $a \sqcap y \le b$; so

$$y \le y \sqcup a' = (a \sqcap y) \sqcup a' \le b \sqcup a'.$$

So $y = (b \sqcup a') \sqcap y$. This shows that $y = (b \sqcup a') \sqcap v$ for $v = y$. Since $x' \sqcap (a' \sqcup y') = b'$, we have

$$x \sqcup (x' \sqcap (a' \sqcup y')) = x \sqcup b'.$$

So

$$x \sqcup a' \sqcup y' = x \sqcup b',$$

$$b \sqcap (x \sqcup a' \sqcup y') = b \sqcap x.$$

Since $x \sqcup (a \sqcap y) = b$, $x \le b$, so $b \sqcap x = x$ proving

$$x = (u \sqcup (a' \sqcup y')) \sqcap b \quad \text{for any } u = x.$$

Chapter 2, §3

1. (1) $\mathcal{P}(M)$ is complete since if $A_\alpha \subseteq M$ for $\alpha \in I$ we have $\bigcup_{\alpha \in I} A_\alpha \subseteq M$ and $\bigcap_{\alpha \in I} A_\alpha \subseteq M$.

   (2) Let $A_i = \{i\}$ for each $i \in \mathbb{N}$. Then $\bigcup_{i \in \mathbb{N}} A_i = \mathbb{N}$ which is not finite.

   (3) This follows immediately since inf L and sup L exist by completeness.

   (4) The only subsets of $\{a\}$ are $\{a\}$ and $\varnothing$.

   (5) Clearly the one-dimensional subspaces are atoms because they contain no proper subspaces. Every subspace of dimension greater than one contains the subspaces generated by nonzero elements. Hence the one-dimensional subspaces are precisely the atoms.

   (6) The only divisors of primes are one and the prime itself. Every nonprime has a prime divisor hence the primes are the atoms.

   (7) From what has already been said it is clear that all the lattices mentioned above are atomic.

   (8) $\{A \subseteq \mathbb{N} | A' \text{ finite}\}$ has no atoms. Suppose A is an atom. Let $B = A \backslash \{a\} \ne \varnothing$ for any $a \in A$ then $B \subseteq A$, $B' = A' \cup \{a\}$ is finite, giving a contradiction.

   (9) From the above $\mathcal{P}(M)$ is complete and atomic. Conversely let L be a complete atomic Boolean algebra. Let M be the set of atoms of L and define $h: L \to \mathcal{P}(M)$ by $h(x) = \{a \in A | a \le x\}$ for each $x \in L$. For all $x, y \in L$, $a \in A$. Then

$$a \in h(x \wedge y) \quad \text{iff} \quad a \le x \wedge y,$$
$$\text{iff} \quad a \le x \text{ and } a \le y,$$
$$\text{iff} \quad a \in h(x) \cap h(y).$$

So $h(x \wedge y) = h(x) \cap h(y)$. Also

$$a \in h(x') \quad \text{iff} \quad a \le x',$$
$$\text{iff } a \not\le x,$$
$$\text{iff } a \in (h(x))'.$$

So $h(x') = (h(x))'$. Hence h is a homomorphism. h is injective: for $x \ne y$ we have $x \not\le y$ or $y \not\le x$. Without loss of generality suppose $y \not\le x$ then $x' \wedge y \ne 0$. L is atomic so $a \in A$ exists with $a \le x' \wedge y$ which means that $a \not\le x$ and $a \le y$. Therefore $a \in h(y) - h(x)$. Hence $h(x) \ne h(y)$ and h is injective. h is surjective: Let $X \in \mathcal{P}(L)$ and $x = VX$. Then $X \subseteq h(x)$ since if $a \in X$ then $a \leq x$, so $a \in h(x)$. If $a \in A \setminus X$ then $a \wedge b = 0$ for all $b \in X$, therefore $b \leq a'$. It follows that $x \leq a'$ and since $a \neq 0$ we have $a \not\leq x$. Therefore $a \notin h(x)$, i.e. $h(x) \subseteq X$ and it follows that $h(x) = X$.

2. (a) B.L.: No, since complements don't exist.
    Atomic: In general atoms don't exist, e.g. if $M = \mathbb{R}$.
    Complete: Yes.
   (b) B.L.: $\Big\}$
       Atomic: $\Big\}$ as above.
       Complete: $\Big\}$
   (c) B.L.: Yes.
       Atomic: Singletons are the atoms.
       Complete: Yes since arbitrary unions and intersection of sets in $\mathcal{P}(M)$ are
                 sets in $\mathcal{P}(M)$.
   (d) B.L.: No.
       Atomic: The primes are the atoms.
       Complete: No, the supremum of arbitrary sets does not always exist.
   (e) B.L.: No.
       Atomic: Yes, the atoms are the subgroups generated by single elements.
       Complete: Yes.
       V_1^1: B.L.: Yes, by inspection.
            Atomic: 1 is the only atom.
            Complete: Yes, by inspection.
       V_1^1: B.L.: Yes; Atomic: 1 is the atom; Complete: yes.
       V_1^2: Yes; Yes; Yes.
       V_1^3: No: Yes; Yes, a has no complement.
       V_1^4: Yes; Yes; Yes.
       V_2^4: No (a, b don't have complements); Yes; Yes.
       V_1^5: (c has no complement); Yes; Yes.
       V_2^5: No; Yes; Yes.
       V_3^5: No (not distributive); Yes; Yes.
       V_4^5: No (not distributive); Yes; Yes.
       V_5^5: No; Yes; Yes.
       V_2^6: No (complements not unique); Yes; Yes.
       V_3^6: No, Yes, Yes.
       V_4^6: No, Yes, Yes.
       V_5^6: No, Yes, Yes.
       V_6^6: No, Yes, Yes.
       V_7^6: No, Yes, Yes.

3. The Hausdorff spaces that are Boolean spaces are precisely those which are totally disconnected.

4. A model is (Ω, A) where $\Omega = \mathbb{R}$ and $A = \mathcal{P}(\mathbb{R})$. A more suitable model would be $\Omega = \mathbb{R}$ and A, the σ-algebra of Borel sets in $\mathbb{R}$.

5. Let the table have dimension $w \times l$, $l \geq w$. Then a model would be (Ω, A) where $\Omega = \{r \in \mathbb{R} | 0 \leq r \leq w/2\}$ and $A = \mathcal{P}(\Omega)$. We could also take A to be the σ-algebra of Borel sets on $\Omega := \{B \cap \Omega | B \in \mathcal{B}\}$.

6. In both examples, if we take Ω to be the σ-algebra of Borel sets on Ω, A is generated by all the open intervals of $\mathbb{R}$ (in Exercise 4) and $\mathbb{R} \cap \Omega$. So these constitute the elementary events.

7. (a) Define $\mu: A \to |A|$. Clearly $\mu(A) \geq 0$, $\mu(\{a\}) = 1 < \infty$. If $\{A_1, A_2, \ldots\}$ is a countable set of disjoint sets then

$$\mu\left(\bigcup_n A_n\right) = \left|\bigcup_n A_n\right|$$

$$= \sum_n |A_n|$$

$$= \sum_n \mu(A_n) \quad \text{so } \mu \text{ is countably additive.}$$

So μ is a measure on $\mathcal{P}(M)$.

(b) If $|M| \geq 2$ then $\mu(M) = 2$. So μ is not a probability measure.

(c) Clearly $P(M) = |M|/|M| = 1$ and the other properties follow from 1. P is a probability measure.

8. (a) $\mathcal{B}$ is atomic with atoms the singletons $\{a\}$, $a \in \mathbb{R}$.

(b) Every finite subset of $\mathbb{R}$ is a union of atoms.

(c) $(0, \infty) = \bigcup_{n \in \mathbb{N}} (0, n) \in \mathcal{B}$.

(d) $\mathbb{Q} = \bigcup_{a \in \mathbb{Q}} \{a\} \in \mathcal{B}$ since $\mathbb{Q}$ is countable and by (1).

9. $\Omega = \{1, \ldots, 6\} \times \{1, \ldots, 6\}$. A model is $(\Omega, \mathcal{P}(\Omega))$. Define a probability measure P by $P(a, b) = \frac{1}{36}$ and if $A = \bigcup_{i=1}^n \{(a_i, b_i) | i = 1(1)n\}$ then $P(A) = \sum_{i=1}^n P(a_i, b_i)$. If $A = \{1\} \times \{1, \ldots, 6\}$ then this is the event that the first dice throws a one. If $B = \{(1, 1), (2, 2), \ldots, (6, 6)\}$, this is the event that both dice are the same number.

$$P(A|B) = \frac{P(A \cap B)}{P(B)} = \frac{\frac{1}{36}}{\frac{1}{6}} = \frac{1}{6},$$

and

$$P(B|A) = \frac{P(A \cap B)}{P(B)}$$

$$= \frac{\frac{1}{36}}{\frac{1}{6}} = \frac{1}{6},$$

$$P(A) = \frac{1}{6} = P(B).$$

Since $P(A|B) = P(A)$, A and B are independent events.

10. $\Omega = \{(y_1, y_2) \in \{AS, \ldots, 1S, AH, \ldots, 1H, AC, \ldots, 1C, AD, \ldots, 1D\}$

$\times \{AS, \ldots, 1D\}, y_1 \neq y_2\}$,

$|\Omega| = 52 \times 51 = 2652$,

i.e. there are 2652 points in the sample space.

$A = \{(y_1, y_2) \in \{AS, \ldots, 1S\} \times \{AS, \ldots, 1S\}, y_1 \neq y_2\}$,

$|A| = 13 \times 12 = 156$,

i.e. there are 156 points in the subset for the event "both are spades".

$B_0 = \{(y_1, y_2) | (y_1, y_2) \notin A_1 \cup A_2\}$,

$B_1 = \{(y_1, y_2) | (y_1, y_2) \in A_1 \text{ or } (y_1, y_2) \in A_2 \text{ and } (y_1, y_2) \notin A_1 \cap A_2\}$

$\quad = \{(y_1, y_2) | (y_1, y_2) \in A_1 \cup A_2 \text{ and } (y_1, y_2) \notin A_1 \cap A_2\}$,

$B_2 = \{(y_1, y_2) | (y_1, y_2) \in A_1 \cap A_2\}$.

So

$$p(B_0) = \frac{|B_0|}{|\Omega|} = \frac{2256}{2652} = \frac{188}{221},$$

$$p(B_1) = \frac{|B_1|}{|\Omega|} = \frac{384}{2652} = \frac{32}{221},$$

$$p(B_2) = \frac{|B_2|}{|\Omega|} = \frac{12}{2652} = \frac{1}{221}.$$

Chapter 3, §1

1. Multiplication is completely determined by $a \cdot a$, if a is a generator for the additive (cyclic) group of a ring R. For arbitrary $x, y \in R$ there are $m, n \in \mathbb{Z}$, such that $x = ma$, $y = na$, and hence $xy = (ma)(na) = mna^2$. Conversely, this last equation can be used to define multiplication for given a^2. Thus there are the following four possibilities to define multiplication to make the given R a ring.

	0	1	2	3
0	0	0	0	0
1	0	0	0	0
2	0	0	0	0
3	0	0	0	0

	0	1	2	3
0	0	0	0	0
1	0	1	2	3
2	0	2	0	2
3	0	3	2	1

	0	1	2	3
0	0	0	0	0
1	0	2	0	2
2	0	0	0	0
3	0	2	0	2

	0	1	2	3
0	0	0	0	0
1	0	3	2	1
2	0	2	0	2
3	0	1	2	3

2. R does not have zero divisors, since if for nonzero $a \in R$ with $axa = a$ we have $ab = 0$, then $a(b + x)a = aba + axa = a$, hence $b + x = x$ and therefore $b = 0$. The proof of the impossibility of $ba = 0$, $b \neq 0$ is similar. Let $e = ax \neq 0$. Then e is idempotent, since $e^2 = axax = ax = e$. For arbitrary $r \in R$ we have $e(er - r) = 0$ and $(re - r)e = 0$. Therefore e is a unit element and a has a right inverse. It is easy to verify the existence of a left inverse of a. Hence any $a \neq 0$ has an inverse.

3. Let $R = \{a_1, \ldots, a_n\}$. Then we verify immediately that $aR = R$ and $Ra = R$. Therefore $ax = b$ and $xa = b$ are solvable in R for arbitrary $b \in R$. Let e be a solution of $ax = a$. Then e is a right unity, since for an arbitrary $r \in R$ this r

can be expressed as $r = sa$ and $re = (sa)e = s(ae) = r$. Similarly, e is a left unity and therefore a unity.

4. $\mathbb{Z}_6$ has subrings $\{[0], [2], [4]\}$ and $\{[0], [3]\}$ with unit elements $[4]$ and $[3]$, respectively, but $\mathbb{Z}_6$ has unit element $[1]$.

5. The set $F = \{n1 | n = 0, 1, \ldots, p - 1\}$ is a subfield of R with p elements and R can be regarded as a finite-dimensional vector space over F. Let $\{a_1, \ldots, a_n\}$ be a basis of n elements then any element $r \in R$ is of the form $r = c_1 a_1 + \ldots + c_n a_n, c_i \in F$.

6. Let I be an ideal of $\mathbb{Z}$. If $I = \{0\}$ then $I = (0)$; if $I \neq \{0\}$ then I contains a smallest natural number k and $(k) \subseteq I$. Let $i \in I, i = qk + r, 0 \le r \le k - 1$. Since $r = i - qk \in I$ it follows that $r = 0$, that is $i = qk \in (k)$. Hence $I = (k)$.

7. Let R be a field, $\{0\} \neq I \trianglelefteq R$ and $i \in I$. Then $r = r1 = ri^{-1}i \in RI \subseteq I$ for all $r \in R$, hence $I = R$. Therefore R is simple. Conversely, let R be a simple, commutative ring with unity. It suffices to show that every $r \in R^*$ is invertible. Since $r \neq 0$ and R is simple we have $rR = (r) = R$. Therefore there exists an $s \in R$ such that $rs = 1$.

8. The criterion follows from 1.11(iv), since there is no ideal strictly between I and R if and only if there is no ideal strictly between $\{0\}$ and R/I.

9. We define $N = \{J \triangleleft R | I \subseteq J$ and $1 \notin J\}$. Then $N \neq \varnothing$ since $I \in N$. By Zorn's Lemma (see Chapter 1, §1), N contains a maximal element J^*. If J^* were not a maximal ideal there would be some ideal J' such that $J^* \subset J' \triangleleft R$. But then $J' \notin N$, hence $1 \in J'$, whence $J' = R$, a contradiction.

10. Let I be a nonzero ideal of $R[[x]]$. The *codegree* of $(a_0, a_1, \ldots)$ is the first index k with $a_k \neq 0$ (whenever such a k exists). In I there is eactly one formal power series p with minimal codegree k and $a_k = 1$,

$$p = (0, \ldots, 0, 1, a_{k+1}, \ldots) = x^k \cdot (1, a_{k+1}, \ldots) := x^k \cdot q.$$

It can be verified that q has an inverse, so $x^k = pq^{-1} \in I$, whence $(x^k) \subseteq I$. The inclusion $I \subseteq (x^k)$ is trivial, hence $I = (x^k)$.

11. (i) Let I be a nonzero ideal of $R[x]$. Pick out some $p \in I$ with minimal degree in I. If $i \in I$ then there are $q, r \in R[x]$ such that $i = p \cdot q + r$ and deg $r <$ deg p. Since $r = i - p \cdot q \in I, r = 0$ and $i \in (p)$, whence $I = (p)$.

   (ii) Since $(p) \subseteq (q)$ is equivalent to $q | p$ (with proper inclusion if and only if q is a proper divisor of p), we get: (p) is maximal $\Leftrightarrow p$ is irreducible. By 1.12(iii), (p) is a prime ideal in this case. It remains to show that if (q) is a prime ideal then q is irreducible. For suppose $q = f \cdot g$, then $f \cdot g \in (q)$, hence $f \in (q)$ or $g \in (q)$; that is, $q | f$ or $q | g$, a contradiction.

   (iii) follows from (ii).

12. We consider the ideal $I := (f) + (g)$. Since I must be principal, there is some monic polynomial d with $I = (d)$. Since $f \in (d)$ and $g \in (d), d | f$ and $d | g$. Also, there are $p, q \in R[x]$ with $d = p \cdot f + q \cdot g$, since $d \in (f) + (g)$. If also $d' | f$ and $d' | g$ hold then $f \in (d')$ and $g \in (d')$, whence $d \in (d')$, which means $d' | d$. The uniqueness is now trivial.

13. We show the existence by induction on the degree n of $f = (a_0, a_1, \ldots, a_n)$. If $n \le 0$ the result is trivially true. Now let the result be shown for all polynomials of degree smaller than n. If f is irreducible then $f = a_n \cdot (a_0 a_n^{-1}, a_1 a_n^{-1}, \ldots, 1)$ is the desired representation. If f is reducible, $f = g \cdot h$, then we get

$$g = r p_1 \ldots p_k \quad \text{and} \quad h = r' p_1' \ldots p_m';$$

hence $f = (rr') p_1 \ldots p_k p_1' \ldots p_m'$.

To show uniqueness, suppose that $f = r p_1 \ldots p_k = s q_1 \ldots q_n$ are two decompositions of the described kind. Since $r = a_n = s, r = s$. Since $p_1 | f = s q_1 \ldots q_n, p_1 | q_i$ (i.e. $p_1 = q_i$) for some i. Hence, since $R[x]$ is integral, $p_2 p_3 \ldots p_k = q_1 \ldots q_{i-1} q_{i+1} \ldots q_n$. Continuing this process we get $k = n$ and $\{p_1, \ldots, p_k\} = \{q_1, \ldots, q_n\}$.

14. If $p = (a_0, a_1, \ldots)$ and $q = (b_0, b_1, \ldots)$ we get for all $r \in R$:

$$\overline{p + q}(r) = (a_0 + b_0) + (a_1 + b_1)r + (a_2 + b_2)r^2 + \ldots$$

$$= (a_0 + a_1 r + a_2 r^2 + \ldots) + (b_0 + b_1 r + b_2 r^2 + \ldots) = \bar{p}(r) + \bar{q}(r)$$

$$= (\bar{p} + \bar{q})(r).$$

Hence $\overline{p + q} = \bar{p} + \bar{q}$. $\overline{p \cdot q} = \bar{p} \cdot \bar{q}$ is shown similarly. (ii) now follows from (i).

15. (i) is just a reformulation of 1.19(i). To show (ii) it suffices to show that Ker $h = \{0\}$. So suppose that Ker $h \ne \{0\}$. In this case, Ker h contains a polynomial p with minimal degree $n \ge 0$. Since $n = 0$ is impossible, we get $n \ge 1$. Let $p = (a_0, a_1, \ldots, a_n)$ and choose r arbitrarily. Then

$$0 = \bar{p}(r + 1) - \bar{p}(r) = a_n (r + 1)^n + a_{n-1}(r + 1)^{n-1} + \ldots + a_0$$

$$- a_n r^n - a_{n-1} r^{n-1} - \ldots - a_0$$

$$= n a_n r^{n-1} + \bar{q}(r) \quad \text{with} \quad \deg q \le n - 2.$$

$s := n a_n x^{n-1} + q$ fulfills $\bar{s}(r) = 0$ for all $r \in R$. Hence $s \in$ Ker h which is impossible: $\deg s = n - 1$, because $n a_n \ne 0$ since $(R, +)$ contains no elements of infinite order.

16. Let $f = a_0 + a_1 x + \ldots + a_n x^n$ and $g = b_0 + b_1 x + \ldots + b_m x^m$ be elements in $I[x]$ and $h = c_0 + c_1 x + \ldots + c_k x^k \in R[x]$. We have $a_i - b_i \in I, a_0 c_i + a_1 c_{i-1} + \ldots + a_i c_0 \in I$ for all $i \in \mathbb{N}_0$, where $a_i = 0$ for $i > n$, $b_i = 0$ for $i > m$ and $c_i = 0$ for $i > k$. Then $f - g \in I[x]$ and $fh = hf \in I[x]$.

17. The element $1 \in \mathbb{Z}_5$ is a root of multiplicity one.

18. f is not divisible by g. We have $\bar{f}(3) + \bar{g}(3) = \overline{f + g}(3) = 3$.

19. (i) For $R = \mathbb{R}, f$ is not divisible by g.
    (ii) For $R = \mathbb{Z}_2, f$ is divisible by g.

20. Let $\psi : F[x] \to F(\alpha)$ be defined by the rule $\psi(\alpha) = \bar{f}(\alpha)$. Then ψ is an epimorphism for all α. If ψ is an isomorphism then ker $\psi = \{0\}$ and so no f exists such that $\bar{f}(\alpha) = 0$ with $0 \ne f \in F[x]$. So α is transcendental. If ψ is not an isomorphism then $0 \ne f \in \ker(\psi)$ exists such that $\bar{f}(\alpha) = \psi(f) = 0$. So α is algebraic.

21. (a) If f is reducible then, since it is of degree 3, it must have at least one linear factor, which cannot be the case since $\bar{f}(0) = \bar{f}(1) = 1$.

(b) $\mathbb{Z}_2[x]/(f) = \{a + (f) | a \in \mathbb{Z}_2[x]\}$

$$= \{a_0 + a_1 x + a_2 x^2 + (f) | a_i \in \mathbb{Z}_2\}.$$

(c) $\bar{f}(x + (f)) = 0, \qquad \bar{f}(x^2 + (f)) = 0.$

Hence f must factor into three linear factors in $\mathbb{Z}_2[x]/(f)$, clearly this is the smallest such field and so is the splitting field.

22. Let $x = (\sqrt{2} + i)$ then $(x - \sqrt{2})^2 = -1$, i.e. $x^2 - 2\sqrt{2}x + 3 = 0$ is the minimal polynomial over $\mathbb{R}$ and $[\mathbb{Q}(\sqrt{2} + i):\mathbb{R}] = 2$. Now $x^2 + 3 = 2\sqrt{2}x$, therefore $(x^2 + 3)^2 = 8x^2$. So $x^4 - 2x^2 + 9 = 0$ is the minimal polynomial over $\mathbb{Q}$ and $[\mathbb{Q}(\sqrt{2} + i):\mathbb{Q}] = 4$.

23. $\mathbb{Q}(\sqrt[3]{2})$ as a vector space over $\mathbb{Q}$ has basis $\{1, \sqrt[3]{2}, \sqrt[3]{4}\}$. Let $a_0 + a_1\sqrt[3]{2} + a_2\sqrt[3]{4}$ be the inverse of $1 + \sqrt[3]{2} + \sqrt[3]{4}$. Then

$$(a_0 + a_1\sqrt[3]{2} + a_2\sqrt[3]{4})(1 + \sqrt[3]{2} + \sqrt[3]{4})$$

$$= (a_0 + 2a_2 + 2a_1) + \sqrt[3]{2}(a_1 + a_0 + 2a_2) + \sqrt[3]{4}(a_0 + a_1 + a_2)$$

$$= 1.$$

Solving, gives $a_0 = -1$, $a_1 = 1$, $a_2 = 0$. Therefore $-1 + \sqrt[3]{2}$ is the inverse.

24. The elements of $\mathbb{F}_2[x]/(x)$ are $[0]$ and $[1]$, so $\mathbb{F}_2[x]/(x) \cong \mathbb{F}_2$.

25. (i) Let $f(x) = x^2 + 1$. Then $f(0) = 1, f(1) = 2, f(2) = 2$ in $\mathbb{Z}_3$. So f is irreducible over $\mathbb{Z}_3$.

   (ii) Let $\alpha^2 + 1 = 0$. The elements of $\mathbb{Z}_3(\alpha)$ are $0, 1, 2, \alpha, 2\alpha, 1 + \alpha, 1 + 2\alpha, 2 + 2\alpha$. Since addition and multiplication are commutative we calculate the "upper triangle" only:

+	0	1	2	α	2α	$1 + \alpha$	$1 + 2\alpha$	$2 + \alpha$	$2 + 2\alpha$
0	0	1	2	α	2α	$1 + \alpha$	$1 + 2\alpha$	$2 + \alpha$	$2 + 2\alpha$
1		2	0	$1 + \alpha$	$1 + 2\alpha$	$2 + \alpha$	$2 + 2\alpha$	α	2α
2			1	$2 + \alpha$	$2 + 2\alpha$	α	2α	$1 + \alpha$	$1 + 2\alpha$
α				2α	0	$1 + 2\alpha$	1	$2 + 2\alpha$	2
2α					α	1	$1 + \alpha$	2	$2 + \alpha$
$1 + \alpha$						$2 + 2\alpha$	2	2α	0
$1 + 2\alpha$							$2 + \alpha$	0	α
$2 + \alpha$								$1 + 2\alpha$	1
$2 + 2\alpha$									$1 + \alpha$

·	0	1	2	α	2α	$1 + \alpha$	$1 + 2\alpha$	$2 + \alpha$	$2 + 2\alpha$
0	0	0	0	0	0	0	0	0	0
1		1	2	α	2α	$1 + \alpha$	$1 + 2\alpha$	$2 + \alpha$	$2 + 2\alpha$
2			1	2α	α	$2 + 2\alpha$	$2 + \alpha$	$1 + 2\alpha$	$1 + \alpha$
α				2	1	$2 + \alpha$	$1 + \alpha$	$2 + 2\alpha$	$1 + 2\alpha$
2α					2	$1 + 2\alpha$	$2 + 2\alpha$	$1 + \alpha$	$2 + \alpha$
$1 + \alpha$						2α	2	1	α
$1 + 2\alpha$							α	2α	1
$2 + \alpha$								α	2
$2 + 2\alpha$									2α

26. (i) $8, \{1, \sqrt{3}, \sqrt{5}, \sqrt{15}, \sqrt{2}, \sqrt{6}, \sqrt{10}, \sqrt{30}\}$.
    (ii) $2, \{1, \sqrt{2}\}$.
    (iii) $6, \{1, \sqrt{2}, \sqrt[3]{2}, \sqrt{2}\sqrt[3]{2}, (\sqrt[3]{2})^2, \sqrt{2}(\sqrt[3]{2})^2\}$.
    (iv) $2, \{1, \sqrt{2}\}$.

27. (i) The polynomial splits in $\mathbb{C}$ into

$$(x + \sqrt{3})(x - \sqrt{3})(x + 1)\left(x - \frac{-1 + i\sqrt{3}}{2}\right)\left(x - \frac{-1 - i\sqrt{3}}{2}\right).$$

So the splitting field is

$$\mathbb{Q}\left(\sqrt{3}, \frac{-1 + i\sqrt{3}}{2}\right),$$

which is the same as $\mathbb{Q}(\sqrt{3}, i)$.

(ii) The zeros of this polynomial in $\mathbb{C}$ are $1 + \sqrt{3}, \pm i$. A splitting field is given by $\mathbb{Q}(1 + \sqrt{3}, i)$, which is the same as $\mathbb{Q}(\sqrt{3}, i)$.

(iii) Both have $\mathbb{Q}(\sqrt{3})$ as a splitting field over $\mathbb{Q}$.

(iv) Let $\zeta^2 + \zeta + 1 = 0$, form the four elements of $\mathbb{Z}_2(\zeta)$ as:

$$0, 1, \zeta, 1 + \zeta.$$

The polynomial splits over $\mathbb{Z}_2(\zeta)$, namely

$$f = x^2 + x + 1 = (x - \zeta)(x - 1 - \zeta).$$

It splits over no smaller field, so $\mathbb{Z}_2(\zeta)$ is the splitting field of f over $\mathbb{Z}_2$.

(v) $f = x^3 + x + 2 = (x + 1)(x^2 - x + 2)$. $x^2 - x + 2$ is irreducible over $\mathbb{Q}$ but splits into linear factors in $\mathbb{Q}((1 + \sqrt{7}i)/2)$. So f has splitting field $\mathbb{Q}((1 + \sqrt{7}i)/2)$.

28. (a) $\bar{f}(0) = 2; \bar{f}(1) = 3; \bar{f}(2) = 2; \bar{f}(3) = 3; \bar{f}(4) = 1$. Hence f is irreducible over $\mathbb{Z}_5$.

(b) Define $\psi: \mathbb{Z}_5(\alpha) \to \mathbb{Z}_5(\beta)$ by $\psi(a + b\alpha) = a + b\beta$. Since $\{1, \beta\}$ is a basis of $\mathbb{Z}_5(\beta)$ and $\{1, \alpha\}$ is a basis of $\mathbb{Z}_5(\alpha)$ it clearly follows that ψ is surjective. ψ is injective since if $\psi(a + b\alpha) = \psi(a' + b'\alpha)$ then $a + b\beta = a' + b'\beta$, therefore $a = a'$ and $b = b'$. ψ is a homomorphism since $\psi((a + b\alpha) + (a' + b'\alpha)) = \psi((a + a') + (b + b')\alpha) = \ldots = \psi(a + b\alpha) + \psi(a' + b'\alpha)$, and

$$\psi((a + b\alpha)(a' + b'\alpha)) = \psi(aa' + bb' + \alpha(ab' + a'b))$$
$$= aa' + bb' + \beta(ab' + a'b)$$
$$= (a + b\beta)(a' + b'\beta)$$
$$= \psi(a + b\alpha)\psi(a' + b'\alpha).$$

29. $x^2 + 4 = (x + 1)(x - 1)$ in $\mathbb{Z}_5$. Then $\mathbb{Z}_5(\beta) = \mathbb{Z}_5$ since $x^2 + 4$ is reducible in $\mathbb{Z}_5$. So $\mathbb{Z}_5(\alpha) = \mathbb{Z}_5(-\frac{1}{2} + \sqrt{3}i/2)$.

30. If n is a prime then $(x^n - 1)/(x - 1)$ is irreducible. If $n = mk$ then

$$(x^n - 1)(x - 1) = [(x^n - 1)/(x^k - 1)][(x^k - 1)/(x - 1)]$$
$$= ((x^k)^{m-1} + \ldots)(x^{k-1} + \ldots).$$

31. $\cos 4\theta = 8\cos^4\theta - 8\cos^2\theta + 1$. Therefore

$$\cos^2\theta = \frac{8 \pm \sqrt{64 - 32(1 - \cos 4\theta)}}{16}.$$

Since taking square roots is constructible, it is possible to construct $\cos\theta$.

32. (a) No, since this would require $\sqrt[3]{3}$ to be constructible.
    (b) No, since $\sqrt[3]{4} = (\sqrt[3]{2})^2$ is not constructible.

33. If a regular n-gon is constructible then so is the regular p-gon, where $p|n$ is a prime. The $p - 1$ primitive pth roots of unity are the roots of $(x^n - 1)/(x - 1)$, which is irreducible. $p - 1$ must be a power of 2, so $p = 2^m + 1$. This can only be prime if $m = 2^k$. Since, for $m = 1$ take $k = 0$, for $m > 1$ take $m = qr$, q odd then $2^m + 1 = (2^r)^q + 1 = (2^r + 1)((2^r)^{q-1} - \ldots)$ is not a prime.

34. $m^3 = 2s^3$ is not constructible.

35. $(f, f') = x^2 - 4x + 4$, therefore f has multiple roots.

Chapter 3, §2

1. (i) $x^4 + x^3 + 1$ is a primitive irreducible polynomial over $\mathbb{F}_2$. Let $\alpha^4 + \alpha^3 + 1 = 0$. Then $\mathbb{F}_{16} = \{\alpha^i | 0 \le i \le 14\} \cup \{0\}$.
    (ii) Alternatively, $\mathbb{F}_{16} = \{a_0 + a_1\alpha + a_2\alpha^2 + a_3\alpha^3 | a_i \in \mathbb{F}_2\}$.
    (iii) Or use Exercise 18, with

$$A = \begin{pmatrix} 0 & 0 & 0 & -1 \\ 1 & 0 & 0 & 0 \\ 0 & 1 & 0 & 0 \\ 0 & 0 & 1 & -1 \end{pmatrix}.$$

Then

$$\mathbb{F}_{16} = \{a_0 I + a_1 A + a_2 A^2 + a_3 A^3 | a_i \in \mathbb{F}_2\}.$$

2. Let $\mathbb{F}_3(\alpha)$ be the splitting field of $x^3 - x + 1$ over $\mathbb{F}_3$ with basis $\{1, \alpha, \alpha^2\}$ and let $\mathbb{F}_3(\beta)$ be the splitting field of $x^3 - x - 1$ over $\mathbb{F}_3(\alpha)$ with basis $\{1, \beta, \beta^2\}$. Then the map $\psi : \mathbb{F}_3(\alpha) \to \mathbb{F}_3(\beta)$ defined by

$$\psi(a_0 + a_1\alpha + a_2\alpha^2) = b_0 + b_1\beta + b_2\beta^2$$

is an isomorphism.

3. Note that $(\beta - \alpha)^{q-1} = 1$ for $\alpha \ne \beta$ and $= 0$ for $\alpha = \beta$. Therefore $\bar{p}(\beta) = f(\alpha)$.

4. Let $f = x^q - x$ then $f = \prod_{i=1}^{q}(x - \alpha_i)$ if $\mathbb{F}_q = \{\alpha_1, \ldots, \alpha_q\}$. Now by expanding this product and equating the coefficients we find,

$$\alpha_1 + \ldots + \alpha_q = 0,$$

$$\alpha_1\alpha_2 + \alpha_1\alpha_3 + \ldots + \alpha_2\alpha_3 + \ldots + \alpha_{q-1}\alpha_q = 0,$$

$$\vdots$$

$$\sum_{\substack{i,j=1 \\ i \ne j}}^{q} \prod_{k \ne i,j} \alpha_k = 0,$$

and

$$\sum_{i=1}^{q} \prod_{j \neq i} \alpha_j = -1.$$

So by direct computation we have $\sum_{a \in \mathbb{F}_q} a^m = 0$ for $0 < m < q - 1$, and $\sum_{a \in \mathbb{F}_q} a^{q-1} = q - 1$ since $a^{q-1} = 1$. Now if $m > q - 1$ we have two cases to consider. Let $m = d(q - 1)$ then

$$\sum_{a \in \mathbb{F}_q} a^m = \sum_{a \in \mathbb{F}_q} a^{d(q-1)}$$

$$= \sum_{a \in \mathbb{F}_q} (a^{q-1})^d$$

$$= q - 1 \quad \text{since } a^{q-1} = 1.$$

If $m = d(q - 1) + k$ where $q - 1 > k > 0$ then

$$\sum_{a \in \mathbb{F}_q} a^m = \sum_{a \in \mathbb{F}_q} a^{(q-1)d+k}$$

$$= \sum_{a \in \mathbb{F}_q} a^k$$

$$= 0 \quad \text{since } a^{q-1} = 1.$$

5. (i) For $\alpha, \beta \in F$

$$\mathrm{Tr}_{F/K}(\alpha + \beta) = \alpha + \beta + (\alpha + \beta)^q + \ldots + (\alpha + \beta)^{q^{m-1}}$$

$$= \alpha + \beta + \alpha^q + \beta^q + \ldots + \alpha^{q^{m-1}} + \beta^{q^{m-1}}$$

$$= \mathrm{Tr}_{F/K}(\alpha) + \mathrm{Tr}_{F/K}(\beta).$$

(ii) For $c \in K$ we have $c^{q^j} = c$ for all $j \geq 0$. Therefore we obtain for $\alpha \in F$,

$$\mathrm{Tr}_{F/K}(c\alpha) = c\alpha + c^q \alpha^q + \ldots + c^{q^{m-1}} \alpha^{q^{m-1}}$$

$$= c\alpha + c\alpha^q + \ldots + c\alpha^{q^{m-1}} = c\,\mathrm{Tr}_{F/K}(\alpha).$$

(iii) The properties (i) and (ii), together with the fact that $\mathrm{Tr}_{F/K}(\alpha) \in K$ for all $\alpha \in F$, show that $\mathrm{Tr}_{F/K}$ is a linear transformation from F into K. To prove that this mapping is onto, it suffices then to show the existence of an $\alpha \in F$ with $\mathrm{Tr}_{F/K}(\alpha) \neq 0$. Now $\mathrm{Tr}_{F/K}(\alpha) = 0$ if and only if α is a root of the polynomial $x^{q^{m-1}} + \ldots + x^q + x \in K[x]$ in F. But since this polynomial can have at most q^{m-1} roots in F and F has q^m elements, we are done.

(iv) This follows immediately from the definition of the trace function.

(v) For $\alpha \in F$ we have $\alpha^{q^m} = \alpha$ and so $\mathrm{Tr}_{F/K}(\alpha^q) = \alpha^q + \alpha^{q^2} + \ldots + \alpha^{q^m} = \mathrm{Tr}_{F/K}(\alpha)$.

6. Let $f = 1 - \sum_{j=0}^{q-1} x^j a_i^{q-1-j}$. Then f maps $a_i \in \mathbb{F}_q$ onto 1 and the other elements of $\mathbb{F}_q$ onto 0, since

$$f(a_i) = 1 - \sum_{i=0}^{q-1} a_i^{q-1} = 1$$

and

$$xf(x) \equiv a_i f(x) \bmod (x^q - x).$$

The polynomial

$$g = \sum_{i=0}^{q-1} \left(1 - \sum_{j=0}^{q-1} x^j a_i^{q-1-j} \right)$$

maps each element of F_q onto 1 iff each element of F_q is amongst $a_0, \ldots, a_{q-1}$. $g(x) = 1$ holds exactly under the conditions given in the problem.

7. $a \in Z, p \in P$.
   (i) ("Little Fermat".) F_p is the set of roots of $x^p - x$ so because of the isomorphism between F_p and Z_p we have for $a \in Z_p$,

$$a^p - a \equiv 0 \, (\mathrm{mod} \, p),$$

   i.e.

$$a^p \equiv a \, (\mathrm{mod} \, p).$$

   If $b < 0$ or $b \geq p$ then $b = kp + a$ with $a \in Z_p$ and $k \in Z$. Then $b^p \equiv a^p \equiv a \, (\mathrm{mod} \, p)$.
   (ii) ("Wilson's Theorem".) As above $x^{p-1} - 1 = \prod_{\alpha \in F_p^*} (x - \alpha)$, so $-1 = \prod_{\alpha \in F_p^*} (-\alpha)$. And since $F_p \cong Z_p$ we have $(-1)(-2)\ldots(-(p-1)) \equiv -1 \, (\mathrm{mod} \, p)$. If $p = 2$ we clearly have $(p - 1)! \equiv -1 \, (\mathrm{mod} \, p)$. If $p \neq 2$ then

$$-1 \equiv (-1)(-2)\ldots(-(p-1)) \quad (\mathrm{mod} \, p)$$

$$\equiv (-1)^{p-1} 1 \cdot 2 \ldots (p-1)$$

$$\equiv (p-1)! \quad \text{as } p-1 \text{ is even.}$$

8.

(i)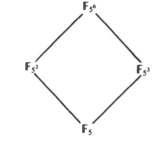

(ii)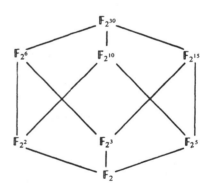

Figure 8.37

9. Let $n = km + r$, $r < m$, then verify

$$x^n - 1 = x^r \left(\sum_i x^{im} \right)(x^m - 1) + (x^r - 1).$$

Therefore

$$(x^m - 1)|(x^n - 1) \quad \text{iff } x^r - 1 = 0; \quad \text{that is, } r = 0.$$

10. 3 and 5 for F_7.

11. Let

$$f(x) = \sum_{i=0}^{m} a_i x^i, \quad \text{then } f(x^q) = \sum a_i (x^q)^i$$

$$= (\sum a_i x^i)^q, \quad \text{since } a_i^q = a_i \text{ and } (\alpha + \beta)^q = \alpha^q + \beta^q.$$

$$(a + b)^p = \sum_{i=0}^{p} \binom{p}{i} a^{p-i} b^i = a^p + b^p,$$

since p divides $\binom{p}{i}$ for $1 \leq i \leq p - 1$.

12. Let $\zeta \neq 1$. Then $1 + \zeta + \ldots + \zeta^{n-1} = (\zeta^n - 1)/(\xi - 1) = 0$, since $\zeta^n - 1 = 0$.

13. (i)

$$Q_{mp^k}(x) = \prod_{d|mp^k} (x^{mp^k/d} - 1)^{\mu(d)} = \prod_{d|mp} ((x^{p^k})^{m/d} - 1)^{\mu(d)} = Q_{mp}(x^{p^{k-1}}).$$

(ii)

$$Q_{pm}(x) = \prod_{d|m} (x^{pm/d} - 1)^{\mu(d)} \prod_{\substack{d|mp \\ d \nmid m}} (x^{pm/d} - 1)^{\mu(d)}$$

$$= \prod_{d|m} ((x^p)^{m/d} - 1)^{\mu(d)} \prod_{d|m} (x^{m/d} - 1)^{-\mu(d)}$$

$$= \frac{Q_m(x^p)}{Q_m(x)}.$$

(iii)

$$Q_n(x) = \prod_{d|n} (x^{n/d} - 1)^{\mu(d)} = (-1)^{\sum_{d|n} \mu(d)} \prod_{d|n} (1 - x^{n/d})^{\mu(d)}$$

$$= \begin{cases} - \prod_{d|n} (1 - x^{n/d})^{\mu(d)} & \text{if } n = 1, \\ \prod_{d|n} (1 - x^{n/d})^{\mu(d)} & \text{if } n > 1. \end{cases}$$

(iv) From (ii) and (iii)

$$Q_{2n}(x) = \frac{Q_n(x^2)}{Q_n(x)} = \prod \left(\frac{1 - x^{2n/d}}{1 - x^{n/d}} \right)^{\mu(d)}$$

$$= \prod_{d|n} (1 + x^{n/d})^{\mu(d)} = Q_n(-x).$$

14. $Q_{36} = 1 - x^6 + x^{12}$,

$$Q_{105} = 1 + x + x^2 - x^5 - x^6 - 2x^7 - x^8 - x^9 + x^{12} + x^{13} + x^{14}$$
$$+ x^{15} + x^{16} + x^{17} - x^{20} - x^{22} - x^{24} - x^{26} - x^{28} + x^{31}$$
$$+ x^{32} + x^{33} + x^{34} + x^{35} + x^{36} - x^{39} - x^{40} - 2x^{41} - x^{42}$$
$$- x^{43} + x^{46} + x^{47} + x^{48}.$$

15. (i)

$$A = \begin{pmatrix} 0 & 2 \\ 1 & 0 \end{pmatrix},$$

$$F_9 = \{aI + bA | a, b \in F_3\}.$$

(ii)

$$A = \begin{pmatrix} 0 & 1 \\ 1 & 2 \end{pmatrix},$$

$$F_9 = \{A^i | 0 \le i \le 8\} \cup \left\{ \begin{pmatrix} 0 & 0 \\ 0 & 0 \end{pmatrix} \right\}.$$

16. Let α be a primitive element of F_r, then $F_q(\alpha) \subseteq F_r$. On the other hand, $F_q(\alpha)$ contains 0 and all powers of α, and therefore all elements *of* F_r. Consequently $F_r = F_q(\alpha)$.

17. $x^{q-1} - 1$ splits in F_q and cannot split in any proper subfield of F_q. So F_q is the splitting field of $x^{q-1} - 1$ over any one of its subfields.

18. Let $f = x^4 + x + 1 \in F_2[x]$. Then the conjugates of α with respect to F_2 are $\alpha, \alpha^2, \alpha^4 = \alpha + 1$, and $\alpha^8 = \alpha^2 + 1$. The conjugates of α with respect to F_4 are α and $\alpha^4 = \alpha + 1$.

19. Let $\alpha \in F_8$ be a root of the irreducible polynomial $x^3 + x^2 + 1$ in $F_2[x]$. Then $\{\alpha, \alpha^2, 1 + \alpha + \alpha^2\}$ is a normal basis of F_8 over F_2, since $1 + \alpha + \alpha^2 = \alpha^4$.

20. If $\operatorname{char} F_q = 2$ then F_q has 2^n elements and each element is a square, thus $\alpha^{-1} = a^2$ for suitable $a \in F_q$. Therefore $1 + \alpha\alpha^{-1} = 0$. Let $\operatorname{char} F_q = p$, $q = p^n$, and let $M_\alpha = \{1 + \alpha x^2 | x \in F_q\}$, $M_\beta = \{-\beta x^2 | x \in F_q\}$. Then $|M_\alpha| = |M_\beta| = (q+1)/2$. Therefore there exists a $c \in M_\alpha \cap M_\beta$, such that $c = 1 + \alpha a^2$, $c = -\beta b^2$. Consequently, $1 + \alpha a^2 = -\beta b^2$.

Chapter 3, §3

1. By Theorem 3.3, the number of (monic) irreducible polynomials over F_2 of degree 4 is

$$I_2(4) = \tfrac{1}{4} \sum_{d|4} \mu\left(\frac{4}{d}\right) 2^d$$

$$= 3.$$

Furthermore, since $x^4 + x + 1$, $x^4 + x^3 + 1$, $x^4 + x^3 + x^2 + 1$ are irreducible (which is shown by direct computation) these are the only ones.

Alternatively, Theorem 3.19 yields $I(q, n, x) = Q_5(x)Q_{15}(x)$. $Q_5(x) = x^4 + x^3 + x^2 + x + 1$ is irreducible in $F_2[x]$. $Q_{15}(x) = (x^4 + x^3 + 1)(x^4 + x + 1)$ gives the remaining two irreducibles polynomials, by first evaluating $Q_5(x + 1) = x^4 + x^3 + 1$, which must divide $Q_{15}(x)$.

2. (i) Let $M = f_1 f_2$, $\deg f_1 > 0$, $\deg f_2 > 0$. Then $M(\alpha) = f_1(\alpha)f_2(\alpha) = 0$ in contradiction to the definition of a minimal polynomial.

(ii) Suppose $f = Mg + r$, $\deg r < \deg M$, $\bar{f}(\alpha) = 0$. Then r is a polynomial of degree less than n with α as a root. Since M is minimal, $r = 0$ and $f \mid M$.

(iii) $1, \alpha, \ldots, \alpha^n$ are linearly independent in $\mathbb{F}_{q^n}$ over $\mathbb{F}_q$. So $\sum_{i=0}^{n} a_i \alpha^i = 0$ for $a_i \in \mathbb{F}_q$ suitably chosen. $F = \sum a_i x^i$ is of degree $\leq n$ having α as a root. So $\deg M \leq n$, i.e. $d \leq n$. M is irreducible over $\mathbb{F}_q$, so the splitting field of M over $\mathbb{F}_q$ has order q^d and contains α. Therefore it contains all of $\mathbb{F}_{q^n}$, so $d \geq n$. Therefore $d = n$.

(iv) Let $M = \sum a_i x^i$, then

$$\bar{M}(\alpha^{q^j}) = \sum a_i \alpha^{iq^j} = (\sum a_i \alpha^i)^{q^j} = \bar{M}(\alpha)^{q^j} = 0 \quad \text{for } 1 \leq j \leq n - 1.$$

Since M is monic and irreducible the result follows.

(v) Follows from (ii).

(vi) Follows from (i) and Theorem 3.1.

3. $f = (x^{19} - 1)/(x - 1)$ is irreducible over $\mathbb{F}_p$ iff p is a primitive eighteenth root of unity. Now $2^{18} - 1 = 262143$ which is divisible by 19 and it can be shown directly that $19 \nmid 2^k - 1$ for $k < 18$. Therefore f is irreducible over $\mathbb{F}_2$ and 2 is clearly the first such prime.

Or: The product over cyclotomic cosets $\neq C_0$ must give $x^{19} - 1$. If the given polynomial is irreducible then the coset must contain all elements, this works for all $p \neq 19$. The smallest one is 2.

4. Let $f \in \mathbb{F}_q[x]$, $\bar{f}(0) \neq 0$ and $\operatorname{ord} f = e \mid m$ then $f \mid x^e - 1$ and $x^e - 1 \mid x^m - 1$ so $f \mid x^m - 1$.

Conversely, if $f \mid x^m - 1$, we have $m \geq e$, so $m = ce + r$, $c \in \mathbb{N}$, $0 \leq r < e$. Since $x^m - 1 = (x^{ce} - 1)x^r + (x^r - 1)$, it follows that $f \mid x^r - 1$. Therefore $r = 0$, and $e \mid m$.

5. Let

$$f = \frac{x^e - 1}{x - 1} = x^{e-1} + x^{e-2} + \ldots + 1.$$

If f is irreducible then there is only one irreducible polynomial of order e and so $\phi(e)/(e - 1) = 1$. Therefore e is a prime number. Furthermore since $\mathbb{F}_{q^{e-1}}$ is the splitting field of f over $\mathbb{F}_q$ we have $e \mid (q^{e-1} - 1)$ and $e \nmid q^k - 1$ for $k < (e - 1)$. In other words q is a primitive $(e - 1)$th root of unity (mod e).

Conversely, if e is prime and q is a primitive $(e - 1)$th root of unity then f is irreducible since a root of f over $\mathbb{F}_q$ must have minimal polynomial of degree $e - 1$.

6. Let N be the set of elements of $\mathbb{F}_{q^n}$ that are of degree $n > 1$ over $\mathbb{F}_q$. Every element α in N has a minimal polynomial over $\mathbb{F}_q$ of degree n. Therefore such an α is a root of $I(q, n; x)$. If β is a root of $I(q, n; x)$ then β is a root of a monic irreducible polynomial over $\mathbb{F}_q$ of degree n. Therefore β is in N and $I(q, n; x) = \prod_{\alpha \in N} (x - \alpha)$. Clearly, for $\alpha \in N$ the order of α is a divisor of $q^n - 1$. It can be shown easily that the order m of an element α of N is such that n is the multiplicative order of q modulo m. This means that $q^n \equiv 1 \pmod{m}$. Let N_m denote the set of elements of N of order m. Then N is the disjoint union of the N_m's. Hence $I(q, n; x) = \prod_m \prod_{\alpha \in N_m} (x - \alpha)$. N_m consists of all elements of $\mathbb{F}_{q^n}^*$ of order m; hence it is the set of primitive mth roots of unity

over $\mathbf{F}_q$. Therefore $\prod_{\alpha \in N_m} (x - \alpha) = Q_m(x)$, by the definition of cyclotomic polynomials.

7 and 8. $$\frac{(x^{64} - x)(x^2 - x)}{(x^8 - x)(x^4 - x)}.$$

9. Let $e = \text{ord}\, f$, $e_i = \text{ord}\, g_i$, $1 \le i \le k$, $m = \text{lcm}(e_1, \ldots, e_k)$ then $g_i \,|\, x^{e_i} - 1$ and $g_i \,|\, x^m - 1$. Therefore $f \,|\, x^m - 1$, since $g_1, \ldots, g_k$ are pairwise coprime. Exercise 4 shows that $e \,|\, m$. Since $f \,|\, x^e - 1$, $g_i \,|\, x^e - 1$, so $e_i \,|\, e$ by Exercise 4. In conclusion: $e = m$.

10. The order is 60.

11. This follows from Theorem 3.13.

Chapter 3, §4

1. There are three distinct irreducible factors of f, since the appropriate matrix $Q - I$ has rank 2 and dim(nullspace of $Q - I$) = 3.

2. *Step* 1. Using Euclid's algorithm f can be shown to be squarefree.
Step 2. Using the approach applied in Example 4.5 we obtain the following matrix:

$$Q = \begin{pmatrix} 1 & 0 & 0 & 0 & 0 & 0 & 0 & 0 & 0 & 0 & 0 & 0 \\ 0 & 0 & 1 & 0 & 0 & 0 & 0 & 0 & 0 & 0 & 0 & 0 \\ 0 & 0 & 0 & 0 & 1 & 0 & 0 & 0 & 0 & 0 & 0 & 0 \\ 0 & 0 & 0 & 0 & 0 & 0 & 1 & 0 & 0 & 0 & 0 & 0 \\ 0 & 0 & 0 & 0 & 0 & 0 & 0 & 0 & 1 & 0 & 0 & 0 \\ 0 & 0 & 0 & 0 & 0 & 0 & 0 & 0 & 0 & 0 & 1 & 0 \\ 1 & 1 & 1 & 0 & 0 & 0 & 1 & 1 & 1 & 0 & 0 & 0 \\ 0 & 0 & 1 & 1 & 1 & 0 & 0 & 0 & 1 & 1 & 1 & 0 \\ 1 & 1 & 1 & 0 & 1 & 1 & 0 & 1 & 1 & 0 & 1 & 1 \\ 1 & 0 & 1 & 0 & 1 & 0 & 0 & 1 & 0 & 0 & 1 & 0 \\ 1 & 1 & 0 & 0 & 1 & 0 & 0 & 1 & 1 & 1 & 0 & 0 \\ 0 & 0 & 1 & 1 & 0 & 0 & 1 & 0 & 0 & 1 & 1 & 1 \end{pmatrix}.$$

Step 3. Reducing $Q - I$ gives: the nullspace has dimension 2 with the independent solutions $v^{(1)} = (1, 0, 0, 0, 0, 0, 0, 0, 0, 0, 0, 0)$ and

$$v^{(2)} = (0, 1, 0, 1, 1, 0, 1, 0, 0, 1, 0, 1).$$

Hence f has two irreducible factors.
Step 4. Let

$$h = \sum_i v_i^{(2)} x^i$$

$$= x^{11} + x^9 + x^6 + x^4 + x^3 + x.$$

Then we use Euclid's algorithm to find the gcd$(f, h - s)$ for all $s \in \mathbf{F}_2$. If $s = 0$, we find the gcd of f and h is $x^5 + x^3 + x^2 + x + 1$ and by dividing this into f we get

$$f = (x^5 + x^3 + x^2 + x + 1)(x^7 + x^5 + x^4 + x^3 + 1),$$

which is the complete factorization over $\mathbf{F}_2$.

3. *Step* 1. Using Euclid's algorithm, $\gcd(f, f') = 1$; so f is squarefree.
   Step 2. Using the method and notation of Example 4.5, we obtain

$$\mathbf{Q} = \begin{pmatrix} 1 & 0 & 0 & 0 & 0 & 0 & 0 & 0 & 0 \\ 0 & 0 & 1 & 0 & 0 & 0 & 0 & 0 & 0 \\ 0 & 0 & 0 & 0 & 1 & 0 & 0 & 0 & 0 \\ 0 & 0 & 0 & 0 & 0 & 0 & 1 & 0 & 0 \\ 0 & 0 & 0 & 0 & 0 & 0 & 0 & 0 & 1 \\ 0 & 1 & 1 & 0 & 0 & 0 & 0 & 0 & 0 \\ 0 & 0 & 0 & 1 & 1 & 0 & 0 & 0 & 0 \\ 0 & 0 & 0 & 0 & 0 & 1 & 1 & 0 & 0 \\ 0 & 0 & 0 & 0 & 0 & 0 & 0 & 1 & 1 \end{pmatrix}.$$

Step 3. By reducing $\mathbf{Q} - \mathbf{I}$ we find the nullspace has dimension 1 which implies that $x^9 + x + 1$ is irreducible over $\mathbf{F}_2$.

4. We have five cases to consider. If $p = 2$ then $f = x^4 + 1 = (x + 1)^4$ is the required factorization and there are four factors. If $p = 8k + 1$ then the $\mathbf{Q}$ matrix defined by

$$x^{qk} \equiv \sum_{i=0}^{3} q_{ki} x^i \pmod{f}, \qquad 0 \le k \le 3,$$

is given by

$$\mathbf{Q} = \begin{pmatrix} 1 & 0 & 0 & 0 \\ 0 & 1 & 0 & 0 \\ 0 & 0 & 1 & 0 \\ 0 & 0 & 0 & 1 \end{pmatrix}.$$

Hence $\mathbf{Q} - \mathbf{I} = \mathbf{0}$, the nullspace has dimension 4 and so there are four factors. If $p = 8k + 3$ then

$$\mathbf{Q} = \begin{pmatrix} 1 & 0 & 0 & 0 \\ 0 & 0 & 0 & 1 \\ 0 & 0 & -1 & 0 \\ 0 & 1 & 0 & 0 \end{pmatrix},$$

So

$$\mathbf{Q} - \mathbf{I} = \begin{pmatrix} 0 & 0 & 0 & 0 \\ 0 & -1 & 0 & 1 \\ 0 & 0 & -2 & 0 \\ 0 & 1 & 0 & -1 \end{pmatrix},$$

which has rank 2. So there are $4 - 2 = 2$ factors. If $p = 8k + 5$ then

$$\mathbf{Q} = \begin{pmatrix} 1 & 0 & 0 & 0 \\ 0 & -1 & 0 & 0 \\ 0 & 0 & 1 & 0 \\ 0 & 0 & 0 & -1 \end{pmatrix}.$$

So

$$Q - I = \begin{pmatrix} 0 & 0 & 0 & 0 \\ 0 & -2 & 0 & 0 \\ 0 & 0 & 0 & 0 \\ 0 & 0 & 0 & -2 \end{pmatrix}.$$

As before f has two factors. If $p = 8k + 7$ then

$$Q = \begin{pmatrix} 1 & 0 & 0 & 0 \\ 0 & 0 & 0 & -1 \\ 0 & 0 & -1 & 0 \\ 0 & -1 & 0 & 0 \end{pmatrix}.$$

So

$$Q - I = \begin{pmatrix} 0 & 0 & 0 & 0 \\ 0 & -1 & 0 & -1 \\ 0 & 0 & -2 & 0 \\ 0 & -1 & 0 & -1 \end{pmatrix}.$$

Again f has two factors.

5. We have

$$\frac{x^4}{3} - x^3 + \frac{x^2}{2} - \frac{x}{2} + \frac{1}{6} \in \mathbb{Q}[x].$$

Consider

$$f = 2x^4 - 6x^3 + 3x^2 - 3x + 1 \in \mathbb{Z}[x].$$

$f(0) = 1$ which has divisors $\{\pm 1\}$,

$f(1) = -3$ which has divisors $\{\pm 1, \pm 3\}$,

$f(-1) = 15$ which has divisors $\{\pm 1, \pm 3, \pm 5, \pm 15\}$.

Selecting $b_0 = 1$, $b_1 = 3$, $b_2 = 3$ we obtain by Lagrange interpolation,

$$\begin{aligned} g(x) &= \frac{1(x-1)(x+1)}{(0-1)(0+1)} + \frac{3(x-0)(x+1)}{(1-0)(1+1)} + \frac{3(x-0)(x-1)}{(-1-0)(-1-1)} \\ &= 2x^2 + 1 \end{aligned}$$

(which satisfies $g(0) = b_0$, $g(1) = b_1$, $g(-1) = b_2$). Using long division $f|g = x^2 - 3x + 1$ which is irreducible over $\mathbb{Q}$. So the complete factorization is $(x^2/3 + 1/6)(x^2 - 3x + 1)$.

6. *Step* 1. By applying Euclid's algorithm it can be shown that $\gcd(f, f') = 1$, i.e. f is squarefree.

Step 2. Following the procedure used in 4.4

$$Q = \begin{pmatrix} 1 & 0 & 0 & 0 & 0 & 0 & 0 \\ 0 & 0 & 0 & 1 & 0 & 0 & 0 \\ 0 & 0 & 0 & 0 & 0 & 0 & 1 \\ 0 & 2 & 0 & 2 & 1 & 1 & 1 \\ 0 & 0 & 1 & 1 & 1 & 1 & 2 \\ 0 & 2 & 1 & 0 & 0 & 0 & 2 \\ 0 & 1 & 0 & 1 & 1 & 0 & 2 \end{pmatrix}.$$

Step 3. Therefore

$$Q - I = \begin{pmatrix} 0 & 0 & 0 & 0 & 0 & 0 & 0 \\ 0 & 2 & 0 & 1 & 0 & 0 & 0 \\ 0 & 0 & 2 & 0 & 0 & 0 & 1 \\ 0 & 2 & 0 & 1 & 1 & 1 & 1 \\ 0 & 0 & 1 & 1 & 0 & 1 & 2 \\ 0 & 2 & 1 & 0 & 0 & 2 & 2 \\ 0 & 1 & 0 & 1 & 1 & 0 & 1 \end{pmatrix}.$$

which reduces to

$$\begin{pmatrix} 0 & 0 & 0 & 0 & 0 & 0 & 0 \\ 0 & 2 & 0 & 1 & 0 & 0 & 0 \\ 0 & 0 & 0 & 0 & 0 & 0 & 1 \\ 0 & 1 & 0 & 0 & 1 & 1 & 0 \\ 0 & 0 & 0 & 1 & 0 & 1 & 2 \\ 0 & 2 & 0 & 0 & 0 & 2 & 2 \\ 0 & 0 & 0 & 0 & 1 & 0 & 0 \end{pmatrix}.$$

So the nullspace of $Q - I$ is

$$\{(a, b, b + c, c, 2b, c + 2b, 2c)^T \,|\, a, b, c \in \mathbf{F}_3\}.$$

Since $r = 3$, f has three irreducible factors.
Step 4. Let $h^{(2)} = x + x^2 + 2x^4 + 2x^5$ and $h^{(3)} = x^2 + x^3 + x^5 + 2x^6$. Then by applying Euclid's algorithm we find that

$$\gcd(f, h^{(2)} - 2) = -x^2 - 1 = f_1,$$

and

$$\gcd(f, h^{(3)}) = x^3 - x^2 + 1 = f_2.$$

So

$$f_1 f_2 = -x^5 + x^4 - x^3 - 1,$$

and

$$f/f_1 f_2 = -x^2 - 2x + 1.$$

So $f = (-x^2 - 1)(x^3 - x^2 + 1)(-x^2 - 2x + 1)$ is the decomposition into irreducible factors.

7. The cyclotomic cosets (mod 21) over F_2 are:

$$C_0 = \{0\},$$
$$C_1 = \{1, 2, 4, 8, 11, 16\},$$
$$C_3 = \{3, 6, 12\},$$
$$C_5 = \{5, 10, 12, 13, 19, 20\},$$
$$C_7 = \{7, 14\},$$
$$C_9 = \{9, 18, 15\}.$$

It follows that

$$x^{21} - 1 = M^{(0)} M^{(1)} M^{(3)} M^{(5)} M^{(7)} M^{(9)}.$$

Now 6 is the smallest integer such that $21 \mid 2^6 - 1$. We choose $\zeta \in F_{2^6}$ with $\zeta^6 + \zeta^4 + \zeta^2 + \zeta + 1 = 0$ and we find:

$$M^{(0)} = 1 + x,$$
$$M^{(1)} = 1 + x + x^2 + x^4 + x^6,$$
$$M^{(3)} = 1 + x^2 + x^3,$$
$$M^{(5)} = 1 + x^2 + x^4 + x^5 + x^6,$$
$$M^{(7)} = 1 + x + x^2,$$
$$M^{(9)} = 1 + x + x^3.$$

To illustrate the procedure let us calculate $M^{(5)}$ in detail. Let $u = \zeta^5$; then calculating modulo $\zeta^6 + \zeta^4 + \zeta^2 + \zeta + 1$ we have:

$$1 = 1,$$
$$u = \zeta^5,$$
$$u^2 = \zeta^2 + \zeta^3 + \zeta^5,$$
$$u^3 = 1 + \zeta^3,$$
$$u^4 = 1 + \zeta + \zeta^3 + \zeta^5,$$
$$u^5 = \zeta^4,$$
$$u^6 = \zeta + \zeta^2 + \zeta^4.$$

Solving

$$[M_0, M_1, M_2, M_3, M_4, M_5, M_6] \begin{bmatrix} 1 & 0 & 0 & 0 & 0 & 0 \\ 0 & 0 & 0 & 0 & 0 & 1 \\ 0 & 0 & 1 & 1 & 0 & 1 \\ 1 & 0 & 0 & 1 & 0 & 0 \\ 1 & 1 & 0 & 1 & 0 & 1 \\ 0 & 0 & 0 & 0 & 1 & 0 \\ 0 & 1 & 1 & 0 & 1 & 0 \end{bmatrix} = 0.$$

We find

$$M_1 = M_3 = 0,$$

$$M_0 = M_2 = M_4 = M_5 = M_6 = 1,$$

i.e. $M^{(5)} = 1 + x^2 + x^4 + x^5 + x^6$.

The other minimal polynomials are calculated in the same manner or by direct computation using

$$M^{(S)} = \prod_{i \in C_S} (x - \zeta^i).$$

8. There are one linear and two factors of degree 8. Since $2^8 - 1 \equiv 0 \pmod{17}$, $\mathbb{F}_{2^8}$ is the smallest field in which $x^{17} - 1$ factors into linear factors.

Chapter 4, §1

1. Clearly $d(x, y) \geq 0$ and $d(x, y) = 0$ iff $x = y$. It is also clear that $d(x, y) = d(y, x)$. It is obvious that $w(x) + w(y) \geq w(x + y)$. Since $d(x, y) = w(x - y)$, we have $d(x, z) + d(z, y) = w(x - z) + w(z - y) \geq w(x - y) = d(x, y)$.

2. We show: a linear code can detect s or fewer errors if and only if its minimum distance is $\geq s + 1$. Suppose C detects all sets of s or fewer errors. Let $x, y \in C$ and $d(x, y) \leq s$. If x is sent and d errors occur in precisely the positions that x and y differ then y will be received and no error will be detected. This is a contradiction, so $d \geq s + 1$.

   Conversely, assume $d_{min} \geq s + 1$. Then if x is transmitted and fewer than s errors occur then the received message will not be a codeword, so transmission errors will be detected.

   The more general result is derived similarly.

3. Proof. (a) $G' = \begin{pmatrix} G_1 & 0 \\ 0 & G_2 \end{pmatrix}$ is a $2k \times (n_1 + n_2)$ matrix and generates a $(n_1 + n_2, 2k)$ code. The codewords are of the form uG' for $u \in \mathbb{F}_q^{2k}$ where

$$uG' = ((u_1, \ldots, u_k)G_1, 0, 0, \ldots, 0) + (0, \ldots, 0, (u_{k+1}, \ldots, u_{2k})G_2).$$

Clearly $d \geq \min\{d_1, d_2\}$ since

$$w(uG') = w((u_1, \ldots, u_k)G_1) + w((u_{k+1}, \ldots, u_{2k})G_2).$$

Suppose without loss of generality that $d_1 \leq d_2$ and that $w((u_1, \ldots, u_k)G_1) = d_1$. Then

$$(u_1, \ldots, u_k, 0, \ldots, 0)G = ((u_1, \ldots, u_k)G_1, 0, 0, \ldots, 0)$$

is a codeword of weight d_1. Therefore $d = \min\{d_1, d_2\}$.

   (b) $G'' = (G_1, G_2)$ is a $k \times (n_1 + n_2)$ matrix and therefore generates a $(n_1 + n_2, k)$ code; the codewords are of the form

$$uG'' = (uG_1, uG_2) \quad \text{for } u \in \mathbb{F}_q^k.$$

So

$$w(\mathbf{u}G'') = w(\mathbf{u}G_1, \mathbf{u}G_2)$$
$$= w(\mathbf{u}G_1) + w(\mathbf{u}G_2)$$
$$\geq d_1 + d_2.$$

Therefore the minimum distance $d \geq d_1 + d_2$.

4. Since elementary row operations leave the row space of G invariant we may take G to be

$$\begin{pmatrix} 1 & 0 & 0 & 1 & 1 \\ 0 & 1 & 0 & 0 & 1 \\ 0 & 0 & 1 & 0 & 1 \end{pmatrix}.$$

The $2^3 = 8$ codewords are u_1 row $1 + u_2$ row $2 + u_3$ row 3 where $u_i \in \mathbf{F}_2$:

$$\begin{array}{ccc}
1\ 0\ 0\ 1\ 1, & 0\ 1\ 0\ 0\ 1, & 0\ 0\ 1\ 0\ 1, \\
1\ 1\ 0\ 1\ 0, & 1\ 0\ 1\ 1\ 0, & 0\ 1\ 1\ 0\ 0, \\
0\ 0\ 0\ 0\ 0, & 1\ 1\ 1\ 1\ 1, &
\end{array}$$

Writing $G = (I_3 \,|\, -A^T)$ we have

$$H = (A\,|\,I_2) = \begin{pmatrix} 1 & 0 & 0 & 1 & 0 \\ 1 & 1 & 1 & 0 & 1 \end{pmatrix}.$$

The minimum distance $= \min\limits_{\substack{t \neq 0 \\ t \in C}} w(t) = 2$ and the rank of G is 3.

5.
$$G = \begin{pmatrix} 01001 \\ 00101 \\ 10011 \end{pmatrix} \quad \text{then} \quad G^\perp = \begin{pmatrix} 10010 \\ 11101 \end{pmatrix} = H,$$

a $(5, 2)$ binary linear code. There are $2^k = 4$ codewords:

$$10010, \quad 11101, \quad 00000, \quad 01111.$$

Coset leader	Coset			Syndrome $= Hc^T$
00000	10010	11101	01111	$(000)^T$
10000	00010	01101	11111	$(001)^T$
01000	11010	10101	00111	$(100)^T$
00100	10110	11001	01011	$(010)^T$
00001	10011	11100	01110	$(111)^T$
11000	01010	00101	10111	$(101)^T$
01100	11110	10001	00011	$(110)^T$
00110	10100	11011	01001	$(011)^T$

There are $2^5/2^2 = 8$ cosets.

If $y = 01001$ then

$$S(\mathbf{y}) = \mathbf{Hy}^T = \begin{pmatrix} 1 \\ 0 \\ 1 \end{pmatrix} = S(11000).$$

Thus the error is 11000 and $\mathbf{y}$ is decoded as $\bar{\mathbf{x}} = \mathbf{y} - \bar{\mathbf{e}} = 01001 - 11000 = 10001$.

6. (a) For Hamming codes $d_{min} = 3$; so the spheres of radius 1 centered on codewords are disjoint. There are $n + 1$ vectors in each sphere and a total of 2^k spheres giving $(n + 1)2^k$ vectors. However, $n + 1 = 2^r$ and $k = 2^r - 1 - r$. So $(n + 1)2^k = 2^{r+k} = 2^{2^r-1} = 2^n$. So every vector of length n is on one of the spheres and the code is perfect.

   (b) Assume we have a binary repetition code of odd block length n. Then the code can correct $(n - 1)/2$ errors and there are only two codewords. So

$$2\Big(1 + (2 - 1)\binom{n}{1} + (2 - 1)^2\binom{n}{2} + \ldots + (2 - 1)^{(n-1)/2}\binom{n}{(n - 1)/2}\Big)$$

$$= 2 \cdot 2^{n-1} = 2^n.$$

   Hence by Theorem 1.16 these codes are perfect.

7. Let $\mathbf{u}$ be a coset leader of weight $> r$. Then $d(\mathbf{c}, \mathbf{u}) = w(\mathbf{u} - \mathbf{c}) > r$ for all $\mathbf{c} \in C$. So $\mathbf{u} \notin S_r(\mathbf{c})$.

   Conversely, let k be the weight of the coset leader with maximum weight. Then we show that spheres $S_k(\mathbf{c})$, $\mathbf{c} \in C$, cover the space. Assume the contrary, then there is an $\mathbf{x}$ with $d(\mathbf{x}, \mathbf{c}) > k$, but $\mathbf{x}$ is in a coset with coset leader $\mathbf{u}$. Hence $\mathbf{x} = \mathbf{u} + \mathbf{c}$ for suitable $\mathbf{c} \in C$. But $d(\mathbf{x}, \mathbf{c}) = d(\mathbf{u} + \mathbf{c}, \mathbf{c}) = w(\mathbf{u}) \leq k$, a contradiction.

8. Let C_1 be the code with generator matrix $\mathbf{G}_1$; then

$$C_1 = \{a(1, 1, 1, 0) + b(0, 1, 1, 0) + c(0, 0, 1, 1) \,|\, a, b, c \in \mathbb{F}_2\};$$

   So

$$C_1 = \{(0, 0, 0, 0), (1, 0, 1, 1), (1, 1, 0, 1), (1, 0, 0, 0),$$

$$(1, 1, 1, 0), (0, 1, 0, 1), (0, 0, 1, 1), (0, 1, 1, 0)\}.$$

   Also if C_2 is the code with generator matrix $\mathbf{G}_2$ then

$$C_2 = \{a(1, 0, 1, 1) + b(0, 1, 1, 1) + c(1, 0, 0, 1) \,|\, a, b, c \in \mathbb{F}_2\},$$

   So

$$C_2 = \{(0, 0, 0, 0), (1, 1, 1, 0), (1, 0, 0, 1), (0, 1, 1, 1),$$

$$(0, 1, 0, 1), (0, 0, 1, 0), (1, 1, 0, 0), (1, 0, 1, 1)\}.$$

   Comparing C_1 and C_2 we see that if we move each of the entries in the code words of C_1 two to the right we get C_2. Hence C_1 and C_2 are equivalent codes.

9. $\mathbf{H} = \begin{pmatrix} 1 & 1 & 1 & 0 \\ 1 & 2 & 0 & 1 \end{pmatrix}$ is in the standard form $(\mathbf{A}|\mathbf{I}_2)$. $\mathbf{G} = (\mathbf{I}_2, -\mathbf{A}^T) =$

$\begin{pmatrix} 1 & 0 & 2 & 2 \\ 0 & 1 & 2 & 1 \end{pmatrix}$. The codewords are $\mathbf{u}G$ for $\mathbf{u} \in \mathbf{F}_3^2$, i.e. 0 0 0 0, 1 1 1 0, 2 2 2 0, 1 2 0 1, 2 0 1 1, 0 1 2 1, 2 1 0 2, 0 2 1 2, 1 0 2 2. The codewords of $C^\perp$ are $\mathbf{u}H$ for $\mathbf{u} \in \mathbf{F}_3^2$. In this case it turns out that $C = C^\perp$.

10. (i) follows immediately, since if G is the generator of the code C with parity-check matrix H then $C^\perp$ has generator H with parity-check matrix G and so $(C^\perp)^\perp$ has generator matrix G. Therefore $(C^\perp)^\perp = C$.

(ii) Let C_1 have generator matrix G_1; C_2 have generator matrix G_2. Then $\mathbf{x} \in C_1 + C_2$ iff $\mathbf{x} = \mathbf{u}G_1 + \mathbf{v}G_2$ for $\mathbf{u} \in \mathbf{F}_p^n$, $\mathbf{v} \in \mathbf{F}_p^m$. So

$$\mathbf{x} \in (C_1 + C_2)^\perp \quad \text{iff} \quad \mathbf{x} \cdot (\mathbf{u}G_1 + \mathbf{v}G_2) = 0 \quad \forall \, \mathbf{u}, \mathbf{v};$$

$$\text{iff} \quad \mathbf{x} \cdot \mathbf{u}G_1 = \mathbf{x} \cdot \mathbf{v}G_2 = 0 \quad \forall \, \mathbf{u}, \mathbf{v};$$

$$\text{iff} \quad \mathbf{x} \in C_1^\perp \text{ and } x \in C_2^\perp, \text{ i.e. } x \in C_1^\perp \cap C_2^\perp.$$

(iii) C_1 has parity-check matrix

$$H = \begin{pmatrix} 1 \\ 1 \\ \vdots & I_{n-1} \\ 1 \end{pmatrix} + G = (I_1 | 1 \ldots 1)$$
$$= (1 \ldots 1).$$

So $C_1^\perp$ has check matrix $(1, 1, \ldots, 1)$ and is the $(n, n-1, 2)$ parity-check code.

11. (i) To show: every vector in C is orthogonal to itself. Any row of G is orthogonal to itself, since it has even weight. If $\mathbf{x}$ and $\mathbf{y}$ are any two words in C then each is a linear combination of the rows of G and $\mathbf{x} \cdot \mathbf{y} = 0$.

(ii) By (i) C is self-orthogonal. To show that all weights in C are divisible by 4, let $\mathbf{g}_1$ and $\mathbf{g}_2$ be rows in G. Then $w(\mathbf{g}_1 + \mathbf{g}_2) = w(\mathbf{g}_1) + w(\mathbf{g}_2) - 2(\mathbf{g}_1 * \mathbf{g}_2)$. $\mathbf{g}_1 * \mathbf{g}_2$ is even, because $\mathbf{g}_1$ and $\mathbf{g}_2$ are orthogonal, so $4 | w(\mathbf{g}_1 + \mathbf{g}_2)$. Finite induction on the number of rows in the expression of a vector as a linear combination of rows of G completes the proof. Here $\mathbf{g}_1 * \mathbf{g}_2$ is the number of 1's common to both $\mathbf{g}_1$ and $\mathbf{g}_2$.

12. The code is self-orthogonal by Exercise 11(i). Its dual code is a $(7, 4)$ code, which is generated by C and the all-one vector $(11 \ldots 1)$.

13. $S(\mathbf{y}) = H\mathbf{y} = \begin{pmatrix} 0 \\ 1 \\ 1 \end{pmatrix}$ = sixth column. So a single error has (probably) occurred in the sixth position, i.e. the codeword sent was

$$(1 \quad 1 \quad 1 \quad 0 \quad 0 \quad 0 \quad 1).$$

14. Let $H = \begin{pmatrix} 0 & 1 & 1 & 1 & 1 & 0 & 0 \\ 1 & 0 & 1 & 1 & 0 & 1 & 0 \\ 1 & 1 & 0 & 1 & 0 & 0 & 1 \end{pmatrix}$ be parity-check matrix of the $(7, 4)$ Hamming code. Then $G = (I, -A^T)$. Generator and parity-check matrix for the

extended $(8, 4)$ code are

$$\hat{G} = \left(G \begin{array}{c} 1 \\ 1 \\ 1 \\ 0 \end{array} \right), \quad \hat{H} = \left(\begin{array}{ccc} 1 & 1 & \dots 1 \\ H & & 0 \\ & & \vdots \\ & & 0 \end{array} \right).$$

For encoding we have:

$$0110G = 0110011, \quad 1011G = 1011010;$$
$$0110\hat{G} = 01100110, \quad 1011\hat{G} = 10110100.$$

Let $y = 11001101$. Its syndrome is $S(y) = 1000^T$. So an error occurred in the last position of y. The codeword was 11001100, the message was 1100.

15.

$$H = \begin{pmatrix} 0 & 0 & 0 & 0 & 1 & 1 & 1 & 1 & 1 & 1 & 1 & 1 & 1 \\ 0 & 1 & 1 & 1 & 0 & 0 & 0 & 1 & 1 & 1 & 2 & 2 & 2 \\ 1 & 0 & 1 & 2 & 0 & 1 & 2 & 0 & 1 & 2 & 0 & 1 & 2 \end{pmatrix}.$$

Permuting the columns $(1, 13), (2, 12), (5, 11)$ we get

$$H^* = \begin{pmatrix} 1 & 1 & 0 & 0 & 1 & 1 & 1 & 1 & 1 & 1 & 1 & 0 & 0 \\ 2 & 2 & 1 & 1 & 2 & 0 & 0 & 1 & 1 & 1 & 0 & 1 & 0 \\ 2 & 1 & 1 & 2 & 0 & 1 & 2 & 0 & 1 & 2 & 0 & 0 & 1 \end{pmatrix}.$$

So

$$G^* = \left(\begin{array}{c|ccc} & 2 & 1 & 1 \\ & 2 & 1 & 2 \\ & 0 & 2 & 2 \\ & 0 & 2 & 1 \\ I_9 & 2 & 1 & 0 \\ & 2 & 0 & 2 \\ & 2 & 0 & 1 \\ & 2 & 2 & 0 \\ & 2 & 2 & 2 \\ & 2 & 2 & 1 \end{array} \right).$$

Reversing the above permutation we obtain:

$$G = \begin{pmatrix} 1 & 1 & 0 & 0 & 2 & 0 & 0 & 0 & 0 & 0 & 0 & 0 & 1 \\ 2 & 1 & 0 & 0 & 2 & 0 & 0 & 0 & 0 & 0 & 0 & 1 & 0 \\ 2 & 2 & 1 & 0 & 0 & 0 & 0 & 0 & 0 & 0 & 0 & 0 & 0 \\ 1 & 2 & 0 & 1 & 0 & 0 & 0 & 0 & 0 & 0 & 0 & 0 & 0 \\ 0 & 1 & 0 & 0 & 2 & 0 & 0 & 0 & 0 & 0 & 1 & 0 & 0 \\ 2 & 0 & 0 & 0 & 2 & 1 & 0 & 0 & 0 & 0 & 0 & 0 & 0 \\ 1 & 0 & 0 & 0 & 2 & 0 & 1 & 0 & 0 & 0 & 0 & 0 & 0 \\ 0 & 2 & 0 & 0 & 2 & 0 & 0 & 1 & 0 & 0 & 0 & 0 & 0 \\ 2 & 2 & 0 & 0 & 2 & 0 & 0 & 0 & 1 & 0 & 0 & 0 & 0 \\ 1 & 2 & 0 & 0 & 2 & 0 & 0 & 0 & 0 & 1 & 0 & 0 & 0 \end{pmatrix}.$$

Now $S(0\ 1\ 1\ 0\ 1\ 0\ 0\ 0\ 0\ 0\ 1\ 1) = (0\ 0\ 1)^T$ = first column of **H**. So **x** = 2 1 1 0 1 0 0 0 0 0 1 1 which decodes to 1 1 1 0 0 0 0 0 0 0.

16. (ii) $A_0 = A_7 = 1$, $A_3 = A_4 = 7$, all other $A_i = 0$.

17. $A_0 = A_8 = 1$, $A_4 = 14$, all other $A_i = 0$.

18.

$$r = 0: \quad \sum_{i=0}^{n} A_i = 2^k,$$

$$r = 1: \quad \sum_{i=0}^{n} iA_i = 2^{k-1}(n - A_1^\perp),$$

$$r = 2: \quad \sum_{i=0}^{n} i^2 A_i = 2^{k-2}n(n+1) - 2^{k-1}nA_1^\perp + 2^{k-1}A_2^\perp.$$

Chapter 4, §2

1. (a) Let $g = 1 + x + x^3$ over $\mathbf{F}_2$. We find

$$\mathbf{G} = \begin{pmatrix} 1 & 1 & 0 & 1 & 0 & 0 & 0 \\ 0 & 1 & 1 & 0 & 1 & 0 & 0 \\ 0 & 0 & 1 & 1 & 0 & 1 & 0 \\ 0 & 0 & 0 & 1 & 1 & 0 & 1 \end{pmatrix}.$$

The codewords are $\mathbf{u}\mathbf{G}$ for $\mathbf{u} \in \mathbf{F}_2^4$, i.e.

1 1 0 1 0 0 0,	0 1 1 0 1 0 0,	0 0 1 1 0 1 0,	0 0 1 1 0 1,
1 0 1 1 1 0 0,	0 1 0 1 1 1 0,	0 0 1 0 1 1 1,	0 0 0 0 0 0,
1 1 1 0 0 1 0,	0 1 1 1 0 0 1,		
1 1 0 0 1 0 1,	0 1 0 0 0 1 1,		
1 0 0 0 1 1 0,			
1 1 1 1 1 1 1,			
1 0 1 1 0 0 1,			
1 0 0 1 0 1 1.			

(b) Let $f = 1 + x^4 + x^5 + x^6$; we find

$$f = g(x^3 + x^2) + x^2 + 1.$$

Since $g \nmid f$, 1000111 is not a codeword; so an error has occurred.

(c) Let $L = x + x^2 + x^5 + x^6$; we find

$$L = g(x^3 + x^2 + x) + x^2.$$

Since $g \nmid L$, an error has occurred. The last word is a codeword.

2. Consider the $(4, 2, 3)$ Hamming code C over $\mathbf{F}_3$ with parity-check matrix

$$\mathbf{H} = \begin{pmatrix} 1 & 1 & 1 & 0 \\ 1 & 2 & 0 & 1 \end{pmatrix}.$$

Now $(1110) \in C$, but $Z(1110) = (0111) \notin C$ since

$$\mathbf{H}\begin{pmatrix} 0 \\ 1 \\ 1 \\ 1 \end{pmatrix} = \begin{pmatrix} 2 \\ 0 \end{pmatrix} \neq \begin{pmatrix} 0 \\ 0 \end{pmatrix} \pmod 3.$$

So C is not cyclic.

3. (a) $g = 1 + x + x^2$, so

$$h = \frac{x^3 + 1}{g} = x + 1.$$

Hence

$$H = \begin{pmatrix} 0 & 1 & 1 \\ 1 & 1 & 0 \end{pmatrix}.$$

We have

Syndrome	Coset leader
00	000
01	100
10	001
11	010

(b) $g = 1 + x^2 + x^3 + x^4$ and $(x^7 + 1)/g = x^3 + x^2 + 1 = h$. So the binary $(7, 3)$ code with generator g has parity-check matrix

$$H = \begin{pmatrix} 0 & 0 & 0 & 1 & 1 & 0 & 1 \\ 0 & 0 & 1 & 1 & 0 & 1 & 0 \\ 0 & 1 & 1 & 0 & 1 & 0 & 0 \\ 1 & 1 & 0 & 1 & 0 & 0 & 0 \end{pmatrix}.$$

Syndromes	Coset leader
0000	0000000
0001	1000000
0010	1100000
0100	0000010
1000	0000001
0011	0100000
0110	0010000
1100	0000011
0101	1000010
1001	1000001
1010	0000100
0111	1010000
1110	0000110
1011	0100001
1101	0001000
1111	0100011

4. (i) Each element of $J_1 + J_2$ is of the form $a_1 g_1 + a_2 g_2$. Let $d = \gcd(g_1, g_2)$, then $J_1 + J_2 \subseteq (d)$. On the other hand, $d \in J_1 + J_2$, therefore $J_1 + J_2 = (d)$.

   (ii) Each element of $J_1 \cap J_2$ is a multiple of $m = \mathrm{lcm}(g_1, g_2)$. Since $m \in J_1, J_2$, we have $m \in J_1 \cap J_2$ and therefore $J_1 \cap J_2 = (m)$.

   (iii) Let $\quad d_1 = a g_1 g_2 + b(x^n - 1) = \gcd(x^n - 1, g_1 g_2)$. $\quad$ Then $\quad d_1 = a g_1 g_2 \bmod (x^n - 1)$ and $d_1 \in J_1 J_2$. Since d_1 divides elements of the form $c g_1 g_2$, $J_1 J_2 = (d_1)$. In general $J_1 J_2 \neq J_1 \cap J_2$. If $(n, q) = 1$, equality holds, since

$$\gcd(g_1 g_2, x^n - 1) = \mathrm{lcm}(g_1, g_2).$$

5. $$g = x^4 + x^2 + x + 1$$

$$= (x + 1)(x^3 + x^2 + 1).$$

$x^3 + x^2 + 1$ is primitive, so if ζ is a primitive seventh root of unity with $\zeta^3 + \zeta^2 + 1 = 0$ we have

$$f \in C \quad \text{iff} \quad f(1) = f(\zeta) = 0.$$

If $f = \sum_{i=1}^{n} a_i x^i$ then

$$f(1) = \sum_{i=1}^{n} a_i = 0,$$

$$f(\zeta) = \sum_{i=1}^{n} a_i \zeta^i = 0.$$

Consequently the parity-check matrix is

$$\mathbf{H} = \begin{bmatrix} \zeta^6 & \zeta^5 & \zeta^4 & \zeta^3 & \zeta^2 & \zeta & 1 \\ 1 & 1 & 1 & 1 & 1 & 1 & 1 \end{bmatrix}.$$

Taking the binary representation for the ζ^i we obtain a parity-check matrix for the extended Hamming code. From Exercise 2 we know that the code generated by $x^3 + x^2 + 1$ has $d_{\min} = 3$. C is obtained by selecting all codewords of even weight, hence $d_{\min} = 4$.

6. Suppose $(g_i) \subset I \subset F_q[x]/(x^n - 1)$. Then $I = (f)$ for a polynomial f which divides g_i and $x^n - 1$. This would only be possible for $f = 1$ or $f = g_i$. Therefore (g_i) is maximal.

7. $\dim C = n - 1$, where $C = (1 + x)$. Then $C^\perp$ has $\dim 1$ and generator polynomial $(x^n - 1)/(x + 1)$. This is its own reciprocal. $C^\perp = \{00\ldots 0, 11\ldots 1\}$. A binary vector is orthogonal to $11\ldots 1$ if and only if it has even weight. A binary cyclic code $\bar{C} = (g)$ contains only even weight vectors $\Leftrightarrow \bar{C} \subset C \Leftrightarrow 1 + x \mid g$.

8. (i) C is self-orthogonal iff $C \subseteq C^\perp$ iff the generator polynomial of $C^\perp$ divides g.

   (ii) Yes, since $(1 + x)(1 + x + x^3) = 1 + x^2 + x^3 + x^4$ and $x^3 h(1/x)$ divides $1 + x^2 + x^3 + x^4$.

9. Suppose $\mathrm{ord}(g) = n$. If a codeword of weight 1 exists it must be of the form x^i for some $0 \le i \le n - 1$. For some $f \in V_n$ we have

$$x_i = f * g.$$

So

$$f(x)g(x) - x^i = q(x)(x^n - 1) \quad \text{for some } q(x)$$

$$= q(x)f(x)g(x).$$

Hence $g \mid x^i$. Therefore $g = 1$ and $g \mid x^j - 1$ for any j, which contradicts $\text{ord}(g) = n$. If a codeword of weight 2 exists it must be of the form $x^i + x^j$ with $0 \leq i < j \leq n - 1$. But $x^i + x^j = x^i(1 + x^{j-i})$ so if $g \mid x^i + x^j$ then $g \mid x^i$ since $j - i < \text{ord}(g)$. This leads to a contradiction as before. Therefore

$$d_{\min} \geq 3.$$

If $\text{ord}(g) = e < n$ then we can write

$$x^e - 1 = a(x)g(x) \quad \text{for some } a(x).$$

So $x^e - 1$ is a codeword of weight 2.

10. C is not cyclic, however it is equivalent to a cyclic code. Consider the binary $(9, 3)$ code C' with generator polynomial $g' = 1 + x^3 + x^6$. We have generator matrix

$$G' = \begin{pmatrix} 100100100 \\ 010010010 \\ 001001001 \end{pmatrix}$$

and the codewords are of the form

$$(v_0 v_1 v_2 v_0 v_1 v_2 v_0 v_1 v_2), \qquad v_i = 0 \text{ or } 1.$$

Now $Z(v_0 v_1 v_2 v_0 v_1 v_2 v_0 v_1 v_2) \in C'$ so C' is cyclic. The map $\sigma = C \to C'$ defined by

$$\sigma: (v_0 v_0 v_0 \quad v_1 v_1 v_1 \quad v_2 v_2 v_2) \mapsto (v_0 v_1 v_2 \quad v_0 v_1 v_2 \quad v_0 v_1 v_2)$$

is clearly bijective and preserves distance since

$$w(v_0 v_0 v_0 \quad v_1 v_1 v_1 \quad v_2 v_2 v_2) = w(v_0 v_1 v_2 \quad v_0 v_1 v_2 \quad v_0 v_1 v_2)$$

for all codewords. Hence C is equivalent to C' which is cyclic. C has generator matrix

$$G = \begin{pmatrix} 111000000 \\ 000111000 \\ 000000111 \end{pmatrix}.$$

11.
$$k = n - \deg(g)$$

$$= 63 - 5$$

$$= 58.$$

So the codewords are $a * g$, $a \in \mathbb{F}_2^{58}$, i.e.

$$a * g = \left(\sum_{i=0}^{58} a_i x^i \right) * (1 + x^4 + x^5).$$

Now $1 * (1 + x^4 + x^5) = 1 + x^4 + x^5$ has weight 3. By inspection there can be no codewords of lesser weight and therefore the minimum distance is 3.

12. We prove that every cyclic code C (i.e. ideal in V_n) has an idempotent generator. Let $C = (g)$, then $x^n - 1 = gh$. Since $\gcd(g, h) = 1$, there are polynomials a and b such that $1 = ag + bh$. Let $e = ag$, then we verify immediately that $e^2 = e \in C$, and $ec = c$ for any $c \in C$. $e^2 = e$ means that $i \in K$ iff $2i \pmod{n} \in K$, where K denotes the set of powers of x that occur with nonzero coefficients in e. This can only occur if K is a union of cyclotomic cosets mod n.

13. The cyclotomic cosets for $n = 7$ are $C_0 = \{0\}$, $C_1 = \{1, 2, 4\}$, $C_3 = \{3, 6, 5\}$. Then we obtain the following cyclic codes generated by g_i, with dimension d_i and idempotent e_i:

(1) $g_1 = 1 + x + x^2 + x^3 + x^4 + x^5 + x^6$, $d_1 = 1$, $e_1 = g_1$.
(2) $g_2 = 1 + x^2 + x^3 + x^4$, $d_2 = 3$, $e_2 = x^3 g_2$.
(3) $g_3 = 1 + x + x^2 + x^4$, $d_3 = 3$, $e_3 = g_3$.
(4) $g_4 = 1 + x$, $d_4 = 6$, $e_4 = e_2 + e_3 = (x + x^3 + x^5)g_4$.
(5) $g_5 = 1 + x + x^3$, $d_5 = 4$, $e_5 = e_1 + e_2 = xg_5$.
(6) $g_6 = 1 + x^2 + x^3$, $d_6 = 4$, $e_6 = e_1 + e_3 = x^3 g_6$.
(7) is the whole vector space
(8) is the 0 space.

14. From the properties of finite fields we know that if $f(\zeta^i) = 0$ then $f(\zeta^{iq^m}) = 0$, $\forall m$. Therefore L must be a union of cyclotomic cosets and therefore if $i \in L$ then $M^{(i)} \mid f$. Let $g = \mathrm{lcm}\{m^{(i)} \mid i \in L\}$ then $g \mid f$. Also $g(\zeta^i) = 0$ iff $f(\zeta^i) = 0$. So by Theorem 2.7, $C = (g)$.

15.
$$\sum_{j=0}^{n-1} v(\zeta^i)\zeta^{-ij} = \sum_{j=0}^{n-1}\sum_{k=0}^{n-1} v_k \zeta^{kj}\zeta^{-ij}$$

$$= \sum_{j=0}^{n-1}\sum_{k=0}^{n-1} v_k \zeta^{(k-i)j}$$

$$= \sum_{k=0}^{n-1} v_k \sum_{j=0}^{n-1} \zeta^{(k-i)j}$$

$$= nv_i \quad \text{for } i = 0(1)n - 1.$$

This follows because if $k = i$ we have

$$\sum_{j=0}^{n-1} \zeta^{(k-i)j} = n.$$

And if $k \neq i$ then

$$\sum_{j=0}^{n-1} \zeta^{(k-1)j} = \frac{1 - \zeta^m}{1 - \zeta} = 0 \quad (\zeta \text{ is primitive}).$$

So

$$v_i = \frac{1}{n}\sum_{j=0}^{n-1} v(\zeta^i)\zeta^{-ij} \quad \text{for } i = 0(1)n - 1.$$

16. (i) By Theorem 2.23

$$v_s = F_v(\zeta^s).$$

But v is a binary vector. So $v_s = 0$ or 1. Therefore $F_v(\zeta^s) = 0$ or 1. And

$$F_v(\zeta^{2s}) = F_v^2(\zeta^s)$$

$$= 0 \text{ or } 1$$

$$= F_v(\zeta^s).$$

If we write $F_v = \sum_{i=0}^{n-1} F_i x^i$ then

$$F_i = \sum_{i=0}^{n-1} F_v(\zeta^i)\zeta^{ij}$$

$$= \sum_{s \in S} \sum_{j \in S_s} \zeta^{-ij},$$

where S is a subset of the cyclotomic cosets. So

$$F_{2i} = F_i$$

and

$$F_v^2 = F_v.$$

(ii) Let $v = u + w \in V_n$ then

$$F_v = \frac{1}{n} \sum_{j=1}^{n} x^{n-j} \sum_{i=0}^{n-1} v_i \zeta^{ij}$$

$$= \frac{1}{n} \sum_{j=1}^{n} x^{n-j} \sum_{j=0}^{n-1} (u_i + w_i)\zeta^{ij}$$

$$= \frac{1}{n} \sum_{j=1}^{n} x^{n-j} \sum_{j=0}^{n-1} u_i \zeta^{ij} + \frac{1}{n} \sum_{j=1}^{n} x^{n-j} \sum_{i=0}^{n-1} w_i \zeta^{ij}$$

$$= F_u + F_w.$$

(iii) Define $G: V_n \to V_n$ by

$$G_v = G(v) = \sum_{i=0}^{n-1} v(\zeta^i) x^i.$$

Now

$$G(F_v) = \sum_{i=0}^{n-1} F_v(\zeta^i) x^i$$

$$= \sum_{i=0}^{n-1} v_i x^i \quad \text{by Theorem 2.20}$$

$$= v,$$

and

$$F(G_v) = F\left(\sum_{i=0}^{n-1} v(\zeta^i) x^i \right)$$

$$= \sum_{j=1}^{n} x^{n-j} \left(\frac{1}{n} \sum_{i=0}^{n-1} v(\zeta^i) \zeta^{-(n-j)i} \right)$$

$$= \sum_{j=1}^{n} x^{n-j} v_{n-j}$$

$$= \sum_{j=0}^{n-1} v_j x^j.$$

So $F^{-1} = G$ and therefore G is bijective. Now if $v = u * w$ then $v(\zeta^j) = u(\zeta^j)w(\zeta^j)$, therefore

$$F_v = \frac{1}{n} \sum_{j=1}^{n} u(\zeta^j)w(\zeta^j)x^{n-j}$$

$$= F_u F_w.$$

Since F is bijective the converse follows immediately.

(iv) Suppose $F_v = nF_u * F_w$ then

$$\frac{1}{n} v(\zeta^j) = \frac{n}{n^2} u(\zeta^j)w(\zeta^j),$$

i.e.

$$v(\zeta^j) = u(\zeta^j)w(\zeta^j);$$

so

$$v_i = \frac{1}{n} \sum_{j=0}^{n-1} v(\zeta^j)\zeta^{-ij}$$

$$= \frac{1}{n} \sum_{j=0}^{n-1} u(\zeta^i)w(\zeta^j)\zeta^{-ij}$$

$$= u_i w_i,$$

so $v = uw$.

17. 127 for $x^7 + x + 1$, 73 for $x^9 + x + 1$. This can be seen from the table of irreducible polynomials with orders e, see Chapter 3, §3.

Chapter 4, §3

1. (a) $S(y) = (10010110)^T \neq 0$ so at least one error has occurred. Now

$$S_1 = (1001)^T = 1 + \zeta^3 = \zeta^{14},$$
$$S_3 = (0110)^T = \zeta + \zeta^2 = \zeta^5,$$

and

$$\sigma = 1 + (1 + \zeta^3)x + \left(1 + \zeta^6 + \frac{\zeta + \zeta^2}{1 + \zeta^3}\right)x^2 = 1 + (1 + \zeta^3)x + x^2.$$

By direct substitution we find σ has zeros ζ^2 and ζ^{13}, so

$$\frac{1}{X_1} = \zeta^2 = \frac{1}{\zeta^{13}}$$

and

$$\frac{1}{X_2} = \zeta^{13} = \frac{1}{\zeta^2}.$$

So the errors occurred in the third and the fourteenth position.

(b) $$g = M^{(1)}M^{(2)}$$

$$= (x^4 + x + 1)(x^4 + x^3 + x^2 + x + 1)$$

$$= x^8 + x^7 + x^6 + x^4 + 1.$$

So the generator matrix is

$$\mathbf{G} = \begin{pmatrix} 1 & 0 & 0 & 0 & 1 & 0 & 1 & 1 & 1 & 0 & 0 & 0 & 0 & 0 & 0 \\ 0 & 1 & 0 & 0 & 0 & 1 & 0 & 1 & 1 & 1 & 0 & 0 & 0 & 0 & 0 \\ 0 & 0 & 1 & 0 & 0 & 0 & 1 & 0 & 1 & 1 & 1 & 0 & 0 & 0 & 0 \\ 0 & 0 & 0 & 1 & 0 & 0 & 0 & 1 & 0 & 1 & 1 & 1 & 0 & 0 & 0 \\ 0 & 0 & 0 & 0 & 1 & 0 & 0 & 0 & 1 & 0 & 1 & 1 & 1 & 0 & 0 \\ 0 & 0 & 0 & 0 & 0 & 1 & 0 & 0 & 0 & 1 & 0 & 1 & 1 & 1 & 0 \\ 0 & 0 & 0 & 0 & 0 & 0 & 1 & 0 & 0 & 0 & 1 & 0 & 1 & 1 & 1 \end{pmatrix}.$$

2. The following BCH code of length 80 over $\mathbb{F}_3$ of designed distance $d = 11$ will correct (at least) five errors. We need some of the cyclotomic cosets (mod 80).

$$C_1 = \{1, 3, 9, 27\},$$

$$C_2 = \{2, 6, 18, 54\},$$

$$C_4 = \{4, 12, 36, 28\},$$

$$C_5 = \{5, 15, 45, 55\},$$

$$C_7 = \{7, 21, 63, 29\},$$

$$C_{10} = \{10, 30\}.$$

Let

$$g = \mathrm{lcm}\{M^{(i)} \,|\, i = 1(1)10\}$$

$$= M^{(1)}M^{(2)}M^{(4)}M^{(5)}M^{(7)}M^{(10)},$$

where $M^{(i)}$ is the minimal polynomial of ζ^i and where ζ is a primitive eightieth root of unity. g has degree 22.

$$\dim(C) = 80 - 22 = 58.$$

3. $S(v) = (1110011101)^T \neq 0$ so at least one error has occurred. Let

$$S_1 = (11100)^T,$$

and

$$S_3 = (11101)^T.$$

Let

$$\sigma = 1 + S_1 z + \left(\frac{S_3}{S_1} + S_1^2\right)z^2 \cdot$$

$$= 1 + \zeta^{11}z + \zeta^6 z^2.$$

By trial and error σ has zeros ζ^{30} and ζ^{26}. So

$$\frac{1}{X_1} = \zeta^{30} = \frac{1}{\zeta},$$

$$\frac{1}{X_2} = \zeta^{26} = \frac{1}{\zeta^5}.$$

So the errors have occurred in the second and sixth positions.

4. In order to correct three errors we take designed distance $d = 7$. A suitable generator polynomial is $g = \mathrm{lcm}\{M^{(i)} | i = 1(1)6\}$. We have cyclotomic cosets (mod 15):

$$C_0 = \{0\},$$
$$C_1 = \{1, 2, 4, 8\},$$
$$C_3 = \{3, 6, 12, 9\},$$
$$C_5 = \{5, 10\},$$
$$C_7 = \{7, 14, 13, 11\}.$$

So $M^{(1)} = M^{(2)} = M^{(4)}$ and $M^{(3)} = M^{(6)}$. Therefore

$$g = M^{(1)} M^{(3)} M^{(5)}$$
$$= (x^4 + x^3 + 1)(x^4 + x^3 + x^2 + x + 1)(x^2 + x + 1)$$
$$= x^{10} + x^9 + x^8 + x^6 + x^5 + x^2 + 1.$$

5. Consider the following binary BCH code of length 31. Let ζ be a primitive thirty-first root of unity (and $\zeta^5 + \zeta^2 + 1 = 0$) and define $g(x) = \mathrm{lcm}\{M^{(1)} M^{(2)} M^{(3)} \ldots M^{(8)}\}$ where $M^{(i)}$ is the minimal polynomial of ζ^i. However,

$$C_1 = \{1, 2, 4, 8, 16\},$$
$$C_3 = \{3, 6, 12, 24, 27\},$$
$$C_5 = \{5, 10, 20, 9, 18\},$$
$$C_7 = \{7, 14, 28, 25, 19\}.$$

Therefore $M^{(1)} = M^{(2)} = M^{(4)} = M^{(8)}$, $M^{(3)} = M^{(6)}$. So

$$g(x) = M^{(1)} M^{(3)} M^{(5)} M^{(7)}$$
$$= (x^5 + x^2 + 1)(x^5 + x^4 + x^3 + x^2 + 1)$$
$$\times (x^5 + x^4 + x^2 + x + 1)(x^5 + x^3 + x^2 + x + 1).$$

Then g generates a binary BCH code of length 31 with designed distance $= 9$. (It is easily seen that $d_{\min} \geq 11$.)

6.
$$g = (x^2 + x + 1)(x^4 + x^3 + x^2 + x + 1)(x^4 + x + 1).$$

Let

$$\mathbf{H} = \begin{pmatrix} 1 & \zeta^1 & \zeta^3 & \cdots & \zeta^{14} \\ 1 & \zeta^3 & \zeta^9 & \cdots & \zeta^{42} \\ 1 & \zeta^5 & \zeta^{15} & \cdots & \zeta^{70} \end{pmatrix}.$$

Now

$$S(\mathbf{v}) = \mathbf{H}\mathbf{v}^T = \begin{pmatrix} \zeta^{11} + \zeta^9 + \zeta^8 + \zeta^5 + \zeta + 1 \\ \zeta^{33} + \zeta^{27} + \zeta^{24} + \zeta^{15} + \zeta^3 + 1 \\ \zeta^{95} \quad \zeta^{45} \quad \zeta^{40} \quad \zeta^{25} \quad \zeta^5 + 1 \end{pmatrix} = \begin{pmatrix} S_1 \\ S_3 \\ S_5 \end{pmatrix}.$$

Using the tables in the text,

$$S(\mathbf{y}) = \begin{pmatrix} S_1 \\ S_3 \\ S_5 \end{pmatrix} = \begin{pmatrix} \zeta^2 \\ 1 + \zeta^2 \\ 1 \end{pmatrix} \neq \mathbf{0}$$

so at least one error has occurred. Suppose two or three errors have occurred and consider the function $\sigma = \sum_{i=0}^{3} \sigma_i z^i$ where $\sigma_0 = 1$ and

$$\begin{pmatrix} 1 & 0 & 0 \\ S_2 & S_1 & 1 \\ S_4 & S_3 & S_2 \end{pmatrix} \begin{pmatrix} \sigma_1 \\ \sigma_2 \\ \sigma_3 \end{pmatrix} = \begin{pmatrix} S_1 \\ S_3 \\ S_5 \end{pmatrix},$$

which has solution

$$\sigma_1 = S_1 = \zeta^2,$$

$$\sigma_2 = \zeta^{12},$$

$$\sigma_3 = 0.$$

We find by trial and error that σ has roots ζ and ζ^2. So

$$\frac{1}{X_1} = \zeta = \frac{1}{\zeta^{14}},$$

$$\frac{1}{X_2} = \zeta^2 = \frac{1}{\zeta^{13}}.$$

So the errors have occurred in the fourteenth and fifteenth positions. Therefore the code word sent was 110001001101011. Decoding is achieved using the division algorithm

$$\frac{x^{14} + x^{13} + x^{11} + x^9 + x^8 + x^5 + x + 1}{x^{10} + x^8 + x^5 + x^4 + x^2 + x + 1} = x^4 + x^3 + x^2 + 1.$$

So the message word sent was 11101.

7. We know $d_{\min} \geq d$. Let ζ be a primitive nth root of unity then $\zeta^{id} \neq 1$ for $i < a$. But

$$x^n - 1 = x^{ad} - 1$$

$$= (x^a - 1)(1 + x^a + x^{2a} + \ldots + \zeta^{(a-1)d}).$$

Since $\zeta, \zeta^2, \ldots, \zeta^{d-1}$ are not zeros of $x^a - 1$ they must be zeros of $x^{(a-1)d} + \ldots + x^a + 1$. Hence $1 + x^a + \ldots + x^{(a-1)d}$ is a codeword of weight d in the code.

8. We know from Chapter 3 that if ζ is a primitive fifteenth root of unity satisfying $\zeta^4 + \zeta + 1 = 0$ then

$$x^{15} - 1 = M^{(0)} M^{(1)} M^{(3)} M^{(5)} M^{(7)}$$

$$= (x + 1)(x^4 + x + 1)(x^4 + x^3 + x^2 + x + 1)(x^2 + x + 1)(x^4 + x^3 + 1).$$

If we let $g(x) = 1$ this gives rise to a $(n, n, 1)$ code. Taking $g = M^{(1)}$ then g has roots $\zeta, \zeta^2, \zeta^4, \zeta^8$.

So the resultant code has designed distance $\delta = 3$ and $k = n - \deg g = 11$. $1 * (1 + x + x^4) = 1 + x + x^4$ is a codeword of weight 3 so $d_{\min} = \delta = 3$. If $g = M^{(1)} M^{(3)}$ then g has roots $\zeta, \zeta^2, \zeta^3, \zeta^4, \zeta^6, \zeta^8, \zeta^9, \zeta^{12}$ so $\delta = 5$, $k = 7$ and $d_{\min} = 5$ since $x^8 + x^7 + x^6 + x^4 + 1$ is a codeword. Hence g generates a $(15, 7, 5)$ code. If $g = M^{(1)} M^{(3)} M^{(5)} = x^{10} + x^8 + x^5 + x^4 + x^2 + x + 1$ then g has roots $\zeta, \zeta^2, \zeta^3, \zeta^4, \zeta^5, \zeta^6, \zeta^8, \zeta^9, \zeta^{10}, \zeta^{12}$, so $\delta = 7$ and $k = 15 - 10 = 5$. $x^{10} + x^8 + x^5 + x^4 + x^2 + x + 1$ is a codeword with weight 7 so g generates a $(15, 5, 7)$ code.
Finally, if

$$g = M^{(1)} M^{(3)} M^{(5)} M^{(7)}$$

$$= x^{14} + x^{13} + \ldots + 1,$$

then g has roots $\zeta, \zeta^2, \ldots, \zeta^{14}$ so $\delta = 9, 10, 11, \ldots, 15$, $k = 1$ and clearly $d_{\min} = 15$. So g generates a $(15, 1, 15)$ code.

9. Using $\zeta^2 + \zeta + 2 = 0$ we obtain the following table for $\mathbb{F}_{3^2}$.

$$0,$$
$$\zeta^0 = 1,$$
$$\zeta^1 = \zeta,$$
$$\zeta^2 = 2\zeta + 1,$$
$$\zeta^3 = 2\zeta + 2,$$
$$\zeta^4 = 2,$$
$$\zeta^5 = 2\zeta,$$
$$\zeta^6 = 2 + \zeta,$$
$$\zeta^7 = 1 + \zeta.$$

The cosets mod 8 are

$$C_0 = \{0\},$$
$$C_1 = \{1, 3\},$$
$$C_2 = \{2, 6\},$$
$$C_4 = \{4\},$$
$$C_5 = \{5, 7\}.$$

In order to construct a double error correcting code we take $\delta = 5$ and

$$g = M^{(1)}M^{(2)}M^{(3)}M^{(4)}$$
$$= M^{(1)}M^{(2)}M^{(4)},$$
$$M^{(1)} = x^2 + x + 2,$$
$$M^{(2)} = (x - \zeta^2)(x - \zeta^6) = x^2 + 1,$$
$$M^{(4)} = (x - \zeta^4) = x - 2.$$

So $g = x^5 - x^4 + x^3 + x^2 - 1$. Notice that $1 * g$ is a codeword of weight 5 so $d_{min} = \delta = 5$. $k = 8 - 5 = 3$.

$$H = \begin{pmatrix} 1 & \zeta & \zeta^2 & \zeta^3 & \zeta^4 & \zeta^5 & \zeta^6 & \zeta^7 \\ 1 & \zeta^2 & \zeta^4 & \zeta^6 & \zeta^8 & \zeta^{10} & \zeta^{12} & \zeta^{14} \\ 1 & \zeta^4 & \zeta^8 & \zeta^{12} & \zeta^{16} & \zeta^{20} & \zeta^{24} & \zeta^{28} \end{pmatrix}$$

and

$$S(\mathbf{v}) = H\mathbf{v}^T = \begin{pmatrix} \zeta^7 + \zeta^4 - \zeta - 1 \\ \zeta^{14} + \zeta^8 - \zeta^2 - 1 \\ \zeta^{28} + \zeta^{16} - \zeta^4 - 1 \end{pmatrix}$$

$$= \begin{pmatrix} -1 \\ 1 - \zeta \\ 0 \end{pmatrix} \quad \text{by using the table above.}$$

$$= \begin{pmatrix} S_1 \\ S_2 \\ S_4 \end{pmatrix}.$$

Assuming two errors have occurred we write

$$\sigma = \sum_{k=0}^{2} \sigma_k z^k \quad \text{with } \sigma_0 = 1$$
$$= 1 + \sigma_1 z + \sigma_2 z^2.$$

We have

$$\begin{pmatrix} S_3 & S_2 & S_1 \\ S_4 & S_3 & S_2 \end{pmatrix} \begin{pmatrix} 1 \\ \sigma_1 \\ \sigma_2 \end{pmatrix} = \begin{pmatrix} 0 \\ 0 \\ 0 \end{pmatrix}$$

and $S_3 = (S_1)^3 = -1$. So,

$$\begin{pmatrix} -1 & 1 - \zeta & -1 \\ 0 & -1 & 1 - \zeta \end{pmatrix} \begin{pmatrix} 1 \\ \sigma_1 \\ \sigma_2 \end{pmatrix} = \begin{pmatrix} 0 \\ 0 \\ 0 \end{pmatrix}.$$

Solving gives $\sigma_1 = 1 - \zeta$ and $\sigma_2 = 1$, so

$$\sigma = 1 + (1 - \zeta)z + z^2.$$

Now by trial and error we find that σ has zeros ζ^3 and ζ^5 so

$$\frac{1}{X_1} = \zeta^3 \Rightarrow X_1 = \zeta^5,$$

$$\frac{1}{X_2} = \zeta^5 \Rightarrow X_2 = \zeta^3.$$

So errors have occurred in the fourth and sixth places. To determine the magnitude of the errors we have, in the notation of 3.13

$$-1 = S_1 = e_3\zeta^3 + e_5\zeta^5,$$

$$1 - \zeta = S_2 = e_3\zeta^6 + e_5\zeta^{10},$$

$$-1 = S_3 = e_3\zeta^9 + e_5\zeta^{15},$$

$$0 = S_4 = e_3\zeta^{12} + e_5\zeta^{20}.$$

Therefore $e_3 = 1$ and $e_5 = -1$. Therefore the word sent was $\mathbf{v} - \mathbf{e} = 10110$–$10$–$1$–$1$, i.e. $x^7 + x^6 + x^4 - x^3 - x - 1$. Upon division by g we find this equals $g(x^2 + x + 1)$. So the original message was 111.

10. (i) No. (ii) Yes.

11. $\alpha = 2$ is a primitive element of F_5, so $g = (x - \alpha)(x - \alpha^2) = x^2 + 4x + 3$.

12. If a codeword $c = c_0 \ldots c_{n-1}$ has weight d, then the minimum weight is increased to $d + 1$ if $c(1) = -c_n = \sum_{i=0}^{n-1} c_i \neq 0$. Now $c = ag$, for suitable a, so $c(1) = a(1)g(1)$. We verify $a(1) \neq 0$, $g(1) \neq 0$ and use the BCH bound to finish the proof.

13. Let ζ be a primitive element in F_{2^m} and let C denote the cyclic code with roots $\zeta, \zeta^2, \ldots \zeta^{2t-1}$. If $M^{(i)}$ is the minimal polynomial of ζ^i then $C = (g)$, with $g = \text{lcm}\{M^{(i)} | i = 1, \ldots, 2t - 1\}$. C is binary so every even power of ζ is a root of a minimal polynomial of some odd power of ζ, e.g. if $M^{(1)}$ is the minimal polynomial of ζ, then $\zeta^2, \zeta^4, \ldots$ are all roots of $M^{(1)}$. So we need consider only odd i. Here $\deg M^{(i)} \leq m$, there are t minimal polynomials for odd i and therefore $\dim C \geq n - mt$.

14. Let $\{0, 1, \ldots, n - 1\}$ be the elements of F_n, if k is a quadratic nonresidue (mod n), then $i \mapsto ik$ (mod n) is a permutation of F_n, such that $kU_0 = U_1$, $kU_1 = U_0$. Thus

$$g_1 = \prod_{j \in U_1} (x - \zeta^j) = \prod_{j \in U_0} (x - \zeta^{kj}).$$

If σ denotes the permutation induced on the components of $\mathbf{c} = (c_0, \ldots, c_{n-1}) \in V^n$, then

$$\sum_{i=0}^{n-1} c_i \zeta^{ij} = \sum_{i=0}^{n-1} c_{\sigma(i)}(\zeta^{kj})^i.$$

Thus ζ^j is a root of $c = \sum c_i x^i$ for all $j \in U_0$ iff ζ^{kj} is a root of $c_\sigma = \sum c_{\sigma(i)} x^i$ for all $kj \in U_1$, that is

$$c \in (g_0) \Leftrightarrow c_\sigma \in (g_1).$$

15. Let $a \in \mathbb{F}_3^6$, let G be the generator matrix and $c = aG$ an element of the $(11, 6, 5)$ Golay code. Then $w(c) = c \cdot c^T$. In $\mathbb{F}_3$ we have

$$aGG^Ta^T = -\left(\sum_{i=2}^{6} a_i\right)^2 \not\equiv 1 \pmod{3},$$

since -1 is a nonsquare in $\mathbb{F}_3$.

16. The dimension of the code is 6, since the generator matrix G is systematic. The rows of G have weights ≥ 5 and any linear combination of two rows has weight ≥ 4, because of Exercise 15 then also weight ≥ 5. Similarly, for combinations of three or four rows of G. Thus the minimum distance is 5, that is, spheres of radius 2 about any two codewords are disjoint. But these spheres cover $\mathbb{F}_3^{11}$, since

$$3^6\left(1 + 2\binom{11}{1} + 2^2\binom{11}{2}\right) = 3^{11}.$$

Therefore the code is perfect.

Chapter 5, §1

1.
$$H_{12} = \begin{pmatrix}
1 & 1 & 1 & 1 & 1 & 1 & 1 & 1 & 1 & 1 & 1 & 1 \\
1 & -1 & -1 & 1 & -1 & -1 & -1 & 1 & 1 & 1 & -1 & 1 \\
1 & -1 & 1 & -1 & -1 & -1 & 1 & 1 & 1 & -1 & 1 & -1 \\
1 & 1 & -1 & -1 & -1 & 1 & 1 & 1 & -1 & 1 & -1 & -1 \\
1 & -1 & -1 & -1 & 1 & 1 & 1 & -1 & 1 & -1 & -1 & 1 \\
1 & -1 & -1 & 1 & 1 & 1 & -1 & 1 & -1 & -1 & 1 & -1 \\
1 & -1 & 1 & 1 & 1 & -1 & 1 & -1 & -1 & 1 & -1 & -1 \\
1 & 1 & 1 & 1 & -1 & 1 & -1 & -1 & 1 & -1 & -1 & -1 \\
1 & 1 & 1 & -1 & 1 & -1 & -1 & 1 & -1 & -1 & -1 & 1 \\
1 & 1 & -1 & 1 & -1 & -1 & 1 & -1 & -1 & -1 & 1 & 1 \\
1 & -1 & 1 & -1 & -1 & 1 & -1 & -1 & -1 & 1 & 1 & 1
\end{pmatrix}.$$

Other Hadamard matrices of order 12 are obtained from H_{12} by permuting rows and columns and multiplying rows and columns by -1.

2. Let $H_m = (h_{ij})$ and define the $mn \times mn$ matrix H_{mn} by

$$H_{mn} = (h_{ij}H_n).$$

The entries of H_{mn} are all $+1$ or -1. The entries of $H_{mn}H_{mn}^T$ are $n \times n$ submatrices with submatrix in the (i, j) position being

$$\sum_{k=1}^{m} h_{ik}h_{jk}H_nH_n^T = \sum_{k=1}^{m} h_{ij}h_{jk}nI_n$$

$$= \begin{cases} mnI_n & \text{if } i = j, \\ 0 & \text{if } i \neq j. \end{cases}$$

So H_{mn} is a Hadmard matrix of order mn.

3. (i) $HH^T = hI_h$. Therefore $h^h = \det(HH^T) = (\det(H))^2$.
   The proofs of (ii) and (iii) are straightforward.

4. An example of a 3×5 Latin rectangle is given

$$
\begin{array}{ccccc}
3 & 1 & 2 & 5 & 4 \\
1 & 3 & 5 & 4 & 2 \\
4 & 5 & 3 & 2 & 1
\end{array}
$$

This can be extended by adding the rows 5 2 4 1 3 and 2 4 1 3 5. In general, an extension is always possible.

5. $L^{(k)} = (a_{ij}^{(k)})$ where $a_{ij} \equiv i + jk \pmod 9$, $i, j = 1, 2, \ldots, 9$. For $k \in \{1, 2, \ldots, 8\}$

$$a_{ij}^{(k)} = a_{lj}^{(k)} \Leftrightarrow i + jk \equiv l + jk \pmod 9$$

$$\Leftrightarrow i \equiv l \pmod 9$$

$$\Leftrightarrow i = l.$$

Each element appears in each column of $L^{(k)}$ exactly once.

$$a_{ij}^{(k)} = a_i^{(n)} \Leftrightarrow i + jk \equiv i + lk \pmod 9$$

$$\Leftrightarrow (j - l)k \equiv 0 \pmod 9$$

$$\Leftrightarrow j = l$$

or $k = 3$ and $j - l = 3$ or 6, or $k = 6$ and $j - l = 3$ or 6.

In $L^{(3)}$ and $L^{(6)}$ an element can appear more than once in the same row. Thus $L^{(3)}$ and $L^{(6)}$ are not Latin squares, whilst the remaining $L^{(k)}$, $k = 1, 2, 4, 5, 7, 8$ are Latin squares.

Suppose $(a_{ij}^{(2)}, a_{ij}^{(5)}) = (a_{mn}^{(2)}, a_{mn}^{(5)})$, then

$$(i + 2j, i + 5j) \pmod 9 = (m + 2n, m + 5n) \pmod 9$$

$$i - m \equiv 2(n - j) \pmod 9 \quad \text{and} \quad i - m \equiv 5(n - j) \pmod 9.$$

$$3(n - j) \equiv 0 \pmod 9, \qquad n - j = 0 \text{ or } 3 \text{ or } 6.$$

Say $n - j = 3$, then $i - m = 6$ so let $(i, j, m, n) = (7, 2, 1, 5)$. Then

$$(a_{72}^{(2)}, a_{72}^{(5)}) \equiv (7 + 2 \cdot 2, 7 + 5 \cdot 2) \pmod 9$$

$$\equiv (2, 8) \pmod 9$$

$$= (1 + 2 \cdot 5, 1 + 5 \cdot 5) \pmod 9$$

$$= (a_{15}^{(2)}, a_{15}^{(5)}).$$

Therefore $L^{(2)}$ and $L^{(5)}$ are not orthogonal.

6. There is one normalized Latin square of order $n = 1, 2, 3$. There are four of order 4 and 56 of order 5. They are

$n = 1$: $= 1$.

$n = 2$: L_2:
$$
\begin{array}{cc}
1 & 2 \\
2 & 1
\end{array}
$$

$n = 3$: L_3:
$$
\begin{array}{ccc}
1 & 2 & 3 \\
2 & 3 & 1 \\
3 & 1 & 2
\end{array}
$$

$n = 4$:

```
1 2 3 4    1 2 3 4
2 4 1 3    2 1 4 3
3 1 4 2    3 4 1 2
4 3 2 1    4 3 2 1

1 2 3 4    1 2 3 4
2 1 4 3    2 3 4 1
3 4 2 1    3 4 1 2
4 3 1 2    4 1 2 3
```

$n = 5$:   We have obtained the following normalized Latin Squares and those which can be obtained from these by relabeling and permuting the rows and columns so that the Latin square is normalized.

```
1 2 3 4 5    1 2 3 4 5    1 2 3 4 5
2 3 4 5 1    2 1 4 5 3    2 1 4 5 3
3 4 5 1 2    3 5 1 2 4    3 4 5 2 1
4 5 1 2 3    4 3 5 1 2    4 5 1 3 2
5 1 2 3 4    5 4 2 3 1    5 3 2 1 4

1 2 3 4 5    1 2 3 4 5    1 2 3 4 5
2 3 4 5 1    2 1 5 3 4    2 1 4 5 3
3 5 2 1 4    3 4 2 5 1    3 4 5 1 2
4 1 5 3 2    4 5 1 2 3    4 5 2 3 1
5 4 1 2 3    5 3 4 1 2    5 3 1 2 4
```

7. Take a set of r orthogonal Latin squares of order n with first row $1, 2, \ldots, n$, i appears in the ith column of each square. i cannot appear in the ith column of any square (except the first row entry), therefore $r \le n - 1$.

8. There are 13 points and 13 lines:

points $(0,0,1), (0,1,1), (0,2,1), (1,0,1), (1,1,1), (1,2,1), (2,0,1), (2,1,1),$
$(2,2,1), (1,0,0), (0,1,0), (1,1,0), (2,1,0),$

lines $x = 0, y = 0, L_\infty, x + 2y = 0, x + 2z = 0, 2y + z = 0, x + z = 0, x + y = 0, y + z = 0, x + y + z = 0, x + y + 2z = 0, x + 2y + z = 0, x + 2y + 2z = 0.$

9. For $d \not\equiv 0 \pmod{13}$ there exists (d_1, d_2) such that $d_1 - d_2 \equiv d \pmod{13}$.

d	(d_1, d_2)	d	(d_1, d_2)
1	$(5,4)$	7	$(7,0)$
2	$(7,5)$	8	$(0,5)$
3	$(7,4)$	9	$(0,4)$
4	$(4,0)$	10	$(4,7)$
5	$(5,0)$	11	$(5,7)$
6	$(0,7)$	12	$(4,5)$

This is a $(13, 4, 1)$ difference set. The blocks $B_t = \{t, 4 + t, 5 + t, 7 + t\}$, $t = 0, 1, \ldots, 12$ satisfy the conditions of points and lines in PG$(2, 3)$.

10. For $d \not\equiv 0$ (mod 19) there are exactly four pairs $(d_1, d_2) \in D \times D$ such that $d_1 - d_2 \equiv d$ (mod 19). For example

d	(d_1, d_2)
1	(1, 0), (2, 1) (3, 2) (13, 12)
2	(2, 0) (3, 1) (5, 3) (7, 5)
3	(0, 16) (5, 2) (3, 0) (16, 13)
4	(1, 16) (5, 1) (7, 3) (16, 12)
5	(2, 16) (7, 2) (12, 7) (5,0)
6	(7, 1) (13, 7) (0, 13) (3, 16)
7	(7, 0) (12, 5) (0, 12) (1, 13)
8	(13,5) (1, 12) (2, 13) (5, 16)
9	(12, 3) (16, 7) (2, 12) (3, 13)
10	(13, 3) (12, 2) (7, 16) (3, 12)
11	(5, 13) (12, 1) (13, 2) (16, 5)
12	same as 7 only multiply each by −1
13	same as 6 only multiply each by −1
14	same as 5 only multiply each by −1
15	same as 4 only multiply each by −1
16	same as 3 only multiply each by −1
17	same as 2 only multiply each by −1
18	same as 1 only multiply each by −1

Here $v = 19$, $k = 9$ and $\lambda = 4$.

11. By direct computation D can be seen to be a difference set with $\lambda = 8$. Also $k = 13$ and $v = 40$, i.e. D is a $(40, 13, 18)$ difference set.

12. Let $P = \{1, 2, \ldots, 9\}$. Each element of P occurs in exactly four blocks, so we have a $(9, 12, 4, 3, 1)$ configuration or a $2\text{-}(9, 3, 1)$ design.

13. The result is true for $i = t$ by definition of a t-design, since any t points are contained in eactly λ blocks. Assume that λ_{i+1} is independent of the choice of $P_1, \ldots, P_{i+1}$. For each block containing $P_1, \ldots, P_i$ and for each point $Q \neq P_j$, $j = 1(1)i$ define

$$\chi(Q, B) = \begin{cases} 1 & \text{if } Q \in B, \\ 0 & \text{if } Q \notin B. \end{cases}$$

Then from the induction hypothesis

$$\sum_Q \sum_B \chi(Q, B) = \lambda_{i+1}(v - i) = \sum_Q \sum_B \chi(Q, B) = \lambda_i(k - i).$$

So λ_i is independent of the choice of $P_1, \ldots, P_i$ and the result follows since $\lambda_i = \lambda_{i+1}[(v - i)/(k - i)]$.

14.

r	0	1	2	3	4	5	6
	0	6	5	4	3	2	1
s	1	0	6	5	4	3	2
	3	2	1	0	6	5	4

is a table of possible values of s, given r, such that point P_r and line L_s (or L_r and P_s) are incident. The diagonal points of the quadrangle $P_2 P_4 P_5 P_6$ are the three points on L_0.

15. Let $\mathbf{M} = (m_{ij}) = (na_{ij} + b_{ij} + 1)$. Then

$$\sum_{i=1}^{n} m_{ij} = n \sum a_{ij} + \sum b_{ij} + n$$

$$= \frac{n^2(n-1)}{2} + \frac{n(n-1)}{2} + n.$$

Similarly for columns and diagonals. Let

$$A = \begin{matrix} 0 & 1 & 2 & 3 \\ 3 & 2 & 1 & 0 \\ 1 & 0 & 3 & 2 \\ 2 & 3 & 0 & 1 \end{matrix} \qquad B = \begin{matrix} 0 & 1 & 2 & 3 \\ 2 & 3 & 0 & 1 \\ 3 & 2 & 1 & 0 \\ 1 & 0 & 3 & 2 \end{matrix}$$

Then

$$M = \begin{matrix} 1 & 6 & 11 & 16 \\ 15 & 12 & 5 & 2 \\ 8 & 3 & 14 & 9 \\ 10 & 13 & 4 & 7 \end{matrix}.$$

16.
$$\begin{matrix} 17 & 24 & 1 & 8 & 15 \\ 23 & 5 & 7 & 14 & 16 \\ 4 & 6 & 13 & 20 & 22 \\ 10 & 12 & 19 & 21 & 3 \\ 11 & 18 & 25 & 2 & 9 \end{matrix}$$

17. Let $\mathbf{H}$ be a normalized Hadamard matrix of order $4t$, $t \geq 2$, then number the rows and columns of $\mathbf{H}$ by $0, 1, 2, \ldots, 4t - 1$ so that the row and column number 0 consists entirely of zeros. With the remaining rows associate varieties a_i, $i = 1, \ldots, 4t - 1$ and remaining columns with blocks B_j, $j = 1, \ldots, 4t - 1$. We say that the variety a_i is incident with the block B_j if the entry b_{ij} in $\mathbf{H}$ is $+1$ and a_i is not incident with B_j if $b_{ij} = -1$. Thus we have an incidence system with $4t - 1$ varieties and blocks. It can be shown because of orthogonality that any row in $\mathbf{H}$, excluding the zero row, has $2t$ entries of $+1$ and in any two rows in $\mathbf{H}$, again excluding the zero row, there are t places where both rows have $+1$'s. Hence if we exclude the zero column and row we have a symmetric block design with $r = b = 4t - 1$, $r = k = 2t - 1$ and $\lambda = t - 1$.

18. The first row has two cells filled and one empty. Interchange rows and permute symbols so that $a_{11} = 0\ \infty$, $a_{12} = 1\ 2$ and a_{13} = blank. The remaining doublet in the first column must be 12. Interchange rows so that $a_{21} = 1\ 2$ so

∞ 0	1 2	–
1 2		
–		

Now row and column 3 cannot contain any more blanks so a_{22} = blank. Now to fulfil the conditions this means that $a_{23} = 0\ \infty$ and $a_{32} = 0\ \infty$,

0 ∞	1 2	–
1 2	–	1 2
–	0 ∞	

Finally to make a room square we have to have $a_{33} = 0\ \infty$ to make column 3 correct $\Rightarrow$ row 3 incorrect; and $a_{33} = 1\ 2$ to make row 3 correct $\Rightarrow$ column 3 incorrect.

19. (i) Denote the nine girls by the integers $1, 2, \ldots, 9$. Place the 9 at the center of a circle and the integers $1, 2, \ldots, 8$ equally spaced around the circumference. Consider the diameter 1 9 5 of the circle and the two triangles 2 3 8 and 6 7 4 as shown in the figure below. These three triplets give a suitable arrangement of the nine girls into three triplets for one day. Now consider the possible rotations of the configuration through angles of 45°, 90° and 135° counterclockwise about the center but leaving the numbers fixed. The four distinct positions including the original one yield suitable arrangements, namely

```
1 9 5    2 9 6    3 9 7    4 9 8
2 3 8    3 4 1    4 5 2    5 6 3
6 4 7    7 5 8    8 6 1    1 7 2
```

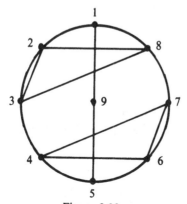

Figure 8.38

(ii) There are 845 solutions, one of the solutions is as follows:

Sunday			Monday			Tuesday			Wednesday			Thursday			Friday			Saturday		
1	2	3	1	4	5	1	6	7	1	8	9	1	10	11	1	12	13	1	14	15
4	8	12	2	8	10	2	9	11	2	12	14	2	13	15	2	4	6	2	5	7
5	10	15	3	13	14	3	12	15	3	5	6	3	4	7	3	9	10	3	8	11
6	11	13	6	9	15	4	10	14	4	11	15	5	9	12	5	11	14	4	9	13
7	9	14	7	11	12	5	8	13	7	10	13	6	8	14	7	8	15	6	10	12

20. From $br = tk$ we have $r = tk/b$ is the number of teams each boy plays in. Since $r(k - 1) = \lambda(b - 1)$, the number of teams on which two boys play together is

$$\lambda = \frac{r(k - 1)}{b - 1} = \frac{tk(k - 1)}{b(b - 1)}.$$

Chapter 5, §2

1. The cipher is decoded by $i \to i + 15 \pmod{26}$. So XTWTEL?J becomes MILITA?Y, which we interpret to be MILITARY.

2. The number of 2×2 involutory matrices $\pmod{26}$ is 736 (from the text). So the number of mappings of the form $b = ka + d$, $a, b, d \in (\mathbb{Z}_{26})_2$ is $736 \cdot 26 = 19{,}136$.

3. (i) We shall encipher the word CRY PTO LOG YZZ. The last two letters being chosen arbitrarily. From 2.8

$$\mathbf{a}_0 = \begin{pmatrix} 1 \\ 2 \\ 3 \end{pmatrix} \quad \text{and} \quad \mathbf{a}_1 = \begin{pmatrix} 3 \\ 18 \\ 25 \end{pmatrix},$$

$$\mathbf{a}_2 = \begin{pmatrix} 15 \\ 19 \\ 14 \end{pmatrix} \quad (P \leftrightarrow 15, T \leftrightarrow 19, 0 \leftrightarrow 14, \text{ etc.}),$$

$$\mathbf{a}_3 = \begin{pmatrix} 11 \\ 14 \\ 6 \end{pmatrix},$$

$$\mathbf{a}_4 = \begin{pmatrix} 24 \\ 25 \\ 25 \end{pmatrix}.$$

So

$$\mathbf{b}_2 = \begin{pmatrix} 208 \\ 289 \\ 174 \end{pmatrix} = \begin{pmatrix} 0 \\ 3 \\ 8 \end{pmatrix} \pmod{26} = \begin{pmatrix} A \\ D \\ S \end{pmatrix},$$

$$\mathbf{b}_3 = \begin{pmatrix} 0 \\ 0 \\ 17 \end{pmatrix} = \begin{pmatrix} A \\ A \\ R \end{pmatrix},$$

and

$$b_4 = \begin{pmatrix} 10 \\ 10 \\ 18 \end{pmatrix} = \begin{pmatrix} K \\ K \\ S \end{pmatrix}.$$

So "CRYPTOLOGY" ZZ is encoded as "TXOADSAARKKS".
(ii) We have $b_i = Ka_i + Ca_{i-1}$, where

$$K = \begin{pmatrix} 1 & 2 & 3 \\ 2 & 5 & 6 \\ 1 & 2 & 4 \end{pmatrix}.$$

Therefore

$$K^{-1} = \begin{pmatrix} 8 & 24 & 23 \\ 24 & 1 & 0 \\ 25 & 0 & 1 \end{pmatrix};$$

when $i = 1$, $b_1 = Ka_1 + Ca_0$. Therefore

$$a_1 = K^{-1}(b_1 - Ca_0) = \begin{pmatrix} 8 & 24 & 23 \\ 24 & 1 & 0 \\ 25 & 0 & 1 \end{pmatrix} b_1 - \begin{pmatrix} 9 \\ 19 \\ 22 \end{pmatrix} (\text{mod } 26)$$

and

$$a_i = K^{-1}(b_i - Ca_{i-1})$$

$$= \begin{pmatrix} 8 & 24 & 23 \\ 24 & 1 & 0 \\ 25 & 0 & 1 \end{pmatrix} b_i - \begin{pmatrix} 25 & 2 & 1 \\ 20 & 24 & 1 \\ 23 & 1 & 25 \end{pmatrix} a_{i-1}.$$

4. Assume that we have a linear transformation cipher of the form $\phi(x) = ax + b \pmod{26}$ where x is the numerical equivalent of a letter. We find the following frequency distribution of letters in the text.

A B C D E F G H I J K L M N O P Q R S T U V W X Y Z
0 3 0 2 1 3 9 7 4 0 1 7 8 2 2 0 0 10 7 2 1 11 1 3 1 7

From inspection of the text Z is most likely an I or an A. It seems the latter is more likely. We guess that V is an E. Assuming this we find

$$a + b = 26 \quad \text{for A} \to \text{Z},$$

$$20a + b = 13 \quad \text{for V} \to \text{E}.$$

Subtracting gives $-4 = 4a$ which has solutions $a \equiv -1, 12 \pmod{26}$, and corresponding values of 1 and -12 for b. Taking the first case we have $\phi(x) = -x + 1$ and $x = 1 - \phi(x)$ and the message deciphers to:
   "A fanatic is a person who is highly enthusiastic about something in which you are not even remotely interested."

5. Let the linear transformation be $C = a\mathrm{P} + b$. We find the following frequency distribution of letters in the cipher.

Letter	Frequency
C	1
E	2
F	7
G	3
L	6
N	8
O	4
P	3
Q	4
R	4
S	3
T	2
W	1
X	2
Y	5
Z	1

It seems reasonable to assume that R should decipher to I or A. If we assume R is I and successively suppose that E has been enciphered as N, F, L, we find when E is assumed to be L,

$$18 = 9a + b,$$
$$12 = 5a + b,$$

$$b = 4a.$$

Now $b \equiv \ldots, -4b, -20, b, 32, 58, 84 \ldots$ (mod 26). If we take $4a = 84$ then $a = 21$ and $b = 11$. So $C = 21\mathrm{P} + 11$. This has inverse $\mathrm{P} = 5C - 3$ and we decipher the message to obtain

I AM NOT AFRAID OF TOMORROW FOR I HAVE SEEN
YESTERDAY AND I LOVE TODAY.

6. Only for $a = 11$ and $a = 24$ we obtain involutory matrices.

7. "The three most difficult things in life are to climb a wall leaning toward you, kiss a woman leaning away from you, and give a lively and well received after dinner speech." (Churchill)

$$\textit{Playfair Square}: \quad \begin{array}{ccccc} W & I & N & S & T \\ O & C & H & U & R \\ L & A & B & D & E \\ F & G & K & M & P \\ Q & V & X & Y & Z \end{array}$$

8. $\mathbf{H}^2 = \begin{pmatrix} a^2 + bc & 2ab \\ 2ac & a^2 + bc \end{pmatrix} \equiv \begin{pmatrix} 1 & 2ab \\ 2ac & 1 \end{pmatrix} \pmod{26}.$

So

$$\mathbf{H}^2 \equiv \mathbf{I} \pmod{26} \quad \text{iff} \quad 2ab \equiv 0 \pmod{26} \text{ and } 2ac \equiv 0 \pmod{26}.$$

$$\text{iff} \quad a \equiv 0, 13, \pmod{26} \text{ or } b \equiv 0, 13 \pmod{26} \text{ and } c$$

$$\equiv 0, 13, \pmod{26}.$$

Case A. If $a \equiv 0 \pmod{26}$ then $bc \equiv 1 \pmod{26}$ which implies that b and c must be one of the pairs 1, 1; 3, 9; 5, 21; 7, 15; 11, 19; 17, 23; or 25, 25.

Case B. If $a \equiv 13 \pmod{26}$ then $bc \equiv 14 \pmod{26}$. Hence b and c must be one of the following pairs 1, 14; 2, 7; 2, 20; 3, 22; 4, 10; 4, 23; 5, 8; 6, 24; 6, 11; 8, 18; 9, 16; 10, 17; 12, 25; 12, 12; 14, 14; 15, 20; 18, 21; 16, 22; 19, 24.

Case C. If b or $c \equiv 0 \pmod{26}$ then $a^2 \equiv 1 \pmod{26}$ so $a \equiv 1, 25 \pmod{26}$.

Case D. If $bc \equiv 13^2 \pmod{26}$ then $a^2 \equiv -168 \pmod{26}$ so $a \equiv 12, 14 \pmod{26}$.

So **H** must be one of the following matrices.

$$\begin{pmatrix} 0 & b \\ c & 0 \end{pmatrix}, \text{ where } b \text{ and } c \text{ are as in Case A.}$$

$$\begin{pmatrix} 13 & b \\ c & 13 \end{pmatrix}, \text{ where } b \text{ and } c \text{ are as in Case B.}$$

$$\begin{pmatrix} 1 & 0 \\ 0 & 1 \end{pmatrix}, \begin{pmatrix} 25 & 0 \\ 0 & 25 \end{pmatrix}, \begin{pmatrix} 1 & 13 \\ 0 & 1 \end{pmatrix}, \begin{pmatrix} 1 & 0 \\ 13 & 1 \end{pmatrix}, \begin{pmatrix} 25 & 13 \\ 0 & 25 \end{pmatrix},$$

$$\begin{pmatrix} 25 & 0 \\ 13 & 25 \end{pmatrix} \text{ and } \begin{pmatrix} 12 & 13 \\ 13 & 12 \end{pmatrix}, \begin{pmatrix} 14 & 13 \\ 13 & 14 \end{pmatrix}$$

giving 56 possible matrices.

9. We write ALGEBRA IS GREAT in the form AL GE BR AX IS XG RE AT where an X denotes a space between words. Numbering the letters $1, \ldots, 26$ and encoding by

$$\begin{pmatrix} b_1 \\ b_2 \end{pmatrix} = \begin{pmatrix} 2 & 23 \\ 3 & 1 \end{pmatrix} \begin{pmatrix} a \\ b \end{pmatrix}$$

we find

$$\begin{pmatrix} A \\ L \end{pmatrix} \rightarrow \begin{pmatrix} 1 \\ 12 \end{pmatrix} \rightarrow \begin{pmatrix} 18 \\ 15 \end{pmatrix} \rightarrow \begin{pmatrix} R \\ O \end{pmatrix} \quad \text{etc.}$$

So the encoded message is ROYZBX . . . . To decipher,

$$\begin{pmatrix} 2 & 23 \\ 3 & 1 \end{pmatrix}^{-1} = \begin{pmatrix} 19 & 5 \\ 21 & 12 \end{pmatrix} \quad \text{so} \quad \begin{pmatrix} a_1 \\ a_2 \end{pmatrix} = \begin{pmatrix} 19 & 5 \\ 21 & 12 \end{pmatrix} \begin{pmatrix} b_1 \\ b_2 \end{pmatrix}.$$

So, e.g.

$$\begin{pmatrix} R \\ O \end{pmatrix} \rightarrow \begin{pmatrix} 18 \\ 15 \end{pmatrix} \rightarrow \begin{pmatrix} 1 \\ 12 \end{pmatrix} = \begin{pmatrix} A \\ L \end{pmatrix}.$$

10.

$$\det \begin{pmatrix} 1 & 0 & 2 \\ 2 & 3 & 2 \\ 0 & 1 & 0 \end{pmatrix} \equiv \begin{cases} 2 \pmod{q}, \\ 0 \pmod{2}. \end{cases}$$

To calculate the inverse for $q > 2$,

$$\left(\begin{array}{ccc|ccc} 1 & 0 & 2 & 1 & 0 & 0 \\ 2 & 3 & 2 & 0 & 1 & 0 \\ 0 & 1 & 0 & 0 & 0 & 1 \end{array}\right) \sim \left(\begin{array}{ccc|ccc} 1 & 0 & 0 & -1 & 1 & -3 \\ 0 & 1 & 0 & 0 & 0 & 1 \\ 0 & 0 & 1 & 1 & -(q+1)/2 & \frac{3}{2}(q+1) \end{array}\right).$$

Since q is odd

$$\frac{q+1}{2}2 \equiv 1 \,(\mathrm{mod}\, q) \quad \text{and} \quad \frac{q+1}{2} \in \mathbf{Z}_q.$$

So

$$\mathbf{A}^{-1} = \left(\begin{array}{ccc} -1 & 1 & -3 \\ 0 & 0 & 1 \\ 1 & -(q+1)/2 & \frac{3}{2}(q+1) \end{array}\right).$$

11. (i)

$$\mathbf{M}^2 = \left(\begin{array}{cc} a^2 + bc & b(a+d) \\ c(a+d) & bc+d^2 \end{array}\right).$$

So if $\mathbf{M}$ is involutory and $\det(\mathbf{M}) \equiv -1 \,(\mathrm{mod}\, 26)$, we obtain $a(a+d) \equiv d(a+d) \equiv 0 \,(\mathrm{mod}\, 26)$. So $(a+d)^2 \equiv 0 \,(\mathrm{mod}\, 26)$. Therefore $a+d \equiv 0 \,(\mathrm{mod}\, 26)$.

Conversely, if $a + d \equiv 0 \,(\mathrm{mod}\, 26)$ then

$$\mathbf{M}^2 = \left(\begin{array}{cc} a^2 + bc & 0 \\ 0 & d^2 + bc \end{array}\right) \equiv \mathbf{I}(\mathrm{mod}\, 26)$$

since $a^2 + bc \equiv a^2 + ad + 1 \equiv 1 \,(\mathrm{mod}\, 26)$ and $d^2 + bc \equiv d^2 + ad + 1 \equiv 1 \,(\mathrm{mod}\, 26)$. So $\mathbf{M}$ is involutory.

(ii) $\mathbf{M} = \left(\begin{array}{cc} 2 & b \\ c & d \end{array}\right)$. By (i), $d \equiv -2 \,(\mathrm{mod}\, 26)$ and $\det(\mathbf{M}) = -4 - bc \equiv -1 \,(\mathrm{mod}\, 26)$. So $bc \equiv -3 \,(\mathrm{mod}\, 26)$ hence b and c must be one of the following pairs, $1, -3; -1, 3; 7, 7; 19, 19; 21, 11; 17, 9; 15, 5$.

12. Since $q - 1$ is the order of the cyclic group $\mathbf{F}_q^*$ the result follows.

13. This follows from Euler's theorem.

14. $1415^{113} \equiv 2440 \,(\mathrm{mod}\, 2803)$.

15. The exact numbers of digits are:

$$2^{127} - 1 \text{ has 39 digits}$$
$$2^{607} - 1 \text{ has 183 digits,}$$
$$2^{3217} - 1 \text{ has 964 digits,}$$
$$2^{19937} - 1 \text{ has 6002 digits,}$$
$$2^{44497} - 1 \text{ has 13,395 digits.}$$

Note that the largest known prime as of January 1984 is claimed to be $2^{132049} - 1$, found by D. Slowinski at Cray Research Laboratories.

16. Using + signs to indicate a space we wish to encode ALGEBRA + IS + GREAT. This is done as follows (reducing modulo 19999).

$$AL = (00000\ 01011) \rightarrow 138,$$

$$GE = (00110\ 00100) \rightarrow 9021,$$

$$BR = (00001\ 10001) \rightarrow 96,$$

$$A+ = (00000\ 11111) \rightarrow 16161,$$

$$IS = (01000\ 10010) \rightarrow 2044,$$

$$+G = (11111\ 00110) \rightarrow 11863,$$

$$RE = (10001\ 00100) \rightarrow 16821,$$

$$AT = (00000\ 10011) \rightarrow 134.$$

So the enciphered message is

$$(138, 9021, 96, 16161, 2044, 11863, 16821, 134).$$

To decipher (14800, 17152, 17162, 2044, 11913, 9718) we first multiply by 100 (mod 19999) and obtain (74, 15285, 16285, 4410, 11359, 11848).
    Now $74 = 1 \cdot 64 + 1 \cdot 10 \Leftrightarrow (01100\ 00000) \leftrightarrow MA$

$$15285 \leftrightarrow (1001100111) \leftrightarrow TH,$$

$$16285 \leftrightarrow (1001011111) \leftrightarrow S+,$$

$$4410 \leftrightarrow (01000\ 10010) \leftrightarrow IS,$$

$$11359 \leftrightarrow (11111\ 00101) \leftrightarrow +F,$$

$$11848 \leftrightarrow (10100\ 01101) \leftrightarrow UN,$$

with "+" signs deleted the message reads "maths is fun".

17. 10110 enciphered is 1327 as knapsack sum. For deciphering we have $1327 \cdot 101 \equiv 293 \pmod{p}$. So $23 + 0 + 91 + 179 + 0 = 293$, giving 10110 as the deciphered number.

18. 0101 enciphered is 197.

19. $a^{560} \equiv 1 \pmod{561}$ for all $(a, 561) = 1$, but $561 = 3.11.17$.

20. Let $n = 257$, then $n - 1 = 2^8$. Then $3^{256} \equiv 1 \pmod{257}$, but $3^{256/2} \equiv 256 \pmod{257}$. So 257 is prime.

21. Consider $y_p = (b^{2p} - 1)/(b^2 - 1)$, p an odd prime and $(p, b^2 - 1) = 1$. Then $y_p = ((b^p - 1)/(b - 1))((b^p + 1)/(b + 1))$ is composite. Also $b^{2p} \equiv 1 \pmod{y_p}$ and $y_p - 1 \equiv 0 \pmod{2p}$.

22. $J(6, 13) = -1$.

24. Answer: -1 or 38 (mod 39).

25. $\phi(n) = 3120$. Then $t = 253$. Let AL = 0116, then

$$116^{253} \equiv 2381 \ (\text{mod } 3233).$$

Therefore AL is enciphered as 2381 (if we use the correspondence A $\leftrightarrow$ 01, B $\leftrightarrow$ 02, etc.).

Chapter 5, §3

1. Define the linear recurring sequence in $\mathbb{F}_2$ by

$$s_{n+2} = s_{n+1} + s_n + 1,$$

so

$$a_0 = a_1 = a_2 = 1, \qquad k = 2.$$

Let

$$s_0 = 1 \quad \text{and} \quad s_1 = 0.$$

We find $\mathbf{s} = 1 \underline{|100|} \underline{|100|} \underline{|1}\ \dots$. $\mathbf{s}$ is not periodic but is ultimately periodic with period 3.

2. (i) $d_{n+5} = d_{n+1} + d_n$.
   (ii) 000010001100101011111100001 ... has period 21.
   (iii)

$$\mathbf{A} = \begin{pmatrix} 0 & 0 & 0 & 0 & 1 \\ 1 & 0 & 0 & 0 & 1 \\ 0 & 1 & 0 & 0 & 0 \\ 0 & 0 & 1 & 0 & 0 \\ 0 & 0 & 0 & 1 & 0 \end{pmatrix}.$$

If $\mathbf{A}^m = \mathbf{A}^n$ then by 3.7, $\mathbf{d}_m = \mathbf{d}_0\mathbf{A}^m = \mathbf{d}_0\mathbf{A}^n = \mathbf{d}_n$. Conversely, suppose $\mathbf{d}_m = \mathbf{d}_n$. Then $\mathbf{d}_0\mathbf{A}^m = \mathbf{d}_0\mathbf{A}^n$. From the linear recurrence for the impulse response sequence $d_0, d_1, \dots$ we obtain $\mathbf{d}_{m+t} = \mathbf{d}_{n+t}$ for all $t \geq 0$. Moreover, $\mathbf{d}_t\mathbf{A}^m = \mathbf{d}_t\mathbf{A}^n$ for all $t \geq 0$. The vectors $\mathbf{d}_0, \mathbf{d}_1, \dots, \mathbf{d}_{k-1}$ form a basis of the vectorspace $\mathbb{F}_q^k$ over $\mathbb{F}_q$, therefore $\mathbf{A}^m = \mathbf{A}^n$.

For the sequence in (i) the order of A in $\text{GL}(5, \mathbb{F}_2)$ is 21, namely the least period of the impulse response sequence.

(α) If the initial state vector of $s_0, s_1, \dots$ is one of the 21 different state vectors in the impulse response sequence, then the least period is 21.

(β) If the initial state vector is $(1, 1, 1, 0, 1)$ we get 111010011101 ... of least period 7. This is also the case if we take any of the seven different state vectors of this sequence as an initial state vector.

(γ) If the initial state vector is $(1, 1, 0, 1, 1)$ we get 11011011 ... of least period 3. This is also the least period for any one of the three different state vectors of this sequence taken as the initial state vector of the sequence with recurrence relation given in (i).

(δ) The all-zero initial state vector yields a sequence of least period 1. This completes consideration of all 32 possibilities.

3.

$$\zeta^4 = 1 + \zeta$$

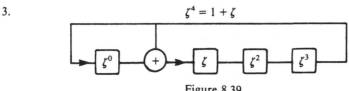

Figure 8.39

4. (i) $d_{n+4} = d_{n+3} + d_{n+1} + d_n$ and $d_0 = d_1 = d_2 = 0$, $d_3 = 1$. So

$$d = \underbrace{\lfloor 000111}_{\text{period} = 6} \underbrace{00011}_{} \ldots .$$

(ii) $G = x^4 + x^5 + x^6 + x^{10} + x^{11} + x^{12} + \ldots .$

5. The impulse response sequence $d_0, d_1, \ldots$ is

0 0 0 0 1 0 0 0 1 1 0 0 1 0 1 0 1 1 1 1 1 0 0 0 0 1...

of least period 21. A feedback shift register is

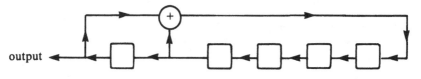

output

Figure 8.40

6. The characteristic polynomial is $f = x^6 - x^4 - x^2 - x - 1 \in \mathbb{F}_2[x]$; it is irreducible and of order 2. The impulse response sequence is

0 0 0 0 0 1 0 1 0 0 1 0 0 1 1 0 0 1 0 1 1 0 0 0 0 0 1...

of least period 21. If 0 0 0 0 1 1 is taken as initial state vectors we obtain the string of binary digits

0 0 0 0 1 1 1 1 0 1 1 0 1 0 1 0 1 1 1 0 1 0 0 0 0 1 1....

7. $f = x^4 + x^3 + 1$. By Theorem 3.3.1, $f|x^{2^n} - x$ iff $4|n$. Therefore $f|x^{16} - x$. So $f|x^{15} - 1 (f(1) \neq 0)$. Now $\text{ord} f|15$, which implies $\text{ord} f = 5$ or 16. Using the division algorithm $f \nmid x^5 - 1$; so it follows that $\text{ord} f = 15$.

8. $2 + x + 4x^2 + 2x^4 + \ldots \in \mathbb{F}_5[[x]]$.

9. $1 + x^2 + x^3 + x^7 + \ldots .$

10. Characteristic polynomial $f = x^5 + x^3 + x^2 + 1$, $\text{ord} f = 12$, minimal polynomial equals f, least period is 12.

11. The characteristic polynomial is $f = x^4 + x^2 + x$, which equals the minimal polynomial. The sequence is not periodic but ultimately periodic with least period $\text{ord} f = 7$.

12. The minimal polynomial of the shifted sequence is $x^3 + x + 1$, which divides $f = x^4 + x^2 + x$.

Chapter 5, §5

1. Two bracelets are distinct if the rotation of one will yield another, but flipping over is not permitted. The number of distinct bracelets is

$$\tfrac{1}{5}(243 + 3 + 3 + 3 + 3) = 51.$$

2. There are seven different chains.

3. There are eight distinct ways of painting the four faces.

4. $\tfrac{1}{24}(n^8 + 17n^4 + 6n^2)$.

5. Three switches are assumed to be not equivalent under permutations of the inputs. There are 2^8 switching functions, 80 are not equivalent.

Chapter 6, §1

1. A = Set of all Theorems. $Z = \mathscr{P}(A)$, where the set $z \in Z$ is the collection of all those Theorems in A "which my brain knows". $\delta(z, a) = z \cup \{a\}$ ("I have learned Theorem a, too").

2. Yes. Inputs = Theorems, z = "Theorems in the storage", $\delta(z, a)$ = "add Theorem a to the Theorems z in the storage".

3. B = list of all Theorems. $\lambda(z, a)$ "tells" that the set $z \cup \{a\}$ of Theorems was learned.

4. For instance, $z_0 = 0$ can mean that you know no Theorem at all.

Chapter 6, §2

1–2. State set $z = \{0, 1, 2, \ldots, 10\}$. Input set $A = \{n, c, w\}$ with n = "no coin inserted", c = "correct coin inserted", w = "wrong coin inserted". x = "no output", s = "a stamp as output", y = "a coin as output".

δ	n	c	w	λ	n	c	w
0	0	0	0	0	x	y	y
1	1	0	1	1	x	s	y
2	2	1	2	2	x	s	y
3	3	2	3	3	x	s	y
4	4	3	4	4	x	s	y
5	5	4	5	5	x	s	y
6	6	5	6	6	x	s	y
7	7	6	7	7	x	s	y
8	8	7	8	8	x	s	y
9	9	8	9	9	x	s	y
10	10	9	10	10	x	s	y

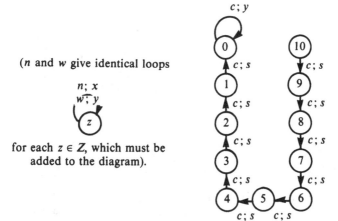

(*n* and *w* give identical loops

$$n; x$$
$$w; y$$

for each $z \in Z$, which must be added to the diagram).

Figure 8.41

3.

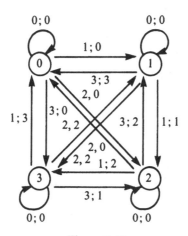

Figure 8.42

4. We just give the tables. $Z = \{0, 2, 4, 6, 8\}$.

δ	0	1	2	3	4	5	6	7	8	9
0	0	0	0	0	0	0	0	0	0	0
2	0	2	4	6	8	0	2	4	6	8
4	0	4	8	2	6	0	4	8	2	6
6	0	6	2	8	4	0	6	2	8	4
8	0	8	6	4	2	0	8	6	4	2

λ	0	1	2	3	4	5	6	7	8	9
0	0	0	0	0	0	0	0	0	0	0
2	2	2	2	2	2	2	2	2	2	2
4	4	4	4	4	4	4	4	4	4	4
6	6	6	6	6	6	6	6	6	6	6
8	8	8	8	8	8	8	8	8	8	8

5. (a) No; ∘ is not necessarily associative.
   (b) Yes, no problem.

6. $A = \{a_1, a_2\}$.

δ	a_1	a_2
z_1	z_2	z_3
z_2	z_3	z_1
z_3	z_1	z_2

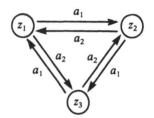

Figure 8.43

Chapter 6, §3

1. If $|M| \le |N|$ then there is some injective map $f: M \to N$. Then $h: \mathcal{P}(M) \to \mathcal{P}(N)$, $A \to f(A) = \{f(a)|a \in A\}$ is injective, too, and a homomorphism. h is an isomorphism iff f is bijective, i.e. if $|M| = |N|$.

2. h is an epimorphism with $\mathbb{Z}/\equiv_n \cong \mathbb{Z}_n$ and $a \equiv_n b \Leftrightarrow a \equiv b \pmod{n}$.

3. It is obvious that $\sim$ is an equivalence relation. Let $x \sim y$, $x' \sim y'$. If $x = y$ and $x' = y'$ then $xx' \sim yy'$. If, e.g. $x = y$ and $x' \in I$, $y' \in I$ then $xx' \in T$ and $yy' \in I$, hence $xx' \sim yy'$. If $x, y, x', y' \in I$ then again $xx' \sim yy'$. $S/\sim$ consists of I and of all singletons $\{x\}$ with $x \notin I$.

4. Let z, i be the adjoined zero (identity)

$\cap$	$\varnothing$	$\{0\}$	$\{1\}$	$\{0,1\}$	z
$\varnothing$	$\varnothing$	$\varnothing$	$\varnothing$	$\varnothing$	z
$\{0\}$	$\varnothing$	$\{0\}$	$\varnothing$	$\{0\}$	z
$\{1\}$	$\varnothing$	0	$\{1\}$	$\{1\}$	z
$\{0,1\}$	$\varnothing$	$\{0\}$	$\{1\}$	$\{1,2\}$	z
z	z	z	z	z	z

$\cap$	$\varnothing$	$\{0\}$	$\{1\}$	$\{0,1\}$	i
$\varnothing$	$\varnothing$	$\varnothing$	$\varnothing$	$\varnothing$	$\varnothing$
$\{0\}$	$\varnothing$	$\{0\}$	$\varnothing$	$\{0\}$	$\{0\}$
$\{1\}$	$\varnothing$	$\varnothing$	$\{1\}$	$\{1\}$	$\{1\}$
$\{0,1\}$	$\varnothing$	$\{0\}$	$\{1\}$	$\{1,2\}$	$\{0,1\}$
i	$\varnothing$	$\{0\}$	$\{1\}$	$\{1,2\}$	i

5. $\P_1((s_1, s_2) \circ (s_1' s_2')) = \P_1(s_1 \circ_1 s_1', s_2 \circ_2 s_2') = s_1 \circ_1 s_1' = \P_1(s_1, s_2) \circ_1 \pi_1(s_1', s_2')$; $\P_1$ is clearly surjective.

6. Number of words of length $1 +$ number of words with length $2 + \dots$ equals

$$7 + 7^2 + 7^3 + \dots + 7^n = \tfrac{7}{6}(7^n - 1).$$

In the free monoid, we get one word more (the empty word). Hence the result is given by $\tfrac{7}{6}(7^n - 1) + 1 = \tfrac{1}{6}(7^{n+1} - 1)$.

7. *Solution 1.* $F =$ free semigroup on $\{1, 2, \dots, n\}$; h sends each word $x_1 x_2 \dots x_k$ onto the product $x_1 \cdot x_2 \cdot \dots \cdot x_k$ in $(\mathbb{Z}_m, \cdot)$.
   Solution 2. Let, e.g. n be 6. Then $E = \{2, 3, 5\}$ generates $(\mathbb{Z}_6, \cdot)$. Let F be free on $\{2, 3, 4\}$. Let h be as before.

8. $(\mathbb{N}, +)$ is free. $(\mathbb{N}_0, +)$ is not free. It is commutative, hence it would be isomorphic to $(\mathbb{N}, +)$, which is impossible $((\mathbb{N}_0, +)$ has an identity, $(\mathbb{N}, +)$ has none, for instance). The same argument applies to $(\mathbb{N}, \cdot)$, $(\mathbb{Q}, \cdot)$ and $(\mathbb{R}, \cdot)$.

9. H is then commutative.

10. Let a_i be the equivalence class of x_i in $F_X/\tilde{R} = H$. We then have $a_1 a_1 = a_1 = a_2 a_2 = a_3 a_3$, $a_1 a_2 = a_1$, $a_1 a_3 = a_1$, $a_2 a_1 = a_1$ and $a_3 a_2 = a_2 a_3$. We get

$\circ$	a_1	a_2	a_3	$a_2 a_3$
a_1	a_1	a_1	a_1	a_1
a_2	a_2	a_1	$a_2 a_3$	a_1
a_3	a_1	$a_2 a_3$	a_1	a_1
$a_2 a_3$	a_2	a_1	a_2	a_1

$H = \{a_1, a_2, a_3, a_2 a_3\}$ has a_1 as a left, but not as a right zero element. Since, e.g. $a_1 a_2 \neq a_2 a_1$; H is not commutative.

Chapter 6, §4

1. (i) $Z = \{z_1, z_2, z_3, z_4\}$, $A = \{a, b\}$, $B = \{0, 1\}$,

δ	a	b
z_1	z_2	z_3
z_2	z_3	z_4
z_3	z_4	z_2
z_4	z_4	z_4

λ	a	b
z_1	0	0
z_2	1	0
z_3	0	1
z_4	0	0

(ii) Final state: z_4; Output sequence: 1000 000.
(iii) *aabb* and *bbaa*.
(iv) If you are in z_4, you'll never come out again.
(v) *a* or *aba* or *ababa* or *abababa* ....
(vi) $z_1 \to z_4$.

2. For $r = 2$, this holds by definition. We show this formula for $r = 3$ (for general r, proceed by induction).

$$\bar{\lambda}(z, a_1 a_2 a_3) = \lambda(z, a_1)\bar{\lambda}(\delta(z, a_1), a_2 a_3)$$
$$= \lambda(z, a_1)\lambda(\delta(z, a_1), a_2)\lambda(\bar{\delta}(z, a_1 a_2), a_3)$$
$$= \bar{\lambda}(z, a_1 a_2)\lambda(\bar{\delta}(z, a_1 a_2), a_3).$$

Interpretations for $\bar{\lambda}(z, a_1 a_2 \ldots a_r)$:
(a) By definition: a_1 gives output $\lambda(z, a_1)$ and changes z into $z' := \delta(z, a_1)$. Then $a_2 a_3 \ldots a_r$ produces the output sequence $\bar{\lambda}(z', a_2 a_3 \ldots a_r)$.
(b) By this formula: $a_1, a_2, \ldots, a_{r-1}$ give (in state z) the output sequence $\bar{\lambda}(z, a_1 a_2 \ldots a_{r-1})$ and a final state $z'' := \delta(z, a_1 a_2 \ldots a_{r-1})$. Then the last input a_r adds another output $\lambda(z'', a_r)$.

3. $\bar{\lambda}(z, a_1 a_2 \ldots a_n)$ gives the last output in the output sequence determined by z and $a_1 a_2 \ldots a_n$. The formula is easily proved by induction.

4. Proof idea. (a) Let f be given by $(Z, A, B, \delta, \lambda)$ and z. Then $f(a_1 \ldots a_n) = \bar{\lambda}(z, a_1 a_2 \ldots a_n)$ is an output word of the same length n as $a_1 a_2 \ldots a_n$. Hence f is sequential.

(b) Conversely, let $f: \bar{A} \to \bar{B}$ be sequential. Take some $z \in Z$ and define $\lambda(z, a) := f(a)$. Adjust δ so that the formula at the beginning of § 4 holds.

5.

δ	a	b	c
z_0	z_2	z_1	z_1
z_1	z_0	z_1	z_1
z_2	z_2	z_2	z_2
z_3	z_2	z_3	z_3

(for instance)

Chapter 6, §5

1.

	f_a	f_b	f_{aa}	f_{ab}	f_{ba}	f_{bb}	f_{aaa}	id
$z_1 \to$	z_2	z_3	z_3	z_2	z_4	z_4	z_4	z_1
$z_2 \to$	z_3	z_2	z_4	z_4	z_3	z_2	z_4	z_2
$z_3 \to$	z_4	z_4	z_4	z_4	z_4	z_4	z_4	z_4
$z_4 \to$	z_4	z_4	z_4	z_4	z_4	z_4	z_4	z_4

M is given by

$\circ$	f_a	f_b	f_{aa}	f_{ab}	f_{ba}	f_{bb}	f_{aaa}	id
f_a	f_{aa}	f_{ba}	f_{aaa}	f_{aa}	f_{aaa}	f_{bb}	f_{aaa}	f_a
f_b	f_{ab}	f_{bb}	f_{aaa}	f_{ab}	f_{aaa}	f_{bb}	f_{aaa}	f_b
f_{aa}	f_{aaa}	f_{aaa}	f_{aaa}	f_{aaa}	f_{aaa}	f_{aaa}	f_{aaa}	f_{aa}
f_{ab}	f_{aaa}	f_{aaa}	f_{aaa}	f_{aaa}	f_{aaa}	f_{aaa}	f_{aaa}	f_{ab}
f_{ba}	f_{aa}	f_{bb}	f_{aaa}	f_{aa}	f_{aaa}	f_{bb}	f_{aaa}	f_{ba}
f_{bb}	f_{ab}	f_{bb}	f_{aaa}	f_{ab}	f_{aaa}	f_{bb}	f_{aaa}	f_{bb}
f_{aaa}	f_{aaa}	f_{aaa}	f_{aaa}	f_{aaa}	f_{aaa}	f_{aaa}	f_{aaa}	f_{aaa}
id	f_a	f_b	f_{aa}	f_{ab}	f_{ba}	f_{bb}	f_{aaa}	id

2. For the parity-check automaton and the trigger flip-flop we get the monoids $\{f_\wedge = \mathrm{id} = f_0, f_1\}$. These are groups of order 2 and hence isomorphic. Since $\{f_\wedge = \mathrm{id}, f_0, f_1\}$ form monoids in the cases of the shift-register and the IR flip-flop, these are their monoids.

3. $(\mathbb{Z}_3, \mathbb{Z}_3, \mathbb{Z}_3, \delta_i, \lambda_i)$ with $\delta_1(z, a) = \lambda_1(z, a) = z + a$ and $\delta_2(z, a) = \lambda_2(z, a) = za$ are suitable automata for the two given semigroups.

4. No; let, for example, $\mathscr{A}$ be the parity-check automaton. Then $M_\mathscr{A} = \{f_0, f_1\} := M$ and $\mathscr{A}_{M_\mathscr{A}} = (M, M, M, \delta, \lambda)$ (as in the proof of 5.7) is not the same as $\mathscr{A}$.

5. See the first case in Exercise 3.

6. Yes. Take, for example, $Z = A = B = \mathbb{N}$, $\lambda(z, a) = z = \delta(z, a)$ for each $z \in Z$ and $a \in A$. Then each $f_a = \mathrm{id}$ and the monoid of this automaton consists only of id.

Chapter 6, §6

1. $Z_1 = \{z_2, z_3, z_4\}$ and $Z_2 = \{z_4\}$ determine subautomata. The monoid of $\mathscr{A}_1 = (Z_1, A, \delta)$ is given by

	f_a	f_b	f_{ab}	id			f_a	f_b	f_{aa}	id
$z_2\to$	z_3	z_2	z_4	z_2		f_a	f_{aa}	f_a	f_{aa}	f_a
$z_3\to$	z_4	z_4	z_4	z_3	and	f_b	f_{aa}	f_b	f_{aa}	f_b
$z_4\to$	z_4	z_4	z_4	z_4		f_{aa}	f_{aa}	f_{aa}	f_{aa}	f_{aa}
						id	f_a	f_b	f_{aa}	id

The map

$$f_a \to f_a, \qquad f_{ba} \to f_a,$$
$$f_b \to f_b, \qquad f_{bb} \to f_b,$$
$$f_{aa} \to f_{aa}, \qquad f_{aaa} \to f_{aa},$$
$$f_{ab} \to f_{aa}, \qquad \text{id} \to \text{id},$$

is the desired epimorphism. The monoid of $\mathscr{A}_2 = (Z_2, A, \delta)$ is $\{\text{id}\}$ and the epimorphism from $M_{\mathscr{A}}$ to $M_{\mathscr{A}_2}$ is trivial: all f_α are mapped onto id.

2. $M_{\mathscr{A}_1} = \{\text{id}, f_{a_2}\}$, $M_{\mathscr{A}_2} = \{\text{id}, f_{a_1}, f_{a_2}, f_{a_2 a_1}, f_{a_1 a_2 a_1}, f_{a_2 a_1 a_2}\}$. Since in $M_{\mathscr{A}_2}$ id and $f_{a_2 a_1 a_2}$ coincide with id when restricted to $\{z_1, z_2\}$, id and $f_{a_2 a_1 a_2}$ are mapped to id, while the remaining four functions in $M_{\mathscr{A}_2}$ are mapped to $f_{a_2} \in M_{\mathscr{A}_1}$.

3. Yes, by Theorem 6.3.

4. Let $\mathscr{A}$ be the parity-check automaton.
   (a) $\mathscr{A} \times \mathscr{A} = (\{z_0, z_1\}^2, \{0, 1\}^2, \{0, 1\}^2, \delta, \lambda)$ with

δ	$(0,0)$	$(0,1)$	$(1,0)$	$(1,1)$	λ	$(0,0)$	$(0,1)$	$(1,0)$	$(1,1)$
(z_0, z_0)	(z_0, z_0)	(z_0, z_1)	(z_1, z_0)	(z_1, z_1)		$(0,0)$	$(0,1)$	$(1,0)$	$(1,1)$
(z_0, z_1)	(z_0, z_1)	(z_0, z_0)	(z_1, z_1)	(z_1, z_0)		$(0,0)$	$(0,1)$	$(1,0)$	$(1,1)$
(z_1, z_0)	(z_1, z_0)	(z_1, z_1)	(z_0, z_0)	(z_0, z_1)		$(0,0)$	$(0,1)$	$(1,0)$	$(1,1)$
(z_1, z_1)	(z_1, z_1)	(z_1, z_0)	(z_0, z_1)	(z_0, z_0)		$(0,0)$	$(0,1)$	$(1,0)$	$(1,1)$

   (b) $\mathscr{A} \parallel \mathscr{A} = (\{z_0, z_1\}^2, \{0, 1\}, \{0, 1\}, \delta, \lambda)$ with

δ	0	1	λ	0	1
(z_0, z_0)	(z_0, z_0)	(z_1, z_1)		0	1
(z_0, z_1)	(z_0, z_1)	(z_1, z_0)		0	1
(z_1, z_0)	(z_1, z_0)	(z_0, z_1)		0	1
(z_1, z_1)	(z_1, z_1)	(z_0, z_0)		0	1

5. We get $(\{0, 1\}^2, \{0, 1\}, \{0, 1\}^2, \delta, \lambda)$ with

$$\delta((z_1, z_2), x) = (x + 1, z_2 + 1) \quad \text{and} \quad \lambda((z_1, z_2), x) = (z_1, z_2);$$

δ	0	1		λ	0	1
$(0,0)$	$(1,1)$	$(0,1)$		$(0,0)$	$(0,0)$	$(0,0)$
$(0,1)$	$(1,0)$	$(0,0)$		$(0,1)$	$(0,1)$	$(0,1)$
$(1,0)$	$(1,1)$	$(0,1)$		$(1,0)$	$(1,0)$	$(1,0)$
$(1,1)$	$(1,0)$	$(0,0)$		$(1,1)$	$(1,1)$	$(1,1)$

6. $(\mathbb{Z}_2, +) \wr (\mathbb{Z}_2, +)$ has eight elements and is isomorphic to the dihedral group (of order 8). Hence this wreath product is a group, but not an abelian one.

7. If $\mathbb{Z}_n$ is the homomorphic image of the subsemigroup H of $(\mathbb{Z}_m, +)$ then H must be isomorphic to $(\mathbb{Z}_k, +)$ with $k|m$. But then $\mathbb{Z}_n \cong \mathbb{Z}_r$ with $r|k$. Hence $n = r|m$.

8. $M_{\mathscr{A}_1}$ is, as a homomorphic image of $M_{\mathscr{A}_2}$ (see Theorem 6.3) itself a group.

9. $M_{\mathscr{A}}$ is the homomorphic image of $M_{\mathscr{A}'}$. A simple group G has, up to isomorphisms, only two homomorphic images: G itself and $\{1\}$. Hence

$$M_{\mathscr{A}} \cong M_{\mathscr{A}'} \quad \text{or} \quad M_{\mathscr{A}} = \{\text{id}\}.$$

10. H is a homomorphic image of a subsemigroup of G. If G is finite, H must be a group, too (see Problem 13 to part B of §3). If G is infinite, H need not be a group (example: $G = (\mathbb{Z}, +)$, $H = (\mathbb{N}, +)$).

11.

Subsemigroups of $(\{a, b, c\}, *)$	Their homomorphic images
$\{a, b, c\}$ $\{a, b\}, \{a, c\}, \{b, c\}$ $\{a\}, \{b\}, \{c\}$	all isomorphic to $(\{a, b, c\}*)$ or $(\{a, b\}, *)$ or $(\{a\}, *)$.

12. $\{0\}, \mathbb{Z}_5$, and all isomorphic copies of these two groups.

Chapter 6, §7

1. Direct and easy.

2. $Z/\sim_1 = \{\{z_1\}, \{z_2, z_3\}\}$, $Z/\sim_2 = \{\{z_1\}, \{z_2\}, \{z_3\}\}$. So $\sim = \sim_2 = \text{id}$. Hence the marriage automaton is already reduced.

3. $Z/\sim_1 = \{\{z_0, z_1\}\} = Z/\sim_2$. So the reduced automaton is trivial and has only one state $\{z_0, z_1\}$.

4. $Z/\sim_1 = \{\{z_1, z_4\}, \{z_2\}, \{z_3\}\}$, $Z/\sim_2 = \{\{z_1\}, \{z_2\}, \{z_3\}, \{z_4\}\}$, so this automaton is reduced.

5. Let $\mathcal{S} = (Z, A, \delta)$ be not irreducible. Enlarge $\mathcal{S}$ to $\mathcal{A} = (Z, A, Z, \delta, \lambda)$ with $\lambda(z, a) = z$. Then $\mathcal{A}$ is reducible, but minimal. Conversely, let $\mathcal{S} = (Z, A, \delta)$ be irreducible with $Z \neq \{0\}$. Then $\mathcal{A} = (Z, A, \{0\}, \delta, \lambda)$ with $\delta(z, a) = 0$ is not minimal.

6. The first part follows the lines indicated in the sketch of the proof. Suppose that $\mathcal{A}_1$ is minimal. Let $\mathcal{A}_1 = (Z_1, A, B, \delta_1, \lambda_1)$. Define $h: Z| \sim \to Z_1$ by $h([z]) =$ the state $z_1 \in Z_1$ which is equivalent to z. h is well defined since $z' \sim z \Rightarrow h([z]) \sim h([z'])$, hence $h([z]) = h([z'])$, since $\mathcal{A}_1$ is minimal; moreover, z_1 is uniquely determined by z. $(h, \text{id}, \text{id})$ is then an isomorphism $\mathcal{A}^* \to \mathcal{A}_1$.

7. A homomorphic image.

8. No, e.g. $\mathcal{A}_i = (Z_i, A_i, B_i, \delta_i, \lambda_i)$ and $|Z_1| < |Z_2|$.

Chapter 7, §1

1. $L(\mathcal{G}) = \{a, aaa, aaaaa, \ldots\}$.

2. $L(\mathcal{G}) = \varnothing$.

3. Both languages are context-free, context-sensitive and ALGOL-like. $L(\mathcal{G})$ of Example 1 is left linear.

4. $L(\mathcal{G})$ is the set of all words containing at least one a_3 such that after the last apparence of a_3 there is no a_2 any more. This means that in order to get an angry husband to be happy, his wife has to cook at least once, and after the last cooking she must not shout any more.

5. $L(\mathcal{G}) = \{b, ab, aab, aaab, \ldots\}$.

6. $L(\mathcal{G}) = \{aa, aaaa, aaaaaa, \ldots\}$.

7. $L(\mathcal{G}) = \{aa, aaaa, aaaaaa, \ldots\}$.

8. The languages in Exercises 6 and 7 are the same. That means that the same language can be generated by different grammars.

9. $A = \{a, b\}$, $G = \{g_0\}$, $\to = \{g_0 \to b, g_0 \to ag_0a\}$ (for instance).

10. $A = \{a, b\}$, $G = \{g_0\}$, $\to = \{g_0 \to \Lambda, g_0 \to ag_0a, g_0 \to b\}$ (for example).

11. This equivalence is certainly reflexive, symmetric and transitive, hence an equivalence relation.

12. (a) Yes. (b) No.

13. Both answers: no (see Theorem 1.23).

14. No for both questions, since T is not closed with respect to forming products and generated submonoids.

15. Yes; $p_4 = 6$ gives $n_0 = 4$ in 1.24. The period is $q = 3$. The corresponding regular language is given by $\{a, a^2, a^3, a^6, a^9, a^{12}, \ldots\}$.

16. Yes, since the exponents n form the periodic set $\{3, 7, 11, 14, \ldots\}$ (see Theorem 1.25).

Chapter 7, §2

1. If one relabels a, b, c, d by $0, 1, 2, 3$, respectively, one gets the table for $(\mathbb{Z}_4, +)$. Hence $\circ_1$ is associative.

2. Relabel a, b, c, d by $(0, 0)$, $(0, 1)$, $(1, 0)$, $(1, 1)$, respectively. Then one gets precisely the (group-) table of $\mathbb{Z}_2 \times \mathbb{Z}_2$. Hence $\circ_2$ is associative.

3. Not associative. Example: $b \circ (b \circ b) \neq (b \circ b) \circ b$. Hence $\circ_3$ is not even "power-associative".

4. No zero element; a is always the identity.

5. By the solutions to Exercises 1–3, the structures in Exercises 1 and 2 are abelian groups, while the one in Exercise 3 is not even a semigroup.

6. In Exercises 1 and 2 the group kernel coincides with $\{a, b, c, d\}$ (since $\{a, b, c, d\}$ form groups in these cases). For Exercise 3 the group kernel is not defined.

7. In Exercise 1 only a is idempotent. In Exercise 2 each element is idempotent. In Exercise 3 a is the only idempotent (if one defines idempotents for groupoids in the obvious way).

8. Since $a \mathbin{\#} (b \mathbin{\#} c) = a \mathbin{\#} z = z = z \mathbin{\#} c = (a \mathbin{\#} b) \mathbin{\#} c$ holds for all $a, b, c \in N$, $(N, \#)$ is a semigroup. The Light test is of no use in this case: every element generates only itself and (z).

9. Only two:

$\circ_1$	e	a	b	c	and	$\circ_2$	e	a	b	c
e	e	a	b	c		e	e	a	b	c
a	a	b	c	e		a	a	e	c	b
b	b	c	e	a		b	b	c	e	a
c	c	e	a	b		c	c	b	a	e

10. Both are abelian groups.

11. This is the number of possibilities to pick out 20 triples (i, j, k) in the $4^3 = 64$-element set $\{1, 2, 3, 4\}^3$; this number is given by $\binom{64}{20} \approx 1.96 \cdot 10^{16}$.

12. $3 \overbrace{}^{(4)}$ $(4)(4)(4)(4)$ can be obtained by
$\quad$ 3
$\quad$ 3 $\qquad g_0 \to (4)(4)g_0 \to (4)(4)(4)(4)g_0 \to (4)(4)(4)(4) = (4)(4)(4)(4)$.
$\quad$ 5
$\quad$ 3 $\overbrace{}^{(4)}$ $(4)(4)(4) \ldots$ cannot be obtained, since it is an infinite string.
$\quad$ 2
$\quad$ 1 $\qquad 12345(4)$ is certainly not in $L(\mathscr{G})$.

Chapter 7, §3

1. S is the free semigroup on the 2-element set X.

2. Yes, since (SiD, D) and (DSi, D) are in R. Hence "sister of daugher", "daughter" and "daughter of sister" are considered to be the same in this case.

3. There is a kinship relationship E such that "E of A" is the same as "A" for each A in S. Also, for each $A \in S$, there is some relationship $B \in S$ such that "A of B", "B of A" and "E" all mean the same.

4. $S = \{[F], [FF], [M], [FM], [FFM]\}$ with the operation table (brackets are omitted)

∘	F	FF	M	FM	FFM
F	FF	F	FM	FFM	FM
FF	F	FF	FFM	FM	FFM
M	FM	FFM	M	FM	FFM
FM	FFM	FM	FM	FFM	FM
FFM	FM	FFM	FF	FM	FFM

5. $S = \{[F], [M], [FM]\}$ with the table (brackets are omitted)

∘	F	M	FM
F	F	FM	FM
M	FM	M	FM
FM	FM	FM	FM

6–7. None of them are monoids, but both are commutative; hence they are not (abelian) groups. This can be seen immediately by using 2.6.

8. $S = \{F^a M^b \mid 0 \le a \le 2, 0 \le b \le 1, ab > 0\}$. The map $h: S \to S'$, $F^a M^b \mapsto F^a M^b$ is a standard epimorphism (F^2 is identified with F in S'). The corresponding partition of S is given by $S = \{F, FF\} \cup \{M\} \cup \{FM, FFM\}$. Hence, according to 6.3.58,

$$\delta(S, S') = \binom{5}{2}^{-1}\left(\binom{2}{2} + \binom{1}{2} + \binom{2}{2}\right) = \frac{1}{10}(1 + 0 + 1) = \frac{1}{5}.$$

9. Let a, b, c be the equivalence classes of P, C and PC, respectively. We then get

S:

	a	b	c
a	a	c	c
b	c	b	c
c	c	c	c

S is not a group (there is no identity, for instance).

10. Same result for most conventions of "few ones". With the convention "weak blocks are ones with no assignment at all" we would get the matrix $\begin{pmatrix} 1 & 1 & 0 \\ 1 & 1 & 0 \\ 1 & 1 & 1 \end{pmatrix}$.

11. Love $L = \begin{pmatrix} 1 & 0 \\ 0 & 0 \end{pmatrix}$, hate $H = \begin{pmatrix} 0 & 1 \\ 1 & 1 \end{pmatrix}$.

$$\langle\{L, H\}\rangle = \left\{ \begin{pmatrix} 0 & 0 \\ 0 & 0 \end{pmatrix}, \begin{pmatrix} 0 & 0 \\ 0 & 1 \end{pmatrix}, \begin{pmatrix} 0 & 0 \\ 1 & 0 \end{pmatrix}, \begin{pmatrix} 0 & 0 \\ 1 & 1 \end{pmatrix}, \begin{pmatrix} 0 & 1 \\ 0 & 0 \end{pmatrix}, \begin{pmatrix} 0 & 1 \\ 0 & 1 \end{pmatrix}, \right.$$

$$\left. \begin{pmatrix} 0 & 1 \\ 1 & 1 \end{pmatrix}, \begin{pmatrix} 1 & 0 \\ 0 & 0 \end{pmatrix}, \begin{pmatrix} 1 & 0 \\ 1 & 0 \end{pmatrix}, \begin{pmatrix} 1 & 1 \\ 0 & 0 \end{pmatrix}, \begin{pmatrix} 1 & 1 \\ 1 & 1 \end{pmatrix} \right\}.$$

12.

$$\text{Love } L = \begin{pmatrix} 1 & 0 & 0 \\ 0 & 1 & 0 \\ 0 & 0 & 1 \end{pmatrix}, \quad \text{hate } H = \begin{pmatrix} 0 & 1 & 1 \\ 1 & 0 & 0 \\ 1 & 0 & 0 \end{pmatrix},$$

$$\langle\{L, H\}\rangle = \left\{ L, H, H^2 = \begin{pmatrix} 1 & 0 & 0 \\ 0 & 1 & 1 \\ 0 & 1 & 1 \end{pmatrix} \right\}.$$

13. $X = \{L, H\}$, $R_1 = \{(LH, H), (HL, H), (H^3, H), (L^2, L)\}$.

14. $X = \{L, H\}$, $R_2 = \{(L^2, L), (H^3, H^2), (LHLH, LH), (HLHL, LHL), (LH^2L, L)\}$. Then the matrices in the solution to Exercise 11 are consecutively given by LHL, HLH, HL, HLH^2, LH, H^2LH, H, L, H^2L, LH^2, H^2.

15. S_1 is a commutative monoid ($e = L$), but not a group.

16. S_2 is not commutative ($LH \neq HL$). There is no identity (but LHL is a (=the) zero element). See 2.6.

17. $\underline{S}$ is presented by $X = \{L, H\}$ and $\underline{R} = R_1 \cup R_2 = \{(LH, H), (HL, H), (H^3, H^2), (L^2, L), (H^3, H), (LHLH, LH), (HLHL, LHL), (LH^2L, L)\}$ (see Exercises 13–14). Then we get (with $\sim$ for $\tilde{R}$) $L \sim LH^2L \sim LHHL \sim HH \sim HHH \sim H$, hence $\underline{S}$ has only one element.

18. By the lines after 6.3.58, $\delta(S_1, \underline{S}) = \delta(S_2, \underline{S}) = 1$.

19. From Exercise 18 we get $\delta(S_1, S_2) = 1$.

20. Yes; each two of S_1, S_2, $\underline{S}$ have distance 1.

21. This is really an open problem.

Appendix

A. Some Fundamental Concepts

A *set* is a collection of different elements; it can be represented in different ways, by enumeration of its elements, say $A = \{a_1, a_2, \dots\}$ or by describing it in terms of a "characteristic property" E which characterizes elements of A, in which case we write $A = \{x \mid x \text{ satisfies } E\}$. Usually, we denote sets by capital letters $A, B, \dots$ and elements by $a, b, \dots$; sets consisting of sets are, in general, denoted by script letters. Some sets which appear frequently have got their own symbols:

$\varnothing$ = the *empty set* (consisting of no elements at all),
$\mathbb{N}$ = the set of all *natural numbers* $1, 2, 3, \dots$ (hence $\mathbb{N}$
  $= \{1, 2, 3, \dots\}$),
$\mathbb{N}_0 = \{0, 1, 2, 3, \dots\}$,
$\mathbb{P} = \{2, 3, 5, 7, 11, \dots\}$ is the set of all *prime numbers*,
$\mathbb{Z} = \{0, 1, -1, 2, -2, 3, -3, \dots\}$ is the set of *integers*,
$\mathbb{Q} = \{x \mid x = a/b, a \in \mathbb{Z}, b \in \mathbb{N}\}$ is the set of *rational numbers*,
$\mathbb{R} = \{x \mid x \text{ is a finite or infinite decimal}$
  $\text{fraction}\}$ is the set of *real numbers*,
$\mathbb{C} = \{x \mid x = a + ib, a, b \in \mathbb{R}, i^2 = -1\}$
  is the set of *complex numbers*.

In order to abbreviate our notation we introduce some logical symbols (without going into the precise definition in terms of mathematical logic). Let p and q stand for so-called propositions (or expressions or statements).

$p \wedge q$ means "p and q";
$p \vee q$ means "p or q (or both)";

$\neg p$ means "not p";

$p \Rightarrow q$ or $q \Leftarrow p$ means "if p then q" or "p implies q";

$p \Leftrightarrow q$ means "p and q are equivalent";

$\forall \, x \in A$: means "for all x in A we have...";

$\exists \, x \in A$: means "there exists at least one x in A such that...";

$\exists | x \in A$: means "there exists exactly one x in A such that...";

$p := q$ or $p :\Leftrightarrow q$ means "p equals q by definition".

In the last line we used the symbol $:=$ in terms of a defining colon.

A set which consists of a single element, say a, is called a *singleton* and denoted by $\{a\}$. A is said to be a *subset of B* if each element of A is an element of B, in symbols: $A \subseteq B$ thus $A \subseteq B :\Leftrightarrow \forall \, x \in A: x \in B$. Further we define equality of sets by $A = B :\Leftrightarrow A \subseteq B \wedge B \subseteq A$. Then $A \neq B :\Leftrightarrow \neg(A = B)$. We say that A is a *proper subset* of B, in symbols $A \subset B$, if $A \subseteq B \wedge A \neq B$.

Let A and B be sets.

(i) $\mathscr{P}(A) := \{S \, | \, S \subseteq A\}$ is called the *power set* of A (the set of all subsets of A).

(ii) $A \times B := \{(a, b) \, | \, a \in A \wedge b \in B\}$ is called the *cartesian product* of A and B, its elements are called *ordered pairs*; equality of pairs is defined by

$$(a, b) = (c, d) :\Leftrightarrow a = c \wedge b = d.$$

We define the cartesian product $\mathsf{X}_{i=1}^{n} A_i$ or $A_1 \times A_2 \times \ldots \times A_n$ of n sets $A_1, \ldots, A_n$ as the set $\{(a_1, \ldots, a_n) \, | \, a_i \in A_i, i = 1, 2, \ldots, n\}$. The elements of $\mathsf{X} A_i$ are called *ordered n-tuples*. We use the notation $A \times A =: A^2$, $A \times A \times A =: A^3$, etc.

(iii) $A \cap B := \{x \, | \, x \in A \wedge x \in B\}$ is the *intersection* of A and B.

(iv) $A \cup B := \{x \, | \, x \in A \vee x \in B\}$ is the *union* of A and B.

(v) More generally we define the intersection and union of sets A_i, $i \in I$, where I is an "index set", as follows:

$$\bigcap_{i \in I} A_i := \{x \, | \, \forall \, i \in I: x \in A_i\} \quad \text{and} \quad \bigcup_{i \in I} A_i := \{x \, | \, \exists \, i \in I: x \in A_i\}.$$

(vi) $A \backslash B := \{x \in A \, | \, x \notin B\}$ is the *difference* of A and B.

(vii) If $A \subseteq S$ then $C(A) := S \backslash A$ is called the *complement* of A in S.

(viii) $A \triangle B := (A \backslash B) \cup (B \backslash A)$ is the *symmetric difference* of A and B.

(ix) If $0 \in A$ then $A^* := A \backslash \{0\}$, e.g. $\mathbb{N}_0^* = \mathbb{N}$.

A *Moore system* $\mathscr{M}$ on a set A is a set of subsets of A which contains A and has the property that any intersection of elements of $\mathscr{M}$ again belongs to $\mathscr{M}$. If $B \subseteq A$ and $\mathscr{M}$ is a Moore system on A then $\mathscr{M}$ contains exactly one smallest subset of A which contains B, namely the intersection of all sets in $\mathcal{M}$ which contain B. We denote this element of $\mathcal{M}$ (= subset of A) by $\langle B \rangle_{\mathcal{M}}$ or briefly by $\langle B \rangle$ and call it the element of $\mathcal{M}$ which is generated by B. Note that $\langle B \rangle \in \mathcal{M}$: B is called a *generating set* for $\langle B \rangle$. If some $M \in \mathcal{M}$ can be written as $M = \langle F \rangle$, where F is a finite set, then M is said to be *finitely generated* (f.g.). Every $M \in \mathcal{M}$ is generated by some subset of A (for instance, by M itself). Subsets of the cartesian product $A \times B$ of two sets A and B are called *relations* from A into B. If $A = B$, we simply speak of a relation on A. Let R be a relation on the set A. We call R to be:

(i) *reflexive* if $(a, a) \in R$ for all $a \in A$;
(ii) *symmetric* if $(a, b) \in R$ implies that $(b, a) \in R$ for all $a, b \in A$;
(iii) *antisymmetric* if $(a, b) \in R$ and $(b, a) \in R$ always implies that $a = b$;
(iv) *transitive* if $(a, b) \in R$ and $(b, c) \in R$ imply that $(a, c) \in R$ for all $a, b, c \in A$.

Quite frequently one writes $a R b$ if $(a, b) \in R$. The negation of this is usually denoted by $a \not{R} b$.

Three types of relations are of fundamental importance: equivalence relations, order relations and functions. A reflexive, symmetric and transitive relation on a set S is called an *equivalence relation*. Usually, an equivalence relation denotes some type of similarity between the elements of A; therefore one frequently uses symbols like $\sim$, $\equiv$, etc. for such a relation rather than R. If $a \in A$ and if $\sim$ is an equivalence relation on A, the set $\{b \in A \, | \, a \sim b\}$ of all elements of A which are equivalent to a is called the *equivalence class of a* and denoted by $[a]$. $A/\sim$ stands for the set of all equivalence classes in A; it is called the *factor set* of A with respect to $\sim$. A reflexive, antisymmetric and transitive relation on a set A is called a *partial order* (on A). These relations are usually denoted by symbols like $\leq$ or $\subseteq$. Then $(A, \leq)$ is called a *partially ordered* set *or poset.*

A relation R from a set A into a set B is called a *function* if for all $a \in A$ there is a unique $b \in B$ such that $(a, b) \in R$. Functions are usually denoted by small letters like f, g, h. If a and b are as above, we denote b by $b = f(a)$ if f is a function from A into B. The whole story is often abbreviated by writing $f: A \to B$, $a \mapsto b = f(a)$. For any set A, the function $\mathrm{id}_A: A \to A$, $a \mapsto a$ is called the *identity function* on A. B^A denotes the set of all functions from A into B. Instead of "function" we also use the terms *mapping* or *map*. If f is a function from A into B, we call f to be:

(i) *injective* (or *one-to-one*) if $f(a) = f(a')$ implies $a = a'$ for all $a, a' \in A$;
(ii) *surjective* (or *onto*) if each $b \in B$ can be written as $b = f(a)$ for some $a \in A$;
(iii) *bijective* if f is injective and surjective.

In a bijective function $f: A \to B$, each $b \in B$ can uniquely be written as $b = f(a)$; hence the assignment $f^{-1}: B \to A$, $b \mapsto a$ (with $f(a) = b$) establishes another function, which is called the *inverse function* to f. It is easily verified that f^{-1} is again bijective and that $(f^{-1})^{-1} = \mathrm{id}_A$ holds. Note that

f^{-1} is only defined if f is bijective. Bijective functions from A to A are called *permutations*. S_A denotes the set of all permutations on A. If $f \in S_A$ and $A = \{1, 2, \ldots, n\}$ then we sometimes write $\begin{pmatrix} 1 & 2 & \ldots & n \\ f(1) & f(2) & \ldots & f(n) \end{pmatrix}$ instead of f.

Let $f: A \to B$ be a function and $A_1 \subseteq A$, $B_1 \subseteq B$. Then $f(A_1) :=$ $\{f(a) \mid a \in A_1\}$ is the *image* of A_1 and $f^{-1}(B_1) := \{a \in A \mid f(a) \in B_1\}$ is the *co-image* of B_1. $f|_{A_1}: A_1 \to B$, $a \mapsto f(a)$ is the *restriction* of f to A_1. Let $f: I \to A$ be a function. f is also called a *family* of elements in A, and instead of $\{(i, f(i)) \mid i \in I\}$ we write $(a_i)_{i \in I}$ with $a_i = f(i)$. Note that, unless f is injective, there exist $i, j \in I$ with $i \neq j$, such that $a_i = a_j$. Let $f: A \to B$ and $g: B \to C$ be functions. Then $g \circ f: A \to C$, $a \mapsto g(f(a))$ is called the *composite* of f and g. Let $f: A \to B$, $g: B \to C$, $h: C \to D$ be functions. Then $(h \circ g) \circ f = h \circ (g \circ f)$.

The composite of injective (surjective, bijective) functions f, g is also injective (surjective, bijective). If f and g are bijective then $(f \circ g)^{-1} = g^{-1} \circ f^{-1}$ holds. Note the change in the order of f, g. Also, if $f: A \to B$ is bijective, then $f^{-1} \circ f = \mathrm{id}_A$ and $f \circ f^{-1} = \mathrm{id}_B$ hold.

Let I be an index set and let A_i be a set for each $i \in I$. Then

$$\underset{i \in I}{\times} A_i := \left\{ f: I \to \bigcup_{i \in I} A_i \mid \forall\, i \in I: f(i) \in A_i \right\}$$

is the cartesian product of the sets A_i ($i \in I$). Each element in $\times_{i \in I} A_i$ is called a *choice function*. If $f(i) = a_i \in A_i$, we usually write $(\ldots, a_i, \ldots)$ instead of f and call this element an $|I|$-tuple. For $I = \{1, 2, \ldots, n\}$, $\times_{i \in I} A_i$ coincides with $\times_{i=1}^n A_i$ as defined earlier. We always require the following:

Axiom ("Axiom of Choice"). *Let I be a nonempty set and let A_i ($i \in I$) be nonempty sets then there is a function $f: I \to \bigcup_{i \in I} A_i$ with $f(i) \in A_i$ for all $i \in I$ (f is a choice function).*

Another way to express the Axiom of Choice is the following: "Any cartesian product of nonempty sets is nonempty." For a set A let $|A|$ denote the number of elements of A (if A is finite), or the symbol ∞ (if A is infinite). Some basic formulas about these "cardinal numbers" are the following (A and B are arbitrary finite sets):

(i) $|A \cup B| = |A| + |B| - |A \cap B|$;
(ii) $|\mathcal{P}(A)| = 2^{|A|}$;
(iii) $|A \times B| = |A| \cdot |B|$;
(iv) $|B^A| = |B|^{|A|}$;
(v) $|S_A| = |A|!$ ($|A|$-factorial);
(vi) $|\varnothing| = 0$.

Let A be a nonempty set. A mapping from A^2 into A is called a (binary)

operation on A. For such operations we use symbols like $\circ$, $+$, $*$, $\cdot$ ....
Instead of $\circ\,(a, b)$ we write $a \circ b$. Here are two examples:

$$+: \mathbb{R} \times \mathbb{R} \to \mathbb{R}, (x, y) \mapsto +(x, y) = x + y,$$

$$\cdot: \mathbb{R} \times \mathbb{R} \to \mathbb{R}, (x, y) \mapsto \cdot(x, y) = x \cdot y.$$

Let $\circ$ be a binary operation on A;

$\circ$ is *associative* $:\Leftrightarrow \forall\, a, b, c \in A: (a \circ b) \circ c = a \circ (b \circ c)$,

$\circ$ is *commutative* $:\Leftrightarrow \forall\, a, b \in A: a \circ b = b \circ a$.

$n \in A$ is called a *neutral element* or *identity* (element) with respect to $\circ$ iff

$$\forall\, a \in A: a \circ n = n \circ a = a.$$

A set A with an associative binary operation $\circ$ is called *semigroup*. A semigroup is called *monoid*, if there exists a neutral element in A. A *group* G is a monoid in which every element has an *inverse*; that is for any g in G there is an h in G such that $g \circ h = h \circ g = n$, where n is the neutral element. These and other structures are extensively studied in this book.

We conclude by introducing some common notations from linear algebra. Examples of vector spaces over a field F are

$$F^n := \{(x_1, \ldots, x_n) \,|\, x_1, \ldots, x_n \in F\},$$

$$F_m := \left\{ \begin{pmatrix} x_1 \\ \vdots \\ x_m \end{pmatrix} =: (x_1, \ldots, x_m)^T \,\bigg|\, x_1, \ldots, x_m \in F \right\},$$

with the usual componentwise operations, i.e.

$$(x_1, \ldots, x_n) + (y_1, \ldots, y_n) = (x_1 + y_1, \ldots, x_n + y_n),$$

$$k(x_1, \ldots, x_n) = (kx_1, \ldots, kx_n),$$

and analogously in F_m.

If $\mathbf{M}$ is an $m \times n$ matrix with entries a_{ij} in a set A, let $\mathbf{M}^T$ denote the *transpose* of $\mathbf{M}$. Let A_n^n be the *set of all square matrices* over A. Then $\mathbf{M} \in A_n^n$ is called *regular* (or *nonsingular*) if there is an $\mathbf{N} \in A_n^n$ such that $\mathbf{MN} = \mathbf{NM} = \mathbf{I}$. The set $\{\mathbf{M} \in A_n^n \,|\, \mathbf{M}$ regular$\}$ is a group with respect to matrix multiplication. It is called the *general linear group* and is denoted by $\mathrm{GL}(n, A)$.

B. Computer Programs

This Appendix contains some examples of computer programs for algorithms relevant to Chapters 3 and 5. The algorithms and programs are presented in a way that allows them to be readily used on several types of micro-computers, such as APPLE II or HP 85. Usually the name Computer

Algebra is used for the discipline that has the study and implementation of algebraic algorithms at its core. The literature in this field is growing rapidly. We refer to a recent survey volume *Computer Algebra*, by J. Calmet (ed.), Lectures Notes in Computer Science, vol. 144, Springer-Verlag, Berlin, 1982. See also LIPSON (references to Chapter 5, §4).

The first few programs in this section are written in BASIC, the algorithms involving polynomials are written in muLISP. These programs should provide sufficient examples on how to implement some of the algorithms given in the text. It is, of course, possible to use other languages for implementation on computers. The book *Mathematical Experiments on the Computer* by U. Grenander (Academic Press, New York, 1982) uses the programming language APL to present a variety of programs, amongst them: root finding, multiplication and division algorithms for polynomials and also Berlekamp's method of factoring a polynomial mod 2.

Other programming languages for computer algebra are REDUCE (by the University of Utah) and MACSYMA (developed by M.I.T.). SAC-2 (of the University of Wisconsin, Madison) is a computer algebra system that is written in the language ALDES. A widely used system for computer algebra on microcomputers is muMATH-80 (from the Microsoft Corporation, Bellevue, Washington). Thomas Beth *et al.* developed several programs, such as factorization, irreducibility test, primitivity test, for polynomials over $\mathbb{F}_p$ (see BETH in Chapter 4). These programs are written in Simula for a Cyber 172 computer. BETH also contains implementations of encoding and decoding algorithms for BCH codes.

We would like to convey our thanks to Stephen Andrewartha (University of Tasmania) for his considerable help in preparing the computer programs.

1. A Probabilistic Primality Test

This first program gives a probabilistic algorithm for checking primality and is due to Rabin (see KNUTH). As a subroutine, calculations $a^s \pmod{n}$ are performed. Rabin's algorithm tests an integer $n = 1 + 2^k q$, q odd, and uses a random integer x in the range $1 < x < n$. If the test is repeated R times and the algorithm reports 20 times in a row that "n is probably prime", then the probability is less than $(\frac{1}{4})^{20}$ that such a 20-times-in-a-row procedure gives the wrong information about n.

The steps of the algorithm are as follows:

(i) Find a random integer x, $1 < x < n$.
(ii) Set $j = 0$ and $y = x^q \pmod{n}$.
(iii) If $j = 0$ and $y = 1$, or if $y = n - 1$, stop and display "n is probably prime". If $j > 0$ and $y = 1$, go to (v).
(iv) Increase j by 1. If $j < k$, set $y = y^2 \pmod{n}$ and return to (iii).
(v) Terminate algorithm and display "n is not prime".

```
10   !  ALGORITHM P
20   !  PROBABILISTIC PRIMALITY
30   !  TEST FOR N=1+Q*2^K
40   !  ----------------------
50   RANDOMIZE
60   INTEGER K,Q,R,N,N1,I,J,Y,A,S
70   DISP "INPUT N";
80   INPUT N
90   DISP "REPETITIONS OF TEST";
100  INPUT R
110  N1=N-1
120  Q=N1 DIV 2
130  K=1
140  IF Q MOD 2=1 THEN 180
150  Q=Q DIV 2
160  K=K+1
170  GOTO 140
180  FOR I=1 TO R
190  X=INT(N-2)*RND)+2
200  J=0
210  A=X
220  S=Q
230  GOSUB 380
240  IF J=0 AND Y=1 OR Y=N1 THEN
     340
250  IF J>0 and Y=1 THEN 320
260  J=J+1
270  IF J>=K THEN 320
280  A=Y
290  S=2
300  GOSUB 380
310  GOTO 240
320  DISP N; "IS NOT PRIME"
330  I=R+2
340  NEXT I
350  IF I<R+2 THEN DISP N; "IS PRO
     BABLY PRIME"
360  STOP
370  !  ----------------------
375  !  ----------------------
380  !  SUBROUTINE TO CALCULATE
390  !  Y=A^S        (MOD N)
400  INTEGER L,M,B(100)
410  L=0
420  IF S=0 THEN 470
430  B(L)=S MOD 2
440  S=S DIV 2
450  L=L+1
460  GOTO 420
```

```
470  Y=1
480  FOR M=L-1 TO 0 STEP -1
490  Y=Y*Y MOD N
500  IF B(M)=1 THEN Y=Y*A MOD N
510  NEXT M
520  RETURN
530  END
```

2. A Primality Test due to Solovay and Strassen (see Exercise 23 in §2 of Chapter 5)

If the test is repeated R times, that is we choose randomly $R = 100$, say, different values of a between 1 and $n - 1$, and if

$$(a, n) = 1 \quad \text{and} \quad J(a, n) \equiv a^{(n-1)/2} \, (\text{mod } n)$$

hold for each a, then the probability that n is prime is approximately $1 - 2^{-100}$.

```
10   !  PROBABILISTIC PRIMALITY
20   !  TEST DUE TO
30   !  SOLOVAY & STRASSEN
40   RANDOMIZE
50   INTEGER N,R,A,I,Y,N1,E1,A0,J
     ,S,G,A1,A2,S1,N2,T
60   DISP "INPUT N";
70   INPUT N
80   DISP "REPETITIONS OF TEST"!
90   INPUT R
100  N1=N-1
110  S1=N1 DIV 2
120  FOR I=1 TO R
130  A=INT(N1*RND)+1
140  GOSUB 380
150  IF G#1 THEN 320
160  J=1
170  A0=A
180  N2=N
190  IF A0=1 THEN 290
200  IF A0 MOD 2=0 THEN 260
210  IF (A0-1)*(N2-1) DIV 4 MOD 2
     =1 THEN J=-J
220  T=A0
230  A0=N2 MOD A0
240  N2=T
250  GOTO 190
```

```
260  IF ((N2-1) DIV 2+1) DIV 2 MO
     D 2=1 THEN J=-J
270  A0=A0 DIV 2
280  GOTO 190
290  S=S1
300  GOSUB 490
310  IF Y=J MOD N THEN 340
320  DISP N; "IS NOT PRIME"
330  I=R+2
340  NEXT I
350  IF I<R+2 THEN DISP N; "IS PRO
     BABLY PRIME"
360  STOP

370  ! --------------------
380  !  EUCLID'S ALGORITHM TO
390  !  CALCULATE G=( A,N )
400  G=A
410  A1=N
420  IF A1=0 THEN 470
430  A2=A1
440  A1=G MOD A1
450  G=A2
460  GOTO 420
470  RETURN
480  ! --------------------
490  !  SUBROUTINE TO CALCULATE
500  !        Y=A^S (MOD N)
510  INTEGER L,M,B(100)
520  L=0
530  IF S=0 THEN 580
540  B(L)=S MOD 2
550  S=S DIV 2
560  L=L+1
570  GOTO 530
580  Y=1
590  FOR M=L-1 TO 0 STEP -1
600  Y=Y*Y MOD N
610  IF B(M)=1 THEN Y=Y*A MOD N
620  NEXT M
630  RETURN
640  END
```

3. A Nullspace Algorithm (due to KNUTH, see Theorem 5.1 in Chapter 4).

Note: The matrix **A** in 4.5.1 is called **D** in the program below, the field K is the field $\mathbb{Z}_p$.

514

The following program contains as a subroutine the *extended Euclidean algorithm* for finding the greatest common divisor G of nonnegative integers A and P (in the subroutine P is not necessarily prime) and expressing G in the form $AS + PT = G$.

```
10  ! ALGORITHM N—CALCULATION
20  ! OF THE NULL SPACE OF D( , )
30  DIM D(20,20),V(20,20),C(20)
40  INTEGER N,N1,R,I,S,J,K,L,T,G
    ,P,A
50  ! INPUT P,N,D( , )
60  N1=N-1
70  R=0
80  FOR I=0 TO N1
90  C(I)=-1
100 NEXT I
110 FOR K=0 TO N1
120 FOR J=0 to N1
130 IF D(K,J)=0 OR C(J)>-1 THEN
    290
140 A=D(K,J)
150 GOSUB 450
160 T=(-S) MOD P
170 FOR L=0 TO N1
180 D(L,J)=T*D(L,J) MOD P
190 NEXT L
200 FOR I=0 TO N1
210 IF I=J THEN 260
220 T=D(K,I)
230 FOR L=0 TO N1
240 D(L,I)=(D(L,I)+T*D(L,J)) MOD
    P
250 NEXT L
260 NEXT I
270 C(J)=K
280 J=100
290 NEXT J
300 IF J>99 THEN 410
310 R=R+1
320 FOR J=0 TO N1
330 FOR I=0 TO N1
340 IF C(I)#J THEN 370
350 V(R,J)=D(K,I)
360 I=100
370 NEXT I
380 IF I>99 THEN 400
390 IF J=K THEN V(R,J)=1 ELSE V(
    R,J)=0
400 NEXT J
```

```
410   NEXT K
420   !  OUTPUT V(,)
430   STOP
440   !
450   !  EUCLID'S EXTENDED
460   !  ALGORITHM A,P->G,S,T
470   INTEGER A1,S1,T1,Q,Q1
480   G=A
490   A1=P
500   S,T1=1
510   S1,T=0
520   IF A1=0 THEN RETURN
530   Q=G DIV A1
540   Q1=G @ G=A1 @ A1=Q1-A1*Q
550   Q1=S @ S=S1 @ S1=Q1-S1*Q
560   Q1=T @ T=T1 @ T1=Q1-T1*Q
570   GOTO 520
580   END
```

4. An Algorithm to Evaluate x^n (mod m) Using the Right-to-Left Binary Method for Exponentiation (see KNUTH)

```
10   !  CALCULATE X^N ( MOD M )
20   INTEGER X,N,M,N1,Y1,Z1,N2
30   DISP "INPUT X,N,M";
40   INPUT X,N,M
50   N1=N
60   Y1=1
70   Z1=X
80   N2=N1
90   N1=N1 DIV 2
100  IF N2 MOD 2=0 THEN 130
110  Y1=Z1*Y1 MOD M
120  IF N1=0 THEN 150
130  Z1=Z1*Z1 MOD M
140  GOTO 80
150  DISP X; "^"; N; "="; Y1; "( MOD";
     M;")"
160  END
```

5. The Chinese Remainder Theorem for Integers

We give a BASIC program for calculations according to the Chinese Remainder Theorem for integers. Let $N_1, \ldots, N_r$ be r positive integers such that $\text{GCD}(N_i, N_j) = 1$ for $i \neq j$. Then the system of linear congruences

$$X \equiv B_1 \;(\text{mod } N_1), \ldots, X \equiv B_r \;(\text{mod } N_r)$$

has a unique simultaneous solution X_0 modulo $N_1 N_2 \ldots N_r$. This solution is obtained by

(1) Form $N = N_1 N_2 \ldots N_r$.
(2) Solve each of

$$(N/N_k)X \equiv 1 \pmod{N_k} \quad \text{for } k = 1, 2, \ldots, r,$$

to obtain a solution T_k (for $k = 1, 2, \ldots, r$) mod N_k. This is done by the extended Euclidean algorithm.
(3) Set $X_0 = B_1(N/N_1)T_1 + \ldots + B_r(N/N_r)T_r \pmod{N}$.

In the program the T_k are written as T for each loop $k = 1, 2, \ldots, r$.

```
10   !  CHINESE REMAINDER
20   !  THEOREM
30   DIM B(100),N(100)
40   INTEGER R,N0,X0,A,P,G,T,I
50   !  INPUT R,B( ),N( )
60   N0=1
61   FOR I=1 TO R
62   N0=N0*N(I)
63   NEXT I
70   FOR I=1 TO R
80   A=N(I)
90   P=N0 DIV A
100  GOSUB 160
110  IF G#1 THEN DISP "N( )'S NOT
     RELATIVELY PRIME" @ STOP
120  X0=(X0+B(I)*P*(T MOD A)) MOD
     N0
130  NEXT I
140  !  OUTPUT X0
145  STOP
150  !
160  !  EUCLID'S EXTENDED
170  !  ALGORITHM A,P->G,T
180  INTEGER A1,T1,Q,Q1
190  G=A
200  A1=P
210  T1=1
220  T=0
230  IF A1=0 THEN RETURN
240  Q=G DIV A1
250  Q1=G @ G=A1 @ A1=Q1-A1*Q
260  Q1=T @ T=T1 @ T1=Q1-T1*Q
270  GOTO 230
280  END
```

We now present some algorithms for polynomials with coefficients in $\mathbb{Z}_p$ (or $\mathbb{F}_p$). Therefore all calculations with integral coefficients are performed modulo a prime P. The programs are written in muLISP.

6. Greatest Common Divisor of Polynomials over $\mathbb{Z}_p$

This algorithm calculates the greatest common divisor of two polynomials A and B modulo a prime number P.

(a) *Usage*

We evaluate (POLYGCDP A B P), where A and B are polynomials represented as lists of coefficients (e.g. represent $x^3 - 3$ as (1 0 0 −3)), and P is a prime number (specifying $\mathbb{Z}_p$).

(b) *Implementation*

```
(DEFUN POLYGCDP (LAMBDA (POLYA POLYB P) % GCD
(POLYA,POLYB) %
    ((NULL POLYB)(MULCOEFF POLYA (INVERSE P (CAR POLYA)
    0 1) P))
    (POLYGCDP POLYB (POLYRMDR POLYA POLYB P) P) ))

(DEFUN POLYRMDR (LAMBDA (POLYA POLYB P) % POLYA mod
POLYB %
    ((GREATERP (DEGREE POLYB)(DEGREE POLYA)) POLYA)
    (POLYRMDR (NOZERO (CASTOUT POLYA (MULCOEFF POLYB
    (TIMES (CAR POLYA)(INVERSE P (CAR POLYB) 0 1 )) P))
    P) POLYB P) ))

(DEFUN MULCOEFF (LAMBDA (POLYA B M) % (POLYA x B) mod
M%
((NULL POLYA) NIL)
(CONS (REMAINDER (TIMES (CAR POLYA) B) M)
    (MULCOEFF (CDR POLYA) B M)) ))

(DEFUN INVERSE (LAMBDA (P B T0 T1) % B⁻¹ mod P %
    ((ZEROP B) T0)
    ((MINUSP B) (INVERSE P (PLUS B P) 0 1))
    (INVERSE B (REMAINDER P B) T1 (DIFFERENCE T0
    (TIMES (QUOTIENT P B) T1))) ))

(DEFUN CASTOUT (LAMBDA (POLYA POLYB)
    ((NULL POLYB) POLYA)
    (CONS (DIFFERENCE (CAR POLYA)(CAR POLYB))
    (CASTOUT (CDR POLYA)(CDR POLYB))) ))
```

```
(DEFUN DEGREE (LAMBDA (POLYA)
  ((NULL POLYA) -1)
  (PLUS 1 (DEGREE (CDR POLYA))) ))

(DEFUN NOZERO (LAMBDA (POLYA P)
  ((NULL POLYA) NIL)
  ((ZEROP (REMAINDER (CAR POLYA) P))(NOZERO (CDR
  POLYA) P))
  POLYA))
```

7. Euclid's Extended Algorithm for Polynomials over $\mathbb{Z}_p$

This algorithm calculates the greatest common divisor of two polynomials A and B over $\mathbb{Z}_p$ as the linear combination of two polynomials S and T over $\mathbb{Z}_p$.

(a) *Usage*

Evaluate (POLYEXTP $A\ B\ P$ (1) NIL NIL (1)), where A and B are polynomials, and P is a prime number as before. POLYEXTP returns the list $(G\ S\ T)$ where $G = \text{GCD}(A, B) = SA + TB$.

(b) *Implementation*

```
(DEFUN POLYEXTP (LAMBDA (POLYA POLYB P POLYS0
          POLYSI POLYT0 POLYT1 RTEMP)
  (NULL POLYB)(LIST POLYA POLYS0 POLYT0))
  (SETQ RTEMP (REVERSE (POLYDIV POLYA POLYB P)))
  (POLYEXTP POLYB (TRANSFORM POLYA POLYB P RTEMP) P
          POLYS1 (TRANSFORM POLYS0 POLYS1 P RTEMP)
          POLYT1 (TRANSFORM POLYT0 POLYT1 P RTEMP)) ))

(DEFUN POLYDIV (LAMBDA (POLYA POLYB P TEMP)
          % POLYA div POLYB %
  ((GREATERP (DEGREE POLYB)(DEGREE POLYA)) NIL)
  (SETQ TEMP (TIMES (CAR POLYA)(INVERSE P (CAR POLYB)
  0 1)))
  (CONS TEMP (POLYDIV (CDR (CASTOUT POLYA
  (MULCOEFF POLYB TEMP P))) POLYB P)) ))

(DEFUN TRANSFORM (LAMBDA (POLYA POLYB P RTEMP)
          % POLYA-POLYB*RTEMP %
  (NOZERO (REVERSE (REVPOLYADD (REVERSE POLYA)
  (MULCOEFF (REVPOLYTIMES (REVERSE POLYB) RTEMP P)
  -1 P) P)) P) ))
```

```
(DEFUN REVPOLYADD (LAMBDA (RPOLYA RPOLYB P)
    % RPOLYA + RPOLYB %
    ((NULL RPOLYA) RPOLYB)
    ((NULL RPOLYB) RPOLYA)
    (CONS (REMAINDER (PLUS (CAR RPOLYA)(CAR RPOLYB)) P)
    (REVPOLYADD (CDR RPOLYA)(CDR RPOLYB) P)) ))

(DEFUN REVPOLYTIMES (LAMBDA (RPOLYA RPOLYB P)
    % RPOLYA*RPOLYB %
    ((NULL RPOLYB) NIL)
    (REVPOLYADD (CONS 0 (REVPOLYTIMES RPOLYA
    (CDR RPOLYB) P))(MULCOEFF RPOLYA (CAR RPOLYB) P)
    P) ))
```

8. The Inverse of a Polynomial modulo a Polynomial over $\mathbb{Z}_p$

This algorithm calculates a polynomial I over $\mathbb{Z}_P$ such that for given polynomials A and M over $\mathbb{Z}_P$

$$AI \equiv 1 \,(\mathrm{mod}\ M).$$

(a) *Usage*

Evaluate (POLYINVERSE $M\,A\,P$), where M and A are polynomials, and P is a prime number. POLYINVERSE returns the polynomial $I = A^{-1}$ (modulo M)(modulo P). If no such inverse exists, NIL (the zero polynomial) is returned.

(b) *Implementation*

```
(DEFUN POLYINVERSE (LAMBDA (POLYM POLYA P TEMP)
    (SETQ TEMP (POLYEXTP POLYM POLYA P (1) NIL NIL (1)))
    ((EQUAL (CAR TEMP)(1))(POLYRMDR (CADDR TEMP) POLYM
    P))
    NIL))
```

9. The Chinese Remainder Theorem for Polynomials over $\mathbb{Z}_P$

This last algorithm calculates a polynomial H over $\mathbb{Z}_P$ which is a solution of a system of certain linear polynomial congruences.

(a) *Usage*

Evaluate (POLYCRT $F\,G\,P$), where F and G are *lists* of polynomials, and P is a prime number. POLYCRT returns a polynomial $H(x)$ modulo P

having the property

$$H(x) \equiv F_i(x) \,(\text{mod } G_i(x)), \qquad i = 1, \ldots n.$$

Here all polynomials are over $\mathbb{Z}_P$ and the polynomials $G_1, \ldots, G_n$ are relatively prime. H is calculated mod P and mod $G_1 \ldots G_n$.

(b) *Implementation*

```
(DEFUN POLYCRT (LAMBDA (FPOLYS GPOLYS P GPROD)
   (SETQ GPROD (POLYPROD GPOLYS P))
   (NOZERO (REVERSE (PCRTREV FPOLYS
      GPOLYS P GPROD)) P) ))

(DEFUN POLYPROD (LAMBDA (POLYS P) % G₁*G₂...*Gₙ %
   ((NULL POLYS) (1))
   (REVERSE (REVPOLYTIMES (REVERSE (CAR POLYS))
      (REVERSE (POLYPROD (CDR POLYS) P)) P)) ))

(DEFUN PCRTREV (LAMBDA (FPOLYS GPOLYS P GPROD)
   ((NULL FPOLYS) NIL)
   (REVPOLYADD (CRTTERM (CAR FPOLYS)(CAR GPOLYS)
      P GPROD)(PCRTREV (CDR FPOLYS)(CDR GPOLYS) P GPROD)
      P) ))

(DEFUN CRTTERM (LAMBDA (POLYF POLYG P GPROD POLYH)
   (SETQ POLYH (POLYDIV GPROD POLYG P))
   (REVPOLYTIMES    (REVERSE    POLYF)(REVPOLYTIMES
   (REVERSE POLYH)
      (REVERSE (CADDR (POLYEXTP POLYG POLYH P (1)
   NIL NIL (1)))) P) P) ))
```

Numerical Examples

The reader might like to use some of the algorithms above for the solution of the following exercises.

1. Calculate $(309)^{253} \,(\text{mod } 3233)$.

2. Solve the system of congruences

$$x \equiv 3 \,(\text{mod } 5), \qquad x \equiv 5 \,(\text{mod } 7), \qquad x \equiv 7 \,(\text{mod } 11).$$

3. Determine the greatest common divisor of

$$f = 3x^6 + x^5 + 4x^4 + 4x^3 + 3x^2 + 4x + 2 \quad \text{and}$$
$$g = 2x^6 + 4x^5 + 3x^4 + 4x^3 + 4x^2 + x + 3$$

over $\mathbb{Z}_7$.

4. Represent the greatest common divisor of f and g as a linear combination of polynomials over $\mathbb{Z}_3$:

$$f = 2x^6 + x^3 + x^2 + 2 \quad \text{and} \quad g = x^4 + x^2 + 2x.$$

5. Find the inverse of $x^3 + x^2 + 1 \pmod{x^5 + x^3 + x^2 + 1}$ over $\mathbb{Z}_2$.

6. Solve the system of polynomial congruences over $\mathbb{Z}_3$:

$$h \equiv x - 1 \pmod{2x^4 + 2}, \qquad h \equiv x + 1 \pmod{x^5 + 2}.$$

Answers to Numerical Examples

1. 1971.

2. $x \equiv 348 \pmod{385}$.

3. $x^3 + 3x^2 + 4x + 6$.

4. $1 = (2x^3 + 2x^2 + x + 2)f + (2x^5 + 2x^4 + 2x^3 + 2x + 2)g$.

5. $x^4 + x^3 + 1$.

6. $h = 2x^8 + x^7 + 2x^6 + x^5 + x^3 + 2x^2 + 2x \ (\mathrm{mod}(2x^4 + 2)(x^5 + 2))$ over $\mathbb{Z}_3$.

Bibliography

General References on Abstract Algebra

BIRKHOFF, G. and MACLANE, S. *A Survey of Modern Algebra*, 4th ed. Macmillan, New York, 1977.
DEAN, R. A. *Elements of Abstract Algebra*. Wiley, New York, 1966.
DUBREIL, P. and DUBREIL-JACOTIN, M. L. *Lectures on Modern Algebra*. Hafner, New York, 1967.
FRALEIGH, J. B. *A First Course in Abstract Algebra*. Addison-Wesley, Reading, Mass., 1967.
GOLDSTEIN, L. J. *Abstract Algebra: A First Course*. Prentice-Hall, Englewood Cliffs, N.J., 1973.
HERSTEIN, I. N. *Topics in Algebra*, 2nd ed. Wiley, New York, 1975.
JACOBSON, N. *Basic Algebra*, vols. I and II. Freeman, San Francisco, 1974, 1980.
KUROSH, A. G. *Lectures on General Algebra*. Chelsea, New York, 1965.
LANG, S. *Algebra*. Addison-Wesley, Reading, Mass., 1967.
MCCOY, N. H. and BERGER, T. R. *Algebra: Groups, Rings and Other Topics*, Allyn and Bacon, Boston, 1977.
MACLANE, S. and BIRKHOFF, G. *Algebra*. Macmillan, New York, 1967.
RÉDEI, L. *Algebra*, vol. I. Pergamon, Oxford, 1967.
VAN DER WAERDEN, B. L. *Modern Algebra*, vols. 1 and 2. Ungar, New York, 1970.

General References on Applied Algebra

BIRKHOFF, G. and BARTEE, T. C. *Modern Applied Algebra*. McGraw-Hill, New York, 1970.
BOBROW, L. S. and ARBIB, M. A. *Discrete Mathematics*. Saunders, Philadelphia, 1974.
CHILDS, L. *A Concrete Introduction to Higher Algebra*. Springer-Verlag, Berlin–Heidelberg–New York, 1979.
DORNHOFF, L. L. and HOHN, F. E. *Applied Modern Algebra*. Macmillan, New York, 1978.
FISHER, J. L. *Application-Oriented Algebra*. Dun-Donnelley, New York, 1977.

GILBERT, W. J. *Modern Algebra with Applications.* Wiley, New York, 1976.
GILL, A. *Applied Algebra for the Computer Sciences.* Prentice-Hall, Englewood Cliffs, N.J., 1976.
LIDL, R. *Algebra für Naturwissenschaftler und Ingenieure.* Walter de Gruyter, Berlin, 1975.
LIDL, R. and PILZ, G. *Angewandte abstrakte Algebra,* vols. I and II. Bibliographisches Institut, Wissenschaftsverlag, Mannheim, 1982.
PRATHER, R. E. *Discrete Mathematical Structures for Computer Science.* Houghton Mifflin, Boston, 1976.
PREPARATA, F. P. and YEH, R. T. *Introduction to Discrete Structures.* Addison-Wesley, Reading, Mass., 1973.
ROY, B. *Modern Algebra and Graph Theory Applied to Management.* Springer-Verlag, New York, 1977.
STONE, H. S. *Discrete Mathematical Structures and Their Applications.* Sci. Res. Associates, Chicago, 1973.

Chapter 1

ABBOTT, J. C. (ed.). *Trends in Lattice Theory.* Van Nostrand, Reinhold, New York, 1970.
BIRKHOFF, G. *Lattice Theory.* American Mathematical Society, Providence, R.I., 1967.
CRAWLEY, P. and DILWORTH, R. P. *Algebraic Theory of Lattices.* Prentice-Hall, Englewood Cliffs, N.J., 1973.
GRÄTZER₁, G. *General Lattice Theory.* Birkhäuser-Verlag, Basel, 1978.
GRÄTZER₂, G. *Lattice Theory.* Freeman, San Francisco, 1971.
HALMOS, P. *Boolean Algebras.* Van Nostrand, Princeton, 1967.
MEHRTENS, H. *Die Entstehung der Verbandstheorie.* Gerstenberg Verlag, Hildesheim, 1979.
REUSCH, B. Generation of prime implicants from subfunctions and a unifying approach to the covering problem. *IEEE Trans. Comput.,* 24 (1975), 924–930.
REUSCH, B. and DETERING, L. On the generation of prime implicants. *Fund. Inform.* 2 (1978/79), 167–186.
RUTHERFORD, D. E. *Introduction to Lattice Theory.* Oliver & Boyd, Edinburgh, 1965.
SZÁSZ, G. *Introduction to Lattice Theory.* Academic Press, New York, 1963.

Chapter 2

§1

DOKTER, F. and STEINHAUER, J. *Digitale Elektronik.* Philips GmbH, Hamburg, 1972.
DWORATSCHEK, S. *Schaltalgebra und digitale Grundschaltungen.* Walter de Gruyter, Berlin, 1970.
FRIEDELL, M. F. Organizations on semilattices. *Amer. Math. Soc. Rev.,* 31 (1967), 46–54.
HARRISON, M. A. *Introduction to Switching and Automata Theory.* McGraw-Hill, New York, 1969.
HOHN, F. E. *Applied Boolean Algebra,* 2nd ed., Macmillan, New York, 1970.
JORDAN, P. Algebraische Betrachtungen zur Theorie des Wirkungsquantums und der Elementarlänge. *Abh. Math. Sem. Hamburg,* 18 (1952), 99–119.
MATSUSHITA, S. Ideals in non-commutative lattices. *Proc. Japan Acad.,* 34 (1958), 407–410.
MENDELSON, E. *Boolean Algebras and Switching Circuits.* Schaum's Outline Series. McGraw-Hill, New York, 1970.

PERRIN, J. P., DENQUETTE, M. and DALCIN, E. *Switching Machines*, vols. I and II. Reidel, Dordrecht, 1972.

PESCHEL, M. *Moderne Anwendungen algebraischer Methoden*. Verlag Technik, Berlin, 1971.

WHITESITT, J. E. *Boolean Algebra and Its Applications*. Addison-Wesley, Reading, Mass., 1961.

§2

BARNES, D. W. and MACK, J. M. *An Algebraic Introduction to Mathematical Logic*. Springer-Verlag, New York–Heidelberg–Berlin, 1975.

HILBERT, D. and ACKERMANN, W. *Grundzüge der theoretischen Logik*. Springer-Verlag, Berlin–Göttingen–Heidelberg, 1959.

RENNIE, M. K. and GIRLE, R. A. *Logic: Theory and Practice*. University of Queensland, Brisbane, 1973.

RUDEANU, S. *Boolean Functions and Equations*. North-Holland, Amsterdam–London, 1974.

§3

BIRKHOFF, G. *Lattice Theory*. American Mathematical Society, Providence, R.I., 1967.

BIRKHOFF, G. and VON NEUMANN, J. The logic of quantum mechanics. *Ann. Math.*, 37 (1936), 823–843.

CHUNG, K. L. *A Course in Probability Theory*. Academic Press, New York, 1974.

DRIESCHNER, M. *Voraussage – Wahrscheinlichkeit – Objekt. Über die begrifflichen Grundlagen der Quantenmechanik*. Lecture Notes in Physics, 99. Springer-Verlag, Berlin–Heidelberg–New York, 1979.

FRASER, D. A. S. *Probability and Statistics*. Duxbury Press, Mass.; Wadsworth, 1976.

FRIEDELL, M. F. Organizations as semilattices. *Amer. Math. Soc. Rev.*, 31 (1967), 46–54.

GNEDENKO, B. V. *The Theory of Probability*. Chelsea, New York, 1968.

GUDDER, S. P. *Axiomatic Quantum Mechanics and Generalized Probability Theory*. Probabilistic Methods in Applied Mathematics, 53–129. Hrsg, Bharucha-Reid; Academic Press, New York, 1970.

HALMOS, P. *Boolean Algebras*. Van Nostrand, Princeton, 1967.

JAUCH, J. M. *Foundations of Quantum Mechanics*. Addison-Wesley, Reading, Mass., 1968.

LOÈVE, M. *Probability Theory*. Van Nostrand, Princeton, N.J., 1963.

MACKEY, G. W. *Mathematical Foundations of Quantum Mechanics*. Benjamin, New York–Amsterdam, 1963.

VARADARAJAN, V. S. *Geometry of Quantum Theory*, vols I and II. Van Nostrand, Princeton, 1968, 1970.

VON NEUMANN, J. *Mathematical Foundations of Quantum Mechanics*. Princeton University Press, Princeton, 1955.

ZELMER, V. and STANCU, A. Mathematical approach on the behavior of bio-systems. *Math. Cluj.*, 15 (1973), 119–128.

Chapter 3

ALANEN, J. D. and KNUTH, D. E. Tables of finite fields. *Sankhya*, Ser. A, 26 (1964), 305–328.

ALBERT, A. A. *Fundamental Concepts of Higher Algebra*. University of Chicago Press, Chicago, 1956.

BERLEKAMP, E. R. *Algebraic Coding Theory*. McGraw-Hill, New York, 1968.

BLAKE, I. F. and MULLIN, R. C. *The Mathematical Theory of Coding*. Academic Press, New York, 1975.

CANTOR, D. G. and ZASSENHAUS, H. A new algorithm for factoring polynomials over finite fields. *Math. Comput.*, **36** (1981), 587–592.

DICKSON, L. E. *History of the Theory of Numbers*, 3 vols. Carnegie Institute, Washington, D.C., 1919, 1920, 1923.

GALOIS, E. Sur la théorie des nombres. *Bull. Sci. Math. M. Férussac*, **13** (1830), 428–435. (See *Oeuvres Math.*, pp. 15–23, Gauthier-Villars, Paris, 1897.)

IRELAND, K. and ROSEN, M. I. *A Classical Introduction to Modern Number Theory*. Springer-Verlag, New York–Heidelberg–Berlin, 1982.

KNUTH, D. E. The Art of Computer Programming. In *Semi-numerical Algorithms*, vol. 2. Addison-Wesley, Reading, Mass., 1981.

KRONECKER, L. Ein Fundamentalsatz der allgemeinen Arithmetik. *J. reine angew. Math.*, **100** (1887), 490–510.

LAZARD, D. On polynomial factorization. In *Computer Algebra*, pp. 126–134. Lecture Notes in Computer Science, 144. Springer-Verlag, Berlin–Heidelberg–New York, 1982.

LIDL, R. and NIEDERREITER, H. *Finite Fields*. Addison-Wesley, Reading, Mass., 1983.

MCCARTHY, P. J. *Algebraic Extensions of Fields*. Chelsea, New York, 1966.

MCDONALD, B. R. *Finite Rings with Identity*. Marcel Dekker, New York, 1974.

MCELIECE, R. J. *Information Theory and Coding Theory*. Addison-Wesley, Reading, Mass., 1977.

MACWILLIAMS, F. J. and SLOANE, N. J. A. *The Theory of Error-Correcting Codes*, vols. I and II. North-Holland, Amsterdam–New York–Oxford, 1977.

MOENCK, R. T. On the efficiency of algorithms for polynomial factoring. *Math. Comput.*, **31** (1977), 235–250.

MOSSIGE, S. *Table of Irreducible Polynomials over GF[2] of Degrees 10 through 20*, 2 vols. University of Bergen, Bergen, 1971.

NAGATA, M. *Field Theory*. Marcel Dekker, New York–Basel, 1977.

PLESS, V. *Introduction to the Theory of Error-Correcting Codes*. Wiley-Interscience, New York, 1982.

RABIN, M. O. Probabilistic algorithms in finite fields. *SIAM J. Comput.*, **9** (1980), 273–280.

STEINITZ, E. Algebraische Theorie der Körper. *J. reine angew. Math.*, **137** (1910), 167–309.

WEBER, H. Die allgemeinen Grundlagen der Galois'schen Gleichungstheorie. *Math. Ann.*, **43** (1893), 521–549.

WINTER, D. J. *The Structure of Fields*. Springer-Verlag, Berlin–Heidelberg–New York, 1974.

Chapter 4

ABRAMSON, N. M. *Information Theory and Coding*. McGraw-Hill, New York, 1963.

BERLEKAMP, E. R. *Algebraic Coding Theory*. McGraw-Hill, New York, 1968.

BERLEKAMP, E. R. The technology of error-correcting codes. *Proc. IEEE*, **68** (1980), 564–593.

BETH, T. and STREHL, V. *Materialien zur Codierungstheorie, Berichte* IMMD, Erlangen, 11 Heft 14, 1978.

BHARGAVA, V. K., HACCOUN, D., MATYAS, R. and NUSPL, P. P. *Digital Communications by Satellite*. Wiley, New York, 1981.

BLAHUT, R. E. *Theory and Practice of Error Control Codes*. Addison-Wesley, Reading, Mass., 1983.

BLAKE, I. F. *Algebraic Coding Theory*: *History and Development*. Dowden-Hutchinson-Ross, Stroudsburg, Penn., 1973.

BLAKE, I. F. and MULLIN, R. C. *The Mathematical Theory of Coding*. Academic Press, New York, 1975.

GALLAGER, R. G. *Information Theory and Reliable Communication*, Wiley, New York, 1968.

GUIASU, S. *Information Theory with Applications*. McGraw-Hill, New York, 1977.

MCELIECE, R. J. *Information Theory and Coding Theory*. Addison-Wesley, Reading, Mass., 1977.

MACWILLIAMS, F. J. and SLOANE, N. J. A. *The Theory of Error-Correcting Codes*, vols. I and II. North-Holland, Amsterdam–New York–Oxford, 1977.

MANN, H. B. *Error Correcting Codes*. Wiley, New York, 1968.

MASSEY, J. *Threshold Decoding*, M.I.T. Press, Cambridge, Mass., 1963.

MATTSON, H. F. Jr. and SOLOMON, G. A new treatment of Bose–Chaudhuri codes. *J. SIAM*, 9 (1961), 654–669.

PETERSON, W. W. and WELDON, E. J. Jr. *Error Correcting Codes*. M.I.T. Press, Cambridge, Mass., 1972.

PLESS, V. *Introduction to the Theory of Error-correcting Codes*. Wiley-Interscience, New York, 1982.

POLLARD, J. M. The fast Fourier transform in a finite field. *Math. Comput.*, 25 (1971), 365–374.

SHANNON, C. E. A mathematical theory of communication. *Bell System Tech. J.*, 27 (1948), 379–423, 623–656.

SLOANE, N. J. A. *A Short Course on Error-correcting Codes*. Springer-Verlag, Wien–New York, 1975.

STREET, A. P. and WALLIS, W. D. *Combinatorics: A First Course*. Winnipeg, Canada, 1982.

VAN LINT, J. H. *Introduction to Coding Theory*. Springer-Verlag, New York–Heidelberg–Berlin, 1982.

ZIERLER, N. Linear recurring sequences and error-correcting codes. In *Error Correcting Codes* (H. B. Mann, ed.), pp. 47–59. Wiley, New York, 1968.

Chapter 5

§1

AIGNER, M. *Combinatorial Theory*. Springer-Verlag, Berlin–Heidelberg–New York, 1979.

BERMAN, G. and FRYER, K. D. *Introduction to Combinatorics*. Academic Press, New York, 1972.

BOGART, K. P. *Introductory Combinatorics*, Pitman, Boston, 1983.

COCHRAN, W. G. and COX, G. M. *Experimental Designs*. Wiley, New York, 1957.

DENES, J. and KEEDWELL, A. D. *Latin Squares and Their Applications*, English Universities Press, London, 1974.

HALL, M. *Combinatorial Theory*. Blaisdell, Waltham, Mass., 1967.

HIRSCHFELD, J. W. P. *Projective Geometries over Finite Fields*. Clarendon Press, Oxford, 1979.

HORADAM, A. F. *A Guide to Undergraduate Projective Geometry*. Pergamon Press (Australia), Rushcutters Bay, 1970.

HUGHES, D. R. and PIPER, F. C. *Projective Planes*. Springer-Verlag, New York–Heidelberg–Berlin, 1973.

MANN, H. B. *Analysis and Design of Experiments*. Dover, New York, 1949.

PILZ, G. *Near-Rings*, 2nd ed. North-Holland, Amsterdam, 1983.

RAGHAVARAO, D. *Constructions and Combinatorial Problems in Design of Experiments*. Wiley, New York, 1971.

RYSER, H. J. *Combinatorial Mathematics.* Mathematical Association of America, New York, 1963.

VAJDA, S. *The Mathematics of Experimental Design.* Hafner, New York, 1967.

VILENKIN, N. Ya. *Combinatorics.* Academic Press, New York–London, 1971.

WALLIS, W. D., STREET, A. P. and WALLIS, J. S. *Combinatorics: Room Squares, Sum-Free Sets, Hadamard Matrices.* Lecture Notes in Mathematics, 292. Springer-Verlag, Berlin–Heidelberg–New York, 1972.

WELSH, D. J. A. *Combinatorial Mathematics and Its Applications.* Academic Press, London–New York, 1971.

WILSON, R. J. (ed.) *Applications of Combinatorics.* Shiva, Nantwich, England; Birkhäuser, Boston–Cambridge, Mass., 1982.

§2

BEKER, H. and PIPER, F. *Cipher Systems, The Protection of Communications.* Northwood Books, London, 1982.

BETH, T., HESS, P. and WIRL, K. *Kryptographie.* B. G. Teubner, Stuttgart, 1983.

BRAWLEY, J. V. and LEVINE₁, J. Involutory commutants with some applications to algebraic cryptography, I. *J. reine angew. Math.,* 224 (1966), 20–43.

BRAWLEY, J. V. and LEVINE₂, J. Some cryptographic applications of permutation polynomials. *Cryptologia,* 1 (1977), 76–92.

DAVIES, D. W., PRICE, W. L. and PARKIN, G. I. An evaluation of public key cryptosystems. NPL Report CTU 1, April 1980.

DEAVOURS, C. A. The Ithaca connection: computer cryptography in the making. *Cryptologia,* 1 (1977), 312–317.

DENNING, D. E. R. *Cryptography and Data Security.* Addison-Wesley, Reading, Mass., 1983.

DIFFIE, W. and HELLMAN, M. E. New directions in cryptography. *IEEE Trans. Inf. Theor.,* 22 (1976), 644–684.

DIFFIE, W. and HELLMAN, M. E. Privacy and authentication—an introduction to cryptography. *Proc. IEEE,* 67 (1979), 397–427.

GALLAND, J. E. *Historical and Analytical Bibliography of the Literature of Cryptology.* American Mathematical Society, Providence, R.I., 1970.

GARDNER, M. A new kind of cipher that would take millions of years to break. *Sci. Amer.,* 237 (1977), 120–124.

HELLMAN, M. E. The mathematics of public-key cryptography. *Sci. Amer.,* 241 (1979), 130–139.

HILL₁, L. S. Cryptography in an algebraic alphabet. *Amer. Math. Mon.,* 36 (1929), 306–312.

HILL₂, L. S. Concerning certain linear transformation apparatus of cryptography. *Amer. Math. Mon.,* 38 (1931), 135–154.

KAHN, D. *The Codebreakers.* Weidenfeld & Nicholson, London, 1967.

KONHEIM, A. G. *Cryptography, A Primer.* Wiley, New York, 1981.

LEISS, E. L. *Principles of Data Security.* Plenum, New York, 1982.

LEVINE, J. Variable matrix substitution in algebraic cryptology. *Amer. Math. Mon.,* 65 (1958), 170–179.

MERKLE, R. C. Secure communications over insecure channels. *Commun. ACM,* 21 (1978), 294–299.

MERKLE, R. C. *Secrecy, Authentication, and Public Key Systems.* UMI Research Press, Ann Arbor, Mich., 1982.

MYER, C. H. and MATYAS, S. M. *Cryptography: A New Dimension in Computer Data Security.* Wiley, New York, 1982.

PRICE, W. L. A fourth annotated bibliography of recent publications on data security and cryptography. NPL Report DNACS 33/1980.

RIVEST, R., SHAMIR, A. and ADLEMAN, L. A method for obtaining digital signatures and public-key cryptosystems. *Commun. ACM,* **21** (1978), 120–126.

SIMMONS, G. J. Cryptology, the mathematics of secure communication. *Math. Intel.,* **1** (1979), 233–246.

SINKOV, A. *Elementary Cryptanalysis, a Mathematical Approach.* Random House, New York, 1978.

SLOANE, N. J. A. Error-correcting codes and cryptography. In *The Mathematical Gardner* (D. A. Klarner, ed.), pp. 346–382. Prindle, Weber and Schmidt, Boston, 1981.

STILLWELL, J. The two faces of coding theory. *Function,* **4** (1980), 8–15.

§3

ASH, R. *Information Theory.* Wiley-Interscience, New York, 1965.

DICKSON, L. E. *History of the Theory of Numbers,* 3 vols. Carnegie Institute, Washington, D.C., 1919, 1920, 1923.

GOLOMB₁, S. W. *Shift Register Sequences.* Holden-Day, San Francisco, 1967.

GOLOMB₂, S. W. (ed.). *Digital Communications with Space Applications.* Prentice-Hall, Englewood Cliffs, N.J., 1964.

LIDL, R. and NIEDERREITER, H. *Finite Fields.* Addison-Wesley, Reading, Mass., 1983.

LIN, S. *An Introduction to Error-Correcting Codes.* Prentice-Hall, Englewood Cliffs, N.J., 1970.

LÜNEBURG, H. *Galoisfelder, Kreisteilungskörper und Schieberegisterfolgen.* Bibliographisches Institut, Mannheim, 1979.

MANN, H. B. (ed.). *Error Correcting Codes.* Wiley, New York, 1968.

SELMER, E. S. *Linear Recurrence Relations over Finite Fields.* University of Bergen, Bergen, 1966.

ZIERLER, N. Linear recurring sequences. *J. SIAM,* **7** (1959), 31–48.

§4

AHO, A. V., HOPCROFT, J. E. and ULLMAN, J. D. *The Design and Analysis of Computer Algorithms.* Addison-Wesley, Reading, Mass., 1974.

BORODIN, A. and MUNRO, I. *The Computational Complexity of Algebraic and Numeric Algorithms.* Elsevier, New York, 1975.

KNUTH, D. E. *The Art of Computer Programming,* vol. 2, 2nd ed. Addison-Wesley, Reading, Mass., 1981.

LIPSON, J. D. *Elements of Algebra and Algebraic Computing.* Addison-Wesley, Reading, Mass., 1981.

LIU, C. L. *Introduction to Combinatorial Mathematics.* McGraw-Hill, New York, 1968.

SZABO, N. S. and TANAKA, R. I. *Residue Arithmetic and Its Applications to Computer Technology.* McGraw-Hill, New York, 1967.

§5

DE BRUIJN, N. G. Pólya's theory of counting. In *Applied Combinatorial Mathematics* (E. F. Beckenbach, ed.), Ch. 5. Robert E. Krieger, Malabar, Florida, 1981.

STONE, H. S. *Discrete Mathematical Structures and Their Applications.* Scientific Research Association, Chicago, 1973.

TUCKER, A. *Applied Combinatorics.* Wiley, New York, 1980.

Chapter 6

ARBIB₁, M. A. *Algebraic Theory of Machines, Languages and Semigroups.* Academic Press, New York–London, 1968.

ARBIB₂, M. A. *Brains, Machines and Mathematics.* McGraw-Hill, New York, 1964.

ARBIB₃, M. A. *Theories of Abstract Automata.* Prentice-Hall, Englewood Cliffs, N.J., 1969.

BOGART, K. P. Preference structures: distances between asymmetric relations. *SIAM J. Appl. Math.,* **29** (1975), 254–262.

CLIFFORD, A. H. and PRESTON, G. B. *The Algebraic Theory of Semigroups,* vols. I and II. American Mathematical Society, Providence, R.I., 1961, 1967.

EILENBERG, S. *Automata, Languages and Machines,* 2 vols. Academic Press, New York, 1974.

GINSBURG, S. *Algebraic and Automata-Theoretic Properties of Formal Languages.* North-Holland, Amsterdam, 1975.

GINZBURG, A. *Algebraic Theory of Automata.* Academic Press, New York, 1968.

HOLCOMBE, W. M. L. *Algebraic Automata Theory.* Cambridge University Press, Cambridge, 1982.

HOWIE, J. M. *An Introduction to Semigroup Theory.* Academic Press, London–New York–San Francisco, 1976.

KALMAN, R. E., FALB, P. L. and ARBIB, M. A. *Topics in Mathematical Systems Theory.* McGraw-Hill, New York, 1969.

LALLEMENT, G. *Semigroups and Combinatorial Applications.* Wiley, New York, 1979.

LJAPIN, E. S. *Semigroups.* American Mathematical Society, Providence, R.I., 1963.

MINSKY, M. *Computation, Finite and Infinite Machines.* Prentice-Hall, Englewood Cliffs, N.J., 1967.

PETRICH, M. *Introduction to Semigroups.* Bell & Howell, Columbus, 1973.

RABIN, M. D. and SCOTT, D. Finite automata and their decision problems. *IBM J.,* **3** (1959), 114–125.

ROSEN, R. *Foundations of Mathematical Biology.* Academic Press, New York, 1972.

SCHADACH, D. J. *Biomathematik,* 2 Bände. Akademie-Verlag, Berlin, 1971.

STARKE, P. H. *Abstract Automata.* North-Holland, Amsterdam, 1972.

TRAKHTENBROT, B. A. and BARZDIN, YA. M. *Finite Automata, Behavior and Synthesis.* North-Holland, Amsterdam, 1973.

WELLS, C. Some applications of the wreath product construction. *Amer. Math. Mon.,* **83** (1976), 317–338.

ZEIGER, H. P. Cascade decomposition of automata. In *Algebraic Theory of Machines, Languages and Semigroups* (M. A. Arbib), pp. 55–80. Academic Press, New York–London, 1968.

Chapter 7

§1

ARBIB₁, M. A. *Algebraic Theory of Machines, Languages and Semigroups.* Academic Press, New York–London, 1968.

ARBIB₂, M. A. *Brains, Machines and Mathematics.* McGraw-Hill, New York, 1964.

ARBIB₃, M. A. *Theories of Abstract Automata.* Prentice-Hall, Englewood Cliffs, N.J. 1969.

BOORMAN, S. A. and WHITE, H. C. Social structure from multiple networks, II: role structures. *Amer. J. Sociol.,* **81** (1976), 1384–1466.

CHOMSKY₁, N. *Syntactic Structures.* Mouton, Den Haag, 1957.

CHOMSKY₂, N. *Aspects of the Theory of Syntax.* Mouton, Den Haag, 1965.

COHN, P. M. Algebra and language theory. *Bull. London Math. Soc.,* **7** (1975), 1–29.

GINSBURG, S. *The Mathematical Theory of Context-Free Languages.* McGraw-Hill, New York, 1966.

SALOMAA₁, A. *Theory of Automata.* Pergamon Press, Oxford, 1969.

SALOMAA₂, A. *Jewels of Formal Language Theory.* Pitman, London, 1981.

§2

BALLONOFF, P. (ed.). *Genetics and Social Structure.* Dowden, Hutchinson, Ross, Stroudsburg, Pennsylvania, 1974.

BOGART, K. P. Preference structures, II: distances between asymmetric relations. *SIAM J. Appl. Math.*, **29** (1975), 254–262.

HERMANN, G. T. and ROSENBERG, G. *Developmental Systems and Languages.* North-Holland, Amsterdam, 1975.

KIERAS, D. Finite-automata and SR models. *J. Math. Psychol.*, **13** (1976), 127–147.

KROHN, K., LANGER, R. and RHODES, J. Algebraic principles for the analysis of a biochemical system. *J. Comput. Syst. Sci.*, **1** (1976), 119–136.

LINDENMAYER, A. Mathematical models for cellular interactions in development, I, II. *J. Theor. Biol.*, **18** (1968), 280–315.

NAHIKIAN, H. M. *A Modern Algebra for Biologists.* University of Chicago Press, Chicago, 1964.

ROSEN₁, R. The DNA-protein coding problem. *Bull. Math. Biophys.*, **21** (1959), 71–95.

ROSEN₂, R. Some further comments on the DNA-protein coding problem. *Bull. Math. Biophys.*, **21** (1959), 289–297.

ROSEN₃, R. *Foundations of Mathematical Biology*, vols. I–III. Academic Press, New York, 1972, 1973.

ROSENBERG, G. and SALOMAA, A. *L-Systems.* Springer-Verlag, Berlin–Heidelberg–New York, 1974.

SUPPES, P. Stimulus-response theory of finite automata. *J. Math. Psychol.*, **6** (1969), 327–355.

§3

BOGART, K. P. Preference structures, II: distances between asymmetric relations. *SIAM J. Appl. Math.*, **29** (1975), 254–262.

BOORMAN, S. A. and WHITE, H. C. Social structure from multiple networks, II: role structures. *Amer. J. Sociol.*, **81** (1976), 1384–1466.

BOYD, J. P., HAEHL, J. H. and SAILER, L. D. Kinship systems and inverse semigroups. *J. Math. Sociol.*, (1972), 37–61.

BOYLE, F. P. Algebraic systems for normal and hierarchical sociograms. *Sociometry*, **32** (1969), 99–119.

BREIGER, R. L., BOORMAN, S. A. and ARABIE, P. An algorithm for clustering relational data with applications to social network analysis and comparison with multi-dimensional scaling. *J. Math. Psychol.*, **12** (1975), 328–383.

CARLSON, R. *An Application of Mathematical Groups to Structures of Human Groups.* UMAP Module 476. Birkhäuser, Boston, 1980.

KIM, K. H. and ROUSH, F. W. *Mathematics for Social Scientists.* Elsevier, New York, 1980.

SAMPSON, S. F. Crisis in a cloister. Dissertation, Cornell University, 1969.

WHITE, H. C. *An Anatomy of Kinship: Mathematical Models for Structures of Cumulated Roles.* Prentice-Hall, Englewood Cliffs, N.J., 1963.

WHITE, H. C., BOORMAN, S. A. and BREIGER, R. L. Social structure from multiple networks, I: blockmodels of roles and positions. *Amer. J. Sociol.*, **81** (1976), 730–780.

History of Mathematics. History of Algebra

BIRKHOFF, G. The rise of modern algebra to 1936. In *Man and Institutions in American Mathematics* (J. D. Tarwater, J. T. White and J. D. Miller, eds.), pp. 41–63. Graduate Studies, No. 13. Texas Technical University, Lubbock, Texas, 1976.

BIRKHOFF, G. The rise of modern Algebra, 1936 to 1950. In *Man and Institutions in American Mathematics* (J. D. Tarwater, J. T. White and J. D. Miller, eds.), pp. 65–85. Graduate Studies, No. 13. Texas Technical University, Lubbock, Texas, 1976.

BOURBAKI, N. *Élements d'Histoire des Mathématiques*, Hermann, Paris, 1974.

DIEUDONNÉ, J. *Abrégé d'Histoire des Mathématiques 1700–1900*. I. Hermann, Paris, 1978.

KLINE, M. *Mathematical Thoughts from Ancient to Modern Times.* Oxford University Press, New York, 1972.

MEHRTENS, H. *Die Entstehung der Verbandstheorie.* Gerstenberg Verlag, Hildesheim, 1979.

NOVÝ, L. *Origins of Modern Algebra*, Noordhoff, Leyden, 1973.

WUSSING, H. *Die Genesis des abstrakten Gruppenbegriffs.* Berlin, 1969.

Applications of Linear Algebra

GOULT, R. Z. *Applied Linear Algebra.* Horwood, New York, 1978.

NOBLE, B. *Applied Linear Algebra.* Prentice-Hall, New York, 1969.

NOBLE, B. and DANIEL, J. W. *Applied Linear Algebra*, 2nd ed. Prentice-Hall, Englewood Cliffs, N.J., 1977.

RORRES, C. and ANTON, H. *Applications of Linear Algebra.* Wiley, New York, 1977.

STRANG, G. *Linear Algebra and Its Applications*, 2nd ed. Academic Press, New York–San Francisco–London, 1980.

Collections of Problems

AYRES, F. *Theory and Problems of Modern Algebra.* Schaum's Outline Series. McGraw-Hill, New York, 1965.

FADDEEV, D. K. and SOMINSKII, I. S. *Problems in Higher Algebra.* Freeman, San Francisco, 1965.

KAISER, H. K., LIDL, R. and WIESENBAUER, J. *Aufgabensammlung zur Algebra.* Akademische Verlagsgesellschaft, Wiesbaden, 1975.

LJAPIN, E. S., AIZENSHTAT, A. YA. and LESOKHIN, M. M. *Exercises in Group Theory.* Plenum, New York, 1972.

References on Further Applications of Algebra

In the limited space of the text we could only consider selected topics of applications of algebra and discrete mathematics. Here we list some references to other areas of applications, with an emphasis on introductory material. Some of the important applications of discrete mathematics use concepts from abstract algebra, combinatorics and probability theory and do not properly fall within one single discipline. For instance, *graph theory* and its many uses are described in several of the applied algebra texts listed in the Bibliography, e.g. BIRKHOFF and BARTEE, DORNHOFF and HOHN, FISHER. Some further interesting examples of applications are given in

ROBERTS, F. S. *Discrete Mathematical Models, with Applications to Social, Biological, and Environmental Problems.* Prentice-Hall, Englewood Cliffs, N.J., 1976.

That book contains also applications in social sciences, for instance rankings, Arrow's Impossibility Theorem, and other preference structures by groups or individuals. For other applied examples and models in the social sciences we refer to

BRAMS, S. J., LUCAS, W. F. and STRAFFIN, Jr., P. D. (eds.). *Political and Related Models.* Springer-Verlag, New York–Heidelberg–Berlin, 1983.

Parts of population genetics can be formulated by using algebraic concepts, such as nonassociative rings, and are summarized under the name *genetic algebra,* see

WÖRZ-BUSEKROS, A. *Algebras in Genetics.* Springer-Verlag, Berlin–Heidelberg–New York, 1980.
BALLONOFF, P. *Genetics and Social Structure: Mathematical Structuralism in Population Genetics and Social Theory.* Dowden, Hutchinson, Ross. Stroudsburg, Pennsylvania, 1974.

Some of the first applications of abstract algebra are connected with the group concept. There is a considerable number of textbooks and monographs on groups and their applications in the physical sciences. We mention a few of them:

BROWN, H., BÜLOW, R., NEUBÜSER, J., WONDRATSCHEK, H. and ZASSENHAUS, H. *Crystallographic Groups of Four-Dimensional Space.* Wiley, New York, 1979.
BUDDEN, F. J. *The Fascination of Groups.* Cambridge University Press, Cambridge, 1972.
COTTON, F. A. *Chemical Applications of Group Theory.* Wiley, New York, 1971.
DORNHOFF, L. *Group Representation Theory,* 2 vols. M. Dekker, New York, 1971.
GELFAND, I. M., MINLOS, R. A. and SHAPIRO, Z. YA. *Representation of Rotation and Lorentz Groups and Their Applications.* Pergamon Press, Oxford, 1963.
LOMONT, J. S. *Applications of Finite Groups.* Academic Press, New York, 1959.
LOEBL, E. M. (ed.). *Group Theory and Its Applications,* 3 vols. Academic Press, New York–London, 1968.

An elementary application of groups to kinship is described in

CARLSON, R. *An Application of Mathematical Groups to Structures of Human Groups,* vol. 1(3). UMAP Module 476. Birkhäuser, Boston–Basel–Stuttgart, 1980.

There are several other examples of the use of algebraic structures. However, these are rather isolated cases and of a very specialized nature. For instance, there is an attempt by Zelmer and Stancu (*Math. Cluj.,* 15 (1973), 119–128) to axiomatically define bio-systems and to give lattice theoretical interpretations of them, e.g. the nervous system as a Boolean algebra. Abstract algebraic structures, such as pseudo-vector spaces, have been used in a description of color theory in psychology (cf. D. H. Krantz, *J. Math. Psychol.,* 12 (1975), 283–303). T. Evans (*Amer. Math. Mon.,* 86 (1979), 466–473) describes an application of universal algebra to Euler's officers problem in combinatorics.

In the main Bibliography we refer to some texts which include applications of linear algebra.

Added in print:

Recently the authors became aware of several books which are related to the topics of this text:

DORNINGER, D. W. and MÜLLER, W. B. *Allgemeine Algebra und Anwendungen.* G. B. Teubner, Stuttgart, 1984.

KOLMAN, B. and BUSBY, R. C. *Discrete Mathematical Structures for Computer Science.* Prentice-Hall, Englewood Cliffs, New Jersey, 1984.

SIMS, C. C. *Abstract Algebra, A Computational Approach.* Wiley, New York, 1984.

Author Index

Subject Index

Undergraduate Texts in Mathematics

continued from ii

Lidl/Pilz: Applied Abstract Algebra.
1984. xiii, 545 pages. 175 illus.

Macki/Strauss: Introduction to Optimal
Control Theory.
1981. xiii, 168 pages. 68 illus.

Malitz: Introduction to Mathematical
Logic: Set Theory - Computable
Functions - Model Theory.
1979. xii, 198 pages. 2 illus.

Martin: The Foundations of Geometry
and the Non-Euclidean Plane.
1975. xvi, 509 pages. 263 illus.

Martin: Transformation Geometry: An
Introduction to Symmetry.
1982. xii, 237 pages. 209 illus.

Millman/Parker: Geometry: A Metric
Approach with Models.
1981. viii, 355 pages. 259 illus.

Owen: A First Course in the
Mathematical Foundations of
Thermodynamics.
1984. xvii, 178 pages. 52 illus.

Prenowitz/Jantosciak: Join Geometrics:
A Theory of Convex Set and Linear
Geometry.
1979. xxii, 534 pages. 404 illus.

Priestly: Calculus: An Historical
Approach.
1979. xvii, 448 pages. 335 illus.

Protter/Morrey: A First Course in Real
Analysis.
1977. xii, 507 pages. 135 illus.

Ross: Elementary Analysis: The Theory
of Calculus.
1980. viii, 264 pages. 34 illus.

Scharlau/Opolka: From Fermat to
Minkowski: Lectures on the Theory of
Numbers and Its Historical Development.
1984. xi, 179 pages. 28 illus.

Sigler: Algebra.
1976. xii, 419 pages. 27 illus.

Simmonds: A Brief on Tensor
Analysis.
1982. xi, 92 pages. 28 illus.

Singer/Thorpe: Lecture Notes on
Elementary Topology and Geometry.
1976. viii, 232 pages. 109 illus.

Smith: Linear Algebra.
Second edition.
1984. vii, 362 pages. 20 illus.

Smith: Primer of Modern Analysis.
1983. xiii, 442 pages. 45 illus.

Thorpe: Elementary Topics in Differential
Geometry.
1979. xvii, 253 pages. 126 illus.

Troutman: Variational Calculus
with Elementary Convexity.
1983. xiv, 364 pages. 73 illus.

Whyburn/Duda: Dynamic Topology.
1979. xiv, 338 pages. 20 illus.

Wilson: Much Ado About Calculus:
A Modern Treatment with Applications
Prepared for Use with the Computer.
1979. xvii, 788 pages. 145 illus.